ADDITIVE COMBINATORICS
A MENU OF RESEARCH PROBLEMS

DISCRETE MATHEMATICS AND ITS APPLICATIONS

Introduction to Combinatorics, Second Edition
W. D. Wallis and J. C. George

Representation Theory of Symmetric Groups
Pierre-Loïc Méliot

Advanced Number Theory with Applications
Richard A. Mollin

A Multidisciplinary Introduction to Information Security
Stig F. Mjølsnes

Combinatorics of Compositions and Words
Silvia Heubach and Toufik Mansour

Handbook of Linear Algebra, Second Edition
Leslie Hogben

Combinatorics, Second Edition
Nicholas A. Loehr

Handbook of Discrete and Computational Geometry, Third Edition
C. Toth, Jacob E. Goodman and Joseph O'Rourke

Handbook of Discrete and Combinatorial Mathematics, Second Edition
Kenneth H. Rosen

Crossing Numbers of Graphs
Marcus Schaefer

Graph Searching Games and Probabilistic Methods
Anthony Bonato and Paweł Prałat

Handbook of Geometric Constraint Systems Principles
Meera Sitharam, Audrey St. John, and Jessica Sidman

Additive Combinatorics:
A Menu of Research Problems
Béla Bajnok

Algorithmics of Nonuniformity:
Tools and Paradigms
Micha Hofri and Hosam Mahmoud

https://www.crcpress.com/Discrete-Mathematics-and-Its-Applications/book-series/CHDISMTHAPP?page=1&order=dtitle&size=12&view=list&status=published,forthcoming

DISCRETE MATHEMATICS AND ITS APPLICATIONS

ADDITIVE COMBINATORICS

A MENU OF RESEARCH PROBLEMS

Béla Bajnok

CRC Press
Taylor & Francis Group
Boca Raton London New York

CRC Press is an imprint of the
Taylor & Francis Group, an **informa** business

CRC Press
Taylor & Francis Group
6000 Broken Sound Parkway NW, Suite 300
Boca Raton, FL 33487-2742

First issued in paperback 2022

Version Date: 20180322

ISBN 13: 978-1-03-247625-4 (pbk)
ISBN 13: 978-0-8153-5301-0 (hbk)

DOI: 10.1201/9781351137621

Publisher's Note
The publisher has gone to great lengths to ensure the quality of this reprint but points out that some imperfections in the original copies may be apparent.

Library of Congress Cataloging-in-Publication Data

Names: Bajnok, Béla, 1961- author.
Title: Additive combinatorics / Béla Bajnok.
Description: Boca Raton : CRC Press, Taylor & Francis Group, 2018. | Includes bibliographical references and author index.
Identifiers: LCCN 2017059342 | ISBN 9780815353010
Subjects: LCSH: Additive combinatorics. | Combinatorial analysis. | Number theory.
Classification: LCC QA164 .B3485 2018 | DDC 511/.6--dc23
LC record available at https://lccn.loc.gov/2017059342

Visit the Taylor & Francis Web site at
http://www.taylorandfrancis.com

and the CRC Press Web site at
http://www.crcpress.com

Contents

Preface

Additive Combinatorics

This book deals with *additive combinatorics*, a vibrant area of current mathematical research. Additive combinatorics—an offspring of combinatorial number theory and additive number theory—can be described as the study of combinatorial properties of *sumsets* (collections of sums with terms from given subsets) in additive structures. For example, given a subset A in an abelian group G, one can consider the sumset $A+A$ consisting of all two-term sums of the elements of A, and then ask how small this sumset may be; furthermore, given that $A+A$ is small, one can examine what can be said about the structure of A.

Additive combinatorics is a rather new field within mathematics that is just now coming to its own; although some of its results have been known for a very long time, many of its most fundamental questions have only been settled recently or are still unsolved. For example, the question about the minimum size of $A+A$ mentioned above was first determined in groups of prime order by Cauchy in 1813 (then re-discovered by Davenport over a hundred years later), but it has only been determined in the general setting in the twenty-first century. For this and many other reasons, additive combinatorics provides an excellent area for research by students of any background: it has intriguing and promising questions for everyone.

Student Research

Student research in mathematics has been increasing dramatically at all levels over the past several decades. Producing research is an expectation at all doctoral programs in mathematics, and it appears that the number of publications before students graduate has increased substantially in recent years. The National Science Foundation (NSF) is sponsoring the popular Research Experiences for Undergraduates (REU) program, and undergraduate research is indeed blossoming at many colleges and universities. There are even organized mathematics research programs for students in high school, such as PRIMES, held at the Massachusetts Institute of Technology and online nationally, and PROMYS, held at Boston University and at the University of Oxford.

The benefits and costs of student research in mathematics have been described elsewhere—see, for example, the report [146] of the Committee on the Undergraduate Program (CUPM) of the Mathematical Association of America (MAA). While the need for an ample inventory of questions for student research is clear, it is often noted how challenging it is to produce an appropriate supply. Indeed, there are quite a few demands on such questions; in the view of this author (see [12]), they are most appropriate when they are:

- based on substantial topics—students should be engaged in the study of non-trivial and not-too-esoteric mathematics;
- challenging at a variety of different levels—students with different backgrounds and

interests should be able to engage in the projects;

- approachable with a variety of different methods—students interested in theoretical, computational, abstract, or concrete work should be able to choose their own approaches;
- incrementally attainable, where at least partial results are within reach yet complete solutions are not easy—no student would want to spend long hours of hard work and not feel productive; and
- new and unsolved—the results attained by the students would have to be, at least in theory, publishable.

I believe that the problems in the book abide by these objectives.

About This Book

In this book we interpret additive combinatorics, somewhat narrowly, as the study of sumsets in finite abelian groups. (A second volume, focusing on the set of integers and other infinite groups, is in the planning phase.) This book offers an extensive menu of research projects to any student interested in pursuing investigations in this area. It contains five parts that we briefly describe, as follows.

Ingredients. We make no assumptions on the backgrounds of students wishing to engage in research projects based on this collection. In order to equip beginners with the necessary background, we provide brief introductions to the relevant branches of number theory, combinatorics, and group theory. The sections in each chapter contain exercises aimed to solidify the understanding of the material discussed.

Appetizers. The short articles in this part are meant to invite everyone to the main entrees of the menu; the appetizers are carefully chosen so that they provide bite-size representative samples of the research projects, as well as make connections to other parts of mathematics that students might have encountered. (One article describes how this author's research in spherical geometry led him to additive combinatorics.)

Sides. Here we present and study some auxiliary functions that appear in several different chapters of the text.

Entrees. Each chapter in this main part of the book discusses a particular family of open questions in additive combinatorics. At the present time, these questions are about

- maximum sumset size,
- spanning sets,
- Sidon sets,
- minimum sumset size,
- critical numbers,
- zero-sum-free sets, and
- sum-free sets.

(We plan to include additional chapters in the near future.) Each chapter is divided into the same four sections, depending on whether the sums may contain repeated terms or not and whether terms can be both added and subtracted or can only be added; each section is then further divided into the same three subsections depending on the number of terms in the sums: whether it is fixed, limited, or is arbitrary. Within most subsections, we investigate both *direct questions* (e.g., what is the maximum size of a set with the required sumset property?) and *inverse questions* (how can we characterize all subsets that achieve this maximum size?).

Many of the questions we discuss have been investigated extensively already, in which case we include all relevant results that are available to the best of our knowledge. We find it essential that our expositions are complete, and the author is committed to continually surveying the literature to assure that the material remains current. (For a variety of reasons, this is not always an easy task: as is often the case with relatively young fields, results appear in a diverse set of outlets, are often presented using different notations and terminology, and, occasionally, are even incorrect.)

While we aim to present each topic as thoroughly as possible, this book is not a historical survey. In particular, in most cases, we only include the best results known on any given question; previously achieved special cases and weaker or partial results can then usually be found in the citations we provide. Furthermore, any reader with an interest in a more contextual treatment of our topics or a desire to see them from different perspectives is encouraged to turn to one of the available books on the subject, such as [87] by Freiman, [98] by Geroldinger and Ruzsa, [103] by Grynkiewicz, [105] by Guy, [109] by Halberstam and Roth, [161] by Nathanson, or [195] by Tao and Vu.

Pudding. The proofs are in the pudding!—well, not quite. The book contains several hundred stated results; most of these results have been published, in which case we give citations where proofs (as well as further material) can be read. In other cases, results have not appeared elsewhere; proofs of these statements—unless rather brief and particularly instructive, in which case they are presented where the results are stated—are separated into this part of the book.

Acknowledgments

The author is most grateful for the immense support he has received from Gettysburg College. The frequent full-year sabbaticals and the many very generous travel grants are much appreciated. Special thanks go to my colleagues in the Mathematics Department who allowed me to teach a research course based on this book in virtually every semester during the past fifteen years.

I am grateful to the many friends and colleagues who have provided useful comments, feedback, and encouragement, including Noga Alon, László Babai, Miklós Bóna, Jill Dietz, Róbert Freud, Ronald Graham, Ben Green, David Grynkiewicz, Mark Kayll, Vsevolod Lev, Ryan Matzke, Steven J. Miller, Wolfgang Schmid, Neil Sloane, Zhi-Wei Sun, James Tanton, Terence Tao, and Paul Zeitz. I am particularly thankful to Samuel Edwards for his help with the bibliography and to Ivaylo Ilinkin for the computer program [120] that he created for this book.

But my biggest appreciation goes to the two hundred or so students at Gettysburg College who have carried out research projects based on this text thus far. Over thirty of these students attained results that are mentioned and cited in this book, at least twenty have given conference presentations on their work, and several have even published their results in refereed journals—I congratulate them full-heartedly! I am convinced, however, that every student benefited from the opportunities for perfecting a variety of skills by engaging in a research experience based on this book.

Notations

Following is the list of the most commonly used notations throughout this book.

Chapter 1: Number theory

Number sets:
Integers: $\mathbb{Z} = \{0, \pm 1, \pm 2, \pm 3, \dots\}$
Nonnegative integers: $\mathbb{N}_0 = \{0, 1, 2, 3, \dots\}$
Positive integers: $\mathbb{N} = \{1, 2, 3, \dots\}$
Interval: $[a, b] = \{c \in \mathbb{Z} \mid a \leq c \leq b\}$ for given $a, b \in \mathbb{Z}$ with $a \leq b$

Divisors:
$D(n) = \{d \in \mathbb{N} \mid d|n\}$ for a given $n \in \mathbb{N}$
$d(n) = |D(n)|$ for a given $n \in \mathbb{N}$
$\gcd(a, b) = \max(D(a) \cap D(b))$ for given $a, b \in \mathbb{N}$
$\gcd(A) = \max\{d \in \mathbb{N} \mid \forall a \in A, d|n\}$ for a given finite $A \subset \mathbb{N}$

Chapter 2: Combinatorics

Counting functions:
$n^{\overline{m}} = n(n+1)\cdots(n+m-1)$ and $n^{\overline{0}} = 1$ for $n \in \mathbb{N}_0$ and $m \in \mathbb{N}$
$n^{\underline{m}} = n(n-1)\cdots(n-m+1)$ and $n^{\underline{0}} = 1$ for $n \in \mathbb{N}_0$ and $m \in \mathbb{N}$
$\binom{n}{m} = \frac{n^{\underline{m}}}{m!} = \frac{n(n-1)\cdots(n-m+1)}{m!}$ for $n, m \in \mathbb{N}_0$
$\left[{n \atop m}\right] = \frac{n^{\overline{m}}}{m!} = \frac{n(n+1)\cdots(n+m-1)}{m!}$ for $n, m \in \mathbb{N}_0$

Some other useful counting functions:

$$a(j,k) = \textstyle\sum_{i\geq 0} \binom{j}{i}\binom{k}{i} 2^i = \begin{cases} 1 & \text{if } j = 0 \text{ or } k = 0 \\ a(j-1,k) + a(j-1,k-1) + a(j,k-1) & \text{if } j \geq 1 \text{ and } k \geq 1 \end{cases}$$

$$c(j,k) = \textstyle\sum_{i\geq 0} \binom{j-1}{i-1}\binom{k}{i} 2^i = \begin{cases} 1 & \text{if } j = 0 \text{ and } k \geq 0 \\ 0 & \text{if } j \geq 1 \text{ and } k = 0 \\ c(j-1,k) + c(j-1,k-1) + c(j,k-1) & \text{if } j \geq 1 \text{ and } k \geq 1 \end{cases}$$

Layers of the integer lattice:
$\mathbb{N}_0^m(h) = \{(\lambda_1, \dots, \lambda_m) \in \mathbb{N}_0^m \mid \Sigma_{i=1}^m \lambda_i = h\}$
$\mathbb{Z}^m(h) = \{(\lambda_1, \dots, \lambda_m) \in \mathbb{Z}^m \mid \Sigma_{i=1}^m |\lambda_i| = h\}$
$\hat{\mathbb{N}}_0^m(h) = \{(\lambda_1, \dots, \lambda_m) \in \{0,1\}^m \mid \Sigma_{i=1}^m \lambda_i = h\}$
$\hat{\mathbb{Z}}^m(h) = \{(\lambda_1, \dots, \lambda_m) \in \{-1,0,1\}^m \mid \Sigma_{i=1}^m |\lambda_i| = h\}$

Chapter 3: Group theory

Finite abelian groups:
$\mathbb{Z}_n$: cyclic group of order n using additive notation
G: (arbitrary) abelian group of order n using additive notation
r: rank of G (number of factors in the invariant decomposition of G)
κ: exponent of G (order of largest factor in the invariant decomposition of G)

Elements and subsets:
$\mathrm{ord}_G(a) = \min\{d \in \mathbb{N} \mid da = 0\}$
$\mathrm{Ord}(G, d) = \{a \in G \mid \mathrm{ord}_G(a) = d\}$
$\langle a \rangle = \{\lambda a \mid \lambda \in [1, \mathrm{ord}_G(a)]\}$
$A = \{a_1, \dots, a_m\}$: (arbitrary) m-subset of G
$\langle A \rangle = \{\Sigma_{i=1}^m \lambda_i a_i \mid \lambda_i \in [1, \mathrm{ord}_G(a_i)]\}$

Unrestricted sumsets:
$hA = \{\Sigma_{i=1}^m \lambda_i a_i \mid \lambda_i \in \mathbb{N}_0, \Sigma\lambda_i = h\}$
$[0,s]A = \cup_{h=0}^s hA = \{\Sigma_{i=1}^m \lambda_i a_i \mid \lambda_i \in \mathbb{N}_0, \Sigma\lambda_i \leq s\}$
$\langle A \rangle = \cup_{h=0}^\infty hA = \{\Sigma_{i=1}^m \lambda_i a_i \mid \lambda_i \in \mathbb{N}_0\}$

Unrestricted signed sumsets:
$h_\pm A = \{\Sigma_{i=1}^m \lambda_i a_i \mid \lambda_i \in \mathbb{Z}, \Sigma|\lambda_i| = h\}$
$[0,s]_\pm A = \cup_{h=0}^s h_\pm A = \{\Sigma_{i=1}^m \lambda_i a_i \mid \lambda_i \in \mathbb{Z}, \Sigma|\lambda_i| \leq s\}$
$\langle A \rangle = \cup_{h=0}^\infty h_\pm A = \{\Sigma_{i=1}^m \lambda_i a_i \mid \lambda_i \in \mathbb{Z}\}$

Restricted sumsets:
$h\hat{}A = \{\Sigma_{i=1}^m \lambda_i a_i \mid \lambda_i \in \{0,1\}, \Sigma\lambda_i = h\}$
$[0,s]\hat{}A = \cup_{h=0}^s h\hat{}A = \{\Sigma_{i=1}^m \lambda_i a_i \mid \lambda_i \in \{0,1\}, \Sigma\lambda_i \leq s\}$
$\Sigma A = \cup_{h=0}^\infty h\hat{}A = \{\Sigma_{i=1}^m \lambda_i a_i \mid \lambda_i \in \{0,1\}\}$
$\Sigma^* A = \cup_{h=1}^\infty h\hat{}A = \{\Sigma_{i=1}^m \lambda_i a_i \mid \lambda_i \in \{0,1\}, \Sigma\lambda_i \geq 1\}$

Restricted signed sumsets:
$h\hat{\pm}A = \{\Sigma_{i=1}^m \lambda_i a_i \mid \lambda_i \in \{-1,0,1\}, \Sigma|\lambda_i| = h\}$
$[0,s]\hat{\pm}A = \cup_{h=0}^s h\hat{\pm}A = \{\Sigma_{i=1}^m \lambda_i a_i \mid \lambda_i \in \{-1,0,1\}, \Sigma|\lambda_i| \leq s\}$
$\Sigma_\pm A = \cup_{h=0}^\infty h\hat{\pm}A = \{\Sigma_{i=1}^m \lambda_i a_i \mid \lambda_i \in \{-1,0,1\}\}$
$\Sigma_\pm^* A = \cup_{h=1}^\infty h\hat{\pm}A = \{\Sigma_{i=1}^m \lambda_i a_i \mid \lambda_i \in \{-1,0,1\}, \Sigma|\lambda_i| \geq 1\}$

Sides

$v_g(n,h) = \max\left\{\left(\left\lfloor \frac{d-1-\gcd(d,g)}{h} \right\rfloor + 1\right) \cdot \frac{n}{d} \mid d \in D(n)\right\}$ for given $n, h, g \in \mathbb{N}$
$v_\pm(n,h) = \max\left\{\left(2 \cdot \left\lfloor \frac{d-2}{2h} \right\rfloor + 1\right) \cdot \frac{n}{d} \mid d \in D(n)\right\}$ for given $n, h \in \mathbb{N}$
$u(n,m,h) = \min\left\{\left(h \cdot \left\lceil \frac{m}{d} \right\rceil - h + 1\right) \cdot d \mid d \in D(n)\right\}$ for given $n, m, h \in \mathbb{N}$

Chapter A: Maximum sumset size

A.1: Unrestricted sumsets:
$\nu(G,m,h) = \max\{|hA| \mid A \subseteq G, |A| = m\}$
$\nu(G,m,[0,s]) = \max\{|[0,s]A| \mid A \subseteq G, |A| = m\}$
$\nu(G,m,\mathbb{N}_0) = \max\{|\langle A \rangle| \mid A \subseteq G, |A| = m\}$

A.2: Unrestricted signed sumsets:
$\nu_{\pm}(G,m,h) = \max\{|h_{\pm}A| \mid A \subseteq G, |A| = m\}$
$\nu_{\pm}(G,m,[0,s]) = \max\{|[0,s]_{\pm}A| \mid A \subseteq G, |A| = m\}$
$\nu_{\pm}(G,m,\mathbb{N}_0) = \max\{|\langle A\rangle| \mid A \subseteq G, |A| = m\}$

A.3: Restricted sumsets:
$\nu\hat{}(G,m,h) = \max\{|h\hat{}A| \mid A \subseteq G, |A| = m\}$
$\nu\hat{}(G,m,[0,s]) = \max\{|[0,s]\hat{}A| \mid A \subseteq G, |A| = m\}$
$\nu\hat{}(G,m,\mathbb{N}_0) = \max\{|\Sigma A| \mid A \subseteq G, |A| = m\}$

A.4: Restricted signed sumsets:
$\nu\hat{_\pm}(G,m,h) = \max\{|h\hat{_\pm}A| \mid A \subseteq G, |A| = m\}$
$\nu\hat{_\pm}(G,m,[0,s]) = \max\{|[0,s]\hat{_\pm}A| \mid A \subseteq G, |A| = m\}$
$\nu\hat{_\pm}(G,m,\mathbb{N}_0) = \max\{|\Sigma_{\pm}A| \mid A \subseteq G, |A| = m\}$

Chapter B: Spanning sets

B.1: Unrestricted sumsets:
$\phi(G,h) = \min\{|A| \mid A \subseteq G, hA = G\}$
$\phi(G,[0,s]) = \min\{|A| \mid A \subseteq G, [0,s]A = G\}$
$\phi(G,\mathbb{N}_0) = \min\{|A| \mid A \subseteq G, \langle A\rangle = G\}$

B.2: Unrestricted signed sumsets:
$\phi_{\pm}(G,h) = \min\{|A| \mid A \subseteq G, h_{\pm}A = G\}$
$\phi_{\pm}(G,[0,s]) = \min\{|A| \mid A \subseteq G, [0,s]_{\pm}A = G\}$
$\phi_{\pm}(G,\mathbb{N}_0) = \min\{|A| \mid A \subseteq G, \langle A\rangle = G\}$

B.3: Restricted sumsets:
$\phi\hat{}(G,h) = \min\{|A| \mid A \subseteq G, h\hat{}A = G\}$
$\phi\hat{}(G,[0,s]) = \min\{|A| \mid A \subseteq G, [0,s]\hat{}A = G\}$
$\phi\hat{}(G,\mathbb{N}_0) = \min\{|A| \mid A \subseteq G, \Sigma A = G\}$

B.4: Restricted signed sumsets:
$\phi\hat{_\pm}(G,h) = \min\{|A| \mid A \subseteq G, h\hat{_\pm}A = G\}$
$\phi\hat{_\pm}(G,[0,s]) = \min\{|A| \mid A \subseteq G, [0,s]\hat{_\pm}A = G\}$
$\phi\hat{_\pm}(G,\mathbb{N}_0) = \min\{|A| \mid A \subseteq G, \Sigma_{\pm}A = G\}$

Chapter C: Sidon sets

C.1: Unrestricted sumsets:
$\sigma(G,h) = \max\left\{|A| \mid A \subseteq G, |hA| = |\mathbb{N}_0^m(h)| = \binom{m+h-1}{h}\right\}$
$\sigma(G,[0,s]) = \max\left\{|A| \mid A \subseteq G, |[0,s]A| = \Sigma_{h=0}^s|\mathbb{N}_0^m(h)| = \binom{m+s}{s}\right\}$

C.2: Unrestricted signed sumsets:
$\sigma_{\pm}(G,h) = \max\left\{|A| \mid A \subseteq G, |h_{\pm}A| = |\mathbb{Z}^m(h)| = c(h,m)\right\}$
$\sigma_{\pm}(G,[0,s]) = \max\left\{|A| \mid A \subseteq G, |[0,s]_{\pm}A| = \Sigma_{h=0}^s|\mathbb{Z}^m(h)| = a(m,s)\right\}$

C.3: Restricted sumsets:
$\sigma\hat{}(G,h) = \max\left\{|A| \mid A \subseteq G, |h\hat{}A| = |\hat{\mathbb{N}}_0^m(h)| = \binom{m}{h}\right\}$
$\sigma\hat{}(G,[0,s]) = \max\left\{|A| \mid A \subseteq G, |[0,s]\hat{}A| = \Sigma_{h=0}^s|\hat{\mathbb{N}}_0^m(h)| = \Sigma_{h=0}^s\binom{m}{h}\right\}$
$\sigma\hat{}(G,\mathbb{N}_0) = \max\left\{|A| \mid A \subseteq G, |\Sigma A| = \Sigma_{h=0}^m|\hat{\mathbb{N}}_0^m(h)| = 2^m\right\}$

C.4: Restricted signed sumsets:
$\sigma\hat{_\pm}(G,h) = \max\left\{|A| \mid A \subseteq G, |h\hat{_\pm}A| = |\hat{\mathbb{Z}}^m(h)| = \binom{m}{h}2^h\right\}$
$\sigma\hat{_\pm}(G,[0,s]) = \max\left\{|A| \mid A \subseteq G, |[0,s]\hat{_\pm}A| = \Sigma_{h=0}^s|\hat{\mathbb{Z}}^m(h)| = \Sigma_{h=0}^s\binom{m}{h}2^h\right\}$
$\sigma\hat{_\pm}(G,\mathbb{N}_0) = \max\left\{|A| \mid A \subseteq G, |\Sigma_\pm A| = \Sigma_{h=0}^m|\hat{\mathbb{Z}}^m(h)| = 3^m\right\}$

Chapter D: Minimum sumset size

D.1: Unrestricted sumsets:
$\rho(G,m,h) = \min\{|hA| \mid A \subseteq G, |A| = m\}$
$\rho(G,m,[0,s]) = \min\{|[0,s]A| \mid A \subseteq G, |A| = m\}$
$\rho(G,m,\mathbb{N}_0) = \min\{|\langle A\rangle| \mid A \subseteq G, |A| = m\}$

D.2: Unrestricted signed sumsets:
$\rho_\pm(G,m,h) = \min\{|h_\pm A| \mid A \subseteq G, |A| = m\}$
$\rho_\pm(G,m,[0,s]) = \min\{|[0,s]_\pm A| \mid A \subseteq G, |A| = m\}$
$\rho_\pm(G,m,\mathbb{N}_0) = \min\{|\langle A\rangle| \mid A \subseteq G, |A| = m\}$

D.3: Restricted sumsets:
$\rho\hat{}(G,m,h) = \min\{|h\hat{}A| \mid A \subseteq G, |A| = m\}$
$\rho\hat{}(G,m,[0,s]) = \min\{|[0,s]\hat{}A| \mid A \subseteq G, |A| = m\}$
$\rho\hat{}(G,m,\mathbb{N}_0) = \min\{|\Sigma A| \mid A \subseteq G, |A| = m\}$

D.4: Restricted signed sumsets:
$\rho\hat{_\pm}(G,m,h) = \min\{|h\hat{_\pm}A| \mid A \subseteq G, |A| = m\}$
$\rho\hat{_\pm}(G,m,[0,s]) = \min\{|[0,s]\hat{_\pm}A| \mid A \subseteq G, |A| = m\}$
$\rho\hat{_\pm}(G,m,\mathbb{N}_0) = \min\{|\Sigma_\pm A| \mid A \subseteq G, |A| = m\}$

Chapter E: The critical number

E.1: Unrestricted sumsets:
$\chi(G,h) = \min\{m \mid A \subseteq G, |A| = m \Rightarrow hA = G\}$
$\chi(G,[0,s]) = \min\{m \mid A \subseteq G, |A| = m \Rightarrow [0,s]A = G\}$
$\chi(G,\mathbb{N}_0) = \min\{m \mid A \subseteq G, |A| = m \Rightarrow \langle A\rangle = G\}$

E.2: Unrestricted signed sumsets:
$\chi_\pm(G,h) = \min\{m \mid A \subseteq G, |A| = m \Rightarrow h_\pm A = G\}$
$\chi_\pm(G,[0,s]) = \min\{m \mid A \subseteq G, |A| = m \Rightarrow [0,s]_\pm A = G\}$
$\chi_\pm(G,\mathbb{N}_0) = \min\{m \mid A \subseteq G, |A| = m \Rightarrow \langle A\rangle = G\}$

E.3: Restricted sumsets:
$\chi\hat{}(G,h) = \min\{m \mid A \subseteq G, |A| = m \Rightarrow h\hat{}A = G\}$
$\chi\hat{}(G,[0,s]) = \min\{m \mid A \subseteq G, |A| = m \Rightarrow [0,s]\hat{}A = G\}$
$\chi\hat{}(G,\mathbb{N}_0) = \min\{m \mid A \subseteq G, |A| = m \Rightarrow \Sigma A = G\}$

E.4: Restricted signed sumsets:
$\chi\hat{_\pm}(G,h) = \min\{m \mid A \subseteq G, |A| = m \Rightarrow h\hat{_\pm}A = G\}$
$\chi\hat{_\pm}(G,[0,s]) = \min\{m \mid A \subseteq G, |A| = m \Rightarrow [0,s]\hat{_\pm}A = G\}$
$\chi\hat{_\pm}(G,\mathbb{N}_0) = \min\{m \mid A \subseteq G, |A| = m \Rightarrow \Sigma\hat{_\pm}A = G\}$

Chapter F: Zero-sum-free sets

F.1: Unrestricted sumsets:
$\tau(G,h) = \max\{|A| \mid A \subseteq G, 0 \not\in hA\}$
$\tau(G,[1,t]) = \max\{|A| \mid A \subseteq G, 0 \not\in [1,t]A\}$

F.2: Unrestricted signed sumsets:
$\tau_\pm(G,h) = \max\{|A| \mid A \subseteq G, 0 \not\in h_\pm A\}$
$\tau_\pm(G,[1,t]) = \max\{|A| \mid A \subseteq G, 0 \not\in [1,t]_\pm A\}$

F.3: Restricted sumsets:
$\tau\hat{}(G,h) = \max\{|A| \mid A \subseteq G, 0 \not\in h\hat{}A\}$
$\tau\hat{}(G,[1,t]) = \max\{|A| \mid A \subseteq G, 0 \not\in [1,t]\hat{}A\}$
$\tau\hat{}(G,\mathbb{N}) = \max\{|A| \mid A \subseteq G, 0 \not\in [1,m]\hat{}A\}$

F.4: Restricted signed sumsets:
$\tau\hat{_\pm}(G,h) = \max\{|A| \mid A \subseteq G, 0 \not\in h\hat{_\pm}A\}$
$\tau\hat{_\pm}(G,[1,t]) = \max\{|A| \mid A \subseteq G, 0 \not\in [1,t]\hat{_\pm}A\}$
$\tau\hat{_\pm}(G,\mathbb{N}) = \max\{|A| \mid A \subseteq G, 0 \not\in [1,m]\hat{_\pm}A\}$

Chapter G: Sum-free sets

G.1: Unrestricted sumsets:
$\mu(G,\{k,l\}) = \max\{|A| \mid A \subseteq G, kA \cap lA = \emptyset\}$
$\mu(G,[0,s]) = \max\{|A| \mid A \subseteq G, 0 \leq l < k \leq s \Rightarrow kA \cap lA = \emptyset\}$

G.2: Unrestricted signed sumsets:
$\mu_\pm(G,\{k,l\}) = \max\{|A| \mid A \subseteq G, k_\pm A \cap l_\pm A = \emptyset\}$
$\mu_\pm(G,[0,s]) = \max\{|A| \mid A \subseteq G, 0 \leq l < k \leq s \Rightarrow k_\pm A \cap l_\pm A = \emptyset\}$

G.3: Restricted sumsets:
$\mu\hat{}(G,\{k,l\}) = \max\{|A| \mid A \subseteq G, k\hat{}A \cap l\hat{}A = \emptyset\}$
$\mu\hat{}(G,[0,s]) = \max\{|A| \mid A \subseteq G, 0 \leq l < k \leq s \Rightarrow k\hat{}A \cap l\hat{}A = \emptyset\}$
$\mu\hat{}(G,\mathbb{N}_0) = \max\{|A| \mid A \subseteq G, 0 \leq l < k \Rightarrow k\hat{}A \cap l\hat{}A = \emptyset\}$

G.4: Restricted signed sumsets:
$\mu\hat{_\pm}(G,\{k,l\}) = \max\{|A| \mid A \subseteq G, kA \cap lA = \emptyset\}$
$\mu\hat{_\pm}(G,[0,s]) = \max\{|A| \mid A \subseteq G, 0 \leq l < k \leq s \Rightarrow k\hat{_\pm}A \cap l\hat{_\pm}A = \emptyset\}$
$\mu\hat{_\pm}(G,\mathbb{N}_0) = \max\{|A| \mid A \subseteq G, 0 \leq l < k \Rightarrow k\hat{_\pm}A \cap l\hat{_\pm}A = \emptyset\}$

Part I

Ingredients

This book deals with *additive combinatorics*, a vibrant area of current mathematical research. Additive combinatorics—which grew out of combinatorial number theory and additive number theory—here is interpreted, somewhat narrowly, as the study of combinatorial properties of *sumsets* in abelian groups. In Chapters 1, 2, and 3 we provide brief introductions to the relevant branches of *number theory*, *combinatorics*, and *group theory*, respectively. The sections in each chapter contain exercises aimed to solidify the understanding of the material discussed.

Chapter 1

Number theory

Number theory, at least in its most traditional form, is the branch of mathematics that studies the set of integers:

$$\mathbb{Z} = \{\ldots, -2, -1, 0, 1, 2, 3, \ldots\},$$

or one of its subsets, such as the set of nonnegative integers:

$$\mathbb{N}_0 = \{0, 1, 2, 3, 4, 5, \ldots\}$$

or the set of positive integers:

$$\mathbb{N} = \{1, 2, 3, 4, 5, 6, \ldots\}.$$

The field of number theory occupies a distinguished spot within mathematics: the German mathematician Carl Friedrich Gauss (1777–1855) even dubbed it the "Queen of Mathematics." Some of the reasons for this distinction is that it (or she?) manages to simultaneously possess the following seemingly contradictory attributes:

- although a substantial amount of the terminology and the methods in number theory, at least at the introductory level, are quite simple and need very little background, number theory reaches into some of the deepest and most complex areas of mathematics;
- even though many of its questions and problems are easy to present, number theory has a cornucopia of impossibly difficult unsolved questions, some hundreds of years old;
- though number theory may be the oldest branch of mathematics and has always had a large number of devotees, it remains one of the most active and perplexing fields within mathematics with new developments and results being published every day.

Below we review some of the foundations of number theory that we will rely on later.

1.1 Divisibility of integers

The most fundamental concept in number theory is probably *divisibility*: Given two integers a and b, we say that a is a *divisor* of b (or b is *divisible* by a) whenever there is an integer c for which $a \cdot c = b$. If a is a divisor of b, we write $a|b$.

For example, $3|6$ and $6|6$, but $6 \not| 3$. (We must be clear with our terminology: 3 is not divisible by 6, but of course 3 can be divided by 6; in the set $\mathbb{Q}$ of rational numbers—fractions of integers—the concept of divisibility is trivial in that every rational number is divisible by every nonzero rational number.) Also, $5|0$, since $5 \cdot 0 = 0$; in fact, $0|0$, since (for example) $0 \cdot 7 = 0$. But $0 \not| 5$, since there is no integer c for which $0 \cdot c = 5$ because for every real number c, $0 \cdot c = 0$. As it is often the case with mathematical definitions, one needs to be careful: saying that a is a divisor of b is not quite equivalent to saying that the fraction b/a is an integer: 0 is a divisor of 0, but $0/0$ is not an integer (it's not even a number)!

In theory, for each integer n, we can easily find the set of its positive divisors, denoted by $D(n)$. For example, we have

$$D(18) = D(-18) = \{1, 2, 3, 6, 9, 18\}$$

and

$$D(19) = D(-19) = \{1, 19\}.$$

Finding the divisors of large positive integers can be very difficult. Cryptography, the study of encoding and decoding information, takes advantage of the dichotomy that multiplying two large integers (as sometimes used in encoding) is easy, but factoring the product without knowing any of the factors (i.e., decoding) could be, if the factors are chosen carefully, very hard!

Given an integer n, we denote the number of its positive divisors by $d(n)$; that is, we set

$$d(n) = |D(n)|.$$

For instance, as the examples above show, we have $d(18) = d(-18) = 6$ and $d(19) = d(-19) = 2$. Clearly, $d(n)$ is always positive as $1|n$ for every integer n. The function $d(n)$ allows us to separate the set of integers into four classes:

- *units* have exactly one positive divisor,
- *prime numbers* have exactly two positive divisors,
- *composite numbers* have three or more positive divisors, and
- *zero* has infinitely many positive divisors.

There are two units among the integers: 1 and -1 divide all integers, but have only one positive divisor. The set of prime numbers

$$P = \{\pm 2, \pm 3, \pm 5, \pm 7, \pm 11, \pm 13, \pm 17, \pm 19, \dots\},$$

as we explain shortly, forms the basic building block of the set of integers. Primes have been studied for thousands of years. They have many intriguing attributes, some of which are not fully understood to this day. For those interested in more on primes may start their investigations with [188], the On-Line Encyclopedia of Integer Sequences; the positive primes appear as the sequence A000040.

Given two positive integers a and b, we define their *greatest common divisor*, denoted by $\gcd(a, b)$, to be the greatest integer that is a divisor of both a and b; in other words,

$$\gcd(a, b) = \max(D(a) \cap D(b)).$$

A pair of integers is said to be relatively prime if their greatest common divisor is 1. The dual concept is the *least common multiple* of two positive integers: it is the smallest positive integer that is a multiple of both of them. The least common multiple of integers a and b is denoted by $\text{lcm}(a, b)$. It is not hard to see that all pairs of positive integers have a unique greatest common divisor as well as a least common multiple.

Exercises

1. Characterize all integers n for which

   (a) $d(n) = 3$,

   (b) $d(n) = 4$,

   (c) $d(n) = 5$.

2. (a) The concepts of greatest common divisor and least common multiple of two positive integers can be extended to three or more positive integers. Find $\gcd(24, 32, 60)$ and $\text{lcm}(24, 32, 60)$.

   (b) Find positive integers a, b, c, and d that are relatively prime, but no two of them are relatively prime (that is, $\gcd(a, b, c, d) = 1$, but $\gcd(a, b) > 1$, $\gcd(a, c) > 1$, etc.).

1.2 Congruences

As a generalization of divisibility, we may consider situations where an integer leaves a remainder when divided by another integer. More precisely, given a positive integer m, we say that an integer a leaves a *remainder* r when divided by m, if there is an integer k for which $a = m \cdot k + r$ and $0 \leq r \leq m - 1$; in this case we sometimes also say that a is *congruent to r mod m* and write $a \equiv r \mod m$. For example, both 13 and 1863 are congruent to 3 mod 10, and so is -57 as it can be written as $10 \cdot (-6) + 3$. (Recall that any remainder mod 10 must be between 0 and 9, inclusive.) It is not hard to see that for any positive integer m and for any integer a, one can determine a unique remainder r that a is congruent to mod m.

A bit more generally, given a positive integer m, we say that two integers are *congruent mod m* if they leave the same remainder when divided by m. For example, 13 and 1863 are congruent mod 10, since they both leave a remainder of 3 when divided by 10; we denote this by writing $13 \equiv 1863 \mod 10$. In fact, all positive integers with a last decimal digit of 3 and all negative integers with a last digit of 7 are congruent to each other, and they form the *congruence class* of

$$[3]_{10} = \{10k + 3 \mid k \in \mathbb{Z}\} = \{\ldots, -27, -17, -7, 3, 13, 23, 33, \ldots\}.$$

Congruence classes allow us to partition the set of integers in a natural way; for example, we have

$$\mathbb{Z} = [0]_{10} \cup [1]_{10} \cup [2]_{10} \cup [3]_{10} \cup [4]_{10} \cup [5]_{10} \cup [6]_{10} \cup [7]_{10} \cup [8]_{10} \cup [9]_{10}.$$

Since, for a given $m \in \mathbb{N}$, the remainder r may be any integer value between 0 and $m - 1$, we have exactly m congruence classes mod m; these classes are disjoint and their union contains every integer. For example, the congruence classes $[0]_2$ and $[1]_2$ contain the even and the odd integers, respectively; the fact that any integer must be either even or odd but not both can be expressed by saying that $[0]_2$ and $[1]_2$ form a *partition* of $\mathbb{Z}$.

Congruence classes play a prominent role in additive combinatorics, and we will return to their study shortly.

Exercises

1. Find all integers between -100 and 100 that are in the congruence class $[7]_{33}$.
2. (a) Find integers a, b, c, and d for which the congruence classes $[a]_2$, $[b]_4$, $[c]_8$, and $[d]_8$ partition $\mathbb{Z}$. (To be a partition, every integer must belong to exactly one congruence class.)

   (b) Partition $\mathbb{Z}$ into exactly five congruence classes where four of the five moduli are distinct. (It is a well-known result that the moduli cannot all be distinct.)

1.3 The Fundamental Theorem of Number Theory

In general, there may be many ways to factor an integer into a product of other integers. In fact, if we allow 1 or -1 to appear as factors, then every integer has infinitely many different factorizations. For example, factorizations of 18 include $1 \cdot 18$, $1 \cdot (-1) \cdot (-18)$, $2 \cdot 3 \cdot 3$, and $3 \cdot 6$. Here and in the next section we briefly discuss two of these factorizations (namely, generalizations of the last two factorizations of 18 that we listed).

First, there is what is referred to as the *prime factorization* of an integer. According to the *Fundamental Theorem of Number Theory*, every integer n with $n \geq 2$ is either a (positive) prime or can be expressed as a product of positive primes; furthermore, this factorization into primes is essentially unique (that is, there is only one factorization if we ignore the order of the prime factors or the possibility of using their negatives). So, for example, the prime factorization of 18 is $2 \cdot 3 \cdot 3$. In general, the prime factorization of an integer n with $n \geq 2$ can be written as

$$n = \underbrace{p_1 \cdot p_1 \cdots p_1}_{\alpha_1} \cdot \underbrace{p_2 \cdot p_2 \cdots p_2}_{\alpha_2} \cdot \cdots \cdot \underbrace{p_k \cdot p_k \cdots p_k}_{\alpha_k},$$

where $p_1, p_2, \ldots, p_k$ are the distinct prime factors of n and $\alpha_1, \alpha_2, \ldots, \alpha_k$ are positive integers. This prime factorization is often turned into the *prime-power factorization*

$$n = p_1^{\alpha_1} \cdot p_2^{\alpha_2} \cdot \cdots \cdot p_k^{\alpha_k};$$

for example, the prime-power factorization of 18 is $2 \cdot 3^2$. (If n has only a single prime factor p that appears α times in its factorization, we simply write $n = p^\alpha$.) Note that $n = 1$, of course, has no prime or prime-power factorization.

The prime and prime-power factorizations of integers are very useful and appear often in discussions. For example, they allow us to quickly see if an integer b is divisible by another integer a: this is the case if, and only if, each of the prime factors of a appears in b as well and at least as many times as it does in a. For example,

$$18 = 2 \cdot 3^2$$

is a factor of

$$252 = 2^2 \cdot 3^2 \cdot 7,$$

but not of

$$840000 = 2^7 \cdot 3 \cdot 5^5 \cdot 7.$$

It also helps us compute the greatest common divisor and the least common multiple of two integers explicitly; for instance, given the prime-power factorizations of 252 and 840000 above, we immediately see that

$$\gcd(252, 840000) = 2^2 \cdot 3 \cdot 7$$

and

$$\text{lcm}(252, 840000) = 2^7 \cdot 3^2 \cdot 5^5 \cdot 7.$$

(We mention, in passing, that

$$\gcd(252, 840000) \cdot \text{lcm}(252, 840000) = 2^9 \cdot 3^3 \cdot 5^5 \cdot 7^2 = 252 \cdot 840000,$$

exemplifying the general fact that the product of the gcd and the lcm of two integers equals the product of the two integers.)

Exercises

1. (a) Find $d(1225)$.
   (b) Suppose that p and k are positive integers and p is a prime. Find, in terms of p and k, $d(p^k)$.
   (c) Find a formula for $d(n)$ for an arbitrary $n \in \mathbb{N}$ in terms of its prime factorization.
2. Let us define, for a given $n \in \mathbb{N}$, $m \in \mathbb{N}$, and $i = 0, 1, \ldots, m-1$, the set

$$D_{m,i}(n) = \{d \in D(n) \mid d \equiv i \ (m)\}$$

   and, if $D_{m,i}(n) \neq \emptyset$, let

$$f_{m,i}(n) = \min D_{m,i}(n).$$

   (a) Find $f_{3,1}(1225)$ and $f_{3,2}(1225)$.
   (b) Explain why $f_{3,2}(n)$ is a prime number for every $n \in \mathbb{N}$ for which $D_{3,2}(n) \neq \emptyset$.
   (c) Is $f_{3,1}(n)$ also a prime number for every $n \in \mathbb{N}$ for which $D_{3,1}(n) \neq \emptyset$?
   (d) For each value of $i = 0, 1, 2, 3$, decide if $f_{4,i}(n)$ must be a prime or not whenever $D_{4,i}(n) \neq \emptyset$.
   (e) For each value of $i = 0, 1, 2, 3, 4$, decide if $f_{5,i}(n)$ must be a prime or not whenever $D_{5,i}(n) \neq \emptyset$.

1.4 Multiplicative number theory

The branch of number theory commonly referred to as *multiplicative number theory* deals with the various ways that integers can be factored into products of other integers. The Fundamental Theorem of Number Theory, discussed above, plays a key role. The prime factorization and the prime-power factorization of an integer are only two of the many different factorizations; here we discuss some others that we will use later.

First, a common generalization of the prime factorization and the prime-power factorization: the so-called *primary factorization*. Indeed, the prime factorization

$$n = \underbrace{p_1 \cdot p_1 \cdots p_1}_{\alpha_1} \cdot \underbrace{p_2 \cdot p_2 \cdots p_2}_{\alpha_2} \cdot \cdots \cdot \underbrace{p_k \cdot p_k \cdots p_k}_{\alpha_k}$$

and prime-power factorization

$$n = p_1^{\alpha_1} \cdot p_2^{\alpha_2} \cdot \cdots \cdot p_k^{\alpha_k}$$

of an integer $n \geq 2$ can be considered the two extremes of the (potentially) many different primary factorizations of the form

$$n = n_1 \cdot n_2 \cdot \cdots \cdot n_k$$

where each factor n_i is a product of (one or more) prime powers with base p_i (here $i = 1, 2, \ldots, k$). For example, the number $n = 18$ has only two primary factorizations, the prime factorization $(2) \cdot (3 \cdot 3)$ and the prime-power factorization $(2) \cdot (3^2)$, but a number such as $n = 840000 = 2^7 \cdot 3 \cdot 5^5 \cdot 7$ has many, for example,

$$840000 = (2 \cdot 2^3 \cdot 2^3) \cdot (3) \cdot (5 \cdot 5^4) \cdot (7)$$

and

$$840000 = (2^2 \cdot 2^2 \cdot 2^3) \cdot (3) \cdot (5 \cdot 5 \cdot 5^3) \cdot (7).$$

(Our parentheses indicate factors n_1, n_2, etc.)

One can enumerate the number of primary factorizations of a given integer n, as follows. Let us first consider the case when n is a prime power itself; for example, let us examine $n = 32 = 2^5$. It is easy to see that 2^5 has seven primary factorizations:

$$2 \cdot 2 \cdot 2 \cdot 2 \cdot 2 = 2 \cdot 2 \cdot 2 \cdot 2^2 = 2 \cdot 2 \cdot 2^3 = 2 \cdot 2^2 \cdot 2^2 = 2 \cdot 2^4 = 2^2 \cdot 2^3 = 2^5.$$

More generally, if $n = p^\alpha$ for some prime p and positive integer α, then the number of primary factorizations of n agrees with the number of ways that α can be written as the sum of positive integers (where the order of the terms is irrelevant). Denoting this quantity by $p(\alpha)$, we find the following values.

α	1	2	3	4	5	6	7	8	9
$p(\alpha)$	1	2	3	5	7	11	15	22	30

For example, $p(5) = 7$ as the ways to write 5 as the sum of positive integers are

$$1+1+1+1+1 = 1+1+1+2 = 1+1+3 = 1+2+2 = 1+4 = 2+3 = 5.$$

Consequently, any number of the form p^5 with prime base p has seven primary factorizations.

The function $p(\alpha)$ is called the *partition function*; it plays an important role in mathematics in various ways, but, unfortunately, there is no closed formula for it. For more values and information, see sequence A000041 in the On-Line Encyclopedia of Integer Sequences [188].

In general, we can see that if the prime-power factorization of n is

$$n = p_1^{\alpha_1} \cdot p_2^{\alpha_2} \cdot \cdots \cdot p_k^{\alpha_k},$$

then the number of primary factorizations of n equals

$$p(\alpha_1) \cdot p(\alpha_2) \cdot \cdots \cdot p(\alpha_k).$$

Therefore,

$$n = 840000 = 2^7 \cdot 3 \cdot 5^5 \cdot 7$$

has

$$p(7) \cdot p(1) \cdot p(5) \cdot p(1) = 15 \cdot 1 \cdot 7 \cdot 1 = 105$$

different primary factorizations.

Related to primary factorizations, we have the so-called invariant factorizations. We say that

$$n = n_1 \cdot n_2 \cdot \cdots \cdot n_r$$

is an *invariant factorization* of the integer $n \geq 2$, if either $r = 1$, or $r \geq 2$ with $n_1 \geq 2$, and $n_i | n_{i+1}$ holds for each $i = 1, 2, \ldots, r-1$. For example, 18 has two invariant factorizations, $3 \cdot 6$ and 18 itself; 840000, however, has many more, for instance,

$$840000 = (2) \cdot (2^3 \cdot 5) \cdot (2^3 \cdot 3 \cdot 5^4 \cdot 7)$$

and

$$840000 = (2^2 \cdot 5) \cdot (2^2 \cdot 5) \cdot (2^3 \cdot 3 \cdot 5^3 \cdot 7).$$

(As before, our parentheses indicate factors n_1, n_2, etc.)

There is a nice one-to-one correspondence between primary factorizations and invariant factorizations. Producing a primary factorization from an invariant factorization is easy: one can simply order and group the prime powers involved according to their prime bases. For example, the primary factorization that we get from

$$(2) \cdot (2^3 \cdot 5) \cdot (2^3 \cdot 3 \cdot 5^4 \cdot 7)$$

is

$$(2 \cdot 2^3 \cdot 2^3) \cdot (3) \cdot (5 \cdot 5^4) \cdot (7).$$

To get an invariant factorization from a primary factorization, start by setting the largest factor of the invariant factorization equal to the product of the largest factors of each of the factors in the primary factorization, follow that by the second largest factors, and so on. For example, for the primary factorization

$$(2^2 \cdot 2^2 \cdot 2^3) \cdot (3) \cdot (5 \cdot 5 \cdot 5^3) \cdot (7),$$

we see that the largest invariant factor equals $2^3 \cdot 3 \cdot 5^3 \cdot 7$, the next one is $2^2 \cdot 5$, and then $2^2 \cdot 5$ again, yielding the invariant factorization

$$(2^2 \cdot 5) \cdot (2^2 \cdot 5) \cdot (2^3 \cdot 3 \cdot 5^3 \cdot 7).$$

Primary factorizations and invariant factorizations will enable us to classify all finite abelian groups—see Chapter 3.

Exercises

1. Find $p(10)$.
2. (a) How many primary factorizations and how many invariant factorizations does 72 have?

   (b) Find all primary factorizations and invariant factorizations of 72.

1.5 Additive number theory

In contrast to multiplicative number theory that deals with ways in which positive integers factor into products of other positive integers, *additive number theory* is concerned with ways that positive integers can be expressed as sums of certain other positive integers.

In a typical setting, one is given a set A of positive integers, and asks whether every positive integer can be written as a sum of terms all in A. In further variations of this question, one may restrict the total number of terms in the sum or the number of times that a particular element of A may occur in a sum.

Regarding this latter restriction: the two most frequent variations are where we allow any element of A to appear an arbitrary number of times and when each element may occur only at most once. For a given $n \in \mathbb{N}$, $A \subseteq \mathbb{N}$ and $H \subseteq \mathbb{N}$, we introduce the notation $p(n, A, H)$ to denote the number of ways that n can be written as a sum of elements of A, where the total number of terms in the sum must be an element of H, but there is no restriction on the number of times that elements may appear; similarly, $\hat{p}(n, A, H)$ will denote the number of those sums where the total number of terms in the sum must be an element of H, but where each element of A may appear at most once.

For instance, one can verify that there are three ways to write 11 as a sum of either three or four terms, where each term is 2, 3, 6, or 7:

$$7+2+2, \quad 6+3+2, \quad 3+3+3+2,$$

thus

$$p(11, \{2,3,6,7\}, \{3,4\}) = 3,$$

but

$$\hat{p}(11, \{2,3,6,7\}, \{3,4\}) = 1,$$

as only the sum $6+3+2$ has distinct terms.

A more familiar example is the case when $A = \mathbb{N}$ and $H = \mathbb{N}$; that is, the terms may be arbitrary positive integers, and we have no restrictions on the number of terms in the sum or the number of times that a particular term may appear. In this case, we get the partition function $p(n)$ introduced on page 10, so

$$p(n, \mathbb{N}, \mathbb{N}) = p(n).$$

For instance,

$$p(5, \mathbb{N}, \mathbb{N}) = p(5) = 7,$$

as 5 can be written as

$$\begin{aligned} 5 &= 5 \\ &= 4+1 \\ &= 3+2 \\ &= 3+1+1 \\ &= 2+2+1 \\ &= 2+1+1+1 \\ &= 1+1+1+1+1. \end{aligned}$$

If we allow only sums where the terms are distinct, we get $\hat{p}(n, \mathbb{N}, \mathbb{N})$; in the case of $n = 5$, we see that $\hat{p}(5, \mathbb{N}, \mathbb{N}) = 3$ as only the first three sums above contain distinct terms. As we noted on page 10, no closed formula exists for $p(n)$, though its values for small n and various estimates for higher n are known. There is also no closed formula for $\hat{p}(n, \mathbb{N}, \mathbb{N})$, although we should mention the remarkable fact that the number of ways to partition a positive integer into distinct parts equals the number of its partitions into odd parts; that is,

$$\hat{p}(n, \mathbb{N}, \mathbb{N}) = p(n, \mathbb{O}, \mathbb{N}).$$

(We denote the set of odd positive integers here by $\mathbb{O}$.) Indeed, we see that 5 has three partitions comprised of only odd terms.

Keeping $A = \mathbb{N}$ and putting no restrictions on the number of times that elements of A may appear in a sum, but limiting the total number of possible terms to at most s (for some $s \in \mathbb{N}$), we get the function $p(n, \mathbb{N}, [1,s])$. (As usual, we denote the set $H = \{1, 2, \ldots, s\}$ by $[1,s]$.) For example, we see that $p(5, \mathbb{N}, [1,3]) = 5$ since five of the sums above contain no more than three terms. We have no closed formula for $p(n, \mathbb{N}, [1,s])$, though we have another remarkable identity:

$$p(n, \mathbb{N}, [1,s]) = p(n, [1,s], \mathbb{N});$$

that is, for all positive integers n and s, the number of partitions of n into at most s parts equals the number of partitions of n into parts that do not exceed s. For example, just like we had five ways to partition 5 into at most three parts, we have five partitions where each part is at most 3. These are just some of the many amazing identities involving the partition function.

In closing this section, we mention some other interesting examples. The famous *Goldbach Conjecture* asserts that every positive even number is the sum of at most two positive primes; denoting the set of positive primes by $\mathbb{P}$ and the set of positive even integers by $\mathbb{E}$, this conjecture can be presented to say that

$$p(n, P, [1,2]) \geq 1$$

holds for all $n \in \mathbb{E}$. While several partial results have been achieved, the Goldbach Conjecture remains as one of the oldest and most famous unsolved problems in mathematics. Very recently, Helfgott solved the related *Weak Goldbach Conjecture*: that

$$p(n, P, [1,3]) \geq 1$$

for all $n \in \mathbb{O}$ and $n > 1$ (see [118]; this work has not been published in a refereed journal yet).

For a fixed positive integer k, let S_k denote the set $\{1^k, 2^k, 3^k, \dots\}$ of all k-th powers of positive integers. *Waring's Problem* asks for the smallest positive integer s, usually denoted by $g(k)$, for which

$$p(n, S_k, [1,s]) \geq 1$$

for all $n \in \mathbb{N}$; that is, the smallest s for which it is true that every positive integer can be written as at most s k-th powers. (Equivalently, we may include 0^k as an element of S_k in which case $g(k)$ is the minimum value of h for which

$$p(n, S_k, \{h\}) \geq 1$$

for all $n \in \mathbb{N}$.) For example, $g(2) = 4$: as the *Four Squares Theorem* or *Lagrange's Theorem* asserts, every positive integer can be written as the sum of at most four positive squares (or, equivalently, as the sum of exactly 4 nonnegative squares). To see that $g(2)$ cannot be less than 4, observe that $n = 7$ (and infinitely many other n) indeed requires four squares. We also know that $g(3) = 9$, $g(4) = 19$, and many other values of $g(k)$, but the question of finding all values is still open today.

Our final famous example is the so-called *Money Changing Problem*. Suppose that we are given a finite set A of relatively prime positive integers; we want to find the largest positive integer n, denoted by $f(A)$, for which

$$p(n, A, \mathbb{N}) = 0;$$

that is, the largest positive integer n that cannot be written as a sum of elements of A. (The reason for the name of the problem should be obvious.) Note that if the elements of A are not relatively prime, then $f(A)$ does not exist; this is the case, for example, if all elements of A are even, as only even numbers can be partitioned into even terms. On the other hand, one can show that if the elements of A are relatively prime, then $f(A)$, called the *Frobenius number of A*, exists. For example, with $A = \{5, 8\}$, we find that $f(A) = 27$ as 27 cannot be written as a sum of 5's and 8's, but every number greater than 27 can be:

$$\begin{aligned}
28 &= 4\cdot 5+1\cdot 8,\\
29 &= 1\cdot 5+3\cdot 8,\\
30 &= 6\cdot 5+0\cdot 8,\\
31 &= 3\cdot 5+2\cdot 8,\\
32 &= 0\cdot 5+4\cdot 8,\\
33 &= 5\cdot 5+1\cdot 8,\\
34 &= 2\cdot 5+3\cdot 8,
\end{aligned}$$

and so on. In general, if $A=\{a,b\}$ (and $\gcd(a,b)=1$), then $f(A)=ab-a-b$; similar formulas for the case when $|A|>2$ are not known.

Exercises

1. (a) Verify that, as listed on page 10, $p(7,\mathbb{N},\mathbb{N})=p(7)=15$.
   (b) Verify that $p\hat{}(7,\mathbb{N},\mathbb{N})=p(7,\mathbb{O},\mathbb{N})$.
   (c) Verify that $p(7,\mathbb{N},[1,3])=p(7,[1,3],\mathbb{N})$.
2. (a) Prove that the Weak Goldbach Conjecture follows from the Goldbach Conjecture.
   (b) Prove that the Weak Goldbach Conjecture implies that
   $$p(n,P,[1,4])\geq 1$$
   holds for all $n\in\mathbb{N}$ and $n>1$.
3. (a) Prove that $g(3)\geq 9$ by finding a positive integer n for which $p(n,S_3,[1,8])=0$.
   (b) Prove that $g(4)\geq 19$ by finding a positive integer n for which $p(n,S_4,[1,18])=0$.
4. Suppose that a fast-food chain sells chicken nuggets in packages of 6, 9, and 20. Find the largest positive integer n for which we are not able to buy exactly n pieces.

Chapter 2

Combinatorics

Explaining what combinatorics is about may be simple since it deals with objects and techniques that are quite familiar to most people. Yet, combinatorics is not easy to define precisely. At the fundamental level, combinatorics deals with the questions related to counting the number of elements in a given set; a bit more precisely, combinatorics deals with *discrete structures*: sets—with, perhaps, some specific characteristics—whose elements can be listed and enumerated, as opposed to sets whose elements vary continuously and cannot be put in a list.

Moreover, beyond its object of study, combinatorics can be characterized by its methods. Typically, by the *combinatorial method* we mean a relatively basic—but, perhaps, surprisingly deep and far-reaching—argument using some relatively elementary tools, rather than the application of sophisticated and elaborately developed machinery. That is what makes combinatorics so highly applicable and why it serves as a very elegant and accessible branch of study in the mathematics curriculum.

In this section we introduce some of the concepts and methods of combinatorics that we will need later.

2.1 Basic enumeration principles

Enumeration—or, simply, counting—is probably one of our earliest intellectual pursuits, and it is a ubiquitous task in everyday life. The principles of enumeration are also what several branches of mathematics are based on, especially probability theory and statistics. In this section we briefly discuss elementary enumeration techniques.

A typical enumeration problem asks us to determine the *size* of a set: the size of a set A, denoted by $|A|$, is the number of elements in A. Clearly, each set has either finite or infinite size. Here we focus on finite sets only.

Most enumeration questions can be reduced to one of two fundamental principles: the Addition Rule and the Multiplication Rule. According to the *Addition Rule*, if A and B are disjoint finite sets, then we have

$$|A \cup B| = |A| + |B|.$$

More generally, if $A_1, A_2, \dots, A_n$ are pairwise disjoint finite sets ($n \in \mathbb{N}$), then we have

$$|A_1 \cup \dots \cup A_n| = |A_1| + \dots + |A_n|.$$

The *Multiplication Rule* says that for arbitrary finite sets A and B, we have

$$|A \times B| = |A| \cdot |B|,$$

and more generally, for arbitrary finite sets $A_1, A_2, \ldots, A_n$ ($n \in \mathbb{N}$), we have

$$|A_1 \times \cdots \times A_n| = |A_1| \cdot \cdots \cdot |A_n|.$$

Observe that the Addition Rule—unlike the Multiplication Rule—requires that the sets be pairwise disjoint. A more general formula treats the case when our sets are not (or not known to be) pairwise disjoint: for finite sets A and B we can verify that

$$|A \cup B| = |A| + |B| - |A \cap B|;$$

indeed, to count the elements in the union of A and B, adding the sizes of A and B together would double-count the elements that are in both A and B, so we need to subtract the number of elements in the intersection of A and B. Similarly, for finite sets A, B, and C, we have

$$|A \cup B \cup C| = |A| + |B| + |C| - |A \cap B| - |A \cap C| - |B \cap C| + |A \cap B \cap C|.$$

The situation gets more complicated as the number of sets increases; while a precise statement (called the *Inclusion–Exclusion Rule*) is readily available, we will not state it here. Instead, we just point out that, in general, for arbitrary sets $A_1, A_2, \ldots, A_n$ we have

$$|A_1 \cup \cdots \cup A_n| \leq |A_1| + \cdots + |A_n|.$$

An often-used consequence of this inequality is the *Pigeonhole Principle*, which says that, if we have

$$|A_1 \cup \cdots \cup A_n| > kn$$

for some nonnegative integer k, then there must be at least one index $i \in \{1, \ldots, n\}$ for which $|A_i| \geq k + 1$. To paraphrase: if more than kn pigeons happen to sit in n holes, then at least one hole must have at least $k + 1$ pigeons in it. Consequently, for example, we see that in a set of 101 positive integers, one can always find 11 (or more) that share their last digits; similarly, among a group of 3000 people there is always a group of at least nine that share the same birthday.

While the counting principles we reviewed here may seem rather elementary, they have far-reaching consequences. We present some in the exercises below.

Exercises

1. Exhibit the Inclusion–Exclusion formula for four sets; that is, find an expression for the size of the union of four sets in terms of the sizes of their various intersections.

2. Prove that however we place seven points inside an 8-by-9 rectangle, we can always find

   (a) a pair whose distance is at most 5, and

   (b) three that form a triangle of area at most 12.

3. Let S be a set of 100 distinct positive integers. Which of the following statements are true?

   (a) If each element of S is at most 198, then S must contain two elements that are relatively prime.

   (b) If each element of S is at most 199, then S must contain two elements that are relatively prime.

   (c) If each element of S is at most 198, then S must contain two elements so that one is divisible by the other.

   (d) If each element of S is at most 199, then S must contain two elements so that one is divisible by the other.

2.2 Counting lists, sequences, sets, and multisets

Before we discuss the four main counting questions in mathematics, we review some familiar terminology and notations, and introduce some new ones. Recall that, for a given set A and positive integer m, an element $(a_1, a_2, \ldots, a_m)$ of A^m is called a sequence of length m. The order of the terms in the sequence matters; for example, the sequence $(2,3,4,5)$ of integers is different from $(3,2,4,5)$. On the other hand, a subset of A of size m is simply a collection of m of its elements, where two subsets are considered equal without regard to the order in which the terms are listed; for example, $\{2,3,4,5\}$ and $\{3,2,4,5\}$ are equal subsets of the set of integers. Recall also that a set remains unchanged if we choose to list some of its elements more than once; for example, the sets $\{2,3,3,5\}$, $\{2,3,5,5\}$, and $\{2,3,5\}$ are all equal, while the sequences $(2,3,3,5)$, $(2,3,5,5)$, and $(2,3,5)$ are all different. Thus, we can consider sets as two-fold relaxations of sequences: we don't care about the order in which the elements are listed, nor do we care how many times the elements are listed.

It will be useful for us to introduce two other objects. First, we say that a sequence $(a_1, a_2, \ldots, a_m)$ of elements of a set A is a *list*, if the m terms are pairwise distinct. Thus, in a list, the order of the elements still matters, but each element is only allowed to appear once. For example, the sequence $(2,3,4,5)$ is a list, but $(2,3,3,5)$ is not. Conversely, in a so-called *multiset* $[a_1, a_2, \ldots, a_m]$ of size m, the order of the elements $a_1, a_2, \ldots, a_m$ of A does not matter (as it is the case with sets), but elements may appear repeatedly (as they may in sequences). For example, the multisets $[2,3,3,5]$, $[2,3,5,5]$, and $[2,3,5]$ are all different, but $[2,3,3,5]$ is still the same as $[2,5,3,3]$.

Given a set A and a positive integer m, we are interested in counting the number of m-sequences (sequences of length m), m-lists (lists of length m), m-multisubsets (multisubsets of size m), and m-subsets (subsets of size m) of A. The schematic summary of these four terms is given in the following table.

	order matters	order does not matter
elements distinct	m-lists	m-sets
elements may repeat	m-sequences	m-multisets

Obviously, if $|A| < m$, then A has neither m-lists nor m-subsets. If $|A| = m$, then the (only) m-subset of A is A itself, while, as we will soon see, if $|A| = m$, then A has $m!$ m-lists. For other situations, we introduce the following notations.

Suppose that n is a nonnegative integer and m is a positive integer. We define the *rising factorial m-th power* and the *falling factorial m-th power* of n to be

$$n^{\overline{m}} = n(n+1)\cdots(n+m-1)$$

and

$$n^{\underline{m}} = n(n-1)\cdots(n-m+1),$$

respectively. For example, we have

$$10^{\overline{3}} = 10 \cdot 11 \cdot 12 = 1320$$

and

$$10^{\underline{3}} = 10 \cdot 9 \cdot 8 = 720.$$

Analogous to $n^0 = 1$ and $0! = 1$, we extend these notations with

$$n^{\underline{0}} = 1 \text{ and } n^{\overline{0}} = 1$$

for arbitrary nonnegative integers n.

Furthermore, we introduce the notations $\binom{n}{m}$ (pronounced "n choose m") and $\left[{n \atop m}\right]$ (pronounced "n multichoose m"): For nonnegative integers m and n,

$$\binom{n}{m} = \frac{n^{\underline{m}}}{m!} = \frac{n(n-1)\cdots(n-m+1)}{m!}$$

and

$$\left[{n \atop m}\right] = \frac{n^{\overline{m}}}{m!} = \frac{n(n+1)\cdots(n+m-1)}{m!}.$$

It is well known that these quantities always denote integers. The values of $\binom{n}{m}$, also known as *binomial coefficients*, are exhibited in *Pascal's Triangle*; here we tabulate some of these values in a table format. (Note that, when $m > n$, the formula above yields $\binom{n}{m} = 0$; keeping the traditional shape of Pascal's Triangle, we omitted these entries from the table below.)

$\binom{n}{m}$	m=0	m=1	m=2	m=3	m=4	m=5	m=6	m=7
n=0	1							
n=1	1	1						
n=2	1	2	1					
n=3	1	3	3	1				
n=4	1	4	6	4	1			
n=5	1	5	10	10	5	1		
n=6	1	6	15	20	15	6	1	
n=7	1	7	21	35	35	21	7	1

Observe that, since

$$\frac{n(n-1)\cdots(n-m+1)}{m!} = \frac{n(n-1)\cdots(m+1)}{(n-m)!}$$

(which we can check by cross-multiplying), we have the identity

$$\binom{n}{m} = \binom{n}{n-m},$$

expressing the fact that the rows in Pascal's Triangle are "palindromic." The explanation for the term "binomial coefficient" will be clear once we discuss the Binomial Theorem below.

The first few values of $\left[{n \atop m}\right]$ are as follows.

$\left[{n \atop m}\right]$	m=0	m=1	m=2	m=3	m=4	m=5	m=6	m=7
n=1	1	1	1	1	1	1	1	1
n=2	1	2	3	4	5	6	7	8
n=3	1	3	6	10	15	21	28	36
n=4	1	4	10	20	35	56	84	120
n=5	1	5	15	35	70	126	210	330
n=6	1	6	21	56	126	252	462	792
n=7	1	7	28	84	210	462	924	1716

As we can see, the two tables contain the same data—values are just shifted: the entries in column m in the first table are moved up by $m-1$ rows in the second table. Indeed, since for integers n and m we clearly have

$$n^{\overline{m}} = n(n+1)\cdots(n+m-1) = (n+m-1)(n+m-2)\cdots n = (n+m-1)^{\underline{m}},$$

we see that values of $\left[{n \atop m} \right]$ can be expressed via the more-often-used binomial coefficients as

$$\left[{n \atop m} \right] = \binom{n+m-1}{m}.$$

We are now ready to "size up" our four main configurations. The *Enumeration Theorem* says that, if A is a set of size n and m is a positive integer, then

- the number of m-sequences of A is n^m,
- the number of m-lists of A is $n^{\underline{m}}$,
- the number of m-multisubsets of A is $\left[{n \atop m} \right]$, and
- the number of m-subsets of A is $\binom{n}{m}$.

Note that, when $n < m$, then $n^{\underline{m}} = 0$ and $\binom{n}{m} = 0$, in accordance with the fact that A has no m-lists and no m-subsets in this case. If $n = m$, then $n^{\underline{m}} = m!$ and $\binom{n}{m} = 1$; indeed, in this case A has $m!$ m-lists while its only m-subset is itself.

The enumeration techniques discussed above are often employed to determine the number of choices one has for selecting or arranging a given number of elements from a given set or collection of sets. For example, the Addition Rule and the Multiplication Rule can be interpreted to say that, given boxes labeled $A_1, A_2, \ldots, A_n$, if box A_i contains m_i distinct objects ($i = 1, 2, \ldots, n$), then there are

$$m_1 + m_2 + \cdots + m_n$$

ways to choose an object from one of the boxes, and there are

$$m_1 \cdot m_2 \cdot \cdots \cdot m_n$$

ways to choose an object from each of the boxes. In a similar manner, the four basic enumeration functions of the Enumeration Theorem are sometimes called "choice functions"; the following table summarizes our results for the number of ways to choose m elements from a given set of n elements.

	order matters	order does not matter
elements distinct	$n^{\underline{m}}$	$\binom{n}{m}$
elements may repeat	n^m	$\left[{n \atop m} \right]$

An important example for enumeration problems, one that we will refer to often, is to count the number of positive integer solutions to an equation of the form

$$x_1 + x_2 + \cdots + x_m = h;$$

that is, to find, for a given $h \in \mathbb{N}$, the number of m-sequences of $\mathbb{N}$ with the property that the entries in the sequence add up to h. (Note that we are counting sequences: the order of the terms does matter.) We can visualize this question by imagining a segment of length h inches with markings at all integer inches (that is, at 1, 2, and so on, all the way to $h-1$); our task is then to find the number of ways this segment can be broken into m pieces at $m-1$ distinct markings: the lengths of the m parts created will correspond, in order, to $x_1, x_2, \ldots, x_m$. By the Enumeration Theorem, the number of ways that this can be done, and therefore the number of positive integer solutions to our equation, is $\binom{h-1}{m-1}$. As a variation, one can easily prove (see one of the exercises below) that the number of nonnegative integer solutions to the same equation equals $\left[{h+1 \atop m-1} \right] = \binom{m+h-1}{h}$.

Exercises

1. Find the number of

   (a) 4-sequences,

   (b) 4-lists,

   (c) 4-multisubsets, and

   (d) 4-subsets

   of a set of size 7. Exhibit one example for each question.

2. Above we have shown that the number of positive integer solutions to an equation

   $$x_1 + x_2 + \cdots + x_m = h$$

   equals $\binom{h-1}{m-1}$. Here we use three different approaches to find the number of nonnegative solutions to the equation.

   (a) Modify the argument used for the number of positive integer solutions to prove that the number of nonnegative integer solutions equals $\left[{h+1 \atop m-1} \right]$.

   (b) Explain why the number of nonnegative integer solutions to

   $$x_1 + x_2 + \cdots + x_m = h$$

   is the same as the number of positive integer solutions to

   $$x_1 + x_2 + \cdots + x_m = h + m,$$

   and use this fact to get that the result is $\binom{h+m-1}{m-1}$.

   (c) Explain why the number of nonnegative integer solutions to

   $$x_1 + x_2 + \cdots + x_m = h$$

   is the same as the number of ways one can place h identical objects into m distinct boxes, and use this fact to get that the result is $\left[{m \atop h} \right]$.

   (d) Verify algebraically that the results of the three previous parts are the same.

2.3 Binomial coefficients and Pascal's Triangle

In this section we discuss some of the many famous and interesting properties of the so-called *binomial coefficients*, that is, the quantities $\binom{n}{m}$. First, let us explain the reason for the name.

A closer look at the rows of the table of entries for $\binom{n}{m}$, exhibited earlier, reveals, in order, the coefficients of the various terms in the expansion of the power $(a+b)^n$ (here a and b are arbitrary real numbers and n is a nonnegative integer). For example, the entries in row 4 of the table are 1, 4, 6, 4, and 1; indeed, the power $(a+b)^4$ expands as

$$(a+b)^4 = a^4 + 4a^3b + 6a^2b^2 + 4ab^3 + b^4.$$

We can easily explain this coincidence as follows. When using the distributive law to expand the expression

$$(a+b)^n = (a+b)\cdots(a+b),$$

we arrive at a sum of products of n factors, where each factor is either a or b. Using the commutative property of multiplication, each term can be arranged so that the a's (if any) all come before the b's (if any). We can then collect "like" terms; that is, terms of the form $a^{n-m}b^m$ for the same $m = 0, 1, \ldots, n$. The number of such terms clearly equals the number of those n-sequences of the set $\{a, b\}$ that contain exactly $n - m$ a's and m b's, which, by the Enumeration Theorem, is exactly $\binom{n}{m}$.

This result is known as (Newton's) Binomial Theorem, and can be stated in general as the identity

$$(a+b)^n = \sum_{m=0}^{n} \binom{n}{m} a^{n-m} b^m.$$

As the name implies, $\binom{n}{m}$ is indeed a "binomial coefficient."

The first few entries for $\binom{n}{m}$ (and, therefore, for $\left[{n \atop m}\right]$ as well) are tabulated in *Pascal's Triangle* below.

$$\begin{array}{ccccccccccccccc}
&&&&&&&1&&&&&&& \\
&&&&&&1&&1&&&&&& \\
&&&&&1&&2&&1&&&&& \\
&&&&1&&3&&3&&1&&&& \\
&&&1&&4&&6&&4&&1&&& \\
&&1&&5&&10&&10&&5&&1&& \\
&1&&6&&15&&20&&15&&6&&1& \\
1&&7&&21&&35&&35&&21&&7&&1
\end{array}$$

We can read off values of $\binom{n}{m}$ as follows. If we label the rows, the "left" diagonals, and the "right" diagonals 0, 1, 2, etc. (we start with 0), then $\binom{n}{m}$ appears as the entry where row n and right diagonal m intersect. For example, we see that $\binom{6}{3} = 20$.

The binomial coefficients possess many interesting properties. We have already mentioned the fact that the rows are "palindromic":

$$\binom{n}{m} = \binom{n}{n-m}.$$

Another important property is known as Pascal's Identity:

$$\binom{n}{m} = \binom{n-1}{m} + \binom{n-1}{m-1}.$$

This identity provides us, actually, with the easiest way to enumerate binomial coefficients: each entry in Pascal's Triangle is simply the sum of the two entries above it. We can, thus, quite quickly find the next row:

$$1 \quad 8 \quad 28 \quad 56 \quad 70 \quad 56 \quad 28 \quad 8 \quad 1$$

Another interesting property of Pascal's Triangle is that the sum of the entries in each row add up to a power of 2:

$$\binom{n}{0} + \binom{n}{1} + \cdots + \binom{n}{n} = 2^n.$$

Note that this identity follows directly from the Binomial Theorem (take $a = b = 1$). Similarly, (by taking $a = 1$ and $b = 2$) we have

$$\binom{n}{0} \cdot 2^0 + \binom{n}{1} \cdot 2^1 + \cdots + \binom{n}{n} \cdot 2^n = 3^n.$$

Of the numerous other interesting properties of Pascal's Triangle, we list only two more:

$$\binom{n-1}{m} + \binom{n-2}{m-1} + \cdots + \binom{n-m-1}{0} = \binom{n}{m},$$

expressing the fact that the entries in each NW-SE diagonal, above a certain row, add to an entry in the next row. Adding up numbers on NE-SW diagonals yields

$$\binom{n-1}{m-1} + \binom{n-2}{m-1} + \cdots + \binom{m-1}{m-1} = \binom{n}{m}.$$

(See the exercises below for proofs.)

We will use each of these identities later.

Exercises

1. What identity of binomial coefficients arises from using the Binomial Theorem for evaluating $(1-1)^n$? Verify your identity for $n = 6$ and $n = 7$.

2. Suppose that n and m are positive integers, and suppose that $m \leq n$.

   (a) We set $A = \{1, 2, \ldots, n\}$; furthermore, we let A_0 be the set of m-subsets of A that do not contain 1, let A_1 be the set of m-subsets of A that contain 1 but do not contain 2, let A_2 be the set of m-subsets of A that contain 1 and 2 but do not contain 3, and so on. Prove the identity

   $$\binom{n-1}{m} + \binom{n-2}{m-1} + \cdots + \binom{n-m-1}{0} = \binom{n}{m}$$

   by considering $A_0 \cup A_1 \cup A_2 \cup \cdots \cup A_m$.

   (b) Prove the identity

   $$\binom{n-1}{m-1} + \binom{n-2}{m-1} + \cdots + \binom{m-1}{m-1} = \binom{n}{m}$$

   using similar techniques as in part (a).

2.4 Some recurrence relations

Let us return to the binomial coefficients discussed in the previous section. Rather than looking at Pascal's Triangle, let's arrange their values in a more convenient table format:

p(j,k)	k=0	k=1	k=2	k=3	k=4	k=5	k=6	k=7
j=0	1	1	1	1	1	1	1	1
j=1	1	2	3	4	5	6	7	8
j=2	1	3	6	10	15	21	28	36
j=3	1	4	10	20	35	56	84	120
j=4	1	5	15	35	70	126	210	330
j=5	1	6	21	56	126	252	462	792
j=6	1	7	28	84	210	462	924	1716
j=7	1	8	36	120	330	792	1716	3432

Here $p(j,k)$ denotes the entry in row j and column k; we have

$$p(j,k) = \left[\begin{matrix} j+1 \\ k \end{matrix} \right] = \binom{j+k}{k}.$$

As we noted before, Pascal's Identity, together with the values in the top row and the left-most column in the table, determine all entries *recursively*: we simply need to add the (previously determined) values directly above and directly to the left of the desired entry. In fact, we can *define* the function $p(j,k)$ recursively by the *recurrence relation*

$$p(j,k) = p(j-1,k) + p(j,k-1)$$

and the *initial conditions* that $p(j,0) = 1$ for all $j \in \mathbb{N}_0$ and $p(0,k) = 1$ for all $k \in \mathbb{N}_0$.

Let us consider a variation where the function $a(j,k)$ is defined by the initial conditions $a(j,0) = 1$ for all $j \in \mathbb{N}_0$ and $a(0,k) = 1$ for all $k \in \mathbb{N}_0$ and by the recursive relation

$$a(j,k) = a(j-1,k) + a(j-1,k-1) + a(j,k-1).$$

The first few values of the function are as follows.

$a(j,k)$	$k=0$	$k=1$	$k=2$	$k=3$	$k=4$	$k=5$	$k=6$
$j=0$	1	1	1	1	1	1	1
$j=1$	1	3	5	7	9	11	13
$j=2$	1	5	13	25	41	61	85
$j=3$	1	7	25	63	129	231	377
$j=4$	1	9	41	129	321	681	1289
$j=5$	1	11	61	231	681	1683	3653
$j=6$	1	13	85	377	1289	3653	8989

The numbers in this table are called *Delannoy numbers*, named after the French amateur mathematician who introduced them in the nineteenth century in [62]. According to its recurrence relation, the Delannoy number $a(j,k)$ is the sum of not only the entries directly above and to the left, but the entry in the "above-left" position as well. Delannoy numbers—like any two-dimensional array—can be turned into a sequence by listing entries by its anti-diagonals; this sequence is given as A008288 in [188]. (The sequence of entries in various columns can be found in [188] as well.) In the next section we shall see an interesting interpretation of Delannoy numbers.

Changing the initial conditions, we next define the function $c(j, k)$ by the initial conditions $c(j, 0) = 0$ for all $j \in \mathbb{N}$ and $c(0, k) = 1$ for all $k \in \mathbb{N}_0$ and by the (same) recursive relation

$$c(j, k) = c(j-1, k) + c(j-1, k-1) + c(j, k-1).$$

The first few values of this function are as follows.

$c(j,k)$	$k=0$	$k=1$	$k=2$	$k=3$	$k=4$	$k=5$	$k=6$
$j=0$	1	1	1	1	1	1	1
$j=1$	0	2	4	6	8	10	12
$j=2$	0	2	8	18	32	50	72
$j=3$	0	2	12	38	88	170	292
$j=4$	0	2	16	66	192	450	912
$j=5$	0	2	20	102	360	1002	2364
$j=6$	0	2	24	146	608	1970	5336

The numbers given by this table can be found in sequence form at A266213 in [188].

The functions $a(j, k)$ and $c(j, k)$ are strongly related; it is not difficult to reduce each one to the other, as we now show.

Consider first the function $c(j, k)$. A quick glance at the tables above suggests that for $j \geq 1$ and $k \geq 1$, the entry $c(j, k)$ is the sum of entries $a(j, k-1)$ and $a(j-1, k-1)$:

$$c(j, k) = a(j, k-1) + a(j-1, k-1).$$

We will prove this by induction. We see that the equation holds for $j = 1$ and for $k = 1$; we then use the defining recursions for both c and a, as well as our inductive hypothesis, for $j \geq 1$ and $k \geq 1$ to write

$$\begin{aligned}
c(j, k) &= c(j-1, k) + c(j-1, k-1) + c(j, k-1) \\
&= a(j-1, k-1) + a(j-2, k-1) + \\
&\quad + a(j-1, k-2) + a(j-2, k-2) + \\
&\quad + a(j, k-2) + a(j-1, k-2) \\
&= a(j, k-1) + a(j-1, k-1),
\end{aligned}$$

as claimed.

Using the recursion for $a(j, k)$ once more, we may rewrite this identity as

$$c(j, k) = a(j, k) - a(j-1, k),$$

from which we get

$$a(j, k) = c(j, k) + a(j-1, k).$$

We can then use this identity to express a in terms of c:

$$\begin{aligned}
a(j, k) &= c(j, k) + a(j-1, k) \\
&= c(j, k) + c(j-1, k) + a(j-2, k) \\
&= c(j, k) + c(j-1, k) + c(j-2, k) + a(j-3, k) \\
&= \ldots \\
&= c(j, k) + c(j-1, k) + \cdots + c(1, k) + a(0, k) \\
&= c(j, k) + c(j-1, k) + \cdots + c(1, k) + c(0, k).
\end{aligned}$$

In summary, we have:

Proposition 2.1 *For the functions $a(j,k)$ and $c(j,k)$, defined recursively above for all nonnegative integers j and k, we have*

$$c(j,k) = a(j,k-1) + a(j-1,k-1) = a(j,k) - a(j-1,k)$$

for all $j,k \in \mathbb{N}$ and

$$a(j,k) = c(j,k) + c(j-1,k) + \cdots + c(1,k) + c(0,k)$$

for all $j,k \in \mathbb{N}_0$.

While recursive expressions are quite helpful, direct formulae, if they exist, would be even more useful, particularly when the variables are large. For example, the binomial coefficients can be easily computed via the function $p(j,k)$ defined above, but it is good to know that

$$p(j,k) = \binom{j+k}{k} = \frac{(j+k)!}{j! \cdot k!}.$$

Although formulae for $a(j,k)$ and $c(j,k)$ are not so direct, we have the following expressions.

Proposition 2.2 *For all nonnegative integers j and k we have*

$$a(j,k) = \sum_{i \geq 0} \binom{j}{i} \binom{k}{i} 2^i$$

and

$$c(j,k) = \sum_{i \geq 0} \binom{j-1}{i-1} \binom{k}{i} 2^i.$$

For a proof of Proposition 2.2, see page 308. Note that, while the summations seem to include infinitely many terms, all but finitely many are zero. Note also that including $i = 0$ in the last sum is only relevant if $j = 0$. (Here we use the convention that $\binom{j-1}{-1}$ equals 1 for $j = 0$ and 0 if $j > 0$.)

It may seem a bit strange that, while $p(j,k)$ and $a(j,k)$ are similarly defined—with the only difference being that p relies on a double recursion while a uses a triple recursion—they yield very different formulae. We gain some insight by the following consideration. Suppose that we are given $j+k$ distinct (for example, numbered) balls, and that j of them are green and k are yellow. Recall that $p(j,k) = \binom{j+k}{k}$, a quantity expressing the number of ways one can select k balls from the collection of these $j+k$ balls. Now if i of the k balls selected are green and the other $k-i$ are yellow, then the number of such choices, by the Multiplication Rule, equals $\binom{j}{i} \cdot \binom{k}{i}$; summing over all possible values of i, we get

$$p(j,k) = \binom{j+k}{j} = \sum_{i \geq 0} \binom{j}{i} \binom{k}{i},$$

a form more closely reminiscent to the one for $a(j,k)$ in Proposition 2.2.

We will make frequent use of the quantities $a(j,k)$ and $c(j,k)$. Without going into detail here, we mention, for example, that

- for an s-spanning set of size m in a group of order n (see Chapter B), we have $n \leq a(m,s)$;
- for a B_h set over $\mathbb{Z}$ of size m in a group of order n (see Chapter C), we have $n \geq c(h,m)$; and

- for a t-independent set of size m in a group of order n (see Chapter F), we have
  - $n \geq c(m, \frac{t+1}{2})$ if t is odd and $t > 1$, and
  - $n \geq a(m, \frac{t}{2})$ if t is even.

We will discuss each of these bounds in the relevant chapters of the book.

Exercises

1. Find $a(7,7)$ and $c(7,7)$ using
   (a) their recursive definitions and
   (b) Proposition 2.2.
2. Suppose that $m \in \mathbb{N}$. Express $a(m,3)$ and $c(m,3)$ as polynomial functions in m.

2.5 The integer lattice and its layers

One of the most often discussed combinatorial objects—and one that we will frequently rely on—is the m-dimensional *integer lattice*

$$\mathbb{Z}^m = \underbrace{\mathbb{Z} \times \mathbb{Z} \times \cdots \times \mathbb{Z}}_{m},$$

consisting of all points (or vectors) $(\lambda_1, \lambda_2, \ldots, \lambda_m)$ with integer coordinates. (Here $m \in \mathbb{N}$ is called the *dimension* of the lattice.) Of course, the 1-dimensional integer lattice is simply the set of integers $\mathbb{Z}$, while the points in $\mathbb{Z}^2$ are arranged in an infinite (2-dimensional) grid in the plane, and $\mathbb{Z}^3$ can be visualized as an infinite grid in 3-space. (For $m \geq 4$, geometric visualization of $\mathbb{Z}^m$ is not convenient.)

A *layer* of the integer lattice is defined as the collection of points with a given fixed *norm*; that is, for a given nonnegative integer h, the hth layer of $\mathbb{Z}^m$ is defined as

$$\mathbb{Z}^m(h) = \{(\lambda_1, \lambda_2, \ldots, \lambda_m) \in \mathbb{Z}^m \mid |\lambda_1| + |\lambda_2| + \cdots + |\lambda_m| = h\}.$$

Obviously, $\mathbb{Z}^m(0)$ consists of a single point, the origin. We can also easily see that $\mathbb{Z}^m(1)$ consists of the points of the m-dimensional lattice with all but one coordinate equal to 0 and the remaining coordinate equal to 1 or -1; there are exactly $2m$ such points.

Describing $\mathbb{Z}^m(h)$ explicitly gets more complicated as h increases, however. For instance, we find that

$$\mathbb{Z}^2(3) = \{(0, \pm 3), (\pm 1, \pm 2), (\pm 2, \pm 1), (\pm 3, 0)\},$$

and

$$\mathbb{Z}^3(2) = \{(0,0,\pm 2), (0, \pm 2, 0), (\pm 2, 0, 0), (0, \pm 1, \pm 1), (\pm 1, 0, \pm 1), (\pm 1, \pm 1, 0)\}.$$

The twelve points of $\mathbb{Z}^2(3)$ lie on the boundary of a square in the plane (occupying the four vertices and two points on each edge), and the eighteen points of $\mathbb{Z}^3(2)$ are on the surface of an octahedron (the six vertices and the midpoints of the twelve edges).

We can derive a formula for the size of $\mathbb{Z}^m(h)$, as follows. For $i = 0, 1, 2, \ldots, m$, let I_i be the set of those elements of $\mathbb{Z}^m(h)$ where exactly i of the m coordinates are nonzero. How many elements are in I_i? We can choose which i of the m coordinates are nonzero in $\binom{m}{i}$ ways. Next, we choose the absolute values of these nonzero coordinates: since the sum of these i positive integers equals h, we have $\binom{h-1}{i-1}$ choices (see page 20). Finally, each of these i coordinates can be positive or negative, and therefore

$$|I_i| = \binom{m}{i}\binom{h-1}{i-1}2^i.$$

Summing now for i yields

$$|\mathbb{Z}^m(h)| = \sum_{i=0}^{h}\binom{m}{i}\binom{h-1}{i-1}2^i;$$

since for $i > h$ the terms vanish, we may write this as

$$|\mathbb{Z}^m(h)| = \sum_{i\geq 0}\binom{m}{i}\binom{h-1}{i-1}2^i.$$

Here we recognize the expression for the size of $\mathbb{Z}^m(h)$ as the quantity $c(h,m)$, discussed in detail in Section 2.4.

In our investigations later, we will also consider certain restrictions of $\mathbb{Z}^m(h)$. In some cases, we will look at the part of $\mathbb{Z}^m(h)$ that is in the "first quadrant;" that is, the subset $\mathbb{N}_0^m(h)$ of $\mathbb{Z}^m(h)$ that contains only those points that contain no negative coordinates:

$$\mathbb{N}_0^m(h) = \{(\lambda_1, \lambda_2, \dots, \lambda_m) \in \mathbb{N}_0^m \mid \lambda_1 + \lambda_2 + \cdots + \lambda_m = h\}.$$

So, for example,

$$\mathbb{N}_0^2(3) = \{(0,3),(1,2),(2,1),(3,0)\}$$

and

$$\mathbb{N}_0^3(2) = \{(0,0,2),(0,2,0),(2,0,0),(0,1,1),(1,0,1),(1,1,0)\}.$$

We can enumerate $\mathbb{N}_0^m(h)$ by observing that it is nothing but the set of m-sequences of $\mathbb{N}_0$ with the property that the entries in the sequence add up to h; as we have seen on page 20, this set has size

$$|\mathbb{N}_0^m(h)| = \binom{m+h-1}{h}.$$

Two other, frequently appearing, cases occur when we restrict $\mathbb{Z}^m(h)$ or $\mathbb{N}_0^m(h)$ to those points where the absolute value of the coordinates are not more than 1. (These points lie within a cube of side length 2 centered at the origin.) We denote these sets by $\hat{\mathbb{Z}}^m(h)$ and $\hat{\mathbb{N}}_0^m(h)$, respectively; namely, we have

$$\hat{\mathbb{Z}}^m(h) = \{(\lambda_1, \lambda_2, \dots, \lambda_m) \in \{-1,0,1\}^m \mid |\lambda_1| + |\lambda_2| + \cdots + |\lambda_m| = h\}$$

and

$$\hat{\mathbb{N}}_0^m(h) = \{(\lambda_1, \lambda_2, \dots, \lambda_m) \in \{0,1\}^m \mid \lambda_1 + \lambda_2 + \cdots + \lambda_m = h\}.$$

So, for example,

$$\hat{\mathbb{Z}}^3(2) = \{(0,\pm 1,\pm 1),(\pm 1,0,\pm 1),(\pm 1,\pm 1,0)\}$$

and

$$\hat{\mathbb{N}}_0^3(2) = \{(0,1,1),(1,0,1),(1,1,0)\},$$

but we have $\hat{\mathbb{Z}}^2(3) = \emptyset$ and $\hat{\mathbb{N}}_0^2(3) = \emptyset$. It is easy to see that the sizes of these sets are given by

$$|\hat{\mathbb{Z}}^m(h)| = \binom{m}{h}2^h$$

and

$$|\hat{\mathbb{N}}_0^m(h)| = \binom{m}{h}.$$

Often, rather than considering a single layer $\mathbb{Z}^m(h)$ of the integer lattice, we will study the union of several of them. Since the layers are pairwise disjoint, for a given range $H \subseteq \mathbb{N}_0$ of norms we have

$$\left| \bigcup_{h \in H} \mathbb{Z}^m(h) \right| = \sum_{h \in H} |\mathbb{Z}^m(h)|;$$

we can similarly just add the sizes of $\mathbb{N}_0^m(h)$, $\hat{\mathbb{Z}}^m(h)$, and $\hat{\mathbb{N}}_0^m(h)$ for all $h \in H$. Most often, we will consider H consisting of

- a single norm h (with $h \in \mathbb{N}_0$),
- a range $[0, s] = \{0, 1, 2, \ldots, s\}$ (with $s \in \mathbb{N}_0$), or
- allow all possible norms (i.e., have $H = \mathbb{N}_0$).

The following table summarizes what we can say, using some of the identities we have seen earlier, about the size of

$$\Lambda^m(H) = \{(\lambda_1, \lambda_2, \ldots, \lambda_m) \in \Lambda^m \mid |\lambda_1| + |\lambda_2| + \cdots + |\lambda_m| \in H\}$$

for these choices of $H \subseteq \mathbb{N}_0$ and our four exemplary sets $\Lambda \subseteq \mathbb{Z}$.

$\lvert\Lambda^m(H)\rvert$	$H = \{h\}$	$H = [0, s]$	$H = \mathbb{N}_0$
$\Lambda = \mathbb{N}_0$	$\binom{m+h-1}{h}$	$\binom{m+s}{s}$	∞
$\Lambda = \mathbb{Z}$	$c(h,m) = \sum_{i \geq 0} \binom{m}{i}\binom{h-1}{i-1}2^i$	$a(m,s) = \sum_{i \geq 0} \binom{m}{i}\binom{s}{i}2^i$	∞
$\Lambda = \{0, 1\}$	$\binom{m}{h}$	$\sum_{h \in H} \binom{m}{h}$	2^m
$\Lambda = \{-1, 0, 1\}$	$\binom{m}{h}2^h$	$\sum_{h \in H} \binom{m}{h}2^h$	3^m

We need to compute the size of an additional set that we use later. Namely, we want to find the number of lattice points that are strictly on one side of one of the coordinate planes of the m-dimensional space; that is, the size of the set

$$\mathbb{Z}^m(h)_{k+} = \{(\lambda_1, \lambda_2, \ldots, \lambda_m) \in \mathbb{Z}^m \mid |\lambda_1| + |\lambda_2| + \cdots + |\lambda_m| = h, \lambda_k > 0\}.$$

For example, the lattice points of the layer $\mathbb{Z}^2(3)$ that are to the right of the y-axis are

$$\mathbb{Z}^2(3)_{1+} = \{(1, \pm 2), (2, \pm 1), (3, 0)\}.$$

Note that $|\mathbb{Z}^m(h)_{k+}|$ is the same for any $k = 1, 2, \ldots, m$; here we calculate $|\mathbb{Z}^m(h)_{1+}|$.

As before, we let, for each $j = 1, \ldots, m$, I_j denote the set of those elements of $\mathbb{Z}^m(h)_{1+}$ where exactly j of the m coordinates are nonzero. (Note that $I_0 = \emptyset$.) How many elements are in I_j? Here we can choose which j of the m coordinates are nonzero in $\binom{m-1}{j-1}$ ways (since we must have $\lambda_1 > 0$). Next, we choose the absolute values of these nonzero coordinates: since the sum of these j positive integers equals h, we have $\binom{h-1}{j-1}$ choices. Finally, $j-1$ of these coordinates can be positive or negative, and therefore

$$|I_j| = \binom{m-1}{j-1}\binom{h-1}{j-1}2^{j-1}.$$

Summing now for j yields

$$|\mathbb{Z}^m(h)_{1+}| = \sum_{j\geq 1}\binom{m-1}{j-1}\binom{h-1}{j-1}2^{j-1}.$$

We can replace $j-1$ by i; this yields

$$|\mathbb{Z}^m(h)_{1+}| = \sum_{i\geq 0}\binom{m-1}{i}\binom{h-1}{i}2^{i},$$

and therefore

$$|\mathbb{Z}^m(h)_{k+}| = \sum_{i\geq 0}\binom{m-1}{i}\binom{h-1}{i}2^{i} = a(m-1, h-1)$$

for every $k = 1, 2, \ldots, m$. Indeed, the set $\mathbb{Z}^2(3)_{1+}$ featured above consists of $a(1,2) = 5$ points.

Exercises

1. We have already evaluated the entries in the column of $H = \{h\}$ in the table on page 28; here we verify the rest. Prove each of the following.

   (a) $|\mathbb{Z}^m([0,s])| = a(m,s)$

   (b) $|\mathbb{N}_0^m([0,s])| = \binom{m+s}{s}$

   (c) $|\hat{\mathbb{Z}}^m(\mathbb{N}_0)| = 3^m$

   (d) $|\hat{\mathbb{N}}_0^m(\mathbb{N}_0)| = 2^m$

2. For each set below, first find the size of the set, then list all its elements.

   (a) $\mathbb{Z}^2([0,3])$

   (b) $\mathbb{N}_0^2([0,3])$

   (c) $\hat{\mathbb{Z}}^2([0,3])$

   (d) $\hat{\mathbb{N}}_0^2([0,3])$

   (e) $\mathbb{Z}^2(3)_{1+}$

Chapter 3

Group theory

A group is arguably the most important structure in abstract mathematics. In essence, a *group* is any set of objects—for example, numbers, functions, vectors, etc.—combined with a binary operation—such as addition, multiplication, composition, etc.—satisfying certain fundamental properties. More precisely, a set G and an operation $*$ form a *group*, if each of the following four properties holds.

- *Closure property*: for any pair of elements a and b of G, $a * b$ is also in G.
- *Associative property*: for any a, b, and c in G, we have $(a * b) * c = a * (b * c)$.
- *Identity property*: there is an element z in G so that $a * z = z * a = a$ holds for any element a of G.
- *Inverse property*: for any element a of G, there exists an element $\overline{a}$, also in G, for which $a * \overline{a} = \overline{a} * a = z$.

Here we discuss a special class of groups: abelian groups. The group is said to be *abelian* if, in addition, the following holds.

- *Commutative property*: for any a and b in G, we have $a * b = b * a$.

One can show that, in a group, the identity element (denoted by z above) is unique, and each element a of the group has its unique inverse $\overline{a}$.

Abelian groups—named after the Norwegian mathematician Niels Abel (1802–1829)—play a central role in most branches of mathematics; our goal here is to investigate some of their fascinating number theoretic properties. Since our focus in this book is on additive combinatorics, we restrict our attention to *additive groups*, where the operation $*$ is addition, denoted as $+$; we will also write 0 for the identity z and $-a$ for the inverse $\overline{a}$ of a.

The additive groups most familiar to us are probably

- the set of integers (nonnegative and negative whole numbers), denoted by $\mathbb{Z}$;
- the set of rational numbers (fractions of integers), denoted by $\mathbb{Q}$; and
- the set of real numbers (finite or infinite decimals), denoted by $\mathbb{R}$.

Other well-studied abelian groups include the set of vectors in n-dimensional space, the set of n-by-m real matrices, and the set of real polynomials. Another example of an abelian group is the set of even integers (including positive and negative even integers and zero); however, the set of odd integers is not a group, since the closure and zero properties fail (neither $1+1$ nor 0 is odd). The sets of positive and nonnegative integers, denoted by $\mathbb{N}$ and $\mathbb{N}_0$, respectively, are also not groups: they both fail the negative inverse property (and $\mathbb{N}$ even fails the zero property).

Our examples of groups above are all *infinite* groups; that is, they have infinitely many elements. The groups we intend to study here, however, are *finite* groups: those that have only a finite number of elements. Thus the title of this chapter really should be an introduction to the theory of finite abelian groups.

3.1 Finite abelian groups

The simplest family of finite abelian groups are the cyclic groups: the *cyclic group* of size (or *order*) n, denoted by $\mathbb{Z}_n$, can be defined as follows. The elements of $\mathbb{Z}_n$ are the nonnegative integers up to $n-1$:

$$\mathbb{Z}_n = \{0, 1, 2, \ldots, n-1\}.$$

Addition is performed "mod n"; that is, for elements a and b of $\mathbb{Z}_n$, the sum $a+b$ is the remainder of $a+b$ when divided by n. For example, in $\mathbb{Z}_{10}$ we have $9+4=3$, representing the fact that when we add two integers whose last decimal digits are 9 and 4, respectively, then their sum will have a last digit of 3. In $\mathbb{Z}_{12}$, we have $9+4=1$: if our evening guests arrive at 9 p.m. and plan to stay for 4 hours, then they will leave at 1 a.m. (Of course, in both $\mathbb{Z}_{10}$ and $\mathbb{Z}_{12}$ we still have $5+3=8$.)

We call $\mathbb{Z}_n$ cyclic because if we add 1 repeatedly to itself, then within n steps we run through each of the elements in the group, after which the cycle repeats itself. The same holds for any other element a that is relatively prime to n; for example, in $\mathbb{Z}_{10}$ we have

$$\{\lambda \cdot 3 \mid \lambda = 0, 1, 2, \ldots, 9\} = \{0, 3, 6, 9, 2, 5, 8, 1, 4, 7\} = \mathbb{Z}_{10}.$$

(The notation $\lambda \cdot 3$, for $\lambda \in \mathbb{N}_0$, stands for the sum of λ terms with each term being 3.) If a and n are not relatively prime, then we still get a cycle, it's just that the cycle will be shorter as we won't run through all n elements; for example, the multiples of 2 in $\mathbb{Z}_{10}$ only yield the set $\{0, 2, 4, 6, 8\}$.

The length of the cycle that an element a of a group G generates is called the *order* of a in G, and is denoted by $\mathrm{ord}_G(a)$ or simply $\mathrm{ord}(a)$. For example, in $\mathbb{Z}_{10}$ we have $\mathrm{ord}(1)=10$, $\mathrm{ord}(2)=5$, $\mathrm{ord}(3)=10$, and so on. According to *Lagrange's Theorem*, in a group of order n, the order of any element is a divisor of n; for example, in $\mathbb{Z}_{10}$, only orders 1, 2, 5, and 10 are possible.

The set of elements in G that have order d is denoted by $\mathrm{Ord}(G,d)$. For example, we have

$$\mathrm{Ord}(\mathbb{Z}_{10}, 1) = \{0\},$$

$$\mathrm{Ord}(\mathbb{Z}_{10}, 2) = \{5\},$$

$$\mathrm{Ord}(\mathbb{Z}_{10}, 5) = \{2, 4, 6, 8\},$$

and

$$\mathrm{Ord}(\mathbb{Z}_{10}, 10) = \{1, 3, 7, 9\}.$$

From cyclic groups, we can build up other finite abelian groups using direct sums. The *direct sum* (which is also called the *direct product*) of the groups G_1 and G_2, denoted by $G_1 \times G_2$, consists of all ordered pairs of the form (a_1, a_2) where a_1 is any element of G_1 and a_2 is any element of G_2; formally,

$$G_1 \times G_2 = \{(a_1, a_2) \mid a_1 \in G_1, a_2 \in G_2\}.$$

If G_1 and G_2 have orders n_1 and n_2, respectively, then the order of $G_1 \times G_2$ is $n_1 n_2$. For example, $\mathbb{Z}_2 \times \mathbb{Z}_5$ has ten elements:

$$\mathbb{Z}_2 \times \mathbb{Z}_5 = \{(0,0),(0,1),(0,2),(0,3),(0,4),(1,0),(1,1),(1,2),(1,3),(1,4)\}.$$

We can also define the direct sum of more than two groups: the direct sum of finite abelian groups $G_1, G_2, \ldots, G_r$ consists of the ordered r-tuples $(a_1, a_2, \ldots, a_r)$ where a_i is any element of G_i (here $i = 1, 2, \ldots, r$). As a special case, if each component in the direct sum is the same group, then we often use exponential notation. For example, $\mathbb{Z}_2 \times \mathbb{Z}_2 \times \mathbb{Z}_2$ is denoted by $\mathbb{Z}_2^3$, and consists of eight elements:

$$\mathbb{Z}_2^3 = \{(0,0,0),(0,0,1),(0,1,0),(0,1,1),(1,0,0),(1,0,1),(1,1,0),(1,1,1)\}.$$

We add and subtract the elements of the direct sum component-wise. For example, in $\mathbb{Z}_2 \times \mathbb{Z}_5$ one has $(1,3)+(1,4)=(0,2)$, and in $\mathbb{Z}_2^3$ we have $(1,0,1)-(1,1,0)=(0,1,1)$.

We must note that not all such direct sum compositions result in new types of groups—as we will see in the next section.

Exercises

1. (a) For each positive integer d, find the set $\mathrm{Ord}(\mathbb{Z}_{18}, d)$.

   (b) For each positive integer d, find the set $\mathrm{Ord}(\mathbb{Z}_3 \times \mathbb{Z}_6, d)$.

2. Exhibit the complete addition table of the groups $\mathbb{Z}_3 \times \mathbb{Z}_4$ and $\mathbb{Z}_2^3$.

3.2 Group isomorphisms

Let us examine again two of the groups mentioned above: $\mathbb{Z}_{10}$ and $\mathbb{Z}_2 \times \mathbb{Z}_5$. Both of these groups have ten elements; furthermore, they form exactly the same structure as we now explain.

Consider the following table.

$\mathbb{Z}_{10}$	0	1	2	3	4	5	6	7	8	9
$\mathbb{Z}_2 \times \mathbb{Z}_5$	$(0,0)$	$(1,1)$	$(0,2)$	$(1,3)$	$(0,4)$	$(1,0)$	$(0,1)$	$(1,2)$	$(0,3)$	$(1,4)$

The table exhibits a correspondence between the elements of the two groups with a very special property: if we add two elements in one group and then add the corresponding elements in the other group, then the two sums will also correspond to each other. For example:

$$\begin{array}{rccccc} \mathbb{Z}_{10}: & 9 & + & 4 & = & 3 \\ & \updownarrow & & \updownarrow & & \updownarrow \\ \mathbb{Z}_2 \times \mathbb{Z}_5: & (1,4) & + & (0,4) & = & (1,3) \end{array}$$

We thus find that each entry in the addition table of $\mathbb{Z}_{10}$ (the ten by ten table that lists all possible pairwise sums of elements) corresponds to the appropriate entry in the addition table of $\mathbb{Z}_2 \times \mathbb{Z}_5$. Therefore, the two groups are essentially the same; using standard terminology, we say that they are *isomorphic*—a fact that we denote as follows:

$$\mathbb{Z}_{10} \cong \mathbb{Z}_2 \times \mathbb{Z}_5.$$

More generally, one can prove that, if n_1 and n_2 are relatively prime integers, then

$$\mathbb{Z}_{n_1} \times \mathbb{Z}_{n_2} \cong \mathbb{Z}_{n_1 n_2}.$$

For example, the groups $\mathbb{Z}_5 \times \mathbb{Z}_{12}$, $\mathbb{Z}_4 \times \mathbb{Z}_{15}$, $\mathbb{Z}_3 \times \mathbb{Z}_{20}$, and $\mathbb{Z}_3 \times \mathbb{Z}_4 \times \mathbb{Z}_5$ are all isomorphic to $\mathbb{Z}_{60}$.

One can also show that, if n_1 and n_2 are not relatively prime, then $\mathbb{Z}_{n_1} \times \mathbb{Z}_{n_2}$ and $\mathbb{Z}_{n_1 n_2}$ are not isomorphic. For example, $\mathbb{Z}_6 \times \mathbb{Z}_{10}$ is an example of a group of order 60 that is not isomorphic to $\mathbb{Z}_{60}$, since 6 and 10 are not relatively prime. (To see why there cannot possibly be an isomorphism between these two groups, note that $\mathbb{Z}_{60}$ is cyclic, but $\mathbb{Z}_6 \times \mathbb{Z}_{10}$ is not cyclic as none of its elements has order 60—see Section 3.3 below.)

Group isomorphism is a very important and useful concept: when two groups are isomorphic, it suffices to study one of them (either one). For example, if $G_1 \cong G_2$, then the number of elements of a certain order in G_1 will be the same as the number of elements of that order in G_2, and the same kind of property holds for other important group characteristics. This reduces our task of studying all finite abelian groups to that of studying only those that have different *isomorphism types* (are pairwise non-isomorphic).

Exercises

1. Show that the groups $\mathbb{Z}_3 \times \mathbb{Z}_4$ and $\mathbb{Z}_{12}$ are isomorphic by finding an explicit correspondence between the elements of the two groups.

2. Explain why the groups $\mathbb{Z}_3 \times \mathbb{Z}_6$ and $\mathbb{Z}_{18}$ are not isomorphic by considering the orders of the various elements in the two groups.

3.3 The Fundamental Theorem of Finite Abelian Groups

Suppose that one wishes to study all abelian groups of order 60, that is, those that have exactly 60 elements. First, we wish to determine how many possible isomorphism types there are for abelian groups of order 60. We have already seen that the groups

$$\mathbb{Z}_3 \times \mathbb{Z}_4 \times \mathbb{Z}_5, \quad \mathbb{Z}_5 \times \mathbb{Z}_{12}, \quad \mathbb{Z}_4 \times \mathbb{Z}_{15}, \quad \mathbb{Z}_3 \times \mathbb{Z}_{20}, \quad \text{and } \mathbb{Z}_{60}$$

are all isomorphic to one another. Considering all other possible factorizations of 60 into a product of positive integers, we see that there are five other direct sums to examine:

$$\mathbb{Z}_2 \times \mathbb{Z}_2 \times \mathbb{Z}_3 \times \mathbb{Z}_5, \quad \mathbb{Z}_2 \times \mathbb{Z}_3 \times \mathbb{Z}_{10}, \quad \mathbb{Z}_2 \times \mathbb{Z}_5 \times \mathbb{Z}_6, \quad \mathbb{Z}_2 \times \mathbb{Z}_{30}, \quad \text{and } \mathbb{Z}_6 \times \mathbb{Z}_{10}.$$

(We should note that, since we only study abelian groups here, changing the order of the terms in a direct sum will not change its isomorphism type.)

It is not hard to determine that these five direct sums are also all isomorphic to each other. For example, to see that

$$\mathbb{Z}_2 \times \mathbb{Z}_{30} \cong \mathbb{Z}_6 \times \mathbb{Z}_{10},$$

note that 5 and 6 are relatively prime, and thus $\mathbb{Z}_2 \times \mathbb{Z}_{30} \cong \mathbb{Z}_2 \times \mathbb{Z}_5 \times \mathbb{Z}_6$, from which, since 2 and 5 are also relatively prime, we get $\mathbb{Z}_6 \times \mathbb{Z}_{10}$. Thus, among the ten possible direct sums, we have two different isomorphism types.

We say that the isomorphism relation is an *equivalence relation*: one can partition the collection of all groups into *equivalence classes* (sometimes called *isomorphism classes*) where all groups within a class are equivalent to each other (but not to those in other classes). As we have just seen, the collection of abelian groups of order 60 form two isomorphism classes.

In general, we can find the number of different isomorphism classes among the groups of order n as follows: if

$$n = p_1^{\alpha_1} \cdots p_k^{\alpha_k}$$

is the prime factorization of n, then the number of different isomorphism classes among the groups of order n equals

$$p(\alpha_1) \cdot \cdots \cdot p(\alpha_k)$$

where p is the partition function introduced in Section 1.4. For example, since

$$60 = 2^2 \cdot 3^1 \cdot 5^1,$$

there are

$$p(2) \cdot p(1) \cdot p(1) = 2$$

different isomorphism types among groups of order 60.

We even have a convenient way of listing representatives of the different isomorphism types for a given order. The *Fundamental Theorem of Finite Abelian Groups* asserts that for any finite abelian group G of order at least 2, there are positive integers r and $n_1, \ldots, n_r$ such that

$$G \cong \mathbb{Z}_{n_1} \times \mathbb{Z}_{n_2} \times \cdots \times \mathbb{Z}_{n_r};$$

furthermore, we can assume that n_{i+1} is divisible by n_i for $i = 1, 2, \ldots, r-1$ and that $n_1 \geq 2$ (and therefore $n_i \geq 2$ for all i). The factorization of G above, which is unique, is called the *invariant decomposition* of G. We say that r is the *rank* of G, and the largest invariant factor, n_r, is called the *exponent* of G; the exponent clearly equals the length of the longest cycle in G. We should also note that our treatment here includes the possibility that $r = 1$, in which case we simply have $G \cong Z_n$ (with $n = n_1$), and the group is cyclic.

Therefore, we get a list of all possible isomorphism types of abelian groups of order n directly from the list of invariant factorizations of n (see Section 1.4). For example, there are two isomorphism classes of abelian groups of order 60, as shown by the two invariant decompositions $\mathbb{Z}_{60}$ and $\mathbb{Z}_2 \times \mathbb{Z}_{30}$. The group $\mathbb{Z}_{60}$ has rank 1 (as does any cyclic group), and the group $\mathbb{Z}_2 \times \mathbb{Z}_{30}$ has rank 2; every cycle in the latter group has length at most 30. From Section 1.4 we also know that there are exactly 105 different isomorphism types among abelian groups of order 840000.

We should also mention two other notable members of the two isomorphism classes for order 60: $\mathbb{Z}_3 \times \mathbb{Z}_4 \times \mathbb{Z}_5$ (which is isomorphic to $\mathbb{Z}_{60}$) and $\mathbb{Z}_2 \times \mathbb{Z}_2 \times \mathbb{Z}_3 \times \mathbb{Z}_5$ (which is isomorphic to $\mathbb{Z}_2 \times \mathbb{Z}_{30}$). These direct sums involve (prime and) prime power orders only, and they are thus called *primary decompositions*. Like the invariant decomposition of G, the primary decomposition is also unique (up to the order of terms, of course).

We can easily convert any decomposition of a finite abelian group into its invariant decomposition or its primary decomposition. Let us consider the example

$$G = \mathbb{Z}_{40} \times \mathbb{Z}_{50} \times \mathbb{Z}_{60} \times \mathbb{Z}_{70}.$$

We start with the primary decomposition as that is easier: since $40 = 8 \cdot 5$, $50 = 2 \cdot 25$, $60 = 4 \cdot 3 \cdot 5$, and $70 = 2 \cdot 5 \cdot 7$, the primary decomposition of G is

$$G \cong \mathbb{Z}_2 \times \mathbb{Z}_2 \times \mathbb{Z}_4 \times \mathbb{Z}_8 \times \mathbb{Z}_3 \times \mathbb{Z}_5 \times \mathbb{Z}_5 \times \mathbb{Z}_5 \times \mathbb{Z}_{25} \times \mathbb{Z}_7,$$

which we can condense as

$$G \cong \mathbb{Z}_2^2 \times \mathbb{Z}_4 \times \mathbb{Z}_8 \times \mathbb{Z}_3 \times \mathbb{Z}_5^3 \times \mathbb{Z}_{25} \times \mathbb{Z}_7.$$

From this, we find that the invariant decomposition is

$$G \cong \mathbb{Z}_{10} \times \mathbb{Z}_{10} \times \mathbb{Z}_{20} \times \mathbb{Z}_{4200},$$

since $8 \cdot 3 \cdot 25 \cdot 7 = 4200$, $4 \cdot 5 = 20$, and $2 \cdot 5 = 10$ (recall from Section 1.4 the procedure of turning a primary decomposition of a positive integer into an invariant decomposition).

Exercises

1. (a) How many different isomorphism types are there among abelian groups of order 72?
   (b) For each isomorphism class of part (a), find the member of the class that is in invariant decomposition and the one that is in primary decomposition.
2. (a) Prove that every abelian group of prime order is cyclic.
   (b) Characterize all positive integers n for which it is true that every abelian group of order n is cyclic.

3.4 Subgroups and cosets

As we have seen before, every element a in a group G determines a cycle

$$\langle a \rangle = \{\lambda \cdot a \mid \lambda = 0, 1, \ldots, d-1\};$$

the length d of the cycle is the order $\text{ord}_G(a)$ of a. We also learned that $\text{ord}_G(a)$ must divide the order $|G|$ of G.

A key feature of the cycle $\langle a \rangle$ is that it is not only a subset, but a *subgroup* of G; that is, $\langle a \rangle$ itself forms a group (for the same operation). In general, if a subset $H \subseteq G$ is itself a group for the same operation, then we call H a subgroup; this fact is denoted by $H \leq G$. For example, $G = \mathbb{Z}_{10}$ has four different subgroups: besides $\{0\}$ and G itself (which are always subgroups), we have

$$\langle 2 \rangle = \{0, 2, 4, 6, 8\} \leq \mathbb{Z}_{10}$$

and

$$\langle 5 \rangle = \{0, 5\} \leq \mathbb{Z}_{10}.$$

The group $\mathbb{Z}_{12}$ has more: $\{0\}$, $\{0, 6\}$, $\{0, 4, 8\}$, $\{0, 3, 6, 9\}$, $\{0, 2, 4, 6, 8, 10\}$, and $\mathbb{Z}_{12}$ itself. It is not hard to prove that the cyclic group $\mathbb{Z}_n$ has exactly $d(n)$ different subgroups (where $d(n)$ is the number of positive divisors of n): each subgroup H of $\mathbb{Z}_n$ must have order d for some divisor d of n, and, conversely, for each divisor d of n, there is a unique subgroup H of $\mathbb{Z}_n$ with order d. Therefore, groups of prime order p—which, by the exercise at the end of the previous section, must be cyclic and thus are isomorphic to $\mathbb{Z}_p$—only have the trivial subgroups: $\{0\}$ and $\mathbb{Z}_p$ itself.

In noncyclic groups, the situation is considerably more complicated. Clearly, if $G = G_1 \times G_2$, then for all subgroups H_1 of G_1 and H_2 of G_2, $H_1 \times H_2$ is a subgroup of G; these kinds of subgroups are called *subproducts*. For example, we may take $H_1 = \{0, 1\} \leq \mathbb{Z}_2$ and $H_2 = \{0, 2\} \leq \mathbb{Z}_4$, with which the subset

$$H_1 \times H_2 = \{(0, 0), (1, 0), (0, 2), (1, 2)\}$$

is a subgroup of $\mathbb{Z}_2 \times \mathbb{Z}_4$. But the collection of subgroups of a noncyclic group is more varied: $G_1 \times G_2$ may have subgroups that are not even in the form $H_1 \times H_2$. In $G = \mathbb{Z}_2 \times \mathbb{Z}_4$, for example,

$$H = \{(0, 0), (1, 2)\}$$

is a subgroup of G and is not of this form. It still holds that any subgroup of a (noncyclic) group of order n must have order d for some divisor d of n, and, for any positive divisor d of n, a group of order n has a subgroup of order d, but we may have more than one such subgroup.

The number of subgroups of finite abelian groups is not yet fully understood. As we mentioned above—and can easily be seen—the number of subgroups of the cyclic group $\mathbb{Z}_n$ is $d(n)$. Regarding groups of rank two, the group $\mathbb{Z}_{n_1} \times \mathbb{Z}_{n_2}$ obviously has $d(n_1) \cdot d(n_2)$ subproducts, but its total number of subgroups is given by

$$\sum_{d_1 \in D(n_1), d_2 \in D(n_2)} \gcd(d_1, d_2).$$

This formula follows from an 1987 paper of Calhoun (cf. [48]) and was recently proved directly by Hampejs et al. (see [116]). Thus, for example, the group $\mathbb{Z}_2 \times \mathbb{Z}_4$ has eight subgroups; of these, six are subproducts and two are not. Similarly simple formulae for groups of rank three or more are not yet known.

Next, we discuss another important concept: cosets of subgroups. Given any subgroup $H \leq G$ and any element $a \in G$, the set

$$a + H = \{a + h \mid h \in H\}$$

is called a *coset* of H in G. For example, if $G = \mathbb{Z}_{10}$, then for $H = \{0, 5\} \leq G$ and $3 \in G$ we have

$$3 + H = \{3, 8\},$$

and if $G = \mathbb{Z}_2 \times \mathbb{Z}_4$, then for $H = \{(0,0), (1,2)\}$ and $(1,1) \in G$ we have

$$(1,1) + H = \{(1,1), (0,3)\}.$$

Observe also that cosets could be represented by any of their other elements. So, for example, with $H = \{0, 5\}$, the coset $\{3, 8\}$ of $\mathbb{Z}_{10}$ can be written both as $3 + H$ and $8 + H$, since in $\mathbb{Z}_{10}$ we have

$$3 + \{0, 5\} = 8 + \{0, 5\};$$

similarly, with $H = \{(0,0), (1,2)\}$, the coset $\{(1,1), (0,3)\}$ of $\mathbb{Z}_2 \times \mathbb{Z}_4$ can be written both as $(1,1) + H$ and $(0,3) + H$ as in $\mathbb{Z}_2 \times \mathbb{Z}_4$ we have

$$(1,1) + \{(0,0), (1,2)\} = (0,3) + \{(0,0), (1,2)\}.$$

Clearly, if H has order d, then every coset of H has size d as well. It turns out that for any two elements a and b of G, the cosets $a + H$ and $b + H$ are either identical (as sets) or entirely disjoint. Therefore, the collection of distinct cosets of H partitions G; that is, if H has order d, then one can find n/d elements $a_1, \ldots, a_{n/d}$ so that

$$G = (a_1 + H) \cup \cdots \cup (a_{n/d} + H).$$

For example, with $H = \{0, 5\}$ in $G = \mathbb{Z}_{10}$, we have

$$\begin{aligned} \mathbb{Z}_{10} &= (0 + H) \cup (1 + H) \cup (2 + H) \cup (3 + H) \cup (4 + H) \\ &= \{0, 5\} \cup \{1, 6\} \cup \{2, 7\} \cup \{3, 8\} \cup \{4, 9\}, \end{aligned}$$

and with $H = \{(0,0), (1,2)\}$ in $G = \mathbb{Z}_2 \times \mathbb{Z}_4$, we have

$$\begin{aligned} \mathbb{Z}_2 \times \mathbb{Z}_4 &= ((0,0) + H) \cup ((0,1) + H) \cup ((1,0) + H) \cup ((1,1) + H) \\ &= \{(0,0), (1,2)\} \cup \{(0,1), (1,3)\} \cup \{(1,0), (0,2)\} \cup \{(1,1), (0,3)\}. \end{aligned}$$

As we pointed out above, the elements $a_1, \ldots, a_{n/d}$ are not unique; for example, we could also write

$$\mathbb{Z}_{10} = (5 + \{0,5\}) \cup (6 + \{0,5\}) \cup (7 + \{0,5\}) \cup (8 + \{0,5\}) \cup (9 + \{0,5\}).$$

Furthermore, we also define the *sum of cosets* $a + H$ and $b + H$ as

$$(a + H) + (b + H) = (a + b) + H.$$

It is easy to see that this addition operation is well-defined (does not depend on which representatives a and b we choose), is closed (the sum of two cosets is also a coset), is associative, has an identity (the coset $0 + H = H$), and each coset has an additive inverse (the inverse of $a + H$ being $-a + H$). Thus, with this operation, the collection of cosets is itself a group, called the *quotient group* of H in G; it is denoted by G/H.

For example, the quotient group $\mathbb{Z}_{10}/\{0,5\}$ consists of the five cosets

$$0 + \{0,5\},\ 1 + \{0,5\},\ 2 + \{0,5\},\ 3 + \{0,5\},\ 4 + \{0,5\};$$

since (for example) $1 + \{0,5\}$ generates all five cosets, we have

$$\mathbb{Z}_{10}/\{0,5\} \cong \mathbb{Z}_5.$$

Similarly, the quotient group $(\mathbb{Z}_2 \times \mathbb{Z}_4)/\{(0,0),(1,2)\}$ consists of the four cosets

$$(0,0) + \{(0,0),(1,2)\},\ (0,1) + \{(0,0),(1,2)\},\ (1,0) + \{(0,0),(1,2)\},\ (1,1) + \{(0,0),(1,2)\};$$

here $(0,1) + \{(0,0),(1,2)\}$ cycles through the four cosets and thus

$$(\mathbb{Z}_2 \times \mathbb{Z}_4)/\{(0,0),(1,2)\} \cong \mathbb{Z}_4.$$

Exercises

1. (a) Find the number of subgroups of $\mathbb{Z}_{100}$. How many of them are cyclic?

   (b) Find the number of subgroups of $\mathbb{Z}_{10}^2$. How many of them are subproducts?

   (c) Using information presented above, prove that the number of subgroups of $\mathbb{Z}_{n_1} \times \mathbb{Z}_{n_2}$ that are not subproducts equals

   $$\sum_{d_1 \in D(n_1), d_2 \in D(n_2)} (\gcd(d_1, d_2) - 1).$$

   (d) Use part (c) to prove that every finite abelian group of rank two has at least one subgroup that is not a subproduct.

   (e) Prove that $\mathbb{Z}_2^2$ is the only finite abelian group of rank two that has exactly one subgroup that is not a subproduct.

2. (a) List all subgroups of $\mathbb{Z}_{18}$. Find the isomorphism type of each subgroup.

   (b) List all subgroups of $\mathbb{Z}_3 \times \mathbb{Z}_6$. Find the isomorphism type of each subgroup.

3. (a) Consider the subgroup $H = \{0,9\}$ of $G = \mathbb{Z}_{18}$. Find each coset of H in G, and find the isomorphism type of G/H.

   (b) Consider the subgroup $H = \{(0,0),(0,3)\}$ of $G = \mathbb{Z}_3 \times \mathbb{Z}_6$. Find each coset of H in G, and find the isomorphism type of G/H.

3.5 Subgroups generated by subsets

Recall that, for any element a of G, the set $\langle a \rangle$ consists of all d multiples of a where $d = \text{ord}_G(a)$ is the order of a:

$$\langle a \rangle = \{\lambda \cdot a \mid \lambda = 0, 1, \ldots, d-1\}.$$

We should note that it makes no difference if we increase the range of λ: since $d \cdot a = 0$, $(d+1) \cdot a = a$, $(d+2) \cdot a = 2 \cdot a$, etc., we may also write

$$\langle a \rangle = \{\lambda \cdot a \mid \lambda \in \mathbb{N}_0\};$$

in fact, since $(-1) \cdot a = (d-1) \cdot a$, $(-2) \cdot a = (d-2) \cdot a$, and so on, we have

$$\langle a \rangle = \{\lambda \cdot a \mid \lambda \in \mathbb{Z}\}.$$

Recall also that $\langle a \rangle$ is a subgroup of G; we call $\langle a \rangle$ the subgroup generated by a as it is the smallest subgroup of G that contains a. More generally, we may look for the smallest subgroup of G that contains all of the set $A = \{a_1, \ldots, a_m\}$; this subgroup, denoted by $\langle A \rangle$, is called the *subgroup generated by* A. It is not hard to see that, with $d_i = \text{ord}(a_i)$ $(i = 1, 2, \ldots, m)$, we have

$$\langle A \rangle = \{\lambda_1 \cdot a_1 + \cdots + \lambda_m \cdot a_m \mid \lambda_i = 0, 1, \ldots, d_i - 1 \text{ for } i = 1, 2, \ldots, m\}.$$

For example, in $G = \mathbb{Z}_{15}$, the subset $A = \{6, 10\}$ generates the subgroup

$$\langle A \rangle = \{\lambda_1 \cdot 6 + \lambda_2 \cdot 10 \mid \lambda_1 = 0, 1, 2, 3, 4; \lambda_2 = 0, 1, 2\},$$

since 6 has order 5 and 10 has order 3 in G. Evaluating the expressions above yields

$$\langle A \rangle = \{0, 6, 12, 3, 9, 10, 1, 7, 13, 4, 5, 11, 2, 8, 14\};$$

so in this case we have $\langle A \rangle = G$. Again, it makes no difference if we let the range of coefficients increase:

$$\langle A \rangle = \{\lambda_1 \cdot a_1 + \cdots + \lambda_m \cdot a_m \mid \lambda_i \in \mathbb{N}_0 \text{ for } i = 1, 2, \ldots, m\}$$

or

$$\langle A \rangle = \{\lambda_1 \cdot a_1 + \cdots + \lambda_m \cdot a_m \mid \lambda_i \in \mathbb{Z} \text{ for } i = 1, 2, \ldots, m\}.$$

In our example above, we have the 2-element set $A = \{6, 10\}$ generating the cyclic group $G = \mathbb{Z}_{15}$; but, like any cyclic group, $G = \mathbb{Z}_{15}$ could be generated by a single element (1 or any other $a \in \mathbb{Z}_{15}$ that is relatively prime to 15). The question may arise: how large does a subset of a group need to be in order to generate the entire group? The general answer is provided by the following proposition.

Proposition 3.1 *An abelian group of rank r cannot be generated by fewer than r elements; that is, if $A = \{a_1, \ldots, a_m\}$ is a subset of G for which $\langle A \rangle = G$, then $m \geq r$.*

For the proof of Proposition 3.1, see page 309. We should note that a subset of size r doesn't always generate a given group of rank r. For example, $\{(1,2), (1,4)\}$ in the group $\mathbb{Z}_6^2$ generates only a subgroup of order 18 (only ordered pairs with an even second component get generated), and the subset $\{(1,2), (3,1)\}$ of $\mathbb{Z}_5^2$ only generates a subgroup of order 5 (each of the two elements alone, in fact, generates the other). It is not easy to see in general how large of a subgroup a given subset A of a group G generates.

Exercises

1. (a) Above, we made the comment that for any $a \in \mathbb{Z}_{15}$ that is relatively prime to 15, a generates the entire group. Verify this statement.
   (b) Prove that, conversely, if an element $a \in \mathbb{Z}_{15}$ is not relatively prime to 15, then a does not generate the entire group.

2. Consider the group $G = \mathbb{Z}_{15}^2$. Find the subgroups of G generated by each of the following subsets, and find the isomorphism type of each subgroup.
   (a) $\{(1,3),(1,5)\}$.
   (b) $\{(1,3),(2,3)\}$.
   (c) $\{(1,4),(4,1)\}$.

3.6 Sumsets

As we have just seen, taking all integer *linear combinations*

$$\lambda_1 \cdot a_1 + \cdots + \lambda_m \cdot a_m$$

of a subset $A = \{a_1, \ldots, a_m\}$ of some finite abelian group G generates a subgroup $\langle A \rangle$ of G; in fact, we arrive at $\langle A \rangle$ by taking only the above linear combinations with coefficients

$$\lambda_i = 0, 1, \ldots, \mathrm{ord}(a_i) - 1$$

$(i = 1, 2 \ldots, m)$.

In many famous and still-investigated problems, we put limitations on the coefficients. For example, while we saw in Section 3.5 that $\{6, 10\}$ generates all of $\mathbb{Z}_{15}$, to generate 7, we may use (for example) 6 twice and 10 once (thus, a total of three terms) to write

$$2 \cdot 6 + 1 \cdot 10,$$

but there is no way to get to 7 without using at least three terms. The question then arises: which terms do we get with, say, at most two terms? In other words, we want to find the linear combinations

$$\lambda_1 \cdot 6 + \lambda_2 \cdot 10$$

with integer coefficients λ_1 and λ_2 satisfying

$$|\lambda_1| + |\lambda_2| \leq 2.$$

The answer then is the set

$$\{0,\ \pm 1 \cdot 6,\ \pm 1 \cdot 10,\ \pm 2 \cdot 6,\ \pm 2 \cdot 10,\ \pm 1 \cdot 6 \pm 1 \cdot 10\} = \{0, 1, 3, 4, 5, 6, 9, 10, 11, 12, 14\}.$$

These eleven elements correspond to the thirteen elements of the layer

$$\mathbb{Z}^2(2) = \{(\lambda_1, \lambda_2) \in \mathbb{Z}^2 \mid |\lambda_1| + |\lambda_2| \leq 2\} = \{(0,0), (\pm 1, 0), (0, \pm 1), (\pm 2, 0), (0, \pm 2), (\pm 1, \pm 1)\}$$

of the 2-dimensional integer lattice (see Section 2.5). (Note that not all elements of $\mathbb{Z}^2(2)$ yield distinct group elements: $1 \cdot 10 = (-2) \cdot 10$ and $2 \cdot 10 = (-1) \cdot 10$.)

More generally, given a subset $\Lambda \subseteq \mathbb{Z}$ and a subset $H \subseteq \mathbb{N}_0$, we consider *sumsets* of $A = \{a_1, \ldots, a_m\}$ corresponding to

$$\Lambda^m(H) = \{(\lambda_1, \lambda_2, \ldots, \lambda_m) \in \Lambda^m \mid |\lambda_1| + |\lambda_2| + \cdots + |\lambda_m| \in H\}$$

(see Section 2.5); namely, we consider the collection

$$\begin{aligned} H_\Lambda A &= \{\lambda_1 a_1 + \cdots + \lambda_m a_m \mid (\lambda_1, \ldots, \lambda_m) \in \Lambda^m(H)\} \\ &= \{\lambda_1 a_1 + \cdots + \lambda_m a_m \mid \lambda_1, \ldots, \lambda_m \in \Lambda, |\lambda_1| + \cdots + |\lambda_m| \in H\} \end{aligned}$$

in G. In our example above we had $G = \mathbb{Z}_{15}$, $A = \{6, 10\}$, $\Lambda = \mathbb{Z}$, and $H = [0, 2] = \{0, 1, 2\}$.

We see that the sumset of A corresponding to Λ and H consists of sums of the elements of A with the conditions that

- the repetition number of each element of A in the sum must be from the set Λ, and
- the total number of terms in each sum must be from the set H.

We need to point out that the order of the elements of A is immaterial: the sumset remains unchanged when the elements $a_1, \ldots, a_m$ are permuted.

There are a number of special cases of sumsets that are most often studied. First, we introduce terms and notations for the case when sums in our sumset all have a fixed number of terms (that is, H consists of a single nonnegative integer h) and when the coefficient-set is $\Lambda = \mathbb{N}_0, \mathbb{Z}, \{0, 1\}$, or $\{-1, 0, 1\}$. More precisely, we introduce the following notations. Suppose that $A = \{a_1, a_2, \ldots, a_m\}$ is a subset of an abelian group G (with $m \in \mathbb{N}$) and that h is a nonnegative integer. We then define:

- the (ordinary) *h-fold sumset* hA of A, consisting of sums of exactly h (not necessarily distinct) terms of A:

$$hA = \{\lambda_1 a_1 + \cdots + \lambda_m a_m \mid \lambda_1, \ldots, \lambda_m \in \mathbb{N}_0,\ \lambda_1 + \cdots + \lambda_m = h\};$$

- the *h-fold signed sumset* $h_\pm A$ of A, consisting of signed sums of exactly h (not necessarily distinct) terms of A:

$$h_\pm A = \{\lambda_1 a_1 + \cdots + \lambda_m a_m \mid \lambda_1, \ldots, \lambda_m \in \mathbb{Z},\ |\lambda_1| + \cdots + |\lambda_m| = h\};$$

- the *restricted h-fold sumset* $h\hat{\ }A$ of A, consisting of sums of exactly h distinct terms of A:

$$h\hat{\ }A = \{\lambda_1 a_1 + \cdots + \lambda_m a_m \mid \lambda_1, \ldots, \lambda_m \in \{0, 1\},\ \lambda_1 + \cdots + \lambda_m = h\};$$

  and

- the *restricted h-fold signed sumset* $h\hat{\pm}A$ of A, consisting of signed sums of exactly h distinct terms of A:

$$h\hat{\pm}A = \{\lambda_1 a_1 + \cdots + \lambda_m a_m \mid \lambda_1, \ldots, \lambda_m \in \{-1, 0, 1\},\ |\lambda_1| + \cdots + |\lambda_m| = h\}.$$

A summary scheme of these four types of sumsets, indicating containment, can be given as follows.

	Terms must be distinct		Repetition of terms is allowed
Terms can be added only	$h\hat{\ }A$	$\subseteq$	hA
	$\rotatebox{-90}{$\subseteq$}$		$\rotatebox{-90}{$\subseteq$}$
Terms can be added or subtracted	$h\hat{\pm}A$	$\subseteq$	$h_\pm A$

We should caution that the h-fold sumset hA of a set A is different from the so-called *h-fold dilation*

$$h \cdot A = \{h \cdot a_1, \dots, h \cdot a_m\}$$

of A, which only contains the multiples of the elements in A. (However, for an element $g \in G$ and an integer $\lambda \in \mathbb{Z}$ we will use the notations $\lambda \cdot g$ and λg interchangeably.) We clearly have $h \cdot A \subseteq hA$ but usually hA is much larger than $h \cdot A$. We will rarely use dilations in this book, except for the negative $-A$ of A, defined, of course, as

$$-A = (-1) \cdot A = \{-a_1, \dots, -a_m\}.$$

Let us now illustrate the spectra of our four types of sumsets as h increases, using the example of $G = \mathbb{Z}_{13}$ and $A = \{2, 3\}$.

The easiest to evaluate are the restricted sumsets; note that since $|A| = 2$, for $h \geq 3$ we have $h\hat{~}A = h\hat{_\pm}A = \emptyset$. The relevant values of

$$\lambda_1 \cdot 2 + \lambda_3 \cdot 3$$

in G are as follows:

$\lambda_2 = 1$	3	5
$\lambda_2 = 0$	0	2
	$\lambda_1 = 0$	$\lambda_1 = 1$

Therefore, we get

$0\hat{~}A$	0												
$1\hat{~}A$			2	3									
$2\hat{~}A$						5							

Similarly, for the restricted signed sumsets we have

$\lambda_2 = 1$	1	3	5
$\lambda_2 = 0$	11	0	2
$\lambda_2 = -1$	8	10	12
	$\lambda_1 = -1$	$\lambda_1 = 0$	$\lambda_1 = 1$

and thus

$0\hat{_\pm}A$	0												
$1\hat{_\pm}A$			2	3							10	11	
$2\hat{_\pm}A$		1				5			8				12

Turning to unrestricted sumsets, we first observe that both hA and $h_\pm A$ equal the entire group $\mathbb{Z}_{13}$ for $h \geq 12$: indeed, we have

$$h_\pm A \supseteq hA = \{(h-i) \cdot 2 + i \cdot 3 \mid i = 0, 1, \dots, h\} = \{2h + i \mid i = 0, 1, \dots, h\},$$

which, if $h \geq 12$, gives the entire group $\mathbb{Z}_{13}$. So we only need to exhibit the h-fold sumset and the h-fold signed sumset of A for $h \leq 12$; a computation similar to the ones above yields the following results:

$0A$	0												
$1A$			2	3									
$2A$					4	5	6						
$3A$							6	7	8	9			
$4A$									8	9	10	11	12
$5A$	0	1	2								10	11	12
$6A$	0	1	2	3	4	5							12
$7A$		1	2	3	4	5	6	7	8				
$8A$				3	4	5	6	7	8	9	10	11	
$9A$	0	1				5	6	7	8	9	10	11	12
$10A$	0	1	2	3	4			7	8	9	10	11	12
$11A$	0	1	2	3	4	5	6	7		9	10	11	12
$12A$	0	1	2	3	4	5	6	7	8	9	10	11	12

$0_{\pm}A$	0												
$1_{\pm}A$			2	3							10	11	
$2_{\pm}A$		1			4	5	6	7	8	9			12
$3_{\pm}A$		1			4	5	6	7	8	9			12
$4_{\pm}A$		1	2	3	4	5	6	7	8	9	10	11	12
$5_{\pm}A$	0	1	2	3		5			8		10	11	12
$6_{\pm}A$	0	1	2	3	4	5	6	7	8	9	10	11	12
$7_{\pm}A$		1	2	3	4	5	6	7	8	9	10	11	12
$8_{\pm}A$		1	2	3	4	5	6	7	8	9	10	11	12
$9_{\pm}A$	0	1	2	3	4	5	6	7	8	9	10	11	12
$10_{\pm}A$	0	1	2	3	4	5	6	7	8	9	10	11	12
$11_{\pm}A$	0	1	2	3	4	5	6	7	8	9	10	11	12
$12_{\pm}A$	0	1	2	3	4	5	6	7	8	9	10	11	12

We can introduce similar notations and terminology for the cases when sums in our sumsets contain a limited number of terms (say $H = [0, s] = \{0, 1, \ldots, s\}$ for some $s \in \mathbb{N}_0$) or an arbitrary number of terms ($H = \mathbb{N}_0$). For instance, with the set $A = \{2, 3\}$ in $G = \mathbb{Z}_{13}$, we get

$$[0,3]A = \cup_{i=0}^{3} hA = \{0\} \cup \{2,3\}, \{4,5,6\} \cup \{6,7,8,9\} = \{0,2,3,4,5,6,7,8,9\},$$

and

$$[0,3]_{\pm}A = \cup_{i=0}^{3} h_{\pm}A = \{0\} \cup \{2,3,10,11\}, \{1,4,5,6,7,8,9,12\} \cup \{1,4,5,6,7,8,9,12\} = G.$$

Our notations and terminology are summarized in the following table.

	$H = \{h\}$	$H = [0,s]$	$H = \mathbb{N}_0$
$\Lambda = \mathbb{N}_0$	hA h-fold sumset	$[0,s]A$ $[0,s]$-fold sumset	$\langle A \rangle$ sumset
$\Lambda = \mathbb{Z}$	$h_{\pm}A$ h-fold signed sumset	$[0,s]_{\pm}A$ $[0,s]$-fold signed sumset	$\langle A \rangle$ signed sumset
$\Lambda = \{0,1\}$	$h\hat{~}A$ restricted h-fold sumset	$[0,s]\hat{~}A$ restricted $[0,s]$-fold sumset	ΣA restricted sumset
$\Lambda = \{-1,0,1\}$	$h\hat{_{\pm}}A$ restricted h-fold signed sumset	$[0,s]\hat{_{\pm}}A$ restricted $[0,s]$-fold signed sumset	$\Sigma_{\pm}A$ restricted signed sumset

Recall that the sumset $\cup_{h=0}^{\infty} hA$ and the signed sumset $\cup_{h=0}^{\infty} h_{\pm}A$ of a subset A both equal $\langle A \rangle$, the subgroup of G generated by A. For example, with our previous example of $A = \{2,3\}$ in $G = \mathbb{Z}_{13}$, we have $\langle A \rangle = \mathbb{Z}_{13}$. (Since 13 is prime, $\mathbb{Z}_{13}$ has no subgroups other than $\{0\}$ and $\mathbb{Z}_{13}$.)

Next, we point out some obvious but useful identities.

Suppose that $A \subseteq G$ and assume, as usual, that $A = \{a_1, \dots, a_m\}$. Let g be an arbitrary element of G. We then define the set $A - g$ as $\{a_1 - g, \dots, a_m - g\}$. For a fixed $h \in \mathbb{N}_0$, we now examine the sumsets $h(A-g)$ and $h\hat{~}(A-g)$. By definition, we have

$$\begin{aligned} h(A-g) &= h\{a_1 - g, \dots, a_m - g\} \\ &= \{\lambda_1(a_1-g) + \cdots + \lambda_m(a_m - g) \mid \lambda_1, \dots, \lambda_m \in \mathbb{N}_0, \lambda_1 + \cdots + \lambda_m = h\} \\ &= \{\lambda_1 a_1 + \cdots + \lambda_m a_m - h \cdot g \mid \lambda_1, \dots, \lambda_m \in \mathbb{N}_0, \lambda_1 + \cdots + \lambda_m = h\} \\ &= hA - h \cdot g. \end{aligned}$$

Similarly,

$$\begin{aligned} h\hat{~}(A-g) &= h\hat{~}\{a_1 - g, \dots, a_m - g\} \\ &= \{\lambda_1(a_1-g) + \cdots + \lambda_m(a_m - g) \mid \lambda_1, \dots, \lambda_m \in \{0,1\}, \lambda_1 + \cdots + \lambda_m = h\} \\ &= \{\lambda_1 a_1 + \cdots + \lambda_m a_m - h \cdot g \mid \lambda_1, \dots, \lambda_m \in \{0,1\}, \lambda_1 + \cdots + \lambda_m = h\} \\ &= h\hat{~}A - h \cdot g. \end{aligned}$$

Therefore, we see that $h(A-g)$ has the same cardinality as hA, and $h\hat{~}(A-g)$ has the same cardinality as $h\hat{~}A$. This is particularly useful when $g \in A$, since then $A - g$ is a subset of G that contains zero. In summary, we have the following:

Proposition 3.2 *For any G, $A \subseteq G$, $g \in G$, and $h \in \mathbb{N}_0$, we have*

$$|h(A-g)| = |hA|$$

and

$$|h\hat{~}(A-g)| = |h\hat{~}A|.$$

In particular, there is a subset A_0 of G so that $0 \in A_0$, $|A_0| = |A|$, $|hA_0| = |hA|$, and $|h\hat{~}A_0| = |h\hat{~}A|$.

We need to point out that the "signed sumset" versions of these identities (when subtraction of elements is allowed) do not necessarily hold.

The following identities are also obvious:

Proposition 3.3 *For any G, $A \subseteq G$, and $s \in \mathbb{N}_0$, we have*

$$[0, s]A = s(A \cup \{0\})$$

and

$$[0, s]_{\pm}A = s_{\pm}(A \cup \{0\}).$$

We should note that the restricted versions of these identities do not necessarily hold.

We also find that the (h-fold, etc.) signed sumset of a subset A is closely related to the (h-fold, etc.) sumset of $A \cup (-A)$. Of course, for $h = 0$, we have

$$0_{\pm}A = 0(A \cup (-A)) = \{0\};$$

for $h = 1$ we get

$$1_{\pm}A = 1(A \cup (-A)) = A \cup (-A).$$

For $h = 2$, we can easily see that $2_{\pm}A$ and $2(A \cup (-A))$ are almost the same, except that $2(A \cup (-A))$ always contains 0 while $2_{\pm}A$ may not, so

$$2(A \cup (-A)) = 2_{\pm}A \cup \{0\} = 2_{\pm}A \cup 0_{\pm}A.$$

More generally, we have the following identities.

Proposition 3.4 *For every subset A of G and every nonnegative integer h we have*

$$h(A \cup (-A)) = h_{\pm}A \cup (h-2)_{\pm}A \cup (h-4)_{\pm}A \cup \cdots;$$

and therefore, for all $s \in \mathbb{N}$,

$$[0, s](A \cup (-A)) = [0, s]_{\pm}A$$

and

$$\langle A \cup (-A) \rangle = \langle A \rangle.$$

We provide the proof on page 310.

Note that when A is *symmetric*, that is, $A = -A$, then an even simpler relation holds:

Proposition 3.5 *For every symmetric subset A of G and every nonnegative integer h we have $h_{\pm}A = hA$.*

The easy proof can be found on page 311.

Regarding restricted sumsets, we always have

$$h\hat{\ }(A \cup (-A)) \subseteq h\hat{_{\pm}}A \cup (h-2)\hat{_{\pm}}A \cup (h-4)\hat{_{\pm}}A \cup \cdots,$$

(see the last paragraph of the proof of Proposition 3.4), but an identity for our two types of restricted sumsets seems a lot more complicated. Note that, for example, when A contains distinct elements a_1 and a_2 for which $a_1 = -a_2$, then $a_1 - a_2$ is definitely an element of $2\hat{_{\pm}}A$, but not necessarily of $2\hat{\ }(A \cup (-A))$. We pose the following vague problem.

Problem 3.6 *Find identities similar to those in Proposition 3.4 for restricted sumsets.*

Exercises

1. Consider the subset $A = \{2, 3\}$ in the cyclic group $G = \mathbb{Z}_{21}$. List the elements of each of the following sumsets.

   (a) $[0,3]_{\pm}A$

   (b) $[0,3]A$

   (c) $[0,3]\hat{_{\pm}}A$

   (d) $[0,3]\hat{}A$

2. Consider the subset $A = \{(0,0,1), (0,1,1), (1,1,1)\}$ in the group $G = \mathbb{Z}_3^3$. List the elements of each of the following sumsets.

   (a) $2_{\pm}A$

   (b) $2A$

   (c) $2\hat{_{\pm}}A$

   (d) $2\hat{}A$

Part II

Appetizers

The short articles in this part are meant to invite everyone to the main entrees of the menu, presented in Part IV of the book. Our appetizers are carefully chosen so that they provide bite-size representative samples of the research projects, as well as make connections to other parts of mathematics that students might have encountered. (The first article describes how this author's research in spherical geometry led him to additive combinatorics.)

Spherical designs

This appetizer tells the (simplified) story of how the author's research in algebraic combinatorics and approximation theory led him to additive combinatorics (cf. [9], [10]).

Imagine that we want to scatter a certain number of points on a sphere—how can we do it in the most "uniformly balanced" way? The answer might be obvious in certain cases, but far less clear in others. Consider, for example, the sphere S^2 in 3-dimensional Euclidean space, and suppose that we need to place six points on it: the most balanced configuration for the points is undoubtedly at the six vertices of a regular octahedron (for example, at the points $(\pm 1, 0, 0), (0, \pm 1, 0), (0, 0, \pm 1)$ in case the sphere is centered at the origin and has radius 1). The answer is perhaps equally clear for four, eight, twelve, or twenty points: place them so that they form a regular tetrahedron, cube, icosahedron, and dodecahedron, respectively. But how should we position five points? Or how about seven or ten or a hundred points? How can one do this in general?

The answer to our question depends, of course, on how we measure the degree to which our pointset is balanced. For example, in the case of a *packing problem*, we want to place our points as far away from each other as possible; more precisely, we may want to maximize the minimum distance between any two of our points. Or, when addressing a *covering problem*, we want to minimize the maximum distance that any place on the sphere has from the closest point of our pointset. There are several other reasonable criteria, but the one that is perhaps the most well known and applicable is the one where we maximize the degree to which our pointset is in *momentum balance*. We make this precise as follows.

As usual, we let S^d denote the sphere in the $(d+1)$-dimensional Euclidean space $\mathbb{R}^{d+1}$; we also assume that S^d is centered at the origin and has radius 1, in which case it can be described as the set of points that have distance 1 from the origin. Given a finite pointset $P \subset S^d$ and a polynomial $f : S^d \to \mathbb{R}$, we define the average of f over P as

$$\overline{f}_P = \frac{1}{|P|} \cdot \sum_{p \in P} f(p);$$

the average of f over the entire sphere is given by

$$\overline{f}_{S^d} = \frac{1}{|S^d|} \cdot \int_{S^d} f(x) \mathrm{d}x$$

(here $|S^d|$ denotes the surface area of S^d). We then say that our finite pointset P is a *spherical t-design on S^d*, if

$$\overline{f}_P = \overline{f}_{S^d}$$

holds for all polynomials f of degree up to t.

Spherical designs were introduced and first studied in 1977 by Delsarte, Goethals, and Seidel in [63]. In this far-reaching paper they also established the following tight lower bound for the number of points needed to form a spherical t-design on S^d:

$$|P| \geq N_t^d = \binom{d + \lfloor t/2 \rfloor}{d} + \binom{d + \lfloor (t-1)/2 \rfloor}{d}.$$

We shall refer to the quantity N_t^d as the *DGS bound*. Spherical designs of this minimum size are called *tight*. Bannai and Damerell (cf. [30], [31]) proved that tight spherical designs for $d \geq 2$ exist only for $t = 1, 2, 3, 4, 5, 7$, or 11. All tight t-designs are known, except possibly for $t = 4, 5$, or 7; in particular, there is a unique 11-design (with d=23 and $|P| = 196560$, coming from the *Leech lattice*).

Clearly, any finite pointset $P \subset S^d$ is a spherical 0-design. Spherical 1-designs and 2-designs have basic interpretations in physics: it is easy to check that P is a spherical 1-design exactly when P is in mass balance (that is, its center of gravity is at the center of the sphere), and we can verify that P is a spherical 2-design if, and only if, P is in both mass balance and inertia balance. Higher degrees of balance find their applications in a variety of areas, including crystallography, coding theory, astronomy, and viral morphology.

According to the definition, to test whether a given pointset P is a spherical t-design, one should check whether $\overline{f}_P$ agrees with $\overline{f}_{S^d}$ for every polynomial f of degree at most t. There is a well-known shortcut: it suffices to do this for non-constant homogeneous harmonic polynomials of degree at most t. A polynomial is called *homogeneous* if all its terms have the same (total) degree; for example,

$$f = x^3z + xz^3 - 6xy^2z$$

is a homogeneous polynomial (of three variables and of degree 4). A polynomial is called *harmonic* if it satisfies the *Laplace equation*; that is, the sum of its unmixed second-order partial derivatives equals zero. We see that the polynomial f just mentioned is harmonic:

$$\frac{\partial^2 f}{\partial x^2} + \frac{\partial^2 f}{\partial y^2} + \frac{\partial^2 f}{\partial z^2} = 6xz - 12xz + 6xz = 0.$$

The reduction to homogeneous harmonic polynomials has two advantages. First, there are a lot fewer of them: the set $\mathrm{Harm}_k(S^d)$ of homogeneous harmonic polynomials over S^d of degree k forms a vector space over $\mathbb{R}$ whose dimension is "only"

$$\dim \mathrm{Harm}_k(S^d) = \binom{d+k}{d} - \binom{d+k-2}{d}.$$

Second, the average of non-constant harmonic polynomials over the sphere is zero. Therefore, a pointset $P \subset S^d$ is a spherical t-design on S^d if, and only if, $\overline{f}_P = 0$ for every $f \in \mathrm{Harm}_k(S^d)$ and $1 \le k \le t$.

So, how can we construct spherical t-designs? We get an inspiration from the case $d = 1$ (when our sphere is a circle). Note that

$$\dim \mathrm{Harm}_k(S^1) = \binom{k+1}{1} - \binom{k-1}{1} = 2,$$

and we can verify that the real and imaginary parts of the polynomial $(x+iy)^k$ (with i denoting the imaginary square root of -1) form a basis of $\mathrm{Harm}_k(S^1)$; for example,

$$\begin{aligned}
\mathrm{Harm}_1(S^1) &= \langle \mathrm{Re}(x+iy)^1, \mathrm{Im}(x+iy)^1 \rangle = \langle x, y \rangle, \\
\mathrm{Harm}_2(S^1) &= \langle \mathrm{Re}(x+iy)^2, \mathrm{Im}(x+iy)^2 \rangle = \langle x^2 - y^2, 2xy \rangle, \\
\mathrm{Harm}_3(S^1) &= \langle \mathrm{Re}(x+iy)^3, \mathrm{Im}(x+iy)^3 \rangle = \langle x^3 - 3xy^2, 3x^2y - y^3 \rangle.
\end{aligned}$$

As it is probably easy to guess and as we now verify: regular polygons yield spherical designs on the circle. More precisely, we show that the set of vertices

$$P = \{(\cos(2\pi j/n), \sin(2\pi j/n)) \mid j = 0, 1, 2, \ldots, n-1\}$$

of the regular n-gon form a spherical t-design on S^1 for all $n \ge t+1$. (Note that, by the DGS bound, a spherical t-design on S^1 must have at least $N_t^1 = t+1$ points.)

Indeed, by identifying S^1 with complex numbers of norm 1, our pointset can be described as

$$P = \{z^j \mid j = 0, 1, 2, \ldots, n-1\},$$

where z is the first n-th root of unity:

$$z = \cos(2\pi/n) + i \sin(2\pi/n).$$

We see that, when k is not a multiple of n, then $z^k \neq 1$, and thus

$$\sum_{j=0}^{n-1} \left(z^j\right)^k = \sum_{j=0}^{n-1} \left(z^k\right)^j = \frac{\left(z^k\right)^n - 1}{z^k - 1} = \frac{\left(z^n\right)^k - 1}{z^k - 1} = \frac{1^k - 1}{z^k - 1} = 0,$$

and thus the average of $\mathrm{Re}(x+iy)^k$ and $\mathrm{Im}(x+iy)^k$ over P both equal zero. (When k is a multiple of n, then each term in the sum equals 1, so only the average of $\mathrm{Im}(x+iy)^k$ equals zero over P.) Therefore, $\overline{f}_P = 0$ for all $f \in \mathrm{Harm}_k(S^1)$ and all $1 \leq k \leq n-1$, and thus P is a spherical t-design for every $t \leq n-1$, as claimed.

Let us see now how we can generalize this construction to higher dimensions. We will assume that d is odd; the case when d is even can be reduced to this case. Let $m = (d+1)/2$, and suppose that $A = \{a_1, \ldots, a_m\}$ is a set of integers. We construct n points on S^d as follows: for each $j = 0, 1, 2, \ldots, n-1$, we let

$$z_j = \left(\cos(2\pi j a_1/n), \sin(2\pi j a_1/n), \ldots, \cos(2\pi j a_m/n), \sin(2\pi j a_m/n)\right);$$

we then set

$$P(A) = \left\{1/\sqrt{m} \cdot z_j \mid j = 0, 1, 2, \ldots, n-1\right\}$$

(the factor $1/\sqrt{m}$ is needed so that $P(A) \subset S^d$). The question that we ask is then the following: for what values of n and for which sets A is $P(A)$ a spherical t-design on S^d?

We answer this question for $t = 1$ first. From our formula on page 51, we see that $\dim \mathrm{Harm}_1(S^d) = d+1$; clearly, the polynomials x_i with $i = 1, 2, \ldots, d+1$ form a basis for $\mathrm{Harm}_1(S^d)$. Consequently, $P(A)$ is a spherical 1-design if, and only if,

$$\sum_{j=0}^{n-1} \cos(2\pi j a_i/n) = \sum_{j=0}^{n-1} \sin(2\pi j a_i/n) = 0$$

for each $a_i \in A$; as we verified above, these equations hold whenever a_i is not a multiple of n. Therefore, with

$$a_1 = a_2 = \cdots = a_m = 1,$$

$P(A)$ is a spherical 1-design for every $n \geq 2$. (Note that the DGS bound requires that a spherical 1-design on S^d has size at least $N_1^d = 2$; obviously, a single point is never in mass balance on the sphere, but two or more points may be.)

Let us turn to $t = 2$. This time our formula from page 51 yields

$$\dim \mathrm{Harm}_2(S^d) = \binom{d+2}{d} - \binom{d}{d} = \binom{d+1}{2} + d,$$

and we can verify that the set

$$\{x_{i_1} x_{i_2} \mid 1 \leq i_1 < i_2 \leq d+1\} \cup \{x_{i+1}^2 - x_i^2 \mid 1 \leq i \leq d\}$$

forms a basis for $\mathrm{Harm}_2(S^d)$. In order to compute the average values of these functions over $P(A)$, we need to use the trigonometric identities

$$\begin{aligned}\sin\alpha\cdot\sin\beta &= 1/2\cdot(\cos(\alpha-\beta)-\cos(\alpha+\beta)),\\ \cos\alpha\cdot\cos\beta &= 1/2\cdot(\cos(\alpha-\beta)+\cos(\alpha+\beta)),\\ \sin\alpha\cdot\cos\beta &= 1/2\cdot(\sin(\alpha-\beta)+\sin(\alpha+\beta)).\end{aligned}$$

We consider first $f = x_{i_1}x_{i_2}$ with $1 \leq i_1 < i_2 \leq d+1$. Let $i_1' = \lceil i_1/2 \rceil$ and $i_2' = \lceil i_2/2 \rceil$, and note that $1 \leq i_1' \leq i_2' \leq m$. Assume first that $i_1' < i_2'$. With our notations, $f(z_j)$ equals the product of the cosine or sine of $2\pi j a_{i_1'}/n$ and the cosine or sine of $2\pi j a_{i_2'}/n$, and with our identities we can turn this into a linear combination of a cosine or sine of $2\pi j(a_{i_1'} - a_{i_2'})/n$ and of $2\pi j(a_{i_1'} + a_{i_2'})/n$. As we have seen above, if neither $a_{i_1'} - a_{i_2'}$ nor $a_{i_1'} + a_{i_2'}$ is a multiple of n, then

$$\sum_{j=0}^{n-1}\cos\left(2\pi j(a_{i_1'} - a_{i_2'})/n\right) = \sum_{j=0}^{n-1}\sin\left(2\pi j(a_{i_1'} - a_{i_2'})/n\right) = 0$$

and

$$\sum_{j=0}^{n-1}\cos\left(2\pi j(a_{i_1'} + a_{i_2'})/n\right) = \sum_{j=0}^{n-1}\sin\left(2\pi j(a_{i_1'} + a_{i_2'})/n\right) = 0,$$

and thus the average of f over $P(A)$ is zero. Turning to the case when $i_1' = i_2'$: note that we then must have $i_2 = i_1 + 1$ with i_1 odd, so

$$f(z_j) = \cos\left(2\pi j a_{i_1}/n\right)\cdot\sin\left(2\pi j a_{i_1}/n\right) = 1/2\cdot\sin\left(2\pi j(2a_{i_1})/n\right) = 1/2\cdot \mathrm{Im} z^j,$$

where z is the $2a_{i_1}$-th n-th root of unity. Since, as noted above, $\sum_{j=0}^{n-1}\mathrm{Im} z^j$ equals zero for any n-th root of unity z, the average of f over $P(A)$ equals zero in this case as well.

Now let $f = x_{i+1}^2 - x_i^2$ with $1 \leq i \leq d$. We then find that, for even values of i we have

$$\begin{aligned} f(z_j) &= \cos^2\left(2\pi j a_{i/2+1}/n\right) - \sin^2\left(2\pi j a_{i/2}/n\right)\\ &= 1/2\cdot\left(\cos\left(2\pi j(2a_{i/2+1})/n\right) + \cos\left(2\pi j(2a_{i/2})/n\right)\right),\end{aligned}$$

so, if $2a$ is not a multiple of n for any element $a \in A$, then f has a zero average over $P(A)$; our computation is similar when i is odd.

Therefore, in summary, $P(A)$ is a spherical 2-design on S^d—that is, every polynomial in $\mathrm{Harm}_1(S^d)$ and $\mathrm{Harm}_2(S^d)$ has a zero average on $P(A)$—if, and only if, A consists of m distinct terms so that

- no element of A,
- the difference of no two distinct elements of A, and
- the sum of no two (not-necessarily-distinct) elements of A

is a multiple of n. Recall that for a subset $A = \{a_1, \ldots, a_m\}$ of an abelian group G we defined $[1,2]_\pm A$ as

$$[1,2]_\pm A = \{\lambda_1 a_1 + \cdots + \lambda_m a_m \mid \lambda_1, \ldots, \lambda_m \in \mathbb{Z},\ 1 \leq |\lambda_1| + \cdots |\lambda_m| \leq 2\},$$

so our three-part condition above can be summarized by saying that we want to assure that A is an m-subset of $\mathbb{Z}_n$ for which $0 \not\in [1,2]_\pm A$; that is, using the terminology of Section F.2.2, A is *2-independent* in $\mathbb{Z}_n$.

It is not hard to find specific 2-independent sets in $\mathbb{Z}_n$; for example, the set

$$A = \{1, 2, \ldots, m\}$$

is clearly 2-independent in $\mathbb{Z}_n$ for every $n \geq 2m + 1$. Recall that we set $m = (d+1)/2$, so we have constructed explicit spherical 2-designs on S^d of size n for every odd d and $n \geq d+2$. (Note that, by the DGS bound, a spherical 2-design on S^d must contain at least $d+2$ points.) As we mentioned before, the case of even d can be reduced to the odd case, but we only get spherical 2-designs of sizes $n = d+2$ or $n \geq d+4$; it can be shown that when d is even, spherical 2-designs of size $d+3$ do not exist on S^d (cf. [158]). So, on S^2, for example, we have 2-designs of size four (the regular tetrahedron) and any size $n \geq 6$, but there is no spherical 2-design formed by five points.

Moving to $t = 3$, we define A to be a *3-independent set* in $\mathbb{Z}_n$ if $0 \notin [1,3]_{\pm}A$, that is, in addition to our three-part condition above,

- the sum of three (not-necessarily-distinct) elements of A and
- the sum of two (not-necessarily-distinct) elements of A minus any element of A

is never a multiple of n. With similar techniques, we can prove that if A is 3-independent in $\mathbb{Z}_n$, then the set $P(A)$ defined above is a spherical 3-design on S^d.

It is an interesting problem to construct 3-independent sets in $\mathbb{Z}_n$. One quickly realizes that the set

$$\{1, 3, 5, \ldots, 2m-1\}$$

works for every $n \geq 6m - 2$, and this bound can be lowered to $n \geq 4m - 1$ when n is even. We can do better yet when n has a divisor p with $p \equiv 5 \bmod 6$:

$$\begin{aligned} A &= \{1, 3, 5, \ldots, (p-2)/3\} + \{0, p, 2p, \ldots, n-p\} \\ &= \{2i + 1 + jp \mid i = 0, 1, \ldots, (p-5)/6,\ j = 0, 1, \ldots, n/p - 1\} \end{aligned}$$

is then 3-independent in $\mathbb{Z}_n$, and this set has size $(p+1)/6 \cdot n/p$. It turns out that we cannot do better: as Bajnok and Ruzsa proved in 2003 in [24], the minimum size $\tau_{\pm}(\mathbb{Z}_n, [1,3])$ of a 3-independent set in $\mathbb{Z}_n$ equals

$$\tau_{\pm}(\mathbb{Z}_n, [1,3]) = \begin{cases} \left\lfloor \frac{n}{4} \right\rfloor & \text{if } n \text{ is even,} \\ \left(1 + \frac{1}{p}\right) \frac{n}{6} & \text{if } n \text{ is odd, has prime divisors congruent to 5 mod 6,} \\ & \text{and } p \text{ is the smallest such divisor,} \\ \left\lfloor \frac{n}{6} \right\rfloor & \text{otherwise.} \end{cases}$$

(The value of $\tau_{\pm}(G, [1,3])$ is not completely known in noncyclic groups G, and quite little is known about $\tau_{\pm}(G, [1,t])$ for $t > 3$ even for cyclic groups—see Section F.2.2.) As a consequence, we get spherical 3-designs on S^d for odd d and every size n with

- $n \geq 2d + 2$ if n is divisible by 4,
- $n \geq \left(1 - \frac{1}{p+1}\right)(3d+3)$ if n has prime divisors congruent to 5 mod 6 and p is the smallest such divisor, or
- $n \geq 3d + 3$.

Explicit constructions of spherical t-designs get increasingly complicated as t grows, and the cases of small sizes n are particularly difficult and largely open (see [7, 8]). For further information on this topic see the extensive survey paper [29] of Bannai and Bannai and its nearly two hundred references. The construction of spherical designs and, more generally, uniformly distributed pointsets on the sphere, was listed by Fields Medalist Steven Smale as number 7 on his list of the most important mathematical problems for the twenty-first century (see [189]).

Caps, centroids, and the game SET

In this appetizer, we explore how one particular question in additive combinatorics connects such diverse topics as centroids of triangles, caps in affine geometry, and the card game SET. We present each of these topics by some relevant puzzles.

Let us start with the popular and award-winning card game SET, published by SET Enterprises (cf. [186]). (The game was invented by the population geneticist Marsha Falco while studying epilepsy patterns in German Shepherds.) The deck consists of 81 cards, each of which features four attributes: shapes (ovals, diamonds, or squiggles); number of shapes (one, two, or three); shadings (empty, striped, or solid); and colors (red, green, or purple). Each of the 81 cards features a different combination of attributes. The game centers on players identifying sets among the cards: three cards form a set if each of the four attributes, when considered individually on the three cards, is either all the same or all different. For example, the three cards

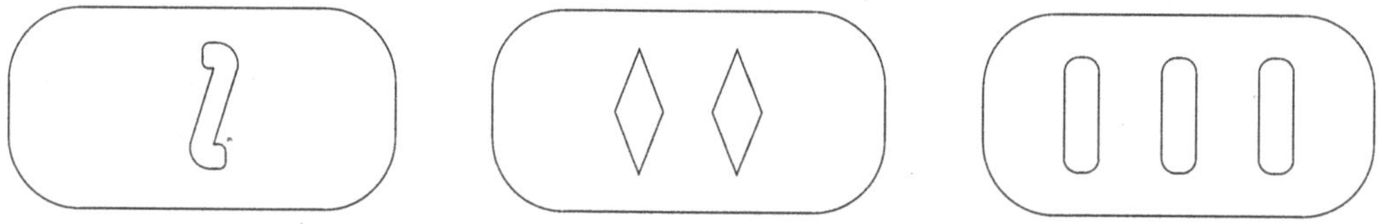

feature all three shapes (squiggle, diamonds, and ovals), all three numbers (one, two, and three), but share the same shading (empty), and they form a set either when they are of the same color (e.g. all are green) or are of different colors (one green, one red, one purple).

In the most popular version of the game, twelve cards are dealt from a shuffled deck and shown face up on a table. When a player sees a set, he or she calls "set," and collects the three cards from the table. The three cards then get replaced from the deck, and the game continues; when the deck runs out and no more sets are left, the game ends. The player with the most sets collected wins. Occasionally—as it turns out, with a probability of slightly more than 3 percent—the twelve cards that were dealt don't contain a set; in this case other cards are dealt from the deck one by one, until a set is present and claimed. The question then arises: what is the maximum number of cards without containing any sets?

We see that sets are quite ubiquitous; in fact, any pair of cards is contained in a (unique) set. Indeed, for each of the four attributes, the two cards either share the attribute, in which case we extend it with a card that also shares that attribute, or they don't, in which case we need a card whose relevant attribute differs from that of both cards. For example, if one card has two striped green diamonds and the other has one striped red squiggle, then we can extend them to form a set with the card containing three striped purple ovals. Therefore, the probability that three cards don't form a set equals 78/79, almost 99 percent.

A similar calculation shows that the probability that four cards don't form a set equals 75/79, slightly under 95 percent. The computations get increasingly difficult as the number of cards gets larger; the exact values have been calculated by Knuth in [134]. It is clear that the probability that there are no sets among k cards decreases rapidly as k increases; as we have mentioned above, the probability that twelve cards don't contain any sets is just over 3 percent. Our question above is equivalent to asking for the largest value of k for which this probability is still not zero. We will reveal the answer shortly.

Let us now move on to a (seemingly) very different topic: affine plane geometry. We start by recalling two properties of the Euclidean plane:

- any two distinct points determine a unique line that contains both points, and
- any point P and any line l that does not contain P determine a unique line that contains P but contains no point of l.

(The second property, which guarantees the existence of a line through P that is *parallel* to l, is referred to as the *Parallel Postulate.*)

Euclidean geometry has many other properties, of course, such as those involving distances. But any structure on a given set S and on a collection $\mathcal{L}$ of its subsets that satisfies the two axioms and is not trivial—that is, S contains at least four points no three of which are *collinear* (are contained in the same element of $\mathcal{L}$)—is worth studying: we may simply identify S with the set of points and $\mathcal{L}$ with the set of lines; if S and $\mathcal{L}$ satisfy the properties we require, we say that they form an *affine plane*. In particular, here we discuss finite affine planes: those that contain a finite number of points.

It turns out that the number of points in a finite affine plane cannot be arbitrary: it must be a square number. In fact, not every square number is possible either: for example, it cannot be 36 or 100, but we still don't know if it can be 144. One can prove that for every finite affine plane there is a positive integer k so that

- there are exactly k^2 points and $k^2 + k$ lines,
- each line contains k points, and
- each point is contained in $k + 1$ lines.

The possible values of k include all prime-powers, and it is one of the most famous open problems to prove that it does not include other values. As we mentioned above, we know that k cannot be 6 or 10, but we still don't know whether it can be 12.

The example that we are focusing on here is the one for $k = 3$; the corresponding affine plane contains nine points and twelve lines. Let us denote the points by $A, B, C, D, E, F, G, H,$ and I; we can then verify that by setting the twelve lines equal to the pointsets

$$\{A, B, C\} \quad \{D, E, F\} \quad \{G, H, I\} \quad \{A, D, G\} \quad \{B, E, H\} \quad \{C, F, I\}$$

$$\{A, F, H\} \quad \{B, D, I\} \quad \{C, E, G\} \quad \{A, E, I\} \quad \{B, F, G\} \quad \{C, D, H\}$$

our required properties hold. For example: the points B and F determine the unique line $\{B, F, G\}$ that contains them both, and the point A and the line $\{C, E, G\}$ determine the unique line $\{A, F, H\}$ that contains A but none of C, E, or G. This affine geometry is denoted by AG(2,3) (with the 2 representing the fact that we are in the plane—affine geometries may be considered in higher dimensions as well), and we can visualize it by the diagram below. (Note that eight of our lines are "straight" and four are "curved.")

Our question then is the following: What is the maximum number of points that one can find so that they do not contain all three points of any line? Pointsets within the geometry that do not contain three collinear points are called *cap-sets*; we can thus rephrase our question to ask for the maximum size of a cap-set in AG$(2, 3)$. We defer answering this question until we first discuss yet another topic: centroids of triangles.

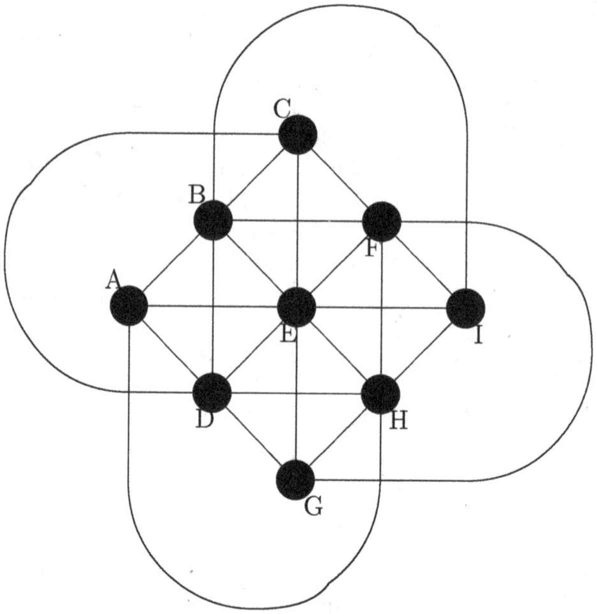

Recall that the *centroid* of a triangle is defined as the point where the three medians of the triangle (lines connecting a vertex with the midpoint of the opposite side) intersect, as illustrated below: the centroid of the triangle with vertices A, B, and C is marked by M. (It is a well-known fact that the three medians go through the same point.) Informally, the centroid of the triangle is the point where the tip of a pin should be if one wants to balance the triangle (made of material with uniform density) on it.

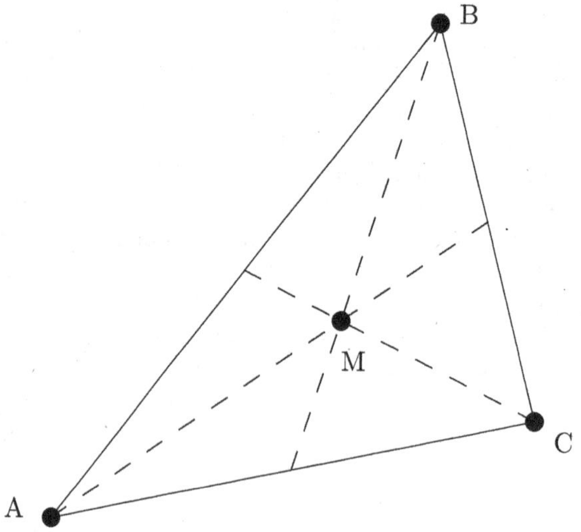

Given the coordinates of the vertices of the triangle in a coordinate system, one can determine the coordinates of the centroid by taking the arithmetic averages of the coordinates of the three vertices; for example, if A=(a_1, a_2), B=(b_1, b_2), and C=(c_1, c_2), then

$$\mathrm{M} = \left(\frac{a_1 + b_1 + c_1}{3}, \frac{a_2 + b_2 + c_2}{3} \right).$$

This formulation allows us to broaden our definition to include the degenerate case when the three points are collinear; for example, the centroid of the collinear points A=$(0,0)$, B=$(2,1)$, and C=$(16,8)$ is the point M=$(6,3)$:

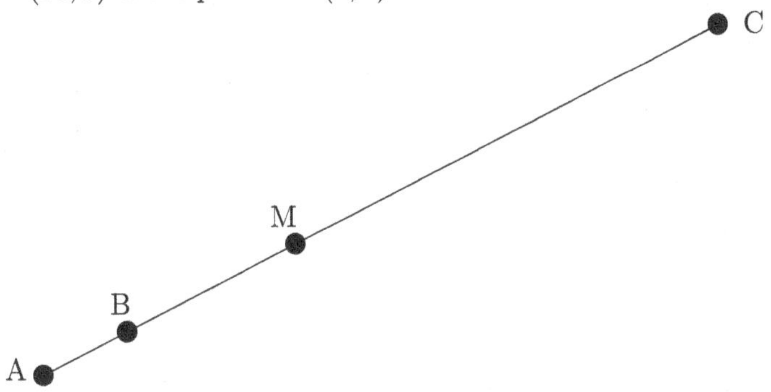

Our question about centroids is the following: What is the maximum number of lattice points that one can find in the integer lattice $\mathbb{Z}^2$ (points in the coordinate plane with integer coordinates) so that no three of them have their centroid at a lattice point?

Well, let us reveal how the three questions we posed relate to one another and how they are special cases of a quantity we investigate in Section F.3.1 of this book. Namely, our interest is in finding the maximum size of a *weakly zero-3-sum-free set* in the group $\mathbb{Z}_3^r$, defined as the maximum size—denoted by $\tau\hat{\ }(\mathbb{Z}_3^r, 3)$—of a subset of $\mathbb{Z}_3^r$ without any three distinct elements adding to zero; that is, the quantity

$$\tau\hat{\ }(\mathbb{Z}_3^r, 3) = \max\{|A| \mid A \subseteq \mathbb{Z}_3^r, 0 \not\in 3\hat{\ }A\}.$$

(The word "weak" signifies that we only disallow the sum of three *distinct* elements to be zero.) The problem of finding $\tau\hat{\ }(\mathbb{Z}_3^r, 3)$ seems forbiddingly difficult at the present time. Exact values are only known for $r \leq 6$:

r	1	2	3	4	5	6
$\tau\hat{\ }(\mathbb{Z}_3^r, 3)$	2	4	9	20	45	112

(See [97] by Gao and Thangadurai and its references for the first five entries and [175] by Potechin for the last.) Here we only present the maximum-sized examples of $r \leq 4$:

$\tau\hat{\ }(\mathbb{Z}_3, 3) = 2:$

$\tau\hat{\ }(\mathbb{Z}_3^2, 3) = 4:$

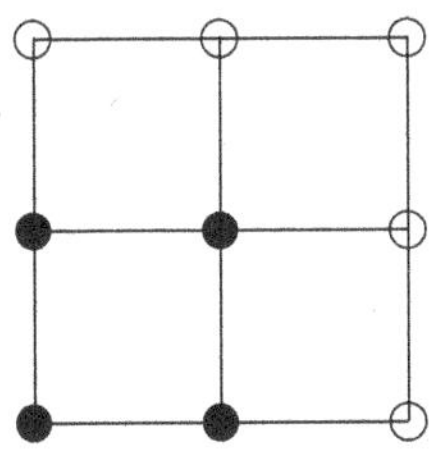

$\tau\hat{\ }(\mathbb{Z}_3^3, 3) = 9:$

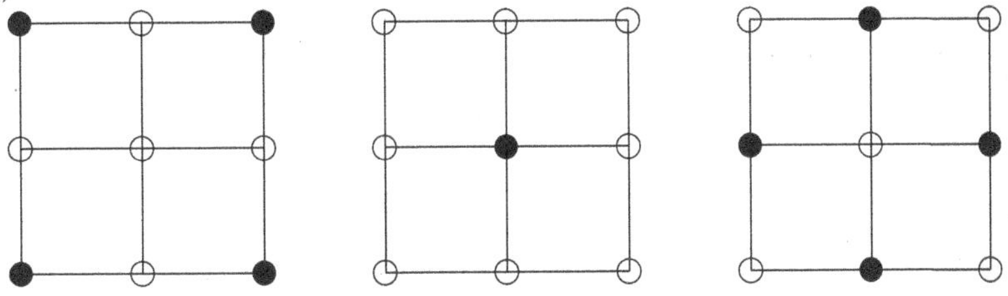

$\tau\hat{}(\mathbb{Z}_3^4, 3) = 20:$

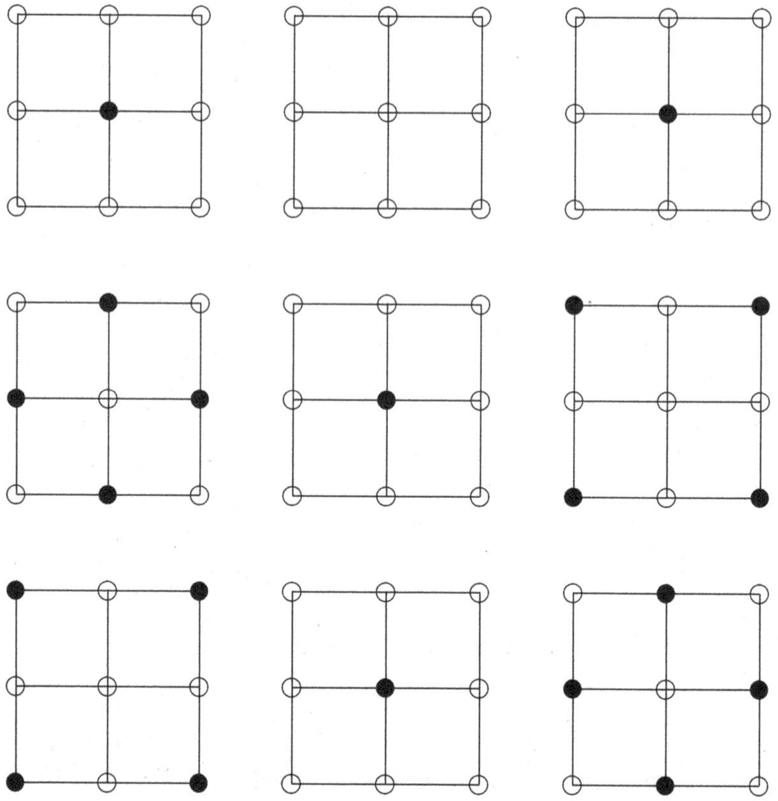

Hopefully, these diagrams are self-explanatory: For example, for rank two, we show the four elements $(0,0), (0,1), (1,0)$, and $(1,1)$ that form a weakly zero-3-sum-free subset in $\mathbb{Z}_3^2$, and within the three grids representing $\mathbb{Z}_3^3$, we feature the weakly zero-3-sum-free subset

$$\{(0,0,0), (0,0,2), (0,2,0), (0,2,2), (1,1,1), (2,0,1), (2,1,0), (2,1,2), (2,2,1)\}.$$

Besides verifying that the sets above are weakly zero-3-sum-free in their respective groups, one would also need to prove that they are of maximum size—these proofs get increasingly difficult as the rank of the group increases.

So how do our three questions above relate to these values? A key is the following observation: three elements of $\mathbb{Z}_3$ add to zero if, and only if, they are all distinct ($0+1+2=0$) or all the same (e.g., $1+1+1=0$). If we assign coordinates to each attribute in the game SET (for example, the first coordinate represents shape: namely, 0, 1, and 2 denote ovals, diamonds, and squiggles, respectively; and similarly with the other three attributes), then each of our 81 cards corresponds to a unique element of $\mathbb{Z}_3^4$, with three cards forming a set exactly when their corresponding group elements add to $(0,0,0,0)$, the zero element of $\mathbb{Z}_3^4$. Therefore, the maximum number of cards without having any sets among them is $\tau\hat{}(\mathbb{Z}_3^4, 3) = 20$.

Now let us turn to the question of finding the maximum size of a cap-set in the geometry AG$(2,3)$. To start, we identify our nine points with the elements of $\mathbb{Z}_3^2$ as follows: $A = (0,0)$, $B = (0,1)$, $C = (0,2)$, $D = (1,0)$, $E = (1,1)$, $F = (1,2)$, $G = (2,0)$, $H = (2,1)$, $I = (2,2)$.

Suppose that X, Y, and Z are three of these points. Recall that they are collinear whenever the vectors from X to Y and from X to Z are parallel, that is, when the vector $Y - X$ is a scalar multiple of the vector $Z - X$. Since our points, and therefore our vectors, are considered here as elements of $\mathbb{Z}_3^2$, that scalar can only be 0, 1, or 2—but it cannot be 0 (since X and Y are distinct) and cannot be 1 (since Y and Z are distinct). Therefore, X, Y, and Z are collinear exactly when $Y - X = 2 \cdot (Z - X)$ in $\mathbb{Z}_3^2$ or, equivalently, $X + Y + Z = (0,0)$. Indeed, it is easy to check that for each of our twelve lines, the sum of the three points within them equals zero; for example, two of our lines were $\{A, D, G\}$ and $\{B, F, G\}$, and indeed we have $A + D + G = (0,0)$ and $B + F + G = (0,0)$. (In fact, the twelve lines exhibit all such 3-subsets: there are three of them where the first coordinates are equal and the second coordinates are all distinct, three where the second coordinates are equal and the first coordinates are all distinct, and six where both the first coordinates and second coordinates are pairwise distinct.) Therefore, a set of points in AG(2, 3) is a cap-set exactly when the corresponding subset of $\mathbb{Z}_3^2$ is a weakly zero-3-sum-free set, and thus the maximum cap-set size in AG(2, 3) equals $\tau\hat{}(\mathbb{Z}_3^2, 3) = 4$. In particular, the cap-set of size four that corresponds to our example of a weakly zero-3-sum-free set illustrated above consists of A, B, D, and E.

Finally, let us consider the maximum size of a subset of $\mathbb{Z}^2$ without the centroid of any three of its points at a lattice point. Note that the centroid of points $A = (a_1, a_2)$, B=(b_1, b_2), and C=(c_1, c_2) is a lattice point if, and only if, $a_1 + b_1 + c_1$ and $a_2 + b_2 + c_2$ are both divisible by 3. Now the sum of three integers is divisible by 3 exactly when they all leave the same remainder mod 3, or when they all leave different remainders mod 3.

Let us map the lattice points of $\mathbb{Z}^2$ to elements of $\mathbb{Z}_3^2$ based on the reminders of their two coordinates. By the previous paragraph, if three points of $\mathbb{Z}^2$ are mapped to the same element of $\mathbb{Z}_3^2$, then the centroid of the three points is a lattice point. Furthermore, the same holds if the three points of $\mathbb{Z}^2$ are mapped to three distinct elements of $\mathbb{Z}_3^2$ whose sum equals $(0, 0)$. Therefore, if our pointset is such that no three of its points have their centroid at a lattice point, then they must be mapped to a subset of $\mathbb{Z}_3^2$ that is weakly zero-3-sum-free, and no more than two points can be mapped to the same element of $\mathbb{Z}_3^2$. But then the maximum size of a subset of $\mathbb{Z}^2$ without the centroid of any three of its points at a lattice point equals $2 \cdot \tau\hat{}(\mathbb{Z}_3^2, 3) = 2 \cdot 4 = 8$. Our argument also shows that an example of such a set of eight points could be

$$\{(0,0), (0,1), (1,0), (1,1), (3,3), (3,7), (4,9), (10,10)\}.$$

Hopefully, our set of examples provided an inviting appetizer for the entrees in this book. One of these entrees is the question of finding $\tau\hat{}(\mathbb{Z}_3^r, 3)$, which (Fields Medalist and Breakthrough Prize Winner) Terence Tao calls "perhaps [his] favorite open question" (see [194]).

How many elements does it take to span a group?

Given a finite abelian group G and a positive integer h, we are interested in finding the minimum possible size of a subset of G so that each element of G can be written as the sum of (exactly) h elements of our set.

Recall that the h-fold sumset of a subset A of G, denoted by hA, consists of all possible h-term sums of (not necessarily distinct) elements of A; more formally, for $A = \{a_1, \ldots, a_m\}$, we have

$$hA = \{\Sigma_{i=1}^{m} \lambda_i a_i \mid \lambda_i \in \mathbb{N}_0, \ \Sigma_{i=1}^{m} \lambda_i = h\}.$$

Thus our task is to find the minimum possible value of m for which G contains an m-subset A so that $hA = G$.

Let us consider an example: suppose that G is the cyclic group $\mathbb{Z}_{10}$ and $h = 2$. After some experimentation, one can find a variety of 5-subsets that yield the whole group; for example, with $A = \{0, 1, 2, 5, 7\}$, we get each element: $0 = 0 + 0$, $1 = 0 + 1$, $2 = 0 + 2$ (or $1 + 1$, or $5 + 7$), and so on, thus $2A = \mathbb{Z}_{10}$. Can we do better? That is, is there a smaller subset that generates the whole group?

Well, it is easy to see that for $2A$ to contain all 10 elements of our group, A must have size at least four. Indeed, more generally, even if all possible h-fold sums of an m-subset of G yield distinct elements, the size of hA is at most $\binom{m+h-1}{h}$, so to get all elements of the group, we must have

$$\binom{m+h-1}{h} \geq |G|.$$

Therefore, in our example, m must be at least four.

Now we prove that, in fact, m has to be at least five. Suppose, indirectly, that there exists a set $A = \{a_1, a_2, a_3, a_4\}$ of size four for which $2A = \mathbb{Z}_{10}$. We then have

$$2A = \{2a_1, 2a_2, 2a_3, 2a_4, a_1 + a_2, a_1 + a_3, a_1 + a_4, a_2 + a_3, a_2 + a_4, a_3 + a_4\}.$$

Therefore, the ten elements listed must all be distinct, and so are, in some order, equal to the elements of $\mathbb{Z}_{10}$; in particular, exactly five of them are even and five are odd. But then exactly one of the last six elements listed in $2A$ is even; say it is $a_1 + a_2$. Now

$$(a_1 + a_2) + (a_3 + a_4)$$

is the sum of an even and an odd number, thus odd. However, this sum also equals

$$(a_1 + a_3) + (a_2 + a_4),$$

the sum of two odd values, which is thus even. This is a contradiction, so the minimum value of m for which $\mathbb{Z}_{10}$ contains an m-subset A with $2A = \mathbb{Z}_{10}$ equals five.

Staying with $\mathbb{Z}_{10}$, let us try to find the answer for $h = 3$. From our general inequality above, we see that no subset of size less than three will have a 3-fold sumset of size ten. While there are numerous sets of size four that yield the entire group—for example, $\{0, 1, 4, 7\}$ does—it seems impossible to find one of size three. We can prove this as follows.

Assume that there is a set $A = \{a_1, a_2, a_3\}$ of size three for which $3A = \mathbb{Z}_{10}$. We see that

$$3A = \{3a_1, 3a_2, 3a_3, 2a_1 + a_2, a_1 + 2a_2, 2a_1 + a_3, a_1 + 2a_3, 2a_2 + a_3, a_2 + 2a_3, a_1 + a_2 + a_3\}.$$

Again, the ten elements listed must all be distinct, so are, in some order, equal to the elements of $\mathbb{Z}_{10}$. Thus, adding the elements in $3A$ and in $\mathbb{Z}_{10}$ should yield the same answer, so

$$10a_1 + 10a_2 + 10a_3 = 0 + 1 + \cdots + 9$$

or $0 = 5$, which is a contradiction.

Letting $\phi(G,h)$ denote the minimum possible size of a subset of G whose h-fold sumset contains each element of G, we just proved that $\phi(\mathbb{Z}_{10},2) = 5$ and $\phi(\mathbb{Z}_{10},3) = 4$.

Let us now consider the variation where we are allowed not only to add terms of our subset but subtract them as well: in particular, we are looking for the smallest value of m for which an m-subset of G exists whose h-fold signed sumset is all of G; we let $\phi_{\pm}(G,h)$ denote this value. Recall that the h-fold signed sumset of an m-subset $A = \{a_1, \dots, a_m\}$ of G is the set

$$h_{\pm}A = \{\Sigma_{i=1}^{m} \lambda_i a_i \mid \lambda_i \in \mathbb{Z},\ \Sigma_{i=1}^{m} |\lambda_i| = h\}.$$

Examining the cyclic group $\mathbb{Z}_{10}$, we run into a bit of uncertainty at $h = 1$ already. We would think that a set $A = \{a_1, a_2, a_3, a_4, a_5\}$ of size five is sufficient as its 1-fold signed sumset is $1_{\pm}A = \{\pm a_1, \pm a_2, \pm a_3, \pm a_4, \pm a_5\}$, which seems to contain ten elements. However, we realize that, in $\mathbb{Z}_{10}$, every element and its inverse have the same parity; therefore, $1_{\pm}A$ cannot contain five even and five odd elements. Since a set of size six with a 1-fold signed sumset of size ten can easily be found (for example, $A = \{0,1,2,3,4,5\}$), we conclude that $\phi_{\pm}(\mathbb{Z}_{10},1) = 6$.

It turns out that the case of $h = 2$ is easy: a set $A = \{a_1, a_2\}$ of size two is clearly insufficient, as its 2-fold sumset,

$$2_{\pm}A = \{\pm 2a_1, \pm 2a_2, \pm a_1 \pm a_2\},$$

will be of size at most eight; but a set of size three whose 2-fold signed sumset is of size ten is not hard to find (e.g., $\{0,1,4\}$). Therefore, $\phi_{\pm}(\mathbb{Z}_{10},2) = 3$.

Now let us attempt to find $\phi_{\pm}(\mathbb{Z}_{10},3)$. Here we see that two elements may be enough: for $A = \{a_1, a_2\}$, we have

$$3_{\pm}A = \{\pm 3a_1, \pm 3a_2, \pm 2a_1 \pm a_2, \pm a_1 \pm 2a_2\};$$

we see twelve (not necessarily different) elements listed. We will prove, however, that this cannot yield all of $\mathbb{Z}_{10}$. (Further analysis would prove that, in fact, $3_{\pm}A$ has size at most eight.)

Note that, if a_1 and a_2 have the same parity (that is, are both even or both odd), then the twelve elements listed all share that parity; in particular, $3_{\pm}A$ has size at most five. Therefore, a_1 and a_2 are of different parity; we will assume that a_1 is even and a_2 is odd, and therefore we see that the elements $\pm 3a_1$ and $\pm a_1 \pm 2a_2$ are even, and $\pm 3a_2$ and $\pm 2a_1 \pm a_2$ are odd. We will need the fact that, if a_1 is even, then $5a_1 = 0$ in $\mathbb{Z}_{10}$.

We also see that if $a_1 = 0$, then

$$3_{\pm}A = \{0, \pm a_2, \pm 2a_2, \pm 3a_2\},$$

which is of size at most seven; so we assume that $a_1 \neq 0$. In that case, $\pm 3a_1 \neq 0$ as well. If 0 were to be an element of $3_{\pm}A$, then it would have to be one of $\pm a_1 \pm 2a_2$; let us assume here that $a_1 + 2a_2 = 0$ and thus $-a_1 - 2a_2 = 0$ as well. (The case when $a_1 - 2a_2 = 0$ and $-a_1 + 2a_2 = 0$ can be examined the same way.) Thus we find that the elements 2, 4, 6, and 8 of $\mathbb{Z}_{10}$ are, in some order, the elements $\pm 3a_1$ and $\pm(a_1 - 2a_2)$; in particular, these four elements must be distinct. But from $-2a_2 = a_1$ and $5a_1 = 0$ we get

$$a_1 - 2a_2 = 2a_1 = 2a_1 - 5a_1 = -3a_1,$$

which is a clear contradiction with the four elements being distinct. Therefore, 0 is not an element of $3_{\pm}A$, and thus $\phi_{\pm}(\mathbb{Z}_{10},3) \geq 3$. Since a 3-subset of $\mathbb{Z}_{10}$ with a 3-fold signed sumset of size ten can be found easily ($\{0,1,2\}$ will do), we conclude that $\phi_{\pm}(\mathbb{Z}_{10},3) = 3$.

Using techniques similar to the ones we just saw, we can find the values of $\phi(\mathbb{Z}_{10}, h)$ and $\phi_{\pm}(\mathbb{Z}_{10}, h)$ for all values of h:

	$h=1$	$h=2$	$h=3$	$h=4$	$h=5$	$h=6$	$h=7$	$h=8$	$h \geq 9$
$\phi(\mathbb{Z}_{10}, h)$	10	5	4	3	3	3	3	3	2
$\phi_{\pm}(\mathbb{Z}_{10}, h)$	6	3	3	3	2	2	2	2	2

The problem of determining $\phi(G, h)$ and $\phi_{\pm}(G, h)$ for a general abelian group G is widely open (not to mention other variations where restricted sums are considered or where the number of terms varies)—Chapter B discusses all that is known (to this author) on this subject.

In pursuit of perfection

We are interested in the "speed" with which a given subset of a group generates the entire group. As a (very nice) example, consider the set $A = \{3, 4\}$ in the group $\mathbb{Z}_{25}$. How "fast" does the set generate the group? Let us explain what we mean by this question.

Note that either element of A alone generates $\mathbb{Z}_{25}$: since 3 and 4 are both relatively prime to 25, their multiples will yield all group elements, but this is not very "fast." For example, to get the element 22, we need to add up 24 terms of 3s. Even if subtraction is also allowed, the fastest way to get to 14 is to subtract 3 (from 0) twelve times. The situation is not better with using the element 4 alone either. However, if both 3s and 4s can be used, and we can both add and subtract them, then we can verify that each element of $\mathbb{Z}_{25}$ can be generated with three or fewer terms:

1	=	$4-3$	7	=	$3+4$	13	=	$-4-4-4$	19	=	$-3-3$
2	=	$3+3-4$	8	=	$4+4$	14	=	$-3-4-4$	20	=	$3-4-4$
3	=	3	9	=	$3+3+3$	15	=	$-3-3-4$	21	=	-4
4	=	4	10	=	$3+3+4$	16	=	$-3-3-3$	22	=	-3
5	=	$4+4-3$	11	=	$3+4+4$	17	=	$-4-4$	23	=	$4-3-3$
6	=	$3+3$	12	=	$4+4+4$	18	=	$-3-4$	24	=	$3-4$

As the table indicates, every element of $\mathbb{Z}_{25}$ can be generated by a signed sum of at most three terms of A. (We consider 0 to be generated trivially.) We therefore call $A = \{3, 4\}$ a *3-spanning set* in $\mathbb{Z}_{25}$.

More generally, given a finite abelian group G, a subset $A = \{a_1, a_2, \ldots, a_m\}$ of G, and a nonnegative integer s, we say that A is an *s-spanning set* in G, if every element of the group can be written as a linear combination

$$\lambda_1 a_1 + \lambda_2 a_2 + \cdots + \lambda_m a_m$$

for some integers $\lambda_1, \lambda_2, \ldots, \lambda_m$ with

$$|\lambda_1| + |\lambda_2| + \cdots + |\lambda_m| \leq s.$$

Using our notations and terminology introduced previously, we can say that A is an s-spanning set of G if the $[0, s]$-fold signed sumset of A is the entire group, that is,

$$[0, s]_{\pm}A = \cup_{h=0}^{s}(h_{\pm}A) = G.$$

In our example above, for $A = \{3, 4\}$ in $\mathbb{Z}_{25}$, we find that

- $0_{\pm}A = \{0\}$,
- $1_{\pm}A = \{\pm 3, \ \pm 4\} = \{3, 4, 21, 22\}$,
- $2_{\pm}A = \{\pm 2 \cdot 3, \ \pm 3 \pm 4, \ \pm 2 \cdot 4\} = \{1, 6, 7, 8, 17, 18, 19, 24\}$, and
- $3_{\pm}A = \{\pm 3 \cdot 3, \ \pm 2 \cdot 3 \pm 4, \ \pm 3 \pm 2 \cdot 4, \ \pm 3 \cdot 4\} = \{2, 5, 9, 10, 11, 12, 13, 14, 15, 16, 20, 23\}$.

Therefore,

$$[0, 3]_{\pm}A = (0_{\pm}A) \cup (1_{\pm}A) \cup (2_{\pm}A) \cup (3_{\pm}A) = \mathbb{Z}_{25}.$$

In fact, the set $A = \{3, 4\}$ has a remarkable property: the

$$1 + 4 + 8 + 12 = 25$$

possible signed sums in $[0,3]_{\pm}A$ are all distinct elements of $\mathbb{Z}_{25}$; in other words, every element of the group can be written *uniquely* as a signed sum of at most three elements of A. We call such a set *perfect*; perfect spanning sets generate the group most efficiently. Unfortunately, they rarely exist!

Just how rare, we do not exactly know in general. If all elements generated are distinct, then $[0,s]_{\pm}A$ must have size

$$a(m,s) = \sum_{i \geq 0} \binom{m}{i} \binom{s}{i} 2^i$$

(see Section 2.5 in Part I for a proof), and if each element of G must be generated, then this quantity must equal the order of G. Therefore, a necessary condition for G to have a perfect s-spanning set of size m is that $|G| = a(m,s)$.

The first few values of the function $a(m,s)$ are as follows:

$a(m,s)$	$s=0$	$s=1$	$s=2$	$s=3$	$s=4$	$s=5$	$s=6$
$m=0$	1	1	1	1	1	1	1
$m=1$	1	3	5	7	9	11	13
$m=2$	1	5	13	25	41	61	85
$m=3$	1	7	25	63	129	231	377
$m=4$	1	9	41	129	321	681	1289
$m=5$	1	11	61	231	681	1683	3653
$m=6$	1	13	85	377	1289	3653	8989

It may be helpful to observe that, besides the formula we gave for $a(m,s)$ above, the function can also be evaluated using the recurrence relation

$$a(m,s) = a(m-1,s) + a(m-1,s-1) + a(m,s-1)$$

(together with the initial conditions $a(m,0) = a(0,s) = 1$). As we see, our function resembles Pascal's Triangle, except here we get to the entry in row m and column s by adding not only the entries directly above and to the left, but the entry in the "above-left" position as well.

We have already seen earlier that the set $A = \{3,4\}$ is a perfect 3-spanning set in $G = \mathbb{Z}_{25}$. Here are some others:

- In any group of odd order, the set of nonzero elements can be partitioned into parts K and $-K$, and both K and $-K$ are perfect 1-spanning sets in the group. For example, the set $\{1,2,\ldots,m\}$ is a perfect 1-spanning set in $\mathbb{Z}_{2m+1}$.

- The set $\{a\}$ is a perfect s-spanning set in $\mathbb{Z}_{2s+1}$ as long as a and $2s+1$ are relatively prime.

- The sets $\{1, 2s+1\}$ and $\{s, s+1\}$ are perfect s-spanning sets in $\mathbb{Z}_{2s^2+2s+1}$.

The first two statements are easy to see. The claim regarding $\{s, s+1\}$ we just demonstrated for $s = 3$ and $n = 25$; we can illustrate the claim about $\{1, 2s+1\}$ for the same parameters by exhibiting the values of

$$\lambda_1 \cdot 1 + \lambda_2 \cdot 7$$

in $\mathbb{Z}_{25}$ for all relevant coefficients λ_1 and λ_2:

	$\lambda_1 = -3$	$\lambda_1 = -2$	$\lambda_1 = -1$	$\lambda_1 = 0$	$\lambda_1 = 1$	$\lambda_1 = 2$	$\lambda_1 = 3$
$\lambda_2 = 3$				21			
$\lambda_2 = 2$			13	14	15		
$\lambda_2 = 1$		5	6	7	8	9	
$\lambda_2 = 0$	22	23	24	0	1	2	3
$\lambda_2 = -1$		16	17	18	19	20	
$\lambda_2 = -2$			10	11	12		
$\lambda_2 = -3$				4			

The table shows a nice pattern for how the elements of the group arise. (The general proofs for $\{1, 2s+1\}$ and $\{s, s+1\}$ are provided on pages 325 and 327, respectively.)

Note that our three types of perfect spanning sets correspond to the cases

- $s = 1$ and $|G| = a(m, 1) = 2m + 1$,
- $m = 1$ and $|G| = a(1, s) = 2s + 1$, and
- $m = 2$ and $|G| = a(2, s) = 2s^2 + 2s + 1$.

respectively. Clearly, the cases of $s = 1$ and $m = 1$ are completely characterized by our list above: the only possibilities are those listed. For $m = 2$, we may have other examples, though none are known. We are also not aware of any perfect spanning sets for $s \geq 2$ and $m \geq 3$. It might be an interesting problem to find and classify all perfect spanning sets (or to prove that no others exist besides the ones listed).

As a modest attempt toward such a classification, we prove that there is neither a perfect 3-spanning set of size two, nor a perfect 2-spanning set of size three in $\mathbb{Z}_5^2$. (Note that $a(2,3) = a(3,2) = 25$. We have already seen that $\mathbb{Z}_{25}$ has perfect 3-spanning sets of size two; we know, via a computer search, that it has no perfect 2-spanning sets of size three. Our proofs below show that, sometimes, it is easier to work with noncyclic groups than with cyclic ones of the same order.)

It is easy to rule out perfect 3-spanning sets of size 2 in $\mathbb{Z}_5^2$: indeed, if an element a of the group were to be in a perfect 3-spanning set, then $2a$ and $-3a$ would need to be distinct elements; however, we have $5a = 0$ for all $a \in \mathbb{Z}_5^2$, so $2a = -3a$, which we just ruled out.

For our second claim, suppose, indirectly, that $A = \{a_1, a_2, a_3\}$ is a perfect 2-spanning set in $\mathbb{Z}_5^2$, in which case

$$\mathbb{Z}_5^2 = \{0,\ \pm a_1,\ \pm a_2,\ \pm a_3,\ \pm 2a_1,\ \pm 2a_2,\ \pm 2a_3,\ \pm a_1 \pm a_2,\ \pm a_1 \pm a_3,\ \pm a_2 \pm a_3\}.$$

In particular, $a_1 + a_2 + a_3$ equals one of the elements listed. Note that it cannot be any of the first seven; if it were, say, equal to $-a_1$, then we would get $a_2 + a_3 = -2a_1$, contradicting the fact that the 25 elements listed above are distinct. Similarly, $a_1 + a_2 + a_3$ cannot equal any of the 12 elements among the last 18 where any of a_1, a_2, or a_3 appears with a positive sign, since cancelling would again result in a repetition. This leaves only two possibilities: $a_1 + a_2 + a_3$ equals $-2a_i$ for some $1 \leq i \leq 3$, or it equals $-a_i - a_j$ for some $1 \leq i < j \leq 3$; without loss of generality,

$$a_1 + a_2 + a_3 = -2a_1$$

or

$$a_1 + a_2 + a_3 = -a_1 - a_2.$$

But neither of these equations can occur: the first leads to

$$a_2 + a_3 = -3a_1 = 2a_1,$$

and the second yields

$$a_3 = -2a_1 - 2a_2,$$

from which we get

$$2a_3 = -4a_1 - 4a_2 = a_1 + a_2,$$

again contradicting the assumption that the 25 elements listed are distinct.

We will study perfect spanning sets later in more detail in Chapter B.

The declaration of independence

Those familiar with linear algebra have undoubtedly heard of "independent sets." A subset $A = \{a_1, \ldots, a_m\}$ of elements in a vector space V (over the set of real numbers) is called *independent*, if the zero element of V cannot be expressed as a *linear combination*

$$\lambda_1 a_1 + \cdots + \lambda_m a_m$$

with some (real number) coefficients $\lambda_1, \ldots, \lambda_m$ non-trivially, that is, without all coefficients $\lambda_1, \ldots, \lambda_m$ being zero. There is an alternate definition: A is independent if no element of V can be expressed in the above form in two different ways. It is a fundamental property of vector spaces that the two definitions are equivalent.

We immediately realize that the analogous concept in finite abelian groups behaves quite differently. In fact, no (nonempty) set in a finite abelian group is independent: if λ is the order of an element a in G, then λa equals 0, so even the 1-element set $\{a\}$ is not independent. Thus to make our concept worth studying in finite abelian groups, we only require that a subset be independent "to a certain degree" rather than that it is "completely" independent. More precisely, rather than considering all linear combinations of the elements of a subset, we limit our attention to those that use only a certain number of terms, that is, where the sum of the absolute values of the coefficients,

$$|\lambda_1| + \cdots + |\lambda_m|,$$

is at most some positive integer t. We will refer to the linear combination above as a *signed sum* of $|\lambda_1| + \cdots + |\lambda_m|$ terms.

Given the two alternatives for defining independence mentioned above, we have two choices: declare a subset $A = \{a_1, \ldots, a_m\}$ of an abelian group G *t-independent* if

- no nontrivial signed sum of at most t terms equals zero (that is, the zero element of the group cannot be expressed as a linear combination

  $$\lambda_1 a_1 + \cdots + \lambda_m a_m$$

  using integer coefficients $\lambda_1, \ldots, \lambda_m$ with

  $$|\lambda_1| + \cdots + |\lambda_m| \leq t,$$

  unless $\lambda_i = 0$ for every $i = 1, 2, \ldots, m$); or

- no element of the group can be written as a signed sum of at most t terms in two different ways (that is, we cannot have

  $$\lambda_1 a_1 + \lambda_2 a_2 + \cdots + \lambda_m a_m = \lambda_1' a_1 + \lambda_2' a_2 + \cdots + \lambda_m' a_m$$

  for some integers $\lambda_1, \lambda_2, \ldots, \lambda_m$ and $\lambda_1', \lambda_2', \ldots, \lambda_m'$ with

  $$|\lambda_1| + |\lambda_2| + \cdots + |\lambda_m| \leq t \text{ and } |\lambda_1'| + |\lambda_2'| + \cdots + |\lambda_m'| \leq t,$$

  unless $\lambda_i = \lambda_i'$ for every $i = 1, 2, \ldots, m$).

It may come as a surprise, but the two possible definitions are not equivalent! Consider, for example, the set $\{1, 3\}$ in the cyclic group $\mathbb{Z}_{10}$: the set is 3-independent according to the first definition, since it is impossible to express zero as a non-trivial signed sum of at most three terms (try!), but the set is not 3-independent using the second definition, as, for example,

$$2 \cdot 1 + (-1) \cdot 3 = 0 \cdot 1 + 3 \cdot 3.$$

So which definition should we declare to be the definition of independence?

It turns out that one definition is more powerful than the other; in fact, one of the two definitions is merely a special case of the other! In particular, we can prove that a subset is t-independent by the second definition if, and only if, it is $2t$-independent by the first definition. What this means is that studying t-independence for all possible t using the second definition is equivalent to studying t-independence just for even t values only using the first definition; therefore, the second definition is indeed superfluous.

To show that if a subset is $2t$-independent by the first definition then it is t-independent by the second definition, assume that two signed sums, each containing at most t terms, equal one another. We can then move all terms to the same side and thereby express zero as a signed sum of at most $2t$ terms. Note that, as a result, some of the terms may get cancelled. In fact, the assumption that our set is $2t$-independent by the first definition means that the signed sumset we created must be the trivial one, thus all terms got cancelled, which means that our two original signed sums were identical.

Conversely, if a signed sum of at most $2t$ terms equals zero, then we can rearrange this equation so that each side contains at most t terms. If we know that our set is t-independent by the second definition, then the two sides are identical, which means that our original signed sum was the trivial one.

So we declare a set to be t-independent following the first definition; recalling our notations and terminology for the collection of signed sums of elements of a set A with exactly h terms being called the h-fold signed sumset of A and denoted by $h_{\pm}A$, we can thus say that a subset A is t-independent in a group G whenever

$$0 \not\in \cup_{h=1}^{t}(h_{\pm}A).$$

Thus, to verify that the subset $A = \{1, 3\}$ of $\mathbb{Z}_{10}$ in our earlier example is 3-independent, we check that none of $1_{\pm}A$, $2_{\pm}A$, or $3_{\pm}A$ contains zero: this is obviously the case for $1_{\pm}A$ and $3_{\pm}A$ as they contain only odd elements, and

$$2_{\pm}A = \{\pm(1+1), \pm(3+3), \pm 1 \pm 3\} = \{2, 4, 6, 8\}$$

doesn't contain zero either. Our set is not 4-independent, since $4_{\pm}A$ does contain zero (e.g., as $1+3+3+3$ or $1+1+1-3$). It is also easy to check that $\mathbb{Z}_{10}$ has no 3-independent sets of size larger than two.

To see a bigger example, consider the set $A = \{1, 4, 6, 9, 11\}$ in $\mathbb{Z}_{25}$. We find that:

- $1_{\pm}A = \{1, 4, 6, 9, 11, 14, 16, 19, 21, 24\}$,
- $2_{\pm}A = \{2, 3, 5, 7, 8, 10, 12, 13, 15, 17, 18, 20, 22, 23\}$, and
- $3_{\pm}A = \{1, 2, 3, 4, 6, 7, 8, 9, 11, 12, 13, 14, 16, 17, 18, 19, 21, 22, 23, 24\}$,

and therefore

$$0 \not\in \cup_{h=1}^{3} h_{\pm}A,$$

implying that A is 3-independent in G. We also see, however, that $1+4+4-9$ (for example) equals zero, so $0 \in 4_{\pm}A$, and therefore A is not 4-independent.

Can we pack more than five elements in $\mathbb{Z}_{25}$ that are 3-independent? Here is a quick argument that shows that we certainly cannot pack seven. Suppose that

$$A = \{a_1, a_2, a_3, a_4, a_5, a_6, a_7\} \subset \mathbb{Z}_{25}.$$

Now consider the following 28 signed sums:

$$0, \ \pm a_1, \ \pm a_2, \ \pm a_3, \ \pm a_4, \ \pm a_5, \ \pm a_6, \ \pm a_7,$$

$$a_1 + a_1,\ a_1 \pm a_2,\ a_1 \pm a_3,\ a_1 \pm a_4,\ a_1 \pm a_5,\ a_1 \pm a_6,\ a_1 \pm a_7.$$

If A were to be 3-independent, then all 28 expressions would be different. (For example, if $-a_2$ were to equal $a_1 - a_4$, then we would have $a_1 + a_2 - a_4 = 0$, contradicting the fact that A is 3-independent.) Since G only contains 25 elements, this cannot happen.

It is also true that $\mathbb{Z}_{25}$ contains no six-element 3-independent sets either, but this is much harder to prove; it follows from the more general result by Bajnok and Ruzsa (see [24]) that the maximum size of a 3-independent set in the cyclic group $\mathbb{Z}_n$ equals

- $\lfloor n/4 \rfloor$ when n is even;
- $(1 + 1/p)\, n/6$ when n is odd, has prime divisors congruent to 5 mod 6, and p is the smallest such divisor; and
- $\lfloor n/6 \rfloor$ otherwise.

(The corresponding results for noncyclic groups or for $t > 3$ are not yet known.)

Our concept of independence is one of the most important ones in this book, and is closely related to several other fundamental concepts, such as the zero-sum-free property, the sum-free property, and the Sidon property, as we now explain.

Recall that we defined a subset A of G to be t-independent in G if no signed sum of t or fewer terms of A (repetition of terms allowed), other than the all-zero sum, equals zero. Observe that any equation expressing that a particular signed sum is zero can be rearranged so that on both sides of the equation the elements of A appear with positive coefficients. Therefore, an equivalent way of saying that A is a t-independent set in G is to say that, for all nonnegative integers k and l with $k + l \leq t$, the sum of k (not necessarily distinct) elements of A can only equal the sum of l (not necessarily distinct) elements of A in a trivial way, that is, $k = l$ and the two sums contain the same terms in some order.

Therefore, we can thus break up our definition of A being t-independent into three conditions:

- the *zero-sum-free property*:
  $$0 \not\in hA,$$
  that is, the sum of h elements of A cannot equal zero—this needs to hold for $1 \leq h \leq t$;
- the *sum-free property*:
  $$(kA) \cap (lA) = \emptyset,$$
  that is, the sum of k elements of A never equals the sum of l elements of A—this needs to hold whenever k and l are distinct positive integers and $k + l \leq t$; and
- the *Sidon property*:
  $$|hA| = \binom{m+h-1}{h},$$
  that is, two h-term sums of elements of A can only be equal if the sums contain the same terms—this needs to hold for $1 \leq h \leq \lfloor t/2 \rfloor$.

(It is enough, in fact, to require these conditions for equations containing a total of t or $t-1$ terms; therefore the total number of equations considered can be reduced to $2 + (t-2) + 1 = t + 1$.)

We can return to our earlier example of the set $A = \{1, 4, 6, 9, 11\}$ in $G = \mathbb{Z}_{25}$ to verify that A is 3-independent in G using the equivalent three-part condition we just gave. We find that

- $1A = \{1,4,6,9,11\}$,
- $2A = \{2,5,7,8,10,12,13,15,17,18,20,22\}$, and
- $3A = \{1,2,3,4,6,8,9,11,12,13,14,16,17,18,19,21,22,23,24\}$.

To conclude that A is 3-independent in G, note that A is zero-sum-free for $h = 1,2,3$ as $0 \not\in hA$ then; sum-free for $(k,l) = (2,1)$ since A and $2A$ are disjoint; and a Sidon set for $h = 1$ (as are all sets). The fact that A is not 4-independent can be seen by the fact that it is not zero-4-sum-free ($1+4+9+11=0$), or that it is not $(3,1)$-sum-free ($1+4+4=9$), or that it is not a Sidon set for $h=2$ ($4+6=1+9$).

We study zero-sum-free sets, sum-free sets, and Sidon sets in detail in Chapters F, G, and C, respectively.

Part III

Sides

Here we introduce and study some functions that prove valuable for several topics in this book.

The function $v_g(n,h)$

Suppose that h and g are fixed positive integers. Since we will only need the cases when $1 \leq g \leq h$, we make that assumption here. Given a positive integer n, recall that $D(n)$ is the set of positive divisors of n. We define

$$v_g(n,h) = \max\left\{\left(\left\lfloor \frac{d-1-\gcd(d,g)}{h} \right\rfloor + 1\right) \cdot \frac{n}{d} \mid d \in D(n)\right\}.$$

(Here we usually think of $v_g(n,h)$ as a function of n for fixed values of g and h.)

For example, we can compute $v_2(18,4)$ by evaluating, for each positive divisor d of 18, the quantity

$$\left(\left\lfloor \frac{d-1-\gcd(d,2)}{4} \right\rfloor + 1\right) \cdot \frac{18}{d};$$

since the maximum occurs at $d=3$ and equals 6, we have $v_2(18,4)=6$.

The evaluation of $v_g(n,h)$ can be quite cumbersome in general. For a prime p, however, $v_g(p,h)$ can be evaluated easily. In this case, there are only two positive divisors to consider: $d=1$ and $d=p$. For $d=1$, the expression yields 0; therefore we have

$$v_g(p,h) = \max\left\{0, \left\lfloor \frac{p-1-\gcd(p,g)}{h} \right\rfloor + 1\right\}.$$

The greatest common divisor of p and g is either p or 1, depending on whether p is a divisor of g or not. This yields the following.

Proposition 4.1 *If p is a positive prime number, then*

$$v_g(p,h) = \begin{cases} 0 & \text{if } p|g, \\ \left\lfloor \frac{p-2}{h} \right\rfloor + 1 & \text{otherwise.} \end{cases}$$

With a bit more work, we can evaluate $v_g(n,h)$ for any prime power value of n, at least in the case when g is not divisible by that prime; since we will use this later, we present the result here:

Proposition 4.2 *If p is a positive prime number that is not a divisor of g, then*

$$v_g(p^r,h) = \begin{cases} \frac{p^r-1}{h} & \text{if } p \equiv 1 \bmod h, \\ \left(\left\lfloor \frac{p-2}{h} \right\rfloor + 1\right) \cdot p^{r-1} & \text{otherwise.} \end{cases}$$

We present the proof on page 312.

Let us now turn to the evaluation of $v_g(n,h)$ when n is arbitrary. For $h=1$ and $h=2$, we can prove the following.

Proposition 4.3 *For all positive integers n we have*

$$v_1(n,1) = n-1$$

and

$$v_g(n,2) = \begin{cases} \left\lfloor \frac{n}{2} \right\rfloor & \text{if } g=1, \\ \left\lfloor \frac{n-1}{2} \right\rfloor & \text{if } g=2. \end{cases}$$

For a proof, see page 313.

For higher values of h, the evaluation of $v_g(n,h)$ becomes more difficult using its definition above. The following table lists the values of $v_1(n,3)$, $v_3(n,3)$, $v_1(n,4)$, $v_2(n,4)$, $v_4(n,4)$, $v_1(n,5)$, $v_3(n,5)$, $v_5(n,5)$ for $n \leq 40$. (These sequences appear in [188] as A211316, A289435, A289436, A289437, A289438, A289439, A289440, and A289441, respectively.)

n	$v_1(n,3)$	$v_3(n,3)$	$v_1(n,4)$	$v_2(n,4)$	$v_4(n,4)$	$v_1(n,5)$	$v_3(n,5)$	$v_5(n,5)$
2	1	1	1	0	0	1	1	1
3	1	0	1	1	1	1	0	1
4	2	2	2	1	0	2	2	2
5	2	2	1	1	1	1	1	0
6	3	3	3	2	2	3	3	3
7	2	2	2	2	2	2	2	2
8	4	4	4	2	1	4	4	4
9	3	2	3	3	3	3	2	3
10	5	5	5	2	2	5	5	5
11	4	4	3	3	3	2	2	2
12	6	6	6	4	4	6	6	6
13	4	4	3	3	3	3	3	3
14	7	7	7	4	4	7	7	7
15	6	6	5	5	5	5	3	5
16	8	8	8	4	3	8	8	8
17	6	6	4	4	4	4	4	4
18	9	9	9	6	6	9	9	9
19	6	6	5	5	5	4	4	4
20	10	10	10	5	4	10	10	10
21	7	6	7	7	7	7	6	7
22	11	11	11	6	6	11	11	11
23	8	8	6	6	6	5	5	5
24	12	12	12	8	8	12	12	12
25	10	10	6	6	6	5	5	4
26	13	13	13	6	6	13	13	13
27	9	8	9	9	9	9	6	9
28	14	14	14	8	8	14	14	14
29	10	10	7	7	7	6	6	6
30	15	15	15	10	10	15	15	15
31	10	10	8	8	8	6	6	6
32	16	16	16	8	7	16	16	16
33	12	12	11	11	11	11	6	11
34	17	17	17	8	8	17	17	17
35	14	14	10	10	10	10	10	10
36	18	18	18	12	12	18	18	18
37	12	12	9	9	9	8	8	8
38	19	19	19	10	10	19	19	19
39	13	12	13	13	13	13	9	13
40	20	20	20	10	9	20	20	20

The next theorem, based in part on work by Butterworth (cf. [46]), simplifies the evaluation of $v_g(n,h)$.

Theorem 4.4 (Bajnok; cf. [19]) *Suppose that n, h, and g are positive integers and that $1 \leq g \leq h$. Let $D(n)$ denote the set of positive divisors of n. For $i = 0, 1, 2, \ldots, h-1$, let*

$$D_i(n) = \{\, d \in D(n) \mid d \equiv i \pmod h \text{ and } \gcd(d,g) < i\}.$$

We let I denote those values of $i = 0, 1, 2, \ldots, h-1$ for which $D_i(n) \neq \emptyset$, and for each $i \in I$, we let d_i be the smallest element of $D_i(n)$.

Then, the value of $v_g(n,h)$ is

$$v_g(n,h) = \begin{cases} \frac{n}{h} \cdot \max\left\{1 + \frac{h-i}{d_i} \mid i \in I\right\} & \text{if } I \neq \emptyset; \\ \left\lfloor \frac{n}{h} \right\rfloor & \text{if } I = \emptyset \text{ and } g \neq h; \\ \left\lfloor \frac{n-1}{h} \right\rfloor & \text{if } I = \emptyset \text{ and } g = h. \end{cases}$$

The proof of Theorem 4.4 can be found on page 314. Observe that this result generalizes Proposition 4.3. Theorem 4.4 makes the computation of $v_g(n,h)$ considerably simpler; for instance, for the example of $v_2(18,4)$ above, we see that $I = \{3\}$, $d_3 = 3$, and thus $v_2(18,4) = 18/4 \cdot (1 + 1/3) = 6$.

Let us now examine what explicit bounds we can deduce from Theorem 4.4. Clearly, $v_g(n,h) \geq \left\lfloor \frac{n-1}{h} \right\rfloor$. To find an upper bound, we assume that $h \geq 2$. First, note that for all n we have $D_0(n) = D_1(n) = \emptyset$. Furthermore, for $i \geq 2$ we have

$$1 + \frac{h-i}{d_i} \leq 1 + \frac{h-2}{2} = \frac{h}{2},$$

with equality if, and only if, $i = 2$ and $d_i = 2$. This gives:

Corollary 4.5 *For all integers n, $h \geq 2$, and $1 \leq g \leq h$, we have*

$$\left\lfloor \frac{n-1}{h} \right\rfloor \leq v_g(n,h) \leq \frac{n}{2},$$

with $v_g(n,h) = \frac{n}{2}$ if, and only if, n is even and g is odd.

We can use Theorem 4.4 to evaluate $v_g(n,h)$ explicitly when h is relatively small. Consider, for example, the case when $h = 3$ and $g = 1$. Then $I = \{2\}$ or $I = \emptyset$ depending on whether n has divisors that are congruent to 2 mod 3 or not. Note that, when an integer has divisors that are congruent to 2 mod 3, then its smallest such divisor must be a prime, since the product of integers that are not congruent to 2 mod 3 will also not be congruent to 2 mod 3. We thus see that

$$v_1(n,3) = \begin{cases} \left(1 + \frac{1}{p}\right)\frac{n}{3} & \text{if } n \text{ has prime divisors congruent to 2 mod 3,} \\ & \text{and } p \text{ is the smallest such divisor,} \\ \left\lfloor \frac{n}{3} \right\rfloor & \text{otherwise.} \end{cases}$$

We can use Theorem 4.4 to evaluate other cases similarly. We find the following expressions.

$$v_2(n,3) = \begin{cases} \left(1+\frac{1}{p}\right)\frac{n}{3} & \text{if } n \text{ has prime divisors congruent to 5 mod 6,} \\ & \text{and } p \text{ is the smallest such divisor,} \\ \left\lfloor\frac{n}{3}\right\rfloor & \text{otherwise;} \end{cases}$$

$$v_3(n,3) = \begin{cases} \left(1+\frac{1}{p}\right)\frac{n}{3} & \text{if } n \text{ has prime divisors congruent to 2 mod 3,} \\ & \text{and } p \text{ is the smallest such divisor,} \\ \left\lfloor\frac{n-1}{3}\right\rfloor & \text{otherwise;} \end{cases}$$

$$v_1(n,4) = \begin{cases} \frac{n}{2} & \text{if } n \text{ is even} \\ \left(1+\frac{1}{p}\right)\frac{n}{4} & \text{if } n \text{ is odd, has prime divisors congruent to 3} \\ & \text{mod 4, and } p \text{ is the smallest such divisor,} \\ \left\lfloor\frac{n}{4}\right\rfloor & \text{otherwise;} \end{cases}$$

$$v_2(n,4) = \begin{cases} \left(1+\frac{1}{p}\right)\frac{n}{4} & \text{if } n \text{ has prime divisors congruent to 3 mod 4,} \\ & \text{and } p \text{ is the smallest such divisor,} \\ \left\lfloor\frac{n}{4}\right\rfloor & \text{otherwise;} \end{cases}$$

$$v_4(n,4) = \begin{cases} \left(1+\frac{1}{p}\right)\frac{n}{4} & \text{if } n \text{ has prime divisors congruent to 3 mod 4,} \\ & \text{and } p \text{ is the smallest such divisor,} \\ \left\lfloor\frac{n-1}{4}\right\rfloor & \text{otherwise;} \end{cases}$$

$$v_1(n,5) = \begin{cases} \max\left\{\left(1+\frac{1}{p_4}\right)\frac{n}{5}, \left(1+\frac{2}{p_3}\right)\frac{n}{5}, \left(1+\frac{3}{p_2}\right)\frac{n}{5}\right\} & \text{if } n \text{ has prime divisors 2, 3, or 4 mod 5, and } p_2, p_3, p_4 \text{ are smallest respectively;} \\ \left\lfloor\frac{n}{5}\right\rfloor & \text{otherwise;} \end{cases}$$

$$v_3(n,5) = \begin{cases} \max\left\{\left(1+\frac{1}{d_4}\right)\frac{n}{5}, \left(1+\frac{2}{p_3}\right)\frac{n}{5}, \left(1+\frac{3}{p_2}\right)\frac{n}{5}\right\} & \text{if } n \text{ has prime divisors 2 mod 5 and } p_2 \text{ is smallest, prime divisors 3 mod 5 other than 3 and } p_3 \text{ is smallest, or divisors 4 mod 5 and } d_4 \text{ smallest,} \\ \left\lfloor\frac{n}{5}\right\rfloor & \text{otherwise;} \end{cases}$$

$$v_5(n,5) = \begin{cases} \max\left\{\left(1+\frac{1}{p_4}\right)\frac{n}{5}, \left(1+\frac{2}{p_3}\right)\frac{n}{5}, \left(1+\frac{3}{p_2}\right)\frac{n}{5}\right\} & \text{if } n \text{ has prime divisors 2, 3, or 4 mod 5, and } p_2, p_3, p_4 \text{ are smallest respectively;} \\ \left\lfloor \frac{n-1}{5} \right\rfloor & \text{otherwise.} \end{cases}$$

Observe that in all the formulas above, with the exception of $v_3(n,5)$, the divisors playing a role can be assumed to be primes (a fact that can be proven easily). In the case of $v_3(n,5)$, however, d_4 is not necessarily prime: for example, when $n = 9$, we have $d_4 = 9$ (and no p_2 or p_3 exist), thus $v_3(9,5) = 2$. We should also point out that the maximum value in the formulae for $h = 5$ above may occur with a prime that is not the smallest prime divisor of n congruent to 2, 3, or 4 mod 5. For example, for $n = 437 = 19 \cdot 23$ we get

$$v_5(437,5) = \max\left\{\left(1+\frac{1}{19}\right)\frac{437}{5}, \left(1+\frac{2}{23}\right)\frac{437}{5}\right\} = \left(1+\frac{2}{23}\right)\frac{437}{5} = 95.$$

Similar expressions for $v_g(n,h)$ get more complicated for some other choices of g and h. It is not true, for example, that in Theorem 4.4 the minimal element d_i of $D_i(n)$ is prime: $v_3(9,5) = 2$ with $d_4 = 9$, and $v_6(16,11) = 4$ with $d_4 = 4$.

The function $v_g(n,h)$ seems to possess many interesting properties; we offer the following rather vague problem.

Problem 4.6 *Investigate some of the properties of the function $v_g(n,h)$.*

The function $v_\pm(n,h)$

Here we introduce and investigate a close relative of the function $v_g(n,h)$ discussed above.

Unlike with $v_g(n,h)$, the function $v_\pm(n,h)$ we discuss here has no g: it depends only on positive integers n and h. (We can think of $v_\pm(n,h)$ as the "plus–minus" version of $v_1(n,h)$.) We define this function as

$$v_\pm(n,h) = \max\left\{\left(2\cdot\left\lfloor\frac{d-2}{2h}\right\rfloor+1\right)\cdot\frac{n}{d} \mid d\in D(n)\right\}.$$

We can immediately see the close relationship between $v_\pm(n,h)$ and $v_1(n,h)$. In particular, note that for every d and h we have

$$2\cdot\left\lfloor\frac{d-2}{2h}\right\rfloor \leq \left\lfloor\frac{d-2}{h}\right\rfloor,$$

since if q and r are, respectively, the quotient and the remainder of $d-2$ when divided by $2h$, then $2\cdot\lfloor(d-2)/(2h)\rfloor = 2q$, while $\lfloor(d-2)/h\rfloor$ equals $2q$ or $2q+1$ (depending on whether r is less than h or not). This yields:

Proposition 4.7 *For all positive integers n and h we have $v_\pm(n,h)\leq v_1(n,h)$.*

The analogue of Proposition 4.1, this time even simpler as we have no g, is:

Proposition 4.8 *If p is a positive prime number, then*

$$v_\pm(p,h) = 2\cdot\left\lfloor\frac{p-2}{2h}\right\rfloor+1.$$

When n is not (necessarily) prime, the evaluation of $v_\pm(n,h)$ is complicated even for small values of h. For $h=1,2,3$, we can prove the following results.

Proposition 4.9 *For all positive integers n we have the following:*

$$v_\pm(n,1) = \begin{cases} n-1 & \text{if } n \text{ is even,}\\ n-2 & \text{if } n \text{ is odd.}\end{cases}$$

$$v_\pm(n,2) = \begin{cases} n/2 & \text{if } n \text{ is even,}\\ (n-1)/2 & \text{if } n\equiv 3 \bmod 4,\\ (n-3)/2 & \text{if } n\equiv 1 \bmod 4.\end{cases}$$

$$v_\pm(n,3) = \begin{cases} n/2 & \text{if } n \text{ is even,}\\ n/3 & \text{if } n\equiv 3 \bmod 6,\\ (n-2)/3 & \text{if } n\equiv 5 \bmod 6,\\ (n-4)/3 & \text{if } n\equiv 1 \bmod 6.\end{cases}$$

For a proof, see page 316.

We go one step further and evaluate $v_\pm(n,4)$:

Proposition 4.10 *For all positive integers n we have*

$$v_{\pm}(n,4) = \begin{cases} \frac{n}{2} & \text{if } n \text{ is even,} \\ \left(1+\frac{1}{d}\right)\frac{n}{4} & \text{if } n \text{ is odd and has divisors congruent to 3 mod 8,} \\ & \text{and } d \text{ is the smallest such divisor,} \\ 2\cdot\left\lfloor\frac{n-2}{8}\right\rfloor+1 & \text{otherwise.} \end{cases}$$

The proof of Proposition 4.10 appears on page 317. We should note that the smallest divisor d of n that is congruent to 3 mod 8 is not necessarily a prime (consider, for example, $n = 35$), unlike in the analogous formula for $v_1(n,4)$ involving the smallest divisor of n that is congruent to 3 mod 4, which is always a prime.

For higher values of h, the evaluation of $v_{\pm}(n,h)$ becomes increasingly difficult. We offer:

Problem 4.11 *Develop the analogue of Theorem 4.4 for $v_{\pm}(n,h)$.*

While we don't yet have an analogue of Theorem 4.4 that helps us evaluate $v_{\pm}(n,h)$ exactly, we can still establish a tight upper bound for $v_{\pm}(n,h)$. By Proposition 4.7 and Corollary 4.5, for $h \geq 2$ we get

$$v_{\pm}(n,h) \leq v_1(n,h) \leq \frac{n}{2}.$$

When n is even, we have $2 \in D(n)$, and thus with

$$g_d(n,h) = \left(2\cdot\left\lfloor\frac{d-2}{2h}\right\rfloor+1\right)\cdot\frac{n}{d},$$

we see that $v_{\pm}(n,h) \geq g_2(n,2) = n/2$; therefore, equality must hold. When n is odd and $h \geq 3$, we can show that $v_{\pm}(n,h) \leq n/3$, since $g_1(n,h) = -n$, and for odd $d \geq 3$, we have

$$g_d(n,h) \leq \left(2\cdot\frac{d-3}{2h}+1\right)\cdot\frac{n}{d} = \left(\frac{h-3}{d}+1\right)\cdot\frac{n}{h} \leq \left(\frac{h-3}{3}+1\right)\cdot\frac{n}{h} = \frac{n}{3},$$

with equality holding if, and only if, $d = 3$. In summary, we get:

Proposition 4.12 *Let n be a positive integer.*

1. *If $h \geq 2$, then $v_{\pm}(n,h) \leq n/2$, with equality if, and only if, $2|n$.*
2. *If n is odd and $h \geq 3$, then $v_{\pm}(n,h) \leq n/3$, with equality if, and only if, $3|n$.*

We should point out that, in contrast, one can have $v_1(n,3) > n/3$ even for odd values of n (see page 78).

As with $v_g(n,h)$, we offer the following rather vague problem:

Problem 4.13 *Investigate some of the properties of the function $v_{\pm}(n,h)$.*

The function $u(n, m, h)$

The next function we discuss here is a variation on the famous Hopf–Stiefel function, introduced in the 1940s to study real division algebras. Since then, the Hopf–Stiefel function has been studied in a variety of contexts, including topology, linear and bilinear algebra, and additive number theory.

Suppose that n, m, and h are fixed positive integers; we will also assume that $m \leq n$. Recall that $D(n)$ is the set of all positive divisors of n. For a fixed $d \in D(n)$, we set

$$f_d = f_d(m, h) = \left(h \cdot \left\lceil \frac{m}{d} \right\rceil - h + 1\right) \cdot d,$$

and then define

$$u(n, m, h) = \min \{f_d(m, h) \mid d \in D(n)\}.$$

The evaluation of $u(n, m, h)$ can become quite cumbersome, particularly when $D(n)$ is relatively large. Since for every positive integer n, 1 and n are divisors of n, we have

$$u(n, m, h) \leq f_1(m, h) = \left(h \cdot \left\lceil \frac{m}{1} \right\rceil - h + 1\right) \cdot 1 = hm - h + 1$$

and

$$u(n, m, h) \leq f_n(m, h) = \left(h \cdot \left\lceil \frac{m}{n} \right\rceil - h + 1\right) \cdot n = n,$$

providing two upper bounds for $u(n, m, h)$. If n is prime, then $D(n) = \{1, n\}$, and thus we have to compare only the two values above; thus we have the following proposition.

Proposition 4.14 *For a prime number p we have*

$$u(p, m, h) = \min\{p, hm - h + 1\}.$$

We can also easily handle the cases of $m = 1$ and $h = 1$. When $m = 1$, we see that $\lceil \frac{m}{d} \rceil = 1$ for every $d \in D(n)$, and thus $u(n, 1, h)$ equals the least element of $D(n)$:

$$u(n, 1, h) = \min \{d \mid d \in D(n)\} = 1.$$

Similarly, when $h = 1$, we see that

$$u(n, m, 1) = \min \left\{ \left\lceil \frac{m}{d} \right\rceil \cdot d \mid d \in D(n) \right\};$$

since the expression $\lceil \frac{m}{d} \rceil \cdot d$ attains its minimum when d is a divisor of m, we can take $d = 1$ and get

$$u(n, m, 1) = m.$$

When n is not prime, evaluating $u(n, m, h)$ is not easy in general. For example, for $n = 15$, we can tabulate the values of $u(15, m, h)$ for all $1 \leq m \leq n$ and for $h = 2, 3, 4, 5$:

m	1	2	3	4	5	6	7	8	9	10	11	12	13	14	15
$u(15, m, 2)$	1	3	3	5	5	9	13	15	15	15	15	15	15	15	15
$u(15, m, 3)$	1	3	3	5	5	12	15	15	15	15	15	15	15	15	15
$u(15, m, 4)$	1	3	3	5	5	15	15	15	15	15	15	15	15	15	15
$u(15, m, 5)$	1	3	3	5	5	15	15	15	15	15	15	15	15	15	15

These values seem to indicate that we always have $m \leq u(n, m, h) \leq n$. The upper bound we have already explained above; in fact, we have seen that

$$u(n, m, h) \leq \min\{n, hm - h + 1\}.$$

We will prove that the lower bound holds as follows. We have seen that $u(n, m, 1) = m$, so let us assume that $h \geq 2$. Consider the expression

$$f_d(m, h) = \left(h \cdot \left\lceil \frac{m}{d} \right\rceil - h + 1\right) \cdot d.$$

Note that $f_m(m, h) = m$. What we will show is that $f_d(m, h)$ is greater than m both when $d > m$ and when $d < m$.

If $d > m$, then $\lceil \frac{m}{d} \rceil = 1$, and thus $f_d(m, h) = d$; since we assumed that $d > m$, we have $f_d(m, h) > m$.

If $d < m$, we see that

$$\begin{aligned} f_d(m, h) &= \left(h \cdot \left\lceil \frac{m}{d} \right\rceil - h + 1\right) \cdot d \\ &\geq \left(h \cdot \frac{m}{d} - h + 1\right) \cdot d \\ &= (h - 1)(m - d) + m; \end{aligned}$$

a value strictly greater than m under the assumptions that $h \geq 2$ and $d < m$.

Therefore, we have the following result.

Proposition 4.15 *For all positive integers n, m, and h, with $m \leq n$, we have*

$$m \leq u(n, m, h) \leq \min\{n, hm - h + 1\};$$

furthermore, $u(n, m, h) = m$ if, and only if, $h = 1$ or n is divisible by m.

In certain cases, it is useful to compare $u(n, m, h)$ to the smallest prime divisor p of n. Suppose first that $m \leq p$. In this case, we have

$$u(n, m, h) = \min\{p, hm - h + 1\},$$

since for $d = 1$ we have $f_d(m, h) = hm - h + 1$, and for all other divisors d of n, we have $d \geq p \geq m$ and, therefore,

$$f_d(m, h) = \left(h \cdot \left\lceil \frac{m}{d} \right\rceil - h + 1\right) \cdot d = d,$$

which attains its minimum when $d = p$. On the other hand, when $p < m$, then $p < u(n, m, h)$ since $u(n, m, h) \geq m$, and, therefore,

$$u(n, m, h) > \min\{p, hm - h + 1\}.$$

In summary, we have shown the following.

Proposition 4.16 *Let n, m, and h be positive integers, $m \leq n$, and let p be the smallest prime divisor of n. We then have*

$$u(n, m, h) \geq \min\{p, hm - h + 1\},$$

with equality if, and only if, $m \leq p$.

Note that Proposition 4.16 is a generalization of Proposition 4.14.

Proposition 4.15 has a complete characterization for situations when $u(n, m, h)$ reaches its lower bound, m—but how about its upper bound? In particular, when is $u(n, m, h) = n$? From its definition, we can observe that the function $u(n, m, h)$ is a nondecreasing function of m and also of h (but not of n). Furthermore, for fixed n and $h \geq 1$, the values of the function as m increases reach n; indeed, noting that if $d \in D(n)$, then $\lceil \frac{n}{d} \rceil = \frac{n}{d}$, we have

$$u(n, n, h) = \min\left\{\left(h \cdot \left\lceil \frac{n}{d} \right\rceil - h + 1\right) \cdot d \mid d \in D(n)\right\} = \min\{hn - d(h-1) \mid d \in D(n)\} = n.$$

We call the minimum value of m for which $u(n, m, h) = n$ the *h-critical number of n*; according to our computation we see that the h-critical number of n is well-defined for every $h \geq 1$, and equals at most n. This value is now known, as we can prove the following result:

Theorem 4.17 *The h-critical number of n equals $v_1(n, h) + 1$.*

The short proof of Theorem 4.17 can be found on page 318. (Campbell in [49] established Theorem 4.17 in some special cases, and provided lower bounds in other cases.) According to Theorem 4.17, the arithmetic function $v_1(n, h)$ is a certain inverse of the Hopf–Stiefel function. In particular, the maximum value of m for which $u(n, m, h) < n$ equals $v_1(n, h)$.

While the function $u(n, m, h)$ seems to be quite mysterious, it may be possible to find its values, at least in certain special cases, more easily, or to determine upper and lower bounds better than those above. We pose the following rather vague problem.

Problem 4.18 *Find some exact formulas or good upper or lower bounds, at least in certain special cases, for $u(n, m, h)$.*

The function $u\hat{}(n,m,h)$

The function $u\hat{}(n,m,h)$ we introduce here is a relative of the Hopf–Stiefel function $u(n,m,h)$ discussed above. (These two functions will express cardinalities of two related sets we study later.)

Suppose that n, m, and h are fixed positive integers; we will also assume that $h < m \leq n$. For a fixed $d \in D(n)$, we let k and r denote the positive remainder of m mod d and h mod d, respectively. That is, we write

$$m = cd + k \quad \text{and} \quad h = qd + r$$

with

$$1 \leq k \leq d \quad \text{and} \quad 1 \leq r \leq d.$$

Note that one can compute these quotients and remainders as

$$c = \left\lceil \frac{m}{d} \right\rceil - 1 \quad \text{and} \quad q = \left\lceil \frac{h}{d} \right\rceil - 1;$$

$$k = m - d\left\lceil \frac{m}{d} \right\rceil + d \quad \text{and} \quad r = h - d\left\lceil \frac{h}{d} \right\rceil + d.$$

We then set

$$f_{\hat{d}}(n,m,h) = \begin{cases} \min\{n, f_d, hm - h^2 + 1\} & \text{if } h \leq \min\{k, d-1\}, \\ \min\{n, hm - h^2 + 1 - \delta_d\} & \text{otherwise;} \end{cases}$$

where f_d is the function

$$f_d(m,h) = \left(h \cdot \left\lceil \frac{m}{d} \right\rceil - h + 1\right) \cdot d$$

defined on page 83, and δ_d is a "correction term" defined as

$$\delta_d(n,m,h) = \begin{cases} (k-r)r - (d-1) & \text{if } r < k, \\ (d-r)(r-k) - (d-1) & \text{if } k < r < d, \\ d-1 & \text{if } k = r = d, \\ 0 & \text{otherwise.} \end{cases}$$

We then define

$$u\hat{}(n,m,h) = \min\{f_{\hat{d}}(n,m,h) \mid d \in D(n)\}.$$

Evaluating $u\hat{}(n,m,h)$ is more complicated than evaluating $u(n,m,h)$; to see some specific values, we compute $u\hat{}(n,m,h)$ for $n = 15$ and for all $h < m \leq n$ and $h = 2,3,4,5$:

m	1	2	3	4	5	6	7	8	9	10	11	12	13	14	15
$u\hat{}(15,m,2)$			3	5	5	9	11	13	15	15	15	15	15	15	15
$u\hat{}(15,m,3)$				4	5	8	13	15	15	15	15	15	15	15	15
$u\hat{}(15,m,4)$					5	9	13	15	15	15	15	15	15	15	15
$u\hat{}(15,m,5)$						6	11	15	15	15	15	15	15	15	15

We pose the following rather vague problem.

Problem 4.19 *Find a simpler way to define $u\hat{}(n,m,h)$.*

Somewhat less ambitiously:

Problem 4.20 *Find a general expression for $u\hat{}(n,m,h)$, at least for some cases.*

As perhaps a step toward Problems 4.19 and 4.20, there may be a simplified expression for δ_d. Indeed, we can easily compute δ_d for small values of d. We get

$$\delta_1 = 0 \text{ for all } m \text{ and } h;$$

$$\delta_2 = \begin{cases} 1 & \text{if } m \text{ and } h \text{ are both even,} \\ 0 & \text{otherwise;} \end{cases}$$

and

$$\delta_3 = \begin{cases} 2 & \text{if } m \text{ and } h \text{ are both divisible by 3,} \\ -1 & \text{if none of } m,\ h, \text{ or } m-h \text{ is divisible by 3,} \\ 0 & \text{otherwise.} \end{cases}$$

As these formulae suggest, one may be able to evaluate δ_d easily.

Problem 4.21 *Find a simpler general expression for $\delta_d(n,m,h)$.*

With a simpler general expression for δ_d, we can hope for an easier way to evaluate $f_{\hat{d}}$. For $d \leq 3$ we find that

$$f_{\hat{1}} = \min\{n, hm - h^2 + 1\};$$

$$f_{\hat{2}} = \begin{cases} \min\{n, hm - h^2\} & \text{if } m \text{ and } h \text{ are both even,} \\ \min\{n, hm - h^2 + 1\} & \text{otherwise;} \end{cases}$$

and

$$f_{\hat{3}} = \begin{cases} \min\{n, 2m-1, 3m-8\} & \text{if } h = 2 \text{ and } 3 | m-2, \\ \min\{n, 2m-3, 3m-8\} & \text{if } h = 2 \text{ and } 3 | m, \\ \min\{n, hm - h^2 - 1\} & \text{if } 3 | m \text{ and } 3 | h, \\ \min\{n, hm - h^2 + 2\} & \text{if } h > 1,\ 3 \nmid m,\ 3 \nmid h, \text{ and } 3 \nmid m-h, \\ \min\{n, hm - h^2 + 1\} & \text{otherwise.} \end{cases}$$

Having explicit formulae for $f_{\hat{d}}$ for all n, m, h, and d would enable us to evaluate $u\hat{}(n,m,h)$. Regarding the case of $h = 1$: we see that either $d \geq 2$ in which case

$$f_{\hat{d}}(n,m,1) = \min\{n, f_d, hm - h^2 + 1\} = \min\{n, \left\lceil \frac{m}{d} \right\rceil \cdot d, m\} = m,$$

or $d = 1$ in which case

$$f_{\hat{1}}(n,m,1) = \min\{n, hm - h^2 + 1\} = \min\{n, m\} = m;$$

and therefore

$$u\hat{}(n,m,1) = m.$$

For $h = 2$ and $h = 3$ we have the following results.

Proposition 4.22 *For $h = 2$, we have*

$$u\hat{}(n,m,2) = \begin{cases} \min\{u(n,m,2), 2m-4\} & \textit{if } n \textit{ and } m \textit{ are both even,} \\ \min\{u(n,m,2), 2m-3\} & \textit{otherwise.} \end{cases}$$

Proposition 4.23 *Let* $d_0 = \gcd(n, m-1)$. *For* $h = 3$, *we have*

$$u\hat{}(n,m,3) = \begin{cases} \min\{u(n,m,3), 3m-3-d_0\} & \text{if } \gcd(n,m-1) \geq 8; \\ \min\{u(n,m,3), 3m-10\} & \text{if } \gcd(n,m-1) = 7, \text{ or} \\ & \gcd(n,m-1) \leq 5,\ 3|n, \text{ and } 3|m; \\ \min\{u(n,m,3), 3m-9\} & \text{if } \gcd(n,m-1) = 6; \\ \min\{u(n,m,3), 3m-8\} & \text{otherwise.} \end{cases}$$

The proof of Propositions 4.22 and 4.23 can be found on pages 319 and 320, respectively. As h increases, the evaluation of $u\hat{}(n,m,h)$ gets more complicated; we offer the following problem.

Problem 4.24 *Find an explicit formula for* $u\hat{}(n,m,h)$ *in terms of* $u(n,m,h)$ *for* $h = 4$, $h = 5$, *and* $h = 6$.

We also know the value of $u\hat{}(n,m,h)$ when n is prime; in that case $D(n) = \{1, n\}$. We have already seen that

$$f_{\hat{1}} = \min\{n, hm - h^2 + 1\};$$

we can similarly verify that

$$f_{\hat{n}} = \min\{n, hm - h^2 + 1\}$$

and thus we have the following proposition.

Proposition 4.25 *For a prime number* p *we have*

$$u\hat{}(p,m,h) = \min\{p, hm - h^2 + 1\}.$$

Other than the results above, we have no general formula for $u\hat{}(n,m,h)$. Below we investigate some upper and lower bounds instead. First, we compare $u\hat{}(n,m,h)$ and $u(n,m,h)$.

Proposition 4.26 *For all* $h < m \leq n$ *we have* $u\hat{}(n,m,h) \leq u(n,m,h)$.

The proof of Proposition 4.26 can be found on page 320.

Let us now see how much less $u\hat{}(n,m,h)$ can get below $u(n,m,h)$. For $h = 1$, we have already seen that $u(n,m,1) = u\hat{}(n,m,1) = m$. For $h = 2$, we use Proposition 4.22 to see that, when $u\hat{}(n,m,2) < u(n,m,2)$ and n and m are both even, then

$$u(n,m,2) - u\hat{}(n,m,2) = u(n,m,2) - (2m-4) \leq f_2(m,2) - (2m-4) = (2m-2) - (2m-4) = 2,$$

and when $u\hat{}(n,m,2) < u(n,m,2)$ and n or m is odd, then

$$u(n,m,2) - u\hat{}(n,m,2) = u(n,m,2) - (2m-3) \leq f_1(m,2) - (2m-3) = (2m-1) - (2m-3) = 2.$$

This proves the following.

Proposition 4.27 *For all* $m \leq n$ *we have*

$$u(n,m,2) - 2 \leq u\hat{}(n,m,2) \leq u(n,m,2).$$

According to Proposition 4.27, the only possible values of $u\hat{}(n,m,2)$ are $u(n,m,2)$, $u(n,m,2)-1$, and $u(n,m,2)-2$.

In contrast to Proposition 4.27, we can show that, when $h \geq 3$, then $u(n,m,h) - u\hat{}(n,m,h)$ can get arbitrarily large: larger than any given (positive) real number C. As an example, take an arbitrary prime $p > h$ and a positive integer $t > 2$ so that

$$(h-2)p^{t-1} \geq C.$$

One can readily verify that for $n = p^t$ and $m = p^{t-1}+1$ we have

$$u(n,m,h) = f_1 = hp^{t-1}+1$$

and for $d = p^{t-1}$ we get

$$u\hat{}(n,m,h) \leq f\hat{d} = 2p^{t-1}.$$

Thus, we have:

Proposition 4.28 *For every $h \geq 3$ and for every positive real number C, one can find positive integers n and m for which*

$$u\hat{}(n,m,h) < u(n,m,h) - C.$$

Recall that by Proposition 4.15, we have

$$m \leq u(n,m,h) \leq \min\{n, hm-h+1\},$$

with $u(n,m,h) = m$ if, and only if, $h=1$ or n is divisible by m. For $u\hat{}(n,m,h)$ we have the following result.

Proposition 4.29 *For all positive integers n, m, and h with $h < m \leq n$ we have*

$$m \leq u\hat{}(n,m,h) \leq \min\{u(n,m,h), hm-h^2+1\} \leq \min\{n, hm-h^2+1\},$$

with $u\hat{}(n,m,h) = m$ if, and only if, (at least) one of the following holds:

(i) $h = 1$,

(ii) $h = m-1$,

(iii) *n is divisible by m, or*

(iv) *$h = 2$, $m = 4$, and n is even.*

The upper bound in Proposition 4.29 follows from the fact that

$$u\hat{}(n,m,h) \leq f\hat{1}(n,m,h) = \min\{n, hm-h^2+1\}$$

and Propositions 4.26 and 4.15. The proof of the lower bound and the classification of its equality can be found on page 321.

We know considerably less about cases when $u\hat{}(n,m,h)$ reaches the upper bound

$$\min\{u(n,m,h), hm-h^2+1\}$$

in Proposition 4.29. Equality clearly always holds when $h=1$, and from Proposition 4.22 we see that, when $h=2$, equality also holds unless n and m are both even and $2m-3 \leq u(n,m,2)$. When comparing the table on page 83 to the table on page 86, we see that, for $n=15$ and $2 \leq h \leq 5$, there is only one case when the upper bound is not reached: $h=3$

and $m = 6$. We may observe that in both these cases we have $\gcd(n, m, h) > 1$; indeed, when $\gcd(n, m, h) = d > 1$ and $hm - h^2 + 1 \leq u(n, m, h)$, then

$$u\hat{}(n, m, h) \leq f\hat{_d}(n, m, h) = hm - h^2 + 1 - (d - 1) < \min\{u(n, m, h), hm - h^2 + 1\}.$$

It is not hard to find other scenarios when the upper bound is not reached, but we do not have a full understanding of all of them. We offer the following challenging problem.

Problem 4.30 *Classify all situations when* $u\hat{}(n, m, h) < \min\{u(n, m, h), hm - h^2 + 1\}$.

As we did for $u(n, m, h)$, we now compare $u\hat{}(n, m, h)$ to the smallest prime divisor p of n. Clearly, if $p < m$ then, by Proposition 4.29,

$$u\hat{}(n, m, h) \geq m > p \geq \min\{p, hm - h^2 + 1\}.$$

On the other hand, if $p \geq m$, then

$$f\hat{_1}(n, m, h) = \min\{n, hm - h^2 + 1\} \geq \min\{p, hm - h^2 + 1\},$$

and for $d \in D(n) \setminus \{1\}$, we have $d \geq p \geq m > h$ and, therefore, $r = h < m = k$, so $h \leq \min\{d - 1, k\}$ and thus

$$\begin{aligned} f\hat{_d}(n, m, h) &= \min\{n, \left(h \cdot \left\lceil \frac{m}{d} \right\rceil - h + 1\right) \cdot d, hm - h^2 + 1\} \\ &= \min\{n, d, hm - h^2 + 1\} \\ &\geq \min\{p, hm - h^2 + 1\}, \end{aligned}$$

with equality when (though not necessarily only when) $d = p$, and thus

$$u\hat{}(n, m, h) = \min\{p, hm - h^2 + 1\}.$$

In summary, we have shown the following.

Proposition 4.31 *Let n, m, and h be positive integers, $h < m \leq n$, and let p be the smallest prime divisor of n. We then have*

$$u\hat{}(n, m, h) \geq \min\{p, hm - h^2 + 1\},$$

with equality if, and only if, $m \leq p$.

Note that Proposition 4.31 is a generalization of Proposition 4.25.

Recall that the function $u(n, m, h)$ is a nondecreasing function of both m and h (but not of n) and, for a fixed h, it reaches n at a certain threshold value of m, called the h-critical number of n. In contrast, as the table on page 86 indicates, $u\hat{}(n, m, h)$ is not a nondecreasing function of h. With some effort, one can prove that $u\hat{}(n, m, h)$ is a nondecreasing function of m and that it reaches n eventually, so we are able to define the *restricted h-critical number of n*. We will only do this here for $h \leq 2$.

Since $u\hat{}(n, m, 1) = m$ is nondecreasing with m reaching n at $m = n$, the restricted 1-critical number of n is clearly well-defined and equals n. For $h = 2$, we use Proposition 4.22 to show that $u\hat{}(n, m + 1, 2) \geq u\hat{}(n, m, 2)$. Indeed, if n is odd, then

$$u\hat{}(n, m+1, 2) = \min\{u(n, m+1, 2), 2(m+1) - 3\} \geq \min\{u(n, m, 2), 2m - 3\} = u\hat{}(n, m, 2);$$

if n is even and m is even, then

$$u\hat{}(n, m+1, 2) = \min\{u(n, m+1, 2), 2(m+1) - 3\} \geq \min\{u(n, m, 2), 2m - 4\} = u\hat{}(n, m, 2);$$

and, if n is even and m is odd, then

$$u\hat{}(n,m+1,2) = \min\{u(n,m+1,2), 2(m+1)-4\} \geq \min\{u(n,m,2), 2m-3\} = u\hat{}(n,m,2).$$

Therefore, $u\hat{}(n,m,2)$ is a nondecreasing function of m. Furthermore, using Theorem 4.17 and Propositions 4.3 and 4.22, we also see that

$$u\hat{}(n,m,2) \begin{cases} < n & \text{if } m \leq \left\lfloor \frac{n}{2} \right\rfloor + 1, \\ = n & \text{if } m \geq \left\lfloor \frac{n}{2} \right\rfloor + 2, \end{cases}$$

and thus we have the following.

Theorem 4.32 *The restricted* 2*-critical number of* n *is* $\lfloor \frac{n}{2} \rfloor + 2$.

Problem 4.33 *For each* $h \geq 3$, *prove that the* h*-critical number of* n *is well-defined and find its value.*

The arithmetic function $u\hat{}(n,m,h)$ seems quite interesting; we offer the following vague problem.

Problem 4.34 *Find some exact formulas or good upper or lower bounds, at least in certain special cases, for* $u\hat{}(n,m,h)$.

Part IV

Entrees

This is the main part of the book where our problems are posed and their backgrounds are explained. At the present time, these problems are about maximum sumset size, spanning sets, Sidon sets, minimum sumset size, critical numbers, zero-sum-free sets, and sum-free sets. (We plan to include additional chapters in the near future.)

Guide to section numbering: In Chapter X (where X $\in$ {A, B, C, ...}), we study, for $A \subseteq G$, $H \subseteq \mathbb{N}_0$, $h \in \mathbb{N}_0$, and $s \in \mathbb{N}_0$:

Section X.1: Unrestricted sumsets: HA

- Section X.1.1: Fixed number of terms: hA
- Section X.1.2: Limited number of terms: $[0, s]A$
- Section X.1.3: Arbitrary number of terms: $\langle A \rangle$

Section X.2: Unrestricted signed sumsets: $H_{\pm}A$

- Section X.2.1: Fixed number of terms: $h_{\pm}A$
- Section X.2.2: Limited number of terms: $[0, s]_{\pm}A$
- Section X.2.3: Arbitrary number of terms: $\langle A \rangle$

Section X.3: Restricted sumsets: $H\hat{\ }A$

- Section X.3.1: Fixed number of terms: $h\hat{\ }A$
- Section X.3.2: Limited number of terms: $[0, s]\hat{\ }A$
- Section X.3.3: Arbitrary number of terms: ΣA

Section X.4: Restricted signed sumsets: $H\hat{_{\pm}}A$

- Section X.4.1: Fixed number of terms: $h\hat{_{\pm}}A$
- Section X.4.2: Limited number of terms: $[0, s]\hat{_{\pm}}A$
- Section X.4.3: Arbitrary number of terms: $\Sigma_{\pm}A$

See the table on page 44 for definitions and terminology.

Chapter A

Maximum sumset size

Recall that for a given finite abelian group G, m-subset $A = \{a_1, \ldots, a_m\}$ of G, $\Lambda \subseteq \mathbb{Z}$, and $H \subseteq \mathbb{N}_0$, we defined the sumset of A corresponding to Λ and H as

$$H_\Lambda A = \{\lambda_1 a_1 + \cdots + \lambda_m a_m \mid (\lambda_1, \ldots, \lambda_m) \in \Lambda^m(H)\}$$

where the index set $\Lambda^m(H)$ is defined as

$$\Lambda^m(H) = \{(\lambda_1, \ldots, \lambda_m) \in \Lambda^m \mid |\lambda_1| + \cdots + |\lambda_m| \in H\}.$$

In this chapter we attempt to answer the following question: Given a finite abelian group G and a positive integer m, how large can a sumset of an m-subset of G be? More precisely, our objective is to determine, for any G, m, Λ, and H the quantity

$$\nu_\Lambda(G, m, H) = \max\{|H_\Lambda A| \mid A \subseteq G, |A| = m\}.$$

We can immediately see two upper bounds for $\nu_\Lambda(G, m, H)$: since $H_\Lambda A$ is a subset of G and since each element of the index set $\Lambda^m(H)$ contributes a unique element toward $H_\Lambda A$ (which may or may not be distinct), we have the following obvious result.

Proposition A.1 *We have*

$$\nu_\Lambda(G, m, H) \leq \min\{|G|, |\Lambda^m(H)|\}.$$

One of the most fascinating questions in additive combinatorics is to investigate the cases when equality occurs in Proposition A.1. In particular, subsets A of G with $H_\Lambda A = G$ (that is, every element of G can be written as an appropriate sum of the elements of A) are called spanning sets, and subsets with $|H_\Lambda A| = |\Lambda^m(H)|$ (i.e., all sums are distinct modulo the rearrangement of the terms) are called Sidon sets. We investigate spanning sets in Chapter B and Sidon sets in Chapter C.

In the following sections we consider $\nu_\Lambda(G, m, H)$ for special $\Lambda \subseteq \mathbb{Z}$ and $H \subseteq \mathbb{N}_0$.

A.1 Unrestricted sumsets

Our goal in this section is to investigate the quantity

$$\nu(G, m, H) = \max\{|HA| \mid A \subseteq G, |A| = m\}$$

where HA is the union of all h-fold sumsets hA for $h \in H$. We consider three special cases: when H consists of a single nonnegative integer h, when H consists of all nonnegative integers up to some value s, and when H is the entire set of nonnegative integers.

A.1.1 Fixed number of terms

We first consider

$$\nu(G,m,h) = \max\{|hA| \mid A \subseteq G, |A| = m\},$$

that is, the maximum size of an h-fold sumset of an m-element subset of G.

In general, we do not know the value of $\nu(G,m,h)$. Obviously, for $h=0$ and $h=1$ we have $hA=\{0\}$ and $hA=A$, respectively, thus

$$\nu(G,m,0) = 1$$

and

$$\nu(G,m,1) = m$$

for every $m \in \mathbb{N}$. If A consists of a single element a, then $hA = \{ha\}$ for every $h \in \mathbb{N}_0$, so

$$\nu(G,1,h) = 1$$

for every $h \in \mathbb{N}_0$.

Before turning to the case of $m=2$, recall that, according to Proposition 3.2, in order to find $\nu(G,m,h)$ (for any G, m, and h), we may assume that the subset A of G of size m yielding $|hA| = \nu(G,m,h)$ will contain 0. Therefore, to find $\nu(G,2,h)$, we only need to look for an element $a \in G$ for which the size of

$$h\{0,a\} = \{0, a, 2a, \ldots, ha\}$$

is maximal. But the size of this set is clearly $\min\{|\langle a \rangle|, h+1\}$. Here $|\langle a \rangle|$ is the order of a; its maximal value is the exponent of G, denoted by κ. Therefore,

$$\nu(G,2,h) = \min\{\kappa, h+1\}$$

for any $h \in \mathbb{N}_0$.

Summarizing our results thus far, we have the following.

Proposition A.2 *In any abelian group G of order n and exponent κ we have*

$$\begin{aligned} \nu(G,m,0) &= 1, \\ \nu(G,m,1) &= m, \\ \nu(G,1,h) &= 1, \\ \nu(G,2,h) &= \min\{\kappa, h+1\}. \end{aligned}$$

For values of $m \geq 3$ and $h \geq 2$, we have no exact values for $\nu(G,m,h)$ in general; as a consequence of Proposition A.1, we have, however, the following upper bound.

Proposition A.3 *In any abelian group G of order n we have*

$$\nu(G,m,h) \leq \min\left\{n, \binom{m+h-1}{h}\right\}.$$

We pose the following very general problem.

Problem A.4 *Find the value of (or, at least, find good bounds for) $\nu(G,m,h)$ for noncyclic groups G and integers m and h.*

The case of cyclic groups is of special interest:

Problem A.5 *For positive integers n, m, and h, find the value of (or, at least, find good bounds for) $\nu(\mathbb{Z}_n, m, h)$.*

The value of $\nu(\mathbb{Z}_n, m, h)$ (or, more generally, the value of $\nu(G, m, h)$) tends to agree with the upper bound in Proposition A.3. Perhaps the most intriguing aspect of Problem A.5 (or, more generally, of Problem A.4) is to analyze the exceptions to this predilection. The number of exceptions seems to vary a great deal. For example, Manandhar in [148] found (using a computer program) that, as n ranges from 2 to 20, the number of such exceptions is as follows.

n	2	3	4	5	6	7	8	9	10	11	12	13	14	15	16	17	18	19	20
exceptions	0	0	0	0	1	0	0	1	2	2	3	1	2	2	2	4	5	4	6

The six exceptions for $n = 20$ are listed in the following table.

h	m	$\min\left\{20, \binom{m+h-1}{h}\right\}$	$\nu(\mathbb{Z}_{20}, m, h)$
2	5	15	14
2	6	20	18
3	4	20	18
4	3	15	14
5	3	20	17
6	3	20	19

These data suggest that Problem A.5 might be difficult. Therefore, we also pose the following special cases.

Problem A.6 *For positive integers n and h, find $\nu(\mathbb{Z}_n, 3, h)$.*

Problem A.7 *For positive integers n and m, find $\nu(\mathbb{Z}_n, m, 2)$.*

Problem A.8 *Find all (or, at least, infinitely many) positive integers n, m, and h, for which*

$$\nu(\mathbb{Z}_n, m, h) < \min\left\{n, \binom{m+h-1}{h}\right\}.$$

A.1.2 Limited number of terms

Here we ought to consider, for a given group G, positive integer m (with $m \leq n = |G|$), and nonnegative integer s,

$$\nu(G, m, [0, s]) = \max\{|[0, s]A| \mid A \subseteq G, |A| = m\},$$

that is, the maximum size of $\cup_{h=0}^{s} hA$ for an m-element subset A of G.

Obviously, we have

$$\nu(G, n, [0, s]) = |[0, s]G| = \begin{cases} 1 & \text{if } s = 0, \\ n & \text{if } s \geq 1; \end{cases}$$

hence we may restrict our attention to the cases when $m \leq n - 1$. However, we have the following result.

Proposition A.9 *For any group G, positive integer $m \leq n - 1$, and nonnegative integer s we have*

$$\nu(G, m, [0, s]) = \nu(G, m + 1, s).$$

We can prove Proposition A.9 as follows. Suppose first that A is a subset of G of size m and that it has maximum-size $[0,s]$-fold sumset:

$$|[0,s]A| = \nu(G,m,[0,s]).$$

Clearly, $|A \cup \{0\}| \leq m+1$, so

$$|s(A \cup \{0\})| \leq \nu(G,m+1,s).$$

But $[0,s]A = s(A \cup \{0\})$ (see Proposition 3.3), and therefore

$$\nu(G,m,[0,s]) \leq \nu(G,m+1,s).$$

For the other direction, choose a subset A of G of size $m+1$ for which

$$|sA| = \nu(G,m+1,s).$$

By Proposition 3.2, we may assume that $0 \in A$; let $A' = A \setminus \{0\}$. Then A' has size m, and we have

$$\nu(G,m+1,s) = |sA| = |s(A' \cup \{0\})| = |[0,s]A'| \leq \nu(G,m,[0,s]).$$

Therefore,

$$\nu(G,m,[0,s]) = \nu(G,m+1,s),$$

as claimed.

Proposition A.9 makes this section superfluous (or makes Section A.1.1 superfluous via this section).

A.1.3 Arbitrary number of terms

Here we consider, for a given group G and positive integer m (with $m \leq n$),

$$\nu(G,m,\mathbb{N}_0) = \max\{|\langle A \rangle| \mid A \subseteq G, |A| = m\}.$$

Recall that $\langle A \rangle$ is the subgroup of G generated by A.

We can easily determine that

$$\nu(\mathbb{Z}_n,m,\mathbb{N}_0) = n$$

for all m and n, since any set A which contains an element of order n will generate all of $\mathbb{Z}_n$.

More generally, consider the invariant decomposition of G,

$$G = \mathbb{Z}_{n_1} \times \mathbb{Z}_{n_2} \times \cdots \times \mathbb{Z}_{n_r},$$

where r and $n_1, \dots, n_r$ are integers all at least 2 and n_{i+1} is divisible by n_i for $i = 1,2,\dots,r-1$. (Here r is the rank of G and $n_r = \kappa$ is the exponent of G.) For each $i = 1,2,\dots,r$, let e_i denote the element $(0,\dots,0,1,0,\dots,0)$ of G where the 1 occurs in position i. When $m = r$, taking $A = \{e_1,\dots,e_m\}$ results in $\langle A \rangle = G$, and thus $\nu(G,m,\mathbb{N}_0) = n$; this holds when $m > r$ for any set A containing $\{e_1,\dots,e_m\}$. When $m < r$, we can take $A = \{e_{r-m+1}, e_{r-m+2}, \dots, e_r\}$, and this results in

$$\langle A \rangle \cong \mathbb{Z}_{n_{r-m+1}} \times \mathbb{Z}_{n_{r-m+2}} \times \cdots \times \mathbb{Z}_{n_r}.$$

We conjecture that we cannot do better:

Conjecture A.10 *Suppose that G is given by its invariant decomposition, as above. Prove that*

$$\nu(G,m,\mathbb{N}_0) = \begin{cases} n_{r-m+1} n_{r-m+2} \cdots n_r & \text{if } m \leq r, \\ n & \text{if } m \geq r. \end{cases}$$

Problem A.11 *Prove (or disprove) Conjecture A.10.*

A.2 Unrestricted signed sumsets

In this section we investigate the quantity

$$\nu_{\pm}(G,m,H) = \max\{|H_{\pm}A| \mid A \subseteq G, |A| = m\}$$

for various $H \subseteq \mathbb{N}_0$.

A.2.1 Fixed number of terms

First, we consider

$$\nu_{\pm}(G,m,h) = \max\{|h_{\pm}A| \mid A \subseteq G, |A| = m\},$$

that is, the maximum size of an h-fold signed sumset of an m-element subset of G.

In general, we do not know the value of $\nu_{\pm}(G,m,h)$, but we can evaluate it for $h=0$, $h=1$, and $m=1$ as follows. Since $0_{\pm}A = \{0\}$ for every $A \subseteq G$, we have

$$\nu_{\pm}(G,m,0) = 1.$$

Furthermore, $1_{\pm}A = A \cup (-A)$, so to find $\nu_{\pm}(G,m,1)$, we need to find a subset A of G with $|A| = m$ for which $|A \cup (-A)|$ is maximal. We can do this as follows.

Let

$$L = \mathrm{Ord}(G,2) \cup \{0\} = \{g \in G \mid 2g = 0\} = \{g \in G \mid g = -g\}.$$

Observe that the elements of $G \setminus L$ are distinct from their inverses, so we have a subset K of $G \setminus L$ with which $G = L \cup K \cup (-K)$, and L, K, and $-K$ are pairwise disjoint. We then see that to maximize the quantity $|A \cup (-A)|$, we can have $A \subseteq K$ when $m \leq |K|$; $K \subseteq A \subseteq K \cup L$ when $|K| \leq m \leq |K \cup L|$; and $K \cup L \subseteq A$ when $m \geq |K \cup L|$. Thus,

$$\nu_{\pm}(G,m,1) = \begin{cases} 2m & \text{if } m \leq |K| \\ m + |K| & \text{if } |K| \leq m \leq |K \cup L| \\ n & \text{if } m \geq |K \cup L|; \end{cases}$$

since $|K \cup L| = n - |K|$, this simplifies to

$$\nu_{\pm}(G,m,1) = \min\{2m, m + |K|, n\}$$

or

$$\nu_{\pm}(G,m,1) = \min\left\{n, 2m, m + \frac{n - |\mathrm{Ord}(G,2)| - 1}{2}\right\}.$$

Let's turn to the case of $m = 1$. If A consists of a single element a, then $h_{\pm}A = \{ha, -ha\}$ for every $h \in \mathbb{N}_0$, so to find $\nu_{\pm}(G,1,h)$, we need to find, if possible, a one-element subset $A = \{a\}$ of G for which the set $\langle A, h \rangle = \{ha, -ha\}$ has size 2. This means that we only need to answer the following question: when does G contain an element a for which $ha \neq -ha$? But, for any $a \in G$, $ha \neq -ha$ is equivalent to $2ha \neq 0$, which is the same as saying that the order of a does not divide $2h$. Since the order of any element in G is a divisor of the exponent κ of G, if the order of a does not divide $2h$, then κ does not divide $2h$ either. Conversely, if κ does not divide $2h$, then choosing a to be an element of order κ will work. Therefore, we get

$$\nu_{\pm}(G,1,h) = \begin{cases} 1 & \text{if } \kappa | 2h; \\ 2 & \text{otherwise.} \end{cases}$$

In summary, we have the following.

Proposition A.12 *In any abelian group G of order n and exponent κ we have*

$$\begin{aligned}
\nu_{\pm}(G,m,0) &= 1; \\
\nu_{\pm}(G,m,1) &= \min\left\{n, 2m, m + \frac{n - |\mathrm{Ord}(G,2)| - 1}{2}\right\}; \\
\nu_{\pm}(G,1,h) &= \begin{cases} 1 & \text{if } \kappa | 2h, \\ 2 & \text{otherwise.} \end{cases}
\end{aligned}$$

For values of $h \geq 2$ and $m \geq 2$, we have no exact values for $\nu_{\pm}(G,m,h)$ in general; as a consequence of Proposition A.1, we have, however, the following upper bound.

Proposition A.13 *In any abelian group G of order n we have*

$$\nu_{\pm}(G,m,h) \leq \min\{n, c(h,m)\}.$$

Recall that

$$c(h,m) = \sum_{i \geq 0} \binom{m}{i} \binom{h-1}{i-1} 2^i.$$

Similar to Section A.1.1, we pose the following problems.

Problem A.14 *Find the value of (or, at least, find good bounds for) $\nu_{\pm}(G,m,h)$ for noncyclic groups G and integers m and h.*

Problem A.15 *Find the value of (or, at least, find good bounds for) $\nu_{\pm}(\mathbb{Z}_n,m,h)$ for all positive integers n, m, and h.*

Problem A.16 *For positive integers n and h, find $\nu_{\pm}(\mathbb{Z}_n,2,h)$.*

Problem A.17 *For positive integers n and m, find $\nu_{\pm}(\mathbb{Z}_n,m,2)$.*

Problem A.18 *Find all (or, at least, infinitely many) positive integers n, m, and h, with $m \geq 2$ and $h \geq 2$, for which*

$$\nu_{\pm}(\mathbb{Z}_n,m,h) < \min\{n, c(h,m)\}.$$

Some of the cases when the inequality of Problem A.31 holds were found by Buell in [44]; namely, when $2 \leq m \leq 5$, the set of n values for which $\nu_{\pm}(\mathbb{Z}_n,m,2)$ is less than $\min\{n, c(2,m)\}$ are as follows:

m	n
2	$6-8$
3	$14-18, 20$
4	$22-33, 36, 40$
5	$34, 36-54, 56-58$

From these data it appears that a complete solution to our problems above may be quite challenging.

A.2.2 Limited number of terms

Here we consider, for a given group G, positive integer m (with $m \leq n = |G|$), and nonnegative integer s,

$$\nu_\pm(G, m, [0, s]) = \max\{|[0, s]_\pm A| \mid A \subseteq G, |A| = m\},$$

that is, the maximum size of $[0, s]_\pm A$ for an m-element subset A of G.

We note that we don't have a version of Proposition 3.2 for signed sumsets, so we are not able to reduce this entire section to Section A.2.1. (However, one may be able to apply similar techniques.)

It is easy to see that, for every m, we have

$$\nu_\pm(G, m, [0, 0]) = 1.$$

Furthermore,

$$[0, 1]_\pm A = A \cup (-A) \cup \{0\},$$

whose maximum size we can find as we did for $A \cup (-A)$ in Section A.2.1, except here we write G as the pairwise disjoint union of four (potentially empty) parts: $\{0\}$, $\mathrm{Ord}(G, 2)$, K, and $-K$. The computation this time yields

$$\nu_\pm(G, m, [0, 1]) = \min\left\{n, 2m + 1, m + \frac{n - |\mathrm{Ord}(G, 2)| + 1}{2}\right\}.$$

Regarding the case of $m = 1$, we see that for $A = \{a\}$ we have

$$[0, s]_\pm A = \{0, \pm a, \pm 2a, \ldots, \pm sa\},$$

thus

$$\nu_\pm(G, 1, [0, s]) = \min\{\kappa, 2s + 1\}.$$

Summarizing our results thus far, we have the following.

Proposition A.19 *In any abelian group G of order n and exponent κ we have*

$$\begin{aligned} \nu_\pm(G, m, [0, 0]) &= 1, \\ \nu_\pm(G, m, [0, 1]) &= \min\left\{n, 2m + 1, m + \frac{n - |\mathrm{Ord}(G, 2)| + 1}{2}\right\}, \\ \nu_\pm(G, 1, [0, s]) &= \min\{\kappa, 2s + 1\}. \end{aligned}$$

For values of $s \geq 2$ and $m \geq 2$, we have no exact values for $\nu_\pm(G, m, [0, s])$ in general; as a consequence of Proposition A.1, we have, however, the following upper bound.

Proposition A.20 *In any abelian group G of order n we have*

$$\nu_\pm(G, m, [0, s]) \leq \min\{n, a(m, s)\}.$$

Recall that

$$a(m, s) = \sum_{i \geq 0} \binom{m}{i} \binom{s}{i} 2^i.$$

Similarly to previous sections, we pose the following problems.

Problem A.21 *Find the value of (or, at least, find good bounds for) $\nu_\pm(G, m, [0, s])$ for noncyclic groups G and integers m and s.*

Problem A.22 *Find the value of (or, at least, find good bounds for)* $\nu_{\pm}(\mathbb{Z}_n, m, [0,s])$ *for all integers* n, m, *and* s.

Problem A.23 *For positive integers* n *and* s, *find* $\nu_{\pm}(\mathbb{Z}_n, 2, [0,s])$.

Problem A.24 *For positive integers* n *and* m, *find* $\nu_{\pm}(\mathbb{Z}_n, m, [0,2])$.

Problem A.25 *Find all (or, at least, infinitely many) positive integers* n, m, *and* s *for which*

$$\nu_{\pm}(\mathbb{Z}_n, m, [0,s]) < \min\{n, a(m,s)\}.$$

A.2.3 Arbitrary number of terms

This subsection is identical to Subsection A.1.3.

A.3 Restricted sumsets

Our goal in this section is to investigate the quantity

$$\nu\hat{}(G, m, H) = \max\{|H\hat{}A| \mid A \subseteq G, |A| = m\}$$

for various $H \subseteq \mathbb{N}_0$.

A.3.1 Fixed number of terms

Here we consider

$$\nu\hat{}(G, m, h) = \max\{|h\hat{}A| \mid A \subseteq G, |A| = m\},$$

that is, the maximum size of a restricted h-fold sumset of an m-element subset of G.

Observe first that for all $h > m$, $h\hat{}A = \emptyset$ and thus $\nu\hat{}(G, m, h) = 0$. Furthermore, note that for every set $A = \{a_1, \dots, a_m\} \subseteq G$ of size m and for every $h \in \mathbb{N}_0$, we have

$$(m-h)\hat{}A = (a_1 + \cdots + a_m) - h\hat{}A;$$

in particular,

$$|(m-h)\hat{}A| = |h\hat{}A|$$

and

$$\nu\hat{}(G, m, m-h) = \nu\hat{}(G, m, h).$$

Therefore, we can restrict our attention to the cases when

$$h \leq \left\lfloor \frac{m}{2} \right\rfloor.$$

It is also quite obvious that we have $0\hat{}A = \{0\}$ and $1\hat{}A = A$; consequently, we have the following.

Proposition A.26 *In any abelian group* G *we have*
$\nu\hat{}(G, m, 0) = 1$,
$\nu\hat{}(G, m, 1) = m$,
$\nu\hat{}(G, m, m-1) = m$,
$\nu\hat{}(G, m, m) = 1$,
$\nu\hat{}(G, m, h) = 0$ *for* $h > m$.

For values of $2 \leq h \leq m-2$, we have no exact values for $\nu\hat{}(G,m,h)$ in general; as a consequence of Proposition A.1, we have, however, the following upper bound.

Proposition A.27 *In any abelian group G of order n we have*

$$\nu\hat{}(G,m,h) \leq \min\left\{n, \binom{m}{h}\right\}.$$

We pose the following problems.

Problem A.28 *Find the value of (or good bounds for) $\nu\hat{}(G,m,h)$ for noncyclic groups G and integers m and h.*

Problem A.29 *Find the value of (or good bounds for) $\nu\hat{}(\mathbb{Z}_n,m,h)$ for all positive integers $n,m,$ and h.*

Problem A.30 *For positive integers n and m, find $\nu\hat{}(\mathbb{Z}_n,m,2)$.*

Problem A.31 *Find all (or, at least, infinitely many) positive integers n, m, and h for which*

$$\nu\hat{}(\mathbb{Z}_n,m,h) < \min\left\{n, \binom{m}{h}\right\}.$$

Manandhar (cf. [148]) found (using a computer program) that for all $n \leq 20$, we have

$$\nu\hat{}(\mathbb{Z}_n,m,h) = \min\left\{n, \binom{m}{h}\right\},$$

with the following exceptions:

n	m	h	$\min\left\{n, \binom{m}{h}\right\}$	$\nu\hat{}(\mathbb{Z}_n,m,h)$
10	5	2, 3	10	9
14	6	2, 4	14	13
15	6	2, 4	15	13
16	6	2, 4	15	14
17	6	2, 4	15	14
18	6	2, 4	15	14
18	7	2, 5	18	17
19	7	2, 5	19	18
20	7	2, 5	20	19

It is interesting to note that all these exceptions occur with $h=2$ or $h=m-2$.

A.3.2 Limited number of terms

Here we consider, for a given group G, positive integer m (with $m \leq n = |G|$), and nonnegative integer s,

$$\nu\hat{}(G,m,[0,s]) = \max\{|[0,s]\hat{}A| \mid A \subseteq G, |A| = m\},$$

that is, the maximum size of $[0,s]\hat{}A$ for an m-element subset A of G. We may assume that $s \leq m$, since for $s > m$ we have

$$\nu\hat{}(G,m,[0,s]) = \nu\hat{}(G,m,[0,m]).$$

It is easy to see that, for every m, we have

$$\nu\hat{}(G, m, [0,0]) = 1$$

and

$$\nu\hat{}(G, m, [0,1]) = \min\{n, m+1\}.$$

For values of $2 \leq s \leq m$, we have no exact values for $\nu\hat{}(G, m, [0,s])$ in general; as a consequence of Proposition A.1, we have, however, the following upper bound.

Proposition A.32 *In any abelian group G of order n we have*

$$\nu\hat{}(G, m, [0,s]) \leq \min\left\{n, \sum_{h=0}^{s} \binom{m}{h}\right\}.$$

As in previous section, we have the following problems.

Problem A.33 *For positive integers n, m, and s, find the value of (or, at least, good bounds for) $\nu\hat{}(\mathbb{Z}_n, m, [0,s])$. In particular, find $\nu\hat{}(\mathbb{Z}_n, m, [0,2])$.*

Problem A.34 *Find all (or, at least, infinitely many) positive integers n, m, and s, for which*

$$\nu\hat{}(\mathbb{Z}_n, m, [0,s]) < \min\left\{n, \sum_{h=0}^{s} \binom{m}{h}\right\}.$$

Problem A.35 *Find the value of (or, at least, good bounds for) $\nu\hat{}(G, m, [0,s])$ for noncyclic groups G and integers m and s.*

A.3.3 Arbitrary number of terms

Here we consider, for a given group G and positive integer m (with $m \leq n$),

$$\nu\hat{}(G, m, \mathbb{N}_0) = \max\{|\Sigma A| \mid A \subseteq G, |A| = m\}.$$

Note that

$$\nu\hat{}(G, m, \mathbb{N}_0) = \nu\hat{}(G, m, [0,s])$$

for every $s \geq m$, thus we could think of the problem of finding $\nu\hat{}(G, m, \mathbb{N}_0)$, the maximum size of a restricted sumset, as a special case of finding $\nu\hat{}(G, m, [0,s])$ when $s = m$ (or any $s \geq m$). However, it may be worthwhile to separate this case as it is of special interest.

As a consequence of Proposition A.1, we have the following upper bound.

Proposition A.36 *In any abelian group G of order n we have*

$$\nu\hat{}(G, m, \mathbb{N}_0) \leq \min\{n, 2^m\}.$$

We can easily prove that, in the case of cyclic groups, equality holds in Proposition A.36. We will consider two cases: when $n \geq 2^m$ and when $n < 2^m$.

If $n \geq 2^m$, then let

$$A = \{1, 2, 2^2, \ldots, 2^{m-1}\}.$$

Clearly, A has size m, and one can also see (recalling the base 2 representation of integers) that

$$|\Sigma A| = 2^m = \min\{n, 2^m\}.$$

Suppose now that $n < 2^m$, and let $k = \lceil \log_2 n \rceil$; note that $2^{k-1} < n \leq 2^k$. Therefore, for the set

$$A' = \{1, 2, 2^2, \ldots, 2^{k-1}\},$$

we get $\Sigma A' = \mathbb{Z}_n$; and this implies that for any set A that contains A' we have $\Sigma A = \mathbb{Z}_n$. But $n < 2^m$ implies that $m > \log_2 n$ and thus $m \geq k$, which means that, again we are able to find an m-subset A of $\mathbb{Z}_n$ for which

$$|\Sigma A| = n = \min\{n, 2^m\}.$$

These constructions, together with Proposition A.36, imply the following:

Proposition A.37 *For positive integers n and m with $m \leq n$ we have*

$$\nu\hat{~}(\mathbb{Z}_n, m, \mathbb{N}_0) = \min\{n, 2^m\}.$$

As we have mentioned above, this also implies that

$$\nu\hat{~}(\mathbb{Z}_n, m, [0, s]) = \min\{n, 2^m\}$$

holds for every integer $s \geq m$.

So, the case of cyclic groups has been settled, which leaves us with the following problem.

Problem A.38 *For every positive integer m and noncyclic group G, find the value of $\nu\hat{~}(G, m, \mathbb{N}_0)$.*

A.4 Restricted signed sumsets

Our goal in this section is to investigate the quantity

$$\nu\hat{_\pm}(G, m, H) = \max\{|H\hat{_\pm}A| \mid A \subseteq G, |A| = m\}$$

for various $H \subseteq \mathbb{N}_0$.

A.4.1 Fixed number of terms

Here we consider

$$\nu\hat{_\pm}(G, m, h) = \max\{|h\hat{_\pm}A| \mid A \subseteq G, |A| = m\},$$

that is, the maximum size of a restricted h-fold signed sumset of an m-element subset of G.

Observe first that, for all $h > m$, $h\hat{_\pm}A = \emptyset$ and thus $\nu\hat{_\pm}(G, m, h) = 0$, so only $h \leq m$ needs to be considered, as was the case in Section A.3.1. On the other hand, here we do not have the "palindromic" property of restricted sumsets that $\nu\hat{~}(G, m, m-h) = \nu\hat{~}(G, m, h)$.

As in Section A.2.1, we get:

Proposition A.39 *In any abelian group G of order n and for every $m \leq n$ we have*

$$\begin{aligned}
\nu\hat{_\pm}(G, m, 0) &= 1, \\
\nu\hat{_\pm}(G, m, 1) &= \min\left\{n, 2m, m + \frac{n - |\mathrm{Ord}(G, 2)| - 1}{2}\right\}.
\end{aligned}$$

For values of $2 \leq h \leq m$, we have no exact values for $\nu\hat{_\pm}(G, m, h)$ in general; as a consequence of Proposition A.1, we have, however, the following upper bound.

Proposition A.40 *In any abelian group G of order n we have*

$$\nu_{\hat{\pm}}(G, m, h) \leq \min\left\{n, \binom{m}{h} \cdot 2^h\right\}.$$

Not much else is known about the value of $\nu_{\hat{\pm}}(G, m, h)$ in general. We thus pose the following.

Problem A.41 *Find the value of (or, at least, good bounds for) $\nu_{\hat{\pm}}(G, m, h)$ for noncyclic groups G and integers m and $2 \leq h \leq m$.*

At least for cyclic groups, we can find the exact answer when $h = m$. As conjectured by Olans in [165], we have the following result:

Proposition A.42 *For all positive integers n and $m \leq n$ we have*

$$\nu_{\hat{\pm}}(\mathbb{Z}_n, m, m) = \begin{cases} \min\{n, 2^m\} & \text{if } n \text{ is odd,} \\[2mm] \min\{n/2, 2^m\} & \text{if } n \text{ is even.} \end{cases}$$

The easy proof is on page 323. This result leaves us with the following open questions.

Problem A.43 *Find the value of $\nu_{\hat{\pm}}(\mathbb{Z}_n, m, h)$ for all positive integers n, m, and $2 \leq h \leq m - 1$.*

The following two special cases of Problem A.43 are worth mentioning separately.

Problem A.44 *For positive integers n and m, find $\nu_{\hat{\pm}}(\mathbb{Z}_n, m, 2)$.*

Problem A.45 *Find all (or, at least, infinitely many) positive integers n, m, and $2 \leq h \leq m - 1$, for which*

$$\nu_{\hat{\pm}}(\mathbb{Z}_n, m, h) < \min\left\{n, \binom{m}{h} \cdot 2^h\right\}.$$

A.4.2   Limited number of terms

Here we consider, for a given group G, positive integer m (with $m \leq n = |G|$), and nonnegative integer s,

$$\nu_{\hat{\pm}}(G, m, [0, s]) = \max\{|[0, s]_{\hat{\pm}} A| \mid A \subseteq G, |A| = m\},$$

that is, the maximum size of $[0, s]_{\hat{\pm}} A$ for an m-element subset A of G. We may assume that $s \leq m$, since for $s > m$ we have

$$\nu_{\hat{\pm}}(G, m, [0, s]) = \nu_{\hat{\pm}}(G, m, [0, m]).$$

As in Section A.2.2, we get:

Proposition A.46 *In any abelian group G of order n and for every $m \leq n$ we have*

$$\nu_{\hat{\pm}}(G, m, [0, 0]) = 1,$$
$$\nu_{\hat{\pm}}(G, m, [0, 1]) = \min\left\{n, 2m + 1, m + \frac{n - |\mathrm{Ord}(G, 2)| + 1}{2}\right\}.$$

For other values of s, we have no exact values for $\nu\hat{_\pm}(G,m,[0,s])$ in general; as a consequence of Proposition A.1, we have, however, the following upper bound.

Proposition A.47 *In any abelian group G of order n we have*

$$\nu\hat{_\pm}(G,m,[0,s]) \leq \min\left\{n, \sum_{h=0}^{s} \binom{m}{h} 2^h\right\}.$$

The following problems are largely unsolved.

Problem A.48 *Find the value of (or, at least, good bounds for) $\nu\hat{_\pm}(\mathbb{Z}_n,m,[0,s])$ for all integers n, m, and s; in particular, find $\nu\hat{_\pm}(\mathbb{Z}_n,m,[0,2])$.*

Problem A.49 *Find all (or, at least, infinitely many) positive integers n, m, and s, for which*

$$\nu\hat{_\pm}(\mathbb{Z}_n,m,[0,s]) < \min\left\{n, \sum_{h=0}^{s} \binom{m}{h} 2^h\right\}.$$

Problem A.50 *Find the value of (or, at least, good bounds for) $\nu\hat{_\pm}(G,m,[0,s])$ for noncyclic groups G and integers m and s.*

A.4.3 Arbitrary number of terms

Here we consider, for a given group G and positive integer m (with $m \leq n$),

$$\nu\hat{_\pm}(G,m,\mathbb{N}_0) = \max\{|\Sigma_\pm A| \mid A \subseteq G, |A| = m\}.$$

Note that

$$\nu\hat{_\pm}(G,m,\mathbb{N}_0) = \nu\hat{_\pm}(G,m,[0,s])$$

for every $s \geq m$, thus we could think of the problem of finding $\nu\hat{_\pm}(G,m,\mathbb{N}_0)$, the maximum size of a restricted sumset, as a special case of finding $\nu\hat{_\pm}(G,m,[0,s])$ when $s = m$ (or any $s \geq m$). Nevertheless, we separate this subsection as it is of special interest.

As a consequence of Proposition A.1, we have the following upper bound.

Proposition A.51 *In any abelian group G of order n we have*

$$\nu\hat{_\pm}(G,m,\mathbb{N}_0) \leq \min\{n, 3^m\}.$$

With an argument similar to the one in Section A.3.3, we show that, when G is cyclic, equality holds in Proposition A.51. We will consider two cases: when $n \geq 3^m$ and when $n < 3^m$.

If $n \geq 3^m$, then let

$$A = \{1, 3, 3^2, \ldots, 3^{m-1}\};$$

clearly, A has size m. Recall that every positive integer up to $3^m - 1$ has a (unique) ternary representation of at most m ternary digits. Subtracting

$$1 + 3 + 3^2 + \cdots + 3^{m-1} = \frac{3^m - 1}{2},$$

every integer between $-(3^m - 1)/2$ and $(3^m - 1)/2$, inclusive, can be written (uniquely) as

$$r_0 + r_1 \cdot 3 + \cdots + r_{m-1} \cdot 3^{m-1}$$

with $r_0, r_1, \ldots, r_{m-1} \in \{-1, 0, 1\}$. Therefore,

$$|\Sigma_{\pm} A| = 3^m = \min\{n, 3^m\}.$$

Suppose now that $n < 3^m$, and let $k = \lceil \log_3 n \rceil$; note that $3^{k-1} < n \leq 3^k$. Therefore, for the set

$$A' = \{1, 3, 3^2, \ldots, 3^{k-1}\},$$

we get $\Sigma_{\pm} A' = \mathbb{Z}_n$; and this implies that for any set A that contains A' we have $\Sigma_{\pm} A = \mathbb{Z}_n$. But $n < 3^m$ implies that $m > \log_3 n$ and thus $m \geq k$, which means that, again we are able to find an m-subset A of $\mathbb{Z}_n$ for which

$$|\Sigma_{\pm} A| = n = \min\{n, 3^m\}.$$

These constructions, together with Proposition A.51, imply the following:

Proposition A.52 *For positive integers n and m with $m \leq n$ we have*

$$\nu_{\pm}\hat{}(\mathbb{Z}_n, m, \mathbb{N}_0) = \min\{n, 3^m\}.$$

As we have mentioned above, this also implies that

$$\nu_{\pm}\hat{}(\mathbb{Z}_n, m, [0, s]) = \min\{n, 3^m\}$$

holds for every integer $s \geq m$.

So, the case of cyclic groups has been settled, which leaves us with the following problem.

Problem A.53 *For every positive integer m and noncyclic group G, find the value of $\nu_{\pm}\hat{}(G, m, \mathbb{N}_0)$.*

Chapter B

Spanning sets

Recall that for a given finite abelian group G, m-subset $A = \{a_1, \ldots, a_m\}$ of G, $\Lambda \subseteq \mathbb{Z}$, and $H \subseteq \mathbb{N}_0$, we defined the sumset of A corresponding to Λ and H as

$$H_\Lambda A = \{\lambda_1 a_1 + \cdots + \lambda_m a_m \mid (\lambda_1, \ldots, \lambda_m) \in \Lambda^m(H)\}$$

where the index set $\Lambda^m(H)$ is defined as

$$\Lambda^m(H) = \{(\lambda_1, \ldots, \lambda_m) \in \Lambda^m \mid |\lambda_1| + \cdots + |\lambda_m| \in H\}.$$

The case when the sumset yields the entire group is of special interest; we say that A is an *H-spanning set over* Λ if $H_\Lambda A = G$. In this chapter we attempt to find the minimum possible size of an H-spanning set over Λ in a given finite abelian group G. Namely, our objective is to determine, for any G, $\Lambda \subseteq \mathbb{Z}$, and $H \subseteq \mathbb{N}_0$ the quantity

$$\phi_\Lambda(G, H) = \min\{|A| \mid A \subseteq G, H_\Lambda A = G\}.$$

If no H-spanning set exists, we put $\phi_\Lambda(G, H) = \infty$.

Note that we have a strong connection between the maximum possible size of sumsets, studied in Chapter A, and the minimum size of spanning sets, studied here. In particular, we have the following obvious proposition.

Proposition B.1 *For any group G of size n, $\Lambda \subseteq \mathbb{Z}$, and $H \subseteq \mathbb{N}_0$, we have*

$$\phi_\Lambda(G, H) = \min\{m \mid \nu_\Lambda(G, m, H) = n\}.$$

Therefore, in theory, the question of finding $\phi_\Lambda(G, H)$ is a special case of the more general problem of finding $\nu_\Lambda(G, m, H)$; however, there is enough interest in spanning sets alone to treat them separately.

We have the following obvious bound.

Proposition B.2 *If A is an H-spanning set over Λ in a group G of order n and $|A| = m$, then*

$$n \leq |\Lambda^m(H)|.$$

Proposition B.2 provides a lower bound for the size of H-spanning sets over Λ in G. The case of equality in Proposition B.2 is of special interest: an H-spanning set over Λ of size m in a group G is called *perfect* when $n = |\Lambda^m(H)|$ holds.

In the following subsections we consider $\phi_\Lambda(G, H)$ for special coefficient sets Λ.

B.1 Unrestricted sumsets

Our goal in this section is to investigate the quantity

$$\phi(G,H) = \min\{|A| \mid A \subseteq G, HA = G\};$$

if no H-spanning set exists, we put $\phi(G,H) = \infty$. Note that we have

$$\phi(G,\{0\}) = \infty$$

for all groups of order at least 2, but if H contains at least one positive integer h, then

$$\phi(G,H) \leq n$$

since we have

$$G = \{(h-1)\cdot 0 + 1 \cdot g \mid g \in G\} \subseteq hG \subseteq HG$$

and thus $HG = G$.

B.1.1 Fixed number of terms

In this subsection we ought to consider

$$\phi(G,h) = \min\{|A| \mid A \subseteq G, hA = G\},$$

that is, the minimum value of m for which G contains a set A of size m with $hA = G$.

However, using Propositions B.1 and A.9, we can easily reduce the study of $\phi(G,h)$ to that of

$$\phi(G,[0,h]) = \min\{|A| \mid A \subseteq G, [0,h]A = G\}.$$

Namely, we have the following result.

Proposition B.3 *For any group G and nonnegative integer h we have*

$$\phi(G,h) = \phi(G,[0,h]) + 1.$$

According to Proposition B.3, it suffices to study only one of $\phi(G,h)$ or $\phi(G,[0,h])$. We elect to study the latter—see Subsection B.1.2 below.

B.1.2 Limited number of terms

A subset A of G for which $[0,s]A = G$ for some nonnegative integer s is called an *s-basis* for G. (The term is somewhat confusing as for a set to be a basis, it only needs to be spanning—no independence property is assumed. Nevertheless, we keep this historical terminology.)

Here we investigate

$$\phi(G,[0,s]) = \min\{|A| \mid A \subseteq G, [0,s]A = G\},$$

that is, the minimum size of an s-basis for G.

As we pointed out above, we have

$$\phi(G,[0,0]) = \infty$$

for all groups of order at least 2, but

$$\phi(G,[0,s]) \leq n$$

for every $s \in \mathbb{N}$. Furthermore, for every G of order 2 or more we get

$$\phi(G, [0,1]) = n - 1,$$

since for every $A \subseteq G$ we have $[0,1]A = \{0\} \cup A$. But for $s \geq 2$, we do not have a formula for $\phi(G, [0,s])$. Therefore we pose the following problem.

Problem B.4 *Find $\phi(G, [0,s])$ for every G and $s \geq 2$.*

From Proposition B.2, we have the following general bound.

Proposition B.5 *If A is an s-basis of size m in G, then*

$$n \leq \binom{m+s}{s}.$$

We are particularly interested in the extremal cases of Proposition B.5. Namely, we would like to find *perfect s-bases*, that is, s-bases of size m with

$$n = \binom{m+s}{s}.$$

It is easy to determine all perfect s-bases for $s = 1$ or $m = 1$:

Proposition B.6 *Let G be a finite abelian group, $A \subseteq G$, and $a \in G$.*

1. *$A \dot{\subseteq} G$ is a perfect 1-basis in G if, and only if, $A = G \setminus \{0\}$.*
2. *$\{a\} \subseteq G$ is a perfect s-basis in G if, and only if, $G \cong \mathbb{Z}_{s+1}$ and $\gcd(a, s+1) = 1$.*

We are not aware of any other perfect bases; in fact, we believe that there are none:

Conjecture B.7 *There are no perfect s-bases of size m in G, unless $s = 1$ or $m = 1$.*

The following result says that Conjecture B.7 holds for $s = 2$ and $s = 3$:

Theorem B.8 *For $s \in \{2,3\}$, there are no perfect s-bases in G of size $m \geq 2$.*

We present the proof starting on page 323. Our proof there can probably be generalized, so we offer:

Problem B.9 *Prove Conjecture B.7.*

Let us return to the question of finding $\phi(G, [0,s])$. As a consequence of Proposition B.5, we have the following lower bound for $\phi(G, [0,s])$.

Proposition B.10 *For any abelian group G of order n and positive integer s we have*

$$\phi(G, [0,s]) > \left\lceil \sqrt[s]{s!n} \right\rceil - s.$$

To find some upper bounds for $\phi(G, [0,s])$, we exhibit an explicit s-basis as follows. Consider first the cyclic group $\mathbb{Z}_n$, and let $a = \lceil \sqrt[s]{n} \rceil$. It is then easy to see that the set

$$A = \{i \cdot a^j \mid i = 1, 2, \ldots, a-1; j = 0, 1, \ldots, s-1\}$$

is an s-basis for $\mathbb{Z}_n$ (this follows from the fact that all nonnegative integers up to $a^s - 1$ have a base a representation of at most s digits, and that $a^s - 1 \geq n - 1$). Since $|A| \leq (a-1) \cdot s$, we get the following result.

Proposition B.11 *For positive integers n and s we have*

$$\phi(\mathbb{Z}_n, [0, s]) \leq s \cdot \left(\lceil \sqrt[s]{n} \rceil - 1\right) < s \cdot \sqrt[s]{n}.$$

Observe that if A_1 and A_2 are s-bases for groups G_1 and G_2, respectively, then

$$A = \{(a_1, a_2) \mid a_1 \in A_1, a_2 \in A_2\}$$

is an s-basis for $G_1 \times G_2$; therefore,

$$\phi(G_1 \times G_2, [0, s]) \leq \phi(G_1, [0, s]) \cdot \phi(G_2, [0, s]).$$

So, from Proposition B.11 we get the following corollary:

Proposition B.12 *Let G be an abelian group of rank r and order n, and let s be a positive integer. We then have*

$$\phi(G, [0, s]) < s^r \cdot \sqrt[s]{n}.$$

A different—and in most cases much lower—upper bound was provided by Jia:

Theorem B.13 (Jia; cf. [122]) *Let G be an abelian group of order n, and let s be a positive integer. We then have*

$$\phi(G, [0, s]) < s \cdot \left(1 + \frac{1}{\sqrt[s]{2}}\right)^{s-1} \cdot \sqrt[s]{n}.$$

Let us now examine the case of cyclic groups in more detail. Krasny (see [137]) developed a computer program that yields the following data.

$$\phi(\mathbb{Z}_n, [0, 2]) = \begin{cases} 1 & \text{if } n = 1, 2, 3; \\ 2 & \text{if } n = 4, 5; \\ 3 & \text{if } n = 6, \dots, 9; \\ 4 & \text{if } n = 10, \dots, 13; \\ 5 & \text{if } n = 14, \dots, 17, \text{ and } n = 19; \\ 6 & \text{if } n = 18, 20, 21; \\ 7 & \text{if } n = 22, \dots, 27,\ n = 29, 30; \\ 8 & \text{if } n = 28,\ n = 31, \dots, 35; \end{cases}$$

$$\phi(\mathbb{Z}_n, [0, 3]) = \begin{cases} 1 & \text{if } n = 1, \dots, 4; \\ 2 & \text{if } n = 5, \dots, 8; \\ 3 & \text{if } n = 9, \dots, 16; \\ 4 & \text{if } n = 17, \dots, 25; \\ 5 & \text{if } n = 26, \dots, 40; \end{cases}$$

and

$$\phi(\mathbb{Z}_n, [0, 4]) = \begin{cases} 1 & \text{if } n = 1, \dots, 5; \\ 2 & \text{if } n = 6, \dots, 11; \\ 3 & \text{if } n = 12, \dots, 27; \\ 4 & \text{if } n = 28, \dots, 49. \end{cases}$$

As these values suggest, $\phi(\mathbb{Z}_n, [0, s])$ behaves rather peculiarly; the following problem is wide open.

Problem B.14 *Find $\phi(\mathbb{Z}_n, [0, s])$ for all n and s.*

Observe that the coefficient of $\sqrt[s]{n}$ in the lower bound of Proposition B.10 and the upper bound of Proposition B.11 are $\sqrt[s]{s!}$ and s, respectively; therefore, even the following problems are of much interest.

Problem B.15 *For a given $s \geq 2$, find a constant $c_1 > \sqrt[s]{s!}$ so that*

$$\phi(\mathbb{Z}_n, [0, s]) > c_1 \cdot \sqrt[s]{n}$$

holds for all n (or, perhaps, all but finitely many n), or prove that such a constant cannot exist.

Problem B.16 *For a given $s \geq 2$, find a constant $c_2 < s$ so that*

$$\phi(\mathbb{Z}_n, [0, s]) < c_2 \cdot \sqrt[s]{n}$$

holds for all n (or, perhaps, all but finitely many n), or prove that such a constant cannot exist.

For example, for $s = 2$, Problems B.15 and B.16 ask for constants $c_1 > \sqrt{2}$ and $c_2 < 2$, if they exist, for which

$$c_1 \cdot \sqrt{n} < \phi(\mathbb{Z}_n, [0, 2]) < c_2 \cdot \sqrt{n}$$

holds for all (but finitely many) n. While there has been no progress toward finding such a c_1, there have been several results for c_2; after a series of papers by Fried in [88], Mrose in [160], and Kohonen in [135], the best such result thus far was found by Jia and Shen who in [123] proved that one can take c_2 to be any real number larger than $\sqrt{3}$. As a special case of Problem B.16, we ask for an improvement of this construction:

Problem B.17 *Find a constant $c_2 < \sqrt{3}$ so that*

$$\phi(\mathbb{Z}_n, [0, 2]) < c_2 \cdot \sqrt{n}$$

holds for all (but finitely many) n, or prove that such a constant cannot exist.

It is also interesting to investigate s-bases from the opposite viewpoint: given positive integers s and m, what are the possible groups G for which $\phi(G, [0, s]) = m$?

For $m = 1$ the answer is clear:

Proposition B.18 *Let s be a positive integer and G be an abelian group of order n. Then G contains an s-basis of size 1 if, and only if, G is cyclic and $n \leq s + 1$.*

For $m \geq 2$, the answer is not known, so we pose the following problem.

Problem B.19 *For each $s \geq 2$ and $m \geq 2$, find all groups G for which $\phi(G, [0, s]) = m$.*

As a special case of Problem B.19, we have:

Problem B.20 *For each $s \geq 2$ and $m \geq 2$, find all values of n for which $\phi(\mathbb{Z}_n, [0, s]) = m$.*

For $m = 2$, Maturo (see [153]) used a computer program to provide the following partial answer to Problem B.20:

s	all n for which $\phi(\mathbb{Z}_n, [0, s]) = 2$
2	$4, 5$
3	$5, \ldots, 8$
4	$6, \ldots, 11$
5	$7, \ldots, 16$
6	$8, \ldots, 19, 21$
7	$9, \ldots, 24, 26$
8	$10, \ldots, 31, 33$
9	$11, \ldots, 40$
10	$12, \ldots, 45, 47$
11	$13, \ldots, 52, 54, 55, 56$
12	$14, \ldots, 61, 63, 65$
13	$15, \ldots, 66, 68, \ldots, 72, 74$
14	$16, \ldots, 81, 84, 85$
15	$17, \ldots, 88, 90, 91, 92, 94, 95, 96$
16	$18, \ldots, 98, 100, 101, 102, 105, 107$
17	$19, \ldots, 120$
18	$20, \ldots, 121, 123, 124, 126, 127, 129, 131, 133$
19	$21, \ldots, 136, 138, 139, 140, 143, 144, 146$
20	$22, \ldots, 155, 157, 159, 161$
21	$23, \ldots, 162, 164, \ldots, 172, 174, 175, 176$
22	$24, \ldots, 180, 182, \ldots, 185, 189, 191$
23	$25, \ldots, 196, 198, \ldots, 208$

The data make Problem B.20 seem challenging even for $m = 2$:

Problem B.21 *For each $s \in \mathbb{N}$, find all values of n for which $\phi(\mathbb{Z}_n, [0, s]) = 2$.*

We can formulate two sub-problems of Problem B.20:

Problem B.22 *For each $s \geq 2$ and $m \geq 2$, find the largest integer $f(m, s)$ for which $\phi(\mathbb{Z}_n, [0, s]) \leq m$ holds for $n = f(m, s)$.*

Problem B.23 *For each $s \geq 2$ and $m \geq 2$, find the largest integer $g(m, s)$ for which $\phi(\mathbb{Z}_n, [0, s]) \leq m$ holds for all $n \leq g(m, s)$.*

For example, from the table above we see that $f(2, 16) = 107$, $g(2, 16) = 98$, and $f(2, 17) = g(2, 17) = 120$.

Regarding $f(2, s)$, it is not hard to prove that

$$f(2, s) \geq \left\lfloor \frac{s^2+4s+3}{3} \right\rfloor,$$

in particular, one can verify that the set $A = \{1, s + r\}$, where r is the positive remainder of s when divided by 3, is an s-basis in $\mathbb{Z}_n$ for

$$n = \left\lfloor \frac{s^2+4s+3}{3} \right\rfloor.$$

For example, for $s = 6$, the set $A = \{1, 9\}$ is a 6-basis in $G = \mathbb{Z}_{21}$, since for nonnegative integer coefficients λ_1 and λ_2 satisfying $\lambda_1 + \lambda_2 \leq 6$, the values of

$$\lambda_1 \cdot 1 + \lambda_2 \cdot 9$$

yield all elements of the group:

	$\lambda_1 = 0$	$\lambda_1 = 1$	$\lambda_1 = 2$	$\lambda_1 = 3$	$\lambda_1 = 4$	$\lambda_1 = 5$	$\lambda_1 = 6$
$\lambda_2 = 0$	0	1	2	3	4	5	6
$\lambda_2 = 1$	9	10	11	12	13	14	
$\lambda_2 = 2$	18	19	20	0	1		
$\lambda_2 = 3$	6	7	8	9			
$\lambda_2 = 4$	15	16	17				
$\lambda_2 = 5$	3	4					
$\lambda_2 = 6$	12						

In fact, Morillo, Fiol, and Fàbrega proved that $f(2,s)$ cannot be larger (this was independently re-proved by Hsu and Jia in [119] and conjectured by Maturo in [153]):

Theorem B.24 (Morillo, Fiol, and Fàbrega; cf. [159]) *Let s be a positive integer. The largest possible value $f(2,s)$ of n for which $\phi(\mathbb{Z}_n, [0,s]) \leq 2$ is*

$$f(2,s) = \left\lfloor \frac{s^2+4s+3}{3} \right\rfloor.$$

Exact values for $f(m,s)$ are not known in general when $m \geq 3$, though we have computational data for small s and $m = 3$ and for small m and $s = 2$. Namely, in the paper [119] of Hsu and Jia we see:

s	2	3	4	5	6	7	8	9	10	11	12	13	14	15
$f(3,s)$	9	16	27	40	57	78	111	138	176	217	273	340	395	462

In [106], Haanpää presents:

m	1	2	3	4	5	6	7	8	9	10	11	12
$f(m,2)$	3	5	9	13	19	21	30	35	43	51	63	67

(Some of these values had been determined earlier by Graham and Sloane in [100].)

More ambitiously than Problem B.22, but more modestly than Problem B.19, we may ask the following:

Problem B.25 *For each $s \geq 2$ and $m \geq 2$, find the largest integer $F(m,s)$ so that there is a group G of order $n = F(m,s)$ for which $\phi(G, [0,s]) \leq m$.*

Since obviously $F(m,s) \geq f(m,s)$, we already know from Theorem B.24 that

$$F(2,s) \geq \left\lfloor \frac{s^2+4s+3}{3} \right\rfloor.$$

In [83], Fiol, Yebra, Alegre, and Valero proved that, when $s \not\equiv 1$ mod 3, then we cannot do better. However, when $s \equiv 1$ mod 3, then the group $G = \mathbb{Z}_k \times \mathbb{Z}_{3k}$ has the s-basis $\{(0,1),(1,3k-1)\}$, where $k = (s+2)/3$. Note that this group has order $\frac{s^2+4s+4}{3}$, and if $s \not\equiv 1$ mod 3, then

$$\left\lfloor \frac{s^2+4s+3}{3} \right\rfloor = \left\lfloor \frac{s^2+4s+4}{3} \right\rfloor.$$

In summary, we have:

Theorem B.26 (Fiol; cf. [82]) *Let s be any positive integer. The largest value $F(2,s)$ for which there is a group G of order $n = F(m,s)$ for which $\phi(G,[0,s]) \leq 2$ is*

$$F(2,s) = \left\lfloor \tfrac{s^2+4s+4}{3} \right\rfloor.$$

Exact values for $F(m,s)$ are not known in general when $m \geq 3$, but in [106], Haanpää computed the values of $F(m,2)$ for $m \leq 12$. He found that $F(m,2) = f(m,2)$ for all $m \leq 12$, except as follows:

- $F(8,2) = 36 = f(8,2)+1$, as shown by the 2-basis
$$\{(1,0,1),(0,0,1),(0,0,2),(1,1,0),(1,2,0),(3,0,2),(3,1,0),(3,2,0)\}$$
in $G = \mathbb{Z}_4 \times \mathbb{Z}_3^2$;
- $F(11,2) = 64 = f(11,2)+1$, as shown by the 2-basis
$$\{(0,1),(0,4),(1,0),(1,2),(2,1),(2,2),(2,6),(4,5),(5,0),(5,2),(6,5)\}$$
in $G = \mathbb{Z}_8^2$; and
- $F(12,2) = 72 = f(12,2)+5$, as shown by the 2-basis
$$\{(0,1,1),(0,0,2),(0,2,1),(0,2,4),(0,2,7),(0,3,1),$$
$$(1,0,3),(1,0,8),(1,1,1),(1,2,5),(1,2,6),(1,3,1)\}$$
in $G = \mathbb{Z}_2 \times \mathbb{Z}_4 \times \mathbb{Z}_9$.

Turning now to $g(m,s)$, first note that, obviously, $g(m,s) \leq f(m,s)$. We can find a lower bound for $g(m,s)$ as follows. Observe that the base a representation of nonnegative integers guarantees that the set

$$A = \{a^j \mid j = 0,1,\ldots,m-1\}$$

is an s-basis for $\mathbb{Z}_n$ for all $n \leq a^m$, as long as $s \geq m \cdot (a-1)$. Thus, with

$$a = \left\lfloor \frac{s}{m} \right\rfloor + 1,$$

we get the following.

Proposition B.27 *Let m and s be positive integers. The largest possible value $g(m,s)$ of n for which $\phi(\mathbb{Z}_n,[0,s]) \leq m$ holds for all $n \leq g(m,s)$ satisfies*

$$g(m,s) \geq \left(\left\lfloor \frac{s}{m} \right\rfloor + 1 \right)^m.$$

In fact, for $g(2,s)$ we can do slightly better:

Proposition B.28 *Let s be a positive integer. The largest possible value $g(2,s)$ of n for which $\phi(\mathbb{Z}_n,[0,s]) \leq 2$ holds for all $n \leq g(2,s)$ satisfies*

$$\left\lfloor \tfrac{s^2+6s+5}{4} \right\rfloor \leq g(2,s) \leq \left\lfloor \tfrac{s^2+4s+3}{3} \right\rfloor.$$

The upper bound follows directly from Theorem B.24 above. To prove the lower bound, we can verify that for each n up to this value, the set

$$A = \{1, \lfloor \tfrac{s+3}{2} \rfloor\}$$

is an s-basis in $\mathbb{Z}_n$—the details can be found on page 325. Note that for $s \in \{2,3,4\}$, the values of $\lfloor \frac{s^2+6s+5}{4} \rfloor$ and $\lfloor \frac{s^2+4s+3}{3} \rfloor$ agree, thus, for these values, $g(2,s)$ is determined and Problem B.21 is answered. However, we still have the following:

Problem B.29 *For each $s \geq 2$, find the largest integer $g(2,s)$ for which $\phi(\mathbb{Z}_n,[0,s]) \leq 2$ holds for all $n \leq g(2,s)$.*

B.1.3 Arbitrary number of terms

B.2 Unrestricted signed sumsets

In this section we investigate the quantity

$$\phi_{\pm}(G,H) = \min\{|A| \mid A \subseteq G, H_{\pm}A = G\};$$

if no H-spanning set exists, we put $\phi_{\pm}(G,H) = \infty$. Note that we have

$$\phi_{\pm}(G,\{0\}) = \infty$$

for all groups of order at least 2, but if H contains at least one positive integer h, then

$$\phi_{\pm}(G,H) \leq n$$

since we have

$$G = \{(h-1)\cdot 0 + 1 \cdot g \mid g \in G\} \subseteq h_{\pm}G \subseteq H_{\pm}G,$$

so $H_{\pm}G = G$.

B.2.1 Fixed number of terms

A subset A of G for which $h_{\pm}A = G$ for some nonnegative integer h is called an *exact h-spanning set* for G. Here we investigate

$$\phi_{\pm}(G,h) = \min\{|A| \mid A \subseteq G, h_{\pm}A = G\},$$

that is, the minimum size of an exact h-spanning set for G.

As we pointed out above, we have

$$\phi_{\pm}(G,0) = \infty$$

for all groups of order at least 2, but

$$\phi_{\pm}(G,h) \leq n$$

for every $h \in \mathbb{N}$.

For $h = 1$, it is clear that A is an exact 1-spanning set if, and only if, for each $g \in G$, A contains at least one of g or $-g$; in particular, A must contain 0, every element of order 2, and half of the elements of order more than 2. Therefore, we have:

Proposition B.30 *For any finite abelian group G we have*

$$\phi_{\pm}(G,1) = \frac{n + |\mathrm{Ord}(G,2)| + 1}{2};$$

in particular,

$$\phi_{\pm}(\mathbb{Z}_n,1) = \lfloor (n+2)/2 \rfloor.$$

But for $h \geq 2$, we do not have a formula for $\phi_{\pm}(G,h)$. Therefore we pose the following general problem.

Problem B.31 *Find $\phi_{\pm}(G,h)$ for every G and $h \geq 2$.*

To find a lower bound for $\phi_\pm(G,h)$, note that an m-subset of G may have an h-fold signed sumset of size at most

$$c(h,m) = \sum_{i\geq 0} \binom{m}{i}\binom{h-1}{i-1} 2^i;$$

if the h-fold signed sumset is in fact G, then the elements that are their own inverses are generated twice, since replacing all coefficients in the linear combination by their negatives yields the same element. Therefore, we have the following general bound:

Proposition B.32 *If A is an exact h-spanning set of size m in G, then*

$$n \leq c(h,m) - |\mathrm{Ord}(G,2)| - 1;$$

in particular, for the cyclic group $\mathbb{Z}_n$ we must have

$$n \leq c(h,m) - 1.$$

(Note that $c(h,m)$ is even for all $n,h \in \mathbb{N}$.)

Proposition B.32 provides us with a lower bound for $\phi_\pm(G,h)$. For $h=1$, it yields

$$\phi_\pm(G,1) \geq \frac{n + |\mathrm{Ord}(G,2)| + 1}{2};$$

in fact, by Proposition B.30, we know that equality holds. For $h=2$, we have:

Proposition B.33 *For all abelian groups G of order n we have*

$$\phi_\pm(G,2) \geq \sqrt{\frac{n + |\mathrm{Ord}(G,2)| + 1}{2}};$$

in particular,

$$\phi_\pm(\mathbb{Z}_n,2) \geq \sqrt{\lfloor (n+2)/2 \rfloor}.$$

For $h \geq 3$ this bound is difficult to state explicitly.

To find some upper bounds for $\phi_\pm(G,h)$, we exhibit an explicit exact h-spanning set as follows. We will only consider the cyclic group $\mathbb{Z}_n$ here (for noncyclic groups we may use the method we discussed on page 114). Let $a = \lceil \sqrt[h]{n+1} \rceil$. We can readily verify that the set

$$A = \{0\} \cup \{i \cdot a^j \mid i = 1,2,\ldots, \lfloor a/2 \rfloor; j = 0,1,\ldots,h-1\}$$

is an exact h-spanning set for $\mathbb{Z}_n$: indeed, all integers with absolute value at most $n/2$ are generated:

$$\left\lfloor \frac{a}{2} \right\rfloor \cdot (1 + a + \cdots + a^{h-1}) = \left\lfloor \frac{a}{2} \right\rfloor \cdot \frac{a^h - 1}{a-1} \geq \frac{a-1}{2} \cdot \frac{a^h-1}{a-1} = \frac{a^h-1}{2} \geq \frac{n}{2}.$$

Since

$$|A| = 1 + \lfloor a/2 \rfloor \cdot h = 1 + h \cdot \lfloor \lceil \sqrt[h]{n+1} \rceil /2 \rfloor \leq 1 + h \cdot \lfloor (\sqrt[h]{n}+1)/2 \rfloor \leq 1 + h \cdot (\sqrt[h]{n}+1)/2,$$

we get the following result.

Proposition B.34 *For positive integers n and h we have*

$$\phi_\pm(\mathbb{Z}_n,h) \leq h/2 \cdot \sqrt[h]{n} + (h+2)/2.$$

The following problem is wide open.

Problem B.35 *For positive integers n and $h \geq 2$, find $\phi_{\pm}(\mathbb{Z}_n, h)$.*

For $h = 2$, from Propositions B.33 and B.34 we get

$$\sqrt{\lfloor (n+2)/2 \rfloor} \leq \phi_{\pm}(\mathbb{Z}_n, 2) \leq \sqrt{n} + 2.$$

We offer the following interesting problem:

Problem B.36 *Find constants $c_1 > \frac{1}{\sqrt{2}}$ and $c_2 < 1$ so that for all (but perhaps finitely many) positive integers n we have*

$$c_1 \cdot \sqrt{n} \leq \phi_{\pm}(\mathbb{Z}_n, 2) \leq c_2 \cdot \sqrt{n},$$

or prove that no such constants exist.

It is also interesting to investigate exact h-spanning sets from the opposite viewpoint: given positive integers h and m, what are the possible groups G for which $\phi_{\pm}(G, h) = m$?

The answer for $m = 1$ follows immediately from Proposition B.32: since $c(h, 1) = 2$, the inequality, and thus $\phi_{\pm}(G, h) = 1$, holds if, and only if, $n = 1$. For $m = 2$, Reckner in [177] has the following results:

Proposition B.37 (Reckner; cf. [177]) *Let n and h be positive integers.*

1. *If*
$$2 \leq n \leq 2h + 1,$$
then $\phi_{\pm}(\mathbb{Z}_n, h) = 2$. In particular, the set $\{0, 1\}$ is an exact h-spanning set in $\mathbb{Z}_n$ for all n in the given range.

2. *If n and h are both odd and*
$$2h + 3 \leq n \leq 3h,$$
then $\phi_{\pm}(\mathbb{Z}_n, h) = 2$. In particular, when h is odd, the set $\{1, 3\}$ is an exact h-spanning set in $\mathbb{Z}_n$ for all odd n in the given range.

Proposition B.37 does not classify all n and h with $\phi_{\pm}(\mathbb{Z}_n, h) = 2$, so we offer the following problem:

Problem B.38 *Find all other values of n and h for which $\phi_{\pm}(\mathbb{Z}_n, h) = 2$.*

Note that, by Proposition B.32, it suffices to investigate Problem B.38 when $n \leq 4h - 1$.

Moving on to $m = 3$, we observe that, as an immediate consequence of Proposition B.54 in the next subsection, we have:

Proposition B.39 *Let n and h be positive integers. If*

$$n \leq 2h^2 + 2h + 1,$$

then $\phi_{\pm}(\mathbb{Z}_n, h) \leq 3$. In particular, the set $\{0, h, h+1\}$ is an exact h-spanning set in $\mathbb{Z}_n$ for all $n \leq 2h^2 + 2h + 1$.

Problem B.40 *Find all other values of n and h for which $\phi_{\pm}(\mathbb{Z}_n, h) \leq 3$.*

Note that, by Proposition B.32, it suffices to investigate Problem B.40 when $n \leq 4h^2 + 1$.

Problem B.41 *For each h and $m \geq 4$, find all n for which $\phi_{\pm}(\mathbb{Z}_n, h) = m$.*

And, more generally:

Problem B.42 *For all positive integers h and m, find all groups G for which $\phi_{\pm}(G, h) = m$.*

B.2.2 Limited number of terms

A subset A of G for which $[0,s]_{\pm}A = G$ for some nonnegative integer s is called an *s-spanning set* for G. Here we investigate

$$\phi_{\pm}(G,[0,s]) = \min\{|A| \mid A \subseteq G, [0,s]_{\pm}A = G\},$$

that is, the minimum size of an s-spanning set for G.

As we pointed out above, we have

$$\phi_{\pm}(G,[0,0]) = \infty$$

for all groups of order at least 2, but

$$\phi_{\pm}(G,[0,s]) \leq n$$

for every $s \in \mathbb{N}$.

For $s = 1$, it is clear that A is 1-spanning if, and only if, for each $g \in G$, A contains at least one of g or $-g$; in particular, A must contain every element of order 2 and half of the elements of order more than 2. Therefore, we have:

Proposition B.43 *For any finite abelian group G we have*

$$\phi_{\pm}(G,[0,1]) = \frac{n + |\mathrm{Ord}(G,2)| - 1}{2};$$

in particular,

$$\phi_{\pm}(\mathbb{Z}_n,[0,1]) = \lfloor n/2 \rfloor.$$

But for $s \geq 2$, we do not have a formula for $\phi_{\pm}(G,[0,s])$. Therefore we pose the following general problem.

Problem B.44 *Find $\phi_{\pm}(G,[0,s])$ for every G and $s \geq 2$.*

As a consequence of Proposition B.2, we have the following general bound.

Proposition B.45 *If A is an s-spanning set of size m in G, then*

$$n \leq a(m,s) = \sum_{i \geq 0} \binom{m}{i}\binom{s}{i} 2^i.$$

The classification of the extremal cases of Proposition B.45 is a particularly intriguing question. Namely, we would like to find *perfect s-spanning sets*, that is, s-spanning sets of size m with $n = a(m,s)$. The values of $a(m,s)$ can be tabulated for small values of m and s (see also Section 2.4):

a(m,s)	s=0	s=1	s=2	s=3	s=4	s=5	s=6
m=1	**1**	**3**	**5**	**7**	**9**	**11**	**13**
m=2	1	**5**	**13**	**25**	**41**	**61**	**85**
m=3	1	**7**	25	63	129	231	377
m=4	1	**9**	41	129	321	681	1289
m=5	1	**11**	61	231	681	1683	3653
m=6	1	**13**	85	377	1289	3653	8989

Cases where there exists a group of size $a(m,s)$ with a known perfect s-spanning set are marked with boldface. The following proposition exhibits perfect spanning sets for all known parameters.

Proposition B.46 *Let m be a positive integer and s be a nonnegative integer, and let G be an abelian group of order n.*

1. *If $n = 2m+1$, then $G \setminus \{0\}$ can be partitioned into parts K and $-K$, and both K and $-K$ are perfect 1-spanning sets in G. For example, the set $\{1, 2, \ldots, m\}$ is a perfect 1-spanning set in $\mathbb{Z}_n$.*
2. *If $n = 2s+1$ and $\gcd(a, n) = 1$, then the set $\{a\}$ is a perfect s-spanning set in $\mathbb{Z}_n$.*
3. *If $n = 2s^2+2s+1$, then the sets $\{1, 2s+1\}$ and $\{s, s+1\}$ are perfect s-spanning sets in $\mathbb{Z}_n$.*

The first two statements are obvious. The fact that the set $\{1, 2s+1\}$ is a perfect s-spanning set in $\mathbb{Z}_n$ is provided on page 325, and the same claim for $\{s, s+1\}$ follows from Proposition B.54 below; note, however, that $\{s, s+1\}$ in Proposition B.54 cannot be replaced by $\{1, 2s+1\}$. (Both claims were demonstrated for $s = 3$ and $n = 25$ in the Appetizer section "In pursuit of perfection") While these two perfect spanning sets of size 2 are known, there has not been a characterization of any others.

Problem B.47 *Find all perfect s-spanning sets of size 2 in the cyclic group $\mathbb{Z}_{2s^2+2s+1}$.*

We could not find perfect spanning sets for $s \geq 2$ and $m \geq 3$ for any n, and neither could we find any noncyclic groups with perfect spanning sets for $s \geq 2$ and $m = 2$. In particular, we definitely know that, other than the ones already mentioned, no perfect spanning sets exist in groups of order up to 100: Laza (cf. [138]) has verified (using a computer program) that no perfect 2-spanning sets exist in $\mathbb{Z}_{25}$, $\mathbb{Z}_{41}$, $\mathbb{Z}_{61}$, or $\mathbb{Z}_{85}$ (note that $a(m, 2) = 25, 41, 61,$ and 85 for $m = 3, 4, 5,$ and 6, respectively) and that no perfect 3-spanning set exists in $\mathbb{Z}_{63}$ (we have $a(3, 3) = 63$); on page 67, we presented simple arguments to prove that there is neither a perfect 3-spanning set of size two, nor a perfect 2-spanning set of size three in $\mathbb{Z}_5^2$; and Jankowski in [121] proved that $\mathbb{Z}_3 \times \mathbb{Z}_{21}$ cannot contain a perfect 3-spanning set of size 3 either.

It might be an interesting problem to find and classify all perfect spanning sets.

Problem B.48 *Find perfect s-spanning sets in $G = \mathbb{Z}_n$ of size m for some values $s \geq 2$ and $m \geq 3$, or prove that such perfect spanning sets do not exist.*

Problem B.49 *Find perfect s-spanning sets in noncyclic groups of size m for some values $s \geq 2$ and $m \geq 2$, or prove that such perfect spanning sets do not exist.*

Let us return to the question of finding $\phi_{\pm}(G, [0, s])$. Proposition B.45 provides us with a lower bound for $\phi_{\pm}(G, [0, s])$, though for $s \geq 3$ this bound is difficult to state explicitly. For $s = 2$ we get:

Proposition B.50 *For all abelian groups G of order n we have*

$$\phi_{\pm}(G, [0, 2]) \geq \frac{\sqrt{2n-1}-1}{2}.$$

As in the subsection B.2.1, we can find some upper bounds for $\phi_{\pm}(\mathbb{Z}_n, [0, s])$ by letting $a = \lceil \sqrt[s]{n+1} \rceil$ and verifying that the set

$$A = \{i \cdot a^j \mid i = 1, 2, \ldots, \left\lfloor \frac{a}{2} \right\rfloor ; j = 0, 1, \ldots, s-1\}$$

is an s-spanning set for $\mathbb{Z}_n$. Since

$$|A| = s \cdot \lfloor a/2 \rfloor = s \cdot \lfloor \lceil \sqrt[s]{n+1} \rceil /2 \rfloor \leq s \cdot \lfloor (\sqrt[s]{n}+1)/2 \rfloor \leq s \cdot (\sqrt[s]{n}+1)/2,$$

we get the following result.

Proposition B.51 *For positive integers n and s we have*

$$\phi_\pm(\mathbb{Z}_n,[0,s]) \leq s/2 \cdot \sqrt[s]{n} + s/2.$$

The computational data of Laza (see [138]) shows that

$$\phi_\pm(\mathbb{Z}_n,[0,2]) = \begin{cases} 1 & \text{if } n = 1,2,3,4,\mathbf{5}; \\ 2 & \text{if } n = 6,7,\ldots,12,\mathbf{13}; \\ 3 & \text{if } n = 14,15,\ldots,21; \\ 4 & \text{if } n = 22,23,\ldots,33, \text{ and } n = 35; \\ 5 & \text{if } n = 34,\ n = 36,37,\ldots,49, \text{ and } n = 51; \end{cases}$$

and

$$\phi_\pm(\mathbb{Z}_n,[0,3]) = \begin{cases} 1 & \text{if } n = 1,2,\ldots,6,\mathbf{7}; \\ 2 & \text{if } n = 8,9\ldots,24,\mathbf{25}; \\ 3 & \text{if } n = 26,27,\ldots,50,\ n = 52, \text{ and } n = 55; \\ 4 & \text{if } n = 51,53,54,\ n = 56,57,\ldots,100, \text{ and } n = 104. \end{cases}$$

(Values marked in boldface indicate perfect spanning sets, as explained above.) The following problem is wide open.

Problem B.52 *For positive integers n and $s \geq 2$, find $\phi_\pm(\mathbb{Z}_n,[0,s])$.*

For $s=2$, from Propositions B.50 and B.51 we get

$$\frac{\sqrt{2n-1}-1}{2} \leq \phi_\pm(\mathbb{Z}_n,[0,2]) \leq \sqrt{n}+1.$$

We offer the following interesting problem:

Problem B.53 *Find constants $c_1 > \frac{1}{\sqrt{2}}$ and $c_2 < 1$ so that for all (but perhaps finitely many) positive integers n we have*

$$c_1 \cdot \sqrt{n} \leq \phi_\pm(\mathbb{Z}_n,[0,2]) \leq c_2 \cdot \sqrt{n},$$

or prove that no such constants exist.

It is also interesting to investigate s-spanning sets from the opposite viewpoint: given positive integers s and m, what are the possible groups G for which $\phi_\pm(G,[0,s]) = m$? Here is what we know:

Proposition B.54 (Bajnok; cf. [11]) *Suppose, as usual, that G is an abelian group of order n. Let $s \geq 1$ be an integer.*

1. We have $\phi_\pm(G,[0,s]) = 1$ if, and only if, G is cyclic and

$$1 \leq n \leq 2s+1.$$

In particular, the set $\{1\}$ is s-spanning in $\mathbb{Z}_n$ for every $n \leq 2s+1$.

2. We have $\phi_\pm(\mathbb{Z}_n,[0,s]) = 2$ if, and only if,

$$2s+2 \leq n \leq 2s^2+2s+1.$$

In particular, the set $\{s, s+1\}$ is s-spanning in $\mathbb{Z}_n$ for every $n \leq 2s^2+2s+1$.

The case of $m=1$ in Proposition B.54 is clear: for any given $s \in \mathbb{N}$, the only groups that contain an s-spanning set of size 1 are the cyclic groups $\mathbb{Z}_n$, and we have $\phi(\mathbb{Z}_n,[0,s]) = 1$ if, and only if, $n \leq 2s+1$ (in which case the set $\{1\}$, for example, is s-spanning in $\mathbb{Z}_n$). For

$m = 2$, the "only if" part follows from the fact that for $n \leq 2s+1$ we have $\phi_\pm(\mathbb{Z}_n, [0,s]) = 1$ (lower bound) and from Proposition B.45 (upper bound). For the rest, see page 327.

Note that the second part of Theorem B.54 treats only cyclic groups. While it is clear from Proposition 3.1 that the only noncyclic groups G possessing s-spanning sets of size two must have rank two, we do not have a characterization of those for which $\phi_\pm(G, [0,s]) = 2$. We have the following obvious conditions:

Proposition B.55 *Let $G = \mathbb{Z}_{n_1} \times \mathbb{Z}_{n_2}$ be of rank two.*

1. *A necessary condition for $\phi_\pm(G, [0,s]) = 2$ is that*
$$n_1 \cdot n_2 \leq 2s^2 + 2s + 1.$$

2. *A sufficient condition for $\phi_\pm(G, [0,s]) = 2$ is that*
$$\lfloor n_1/2 \rfloor + \lfloor n_2/2 \rfloor \leq s.$$

The first claim is due to Proposition B.45, and the second claim follows from the fact that if the stated condition holds, then (for example) the set $\{(0,1),(1,0)\}$ is s-spanning in G.

While the two conditions in Proposition B.55 are generally far away from each other, we can see what conclusions we have for small values of s. For $s = 1$, by the necessary condition, the only possible group with a 1-spanning set of size two is $\mathbb{Z}_2^2$, but (e.g. by Proposition B.43) $\phi_\pm(\mathbb{Z}_2^2, [0,1]) = 3$. For $s = 2$, the only possible groups with a 2-spanning set of size two are $\mathbb{Z}_2^2$, $\mathbb{Z}_2 \times \mathbb{Z}_4$, $\mathbb{Z}_3^2$, and $\mathbb{Z}_2 \times \mathbb{Z}_6$; of these, the first three have one (e.g. the set $\{0,1),(1,1)\}$ works in each), but, as we can easily check, the fourth one does not. For $s = 3$, the necessary condition of Proposition B.55 yields six other possibilities besides the ones mentioned, which we can then be checked individually. We summarize our findings as follows:

Corollary B.56
1. *There is no group G of rank two with $\phi_\pm(G, [0,1]) = 2$.*
2. *There are exactly three groups G of rank two with $\phi_\pm(G, [0,2]) = 2$; namely, $\mathbb{Z}_2^2$, $\mathbb{Z}_3^2$, and $\mathbb{Z}_2 \times \mathbb{Z}_4$.*
3. *There are exactly six groups G of rank two with $\phi_\pm(G, [0,3]) = 2$; namely, $\mathbb{Z}_2 \times \mathbb{Z}_{2k}$ for $k \in \{1,2,3,4\}$, and $\mathbb{Z}_3 \times \mathbb{Z}_{3k}$ for $k \in \{1,2\}$.*

For groups of rank two whose first component is of size two, we have the following more general result:

Proposition B.57 *Suppose that k and s are positive integers satisfying one of the following conditions:*

- *$s = 2$ and $k \in \{1,2\}$,*
- *$s = 3$ and $k \in \{1,2,3,4\}$,*
- *$s = 4$ and $k \in \{1,2,3,4,5,6,8\}$, or*
- *$s \geq 5$ and $k \leq 3s - 4$.*

Then $\phi_\pm(\mathbb{Z}_2 \times \mathbb{Z}_{2k}, [0,s]) = 2$.

The fact that $k = 7$ is not listed for $s = 4$ is no typo: $\phi_\pm(\mathbb{Z}_2 \times \mathbb{Z}_{14}, [0,4]) = 3$. The proof of Proposition B.57 starts on page 327.

We also believe that the converse of Proposition B.57 holds as well:

Conjecture B.58 *All values of s and k for which $\phi_\pm(\mathbb{Z}_2 \times \mathbb{Z}_{2k}, [0,s]) = 2$ are listed in Proposition B.57.*

We pose the following interesting problems.

Problem B.59 *Prove or disprove Conjecture B.58.*

Problem B.60 *For each $s \geq 2$, find all values of k and l for which $\phi_\pm(\mathbb{Z}_3 \times \mathbb{Z}_{3k}, [0,s]) = 2$ and $\phi_\pm(\mathbb{Z}_4 \times \mathbb{Z}_{4l}, [0,s]) = 2$.*

Problem B.61 *For each $s \geq 2$, find all values of k for which $\phi_\pm(\mathbb{Z}_k^2, [0,s]) = 2$.*

Problem B.62 *For each $s \geq 2$, find all groups G of rank two for which $\phi_\pm(G, [0,s]) = 2$.*

Problem B.63 *For each $s \geq 2$ and $m \geq 3$, find all values of n for which $\phi_\pm(\mathbb{Z}_n, [0,s]) = m$.*

Problem B.64 *For each $s \geq 2$ and $m \geq 3$, find all groups G for which $\phi_\pm(G, [0,s]) = m$.*

We can formulate two sub-problems of Problem B.63:

Problem B.65 *For all positive integers s and m, find the largest integer $f_\pm(m,s)$ for which $\phi_\pm(\mathbb{Z}_n, [0,s]) \leq m$ holds for $n = f_\pm(m,s)$.*

Problem B.66 *For all positive integers s and m, find the largest integer $g_\pm(m,s)$ for which $\phi_\pm(\mathbb{Z}_n, [0,s]) \leq m$ holds for all $n \leq g_\pm(m,s)$.*

Recall that Proposition B.45 provides the following upper bound:

$$g_\pm(m,s) \leq f_\pm(m,s) \leq a(m,s) = \sum_{i \geq 0} \binom{m}{i}\binom{s}{i} 2^i.$$

According to Proposition B.54, we have

$$g_\pm(1,s) = f_\pm(1,s) = 2s+1 = a(1,s),$$

and

$$g_\pm(2,s) = f_\pm(2,s) = 2s^2+2s+1 = a(2,s).$$

We can find a general lower bound for $g_\pm(m,s)$, and thus for $f_\pm(m,s)$, using an analogous argument to the one on page 118; we get the following.

Proposition B.67 *Let m and s be a positive integers. The largest possible value $g_\pm(m,s)$ of n for which $\phi_\pm(\mathbb{Z}_n, [0,s]) = m$ holds for all $n \leq g_\pm(m,s)$ satisfies*

$$g_\pm(m,s) \geq \left(2\left\lfloor \frac{s}{m} \right\rfloor + 1\right)^m.$$

For example, for $m = 3$ we get

$$\frac{8}{27}s^3 \sim \left(2\left\lfloor \frac{s}{3} \right\rfloor + 1\right)^3 \leq g_\pm(3,s) \leq f_\pm(3,s) \leq a(3,s) \sim \frac{4}{3}s^3.$$

Problem B.68 *Find (if possible) better bounds for $g_\pm(3,s)$ and $f_\pm(3,s)$ than those above.*

We mention that, as an attempt to find a lower bound for $g_\pm(3,s)$, Doskov, Pokhrel, and Singh in [69] conjectured that the set

$$A(s) = \left\{\left\lceil \frac{s^2}{2} \right\rceil - s + 1, \ \frac{s^2-s}{2} + 2, \ \frac{s^2-s}{2} + 3\right\}$$

is s-spanning in $\mathbb{Z}_n$ for all n up to $s^3 - s^2 + 6s + 1$ and thus $g_\pm(3,s) \geq s^3 - s^2 + 6s + 1$; while this indeed holds for $s \leq 7$, $A(8)$ only generates 493 elements of $\mathbb{Z}_{497}$.

B.2.3 Arbitrary number of terms

B.3 Restricted sumsets

B.3.1 Fixed number of terms

B.3.2 Limited number of terms

A subset A of G for which $[0,s]\hat{}A = G$ for some nonnegative integer s is called a *restricted s-basis* for G. Here we investigate

$$\phi\hat{}(G,[0,s]) = \min\{|A| \mid A \subseteq G, [0,s]\hat{}A = G\},$$

that is, the minimum size of a restricted s-basis for G.

Clearly, we have

$$\phi\hat{}(G,[0,0]) = \infty$$

for all groups of order at least 2, but, since $1\hat{}G = G$,

$$\phi\hat{}(G,[0,s]) \leq n$$

for every $s \in \mathbb{N}$. Furthermore, for every G of order 2 or more we get

$$\phi\hat{}(G,[0,1]) = n-1,$$

since for every $A \subseteq G$ we have $[0,1]\hat{}A = \{0\} \cup A$. But for $s \geq 2$, we do not have a formula for $\phi\hat{}(G,[0,s])$. Therefore we pose the following problem.

Problem B.69 *Find $\phi\hat{}(G,[0,s])$ for every G and $s \geq 2$.*

From Proposition B.2, we have the following general bound.

Proposition B.70 *If A is an s-basis of size m in G, then*

$$n \leq \sum_{h=0}^{s} \binom{m}{h}.$$

We are particularly interested in the extremal cases of Proposition B.70. Namely, we would like to find *perfect restricted s-bases*, that is, restricted s-bases of size m with

$$n = \sum_{h=0}^{s} \binom{m}{h}.$$

It is easy to determine all perfect restricted 1-bases:

Proposition B.71 *A subset A of a finite abelian group is a perfect restricted 1-basis in G if, and only if, $A = G \setminus \{0\}$.*

Let us pursue a search for perfect restricted 2-bases. If A is an m-subset of G that is a perfect restricted 2-basis, then we must have

$$n = \frac{m^2+m+2}{2}.$$

Thus, for $m=1$ we have $n=2$ and thus $G \cong \mathbb{Z}_2$; $\{1\}$ is indeed a perfect restricted 2-basis in $\mathbb{Z}_2$. For $m=2$ we have $n=4$ and thus $G \cong \mathbb{Z}_4$ or $G \cong \mathbb{Z}_2^2$; $\{1,2\}$ is a perfect restricted

2-basis in $\mathbb{Z}_4$, and $\{(0,1),(1,0)\}$ is a perfect restricted 2-basis in $\mathbb{Z}_2^2$. Moving on to $m=3$, we have $n=7$ and thus $G \cong \mathbb{Z}_7$; $\{1,2,4\}$ (for example) is a perfect restricted 2-basis in $\mathbb{Z}_7$. However, as Krasny in [137] verified, there are no perfect restricted 2-bases of size 4, 5, 6, 7, or 8 in the relevant cyclic groups. In fact, we are not aware of any other perfect restricted s-bases.

Problem B.72 *Find restricted perfect s-bases for $s \geq 3$ or for $s=2$ and $m \geq 4$, or prove that they do not exist.*

Let us return to the question of finding $\phi\hat{}(G,[0,s])$. As a consequence of Proposition B.70, we have a lower bound for $\phi\hat{}(G,[0,s])$, although this bound is difficult to exhibit exactly. For $s=2$, though, we have the following.

Proposition B.73 *For any abelian group G of order n we have*

$$\phi\hat{}(G,[0,2]) \geq \left\lceil \frac{\sqrt{8n-7}-1}{2} \right\rceil.$$

Regarding the upper bound, note that our construction for Propositions B.11 and B.12 used distinct terms, thus we again have:

Proposition B.74 *For positive integers n and s we have*

$$\phi\hat{}(\mathbb{Z}_n,[0,s]) \leq s \cdot \left(\lceil \sqrt[s]{n} \rceil - 1\right) < s \cdot \sqrt[s]{n}.$$

Proposition B.75 *Let G be an abelian group of rank r and order n, and let s be a positive integer. We then have*

$$\phi\hat{}(G,[0,s]) < s^r \cdot \sqrt[s]{n}.$$

Krasny (see [137]) developed a computer program that yields the following data.

$$\phi\hat{}(\mathbb{Z}_n,[0,2]) = \begin{cases} 1 & \text{if } n=1,2; \\ 2 & \text{if } n=3,4; \\ 3 & \text{if } n=5,6,7; \\ 4 & \text{if } n=8,9,10; \\ 5 & \text{if } n=11,12,13,14; \\ 6 & \text{if } n=15,16,17,18,19,20; \\ 7 & \text{if } n=21,22,23,24; \\ 8 & \text{if } n=25,26,27,28,29,30; \\ 9 & \text{if } n=31,32,33,34,35,36,37; \end{cases}$$

$$\phi\hat{}(\mathbb{Z}_n,[0,3]) = \begin{cases} 1 & \text{if } n=1,2; \\ 2 & \text{if } n=3,4; \\ 3 & \text{if } n=5,6,7,8; \\ 4 & \text{if } n=9,\dots,15; \\ 5 & \text{if } n=16,\dots,24; \\ 6 & \text{if } n=25,\dots,35; \\ 7 & \text{if } n=36,\dots,50; \end{cases}$$

and

$$\phi\hat{}(\mathbb{Z}_n,[0,4]) = \begin{cases} 1 & \text{if } n=1,2; \\ 2 & \text{if } n=3,4; \\ 3 & \text{if } n=5,6,7,8; \\ 4 & \text{if } n=9,\dots,16; \\ 5 & \text{if } n=17,\dots,31; \\ 6 & \text{if } n=32,\dots,52. \end{cases}$$

As these values suggest, $\phi\hat{}(\mathbb{Z}_n,[0,s])$ behaves rather peculiarly; the following problem is wide open.

Problem B.76 *Find* $\phi\hat{}(\mathbb{Z}_n, [0, s])$ *for all* n *and* s.

In particular, it is worth considering the following special cases:

Problem B.77 *Find* $\phi\hat{}(\mathbb{Z}_n, [0, 2])$ *for all* n.

Problem B.78 *For all positive integers* s *and* m, *find the largest integer* $f\hat{}(m, s)$ *for which* $\phi\hat{}(\mathbb{Z}_n, [0, s]) \leq m$ *holds for* $n = f\hat{}(m, s)$.

Problem B.79 *For all positive integers* s *and* m, *find the largest integer* $g\hat{}(m, s)$ *for which* $\phi\hat{}(\mathbb{Z}_n, [0, s]) \leq m$ *holds for all* $n \leq g\hat{}(m, s)$.

B.3.3 Arbitrary number of terms

B.4 Restricted signed sumsets

B.4.1 Fixed number of terms

B.4.2 Limited number of terms

B.4.3 Arbitrary number of terms

Chapter C

Sidon sets

Recall that for a given finite abelian group G, m-subset $A = \{a_1, \ldots, a_m\}$ of G, $\Lambda \subseteq \mathbb{Z}$, and $H \subseteq \mathbb{N}_0$, we defined the sumset of A corresponding to Λ and H as

$$H_\Lambda A = \{\lambda_1 a_1 + \cdots + \lambda_m a_m \mid (\lambda_1, \ldots, \lambda_m) \in \Lambda^m(H)\},$$

where the index set $\Lambda^m(H)$ is defined as

$$\Lambda^m(H) = \{(\lambda_1, \ldots, \lambda_m) \in \Lambda^m \mid |\lambda_1| + \cdots + |\lambda_m| \in H\}.$$

The cases where the elements of the sumset corresponding to distinct elements of the index set are themselves distinct have generated much interest: we say that a subset A of size m is a *Sidon set over* Λ in G if

$$|H_\Lambda A| = |\Lambda^m(H)|.$$

Sidon sets are named after the Hungarian mathematician Simon Sidon who introduced them in the 1930s to study Fourier series. (Traditionally, only the special case of $\Lambda = \mathbb{N}_0$ and $H = \{2\}$ is being referred to as a Sidon set.)

In this chapter we attempt to find the maximum possible size of a Sidon set over Λ in a given finite abelian group G. Namely, our objective is to determine, for any G, $\Lambda \subseteq \mathbb{Z}$, and $H \subseteq \mathbb{N}_0$, the quantity

$$\sigma_\Lambda(G, H) = \max\{|A| \mid A \subseteq G, |H_\Lambda A| = |\Lambda^{|A|}(H)|\}.$$

If no Sidon set exists, we put $\sigma_\Lambda(G, H) = 0$.

There are strong connections between the maximum possible size of Sidon sets and several other quantities studied in this book—we will point out these connections throughout.

We have the following obvious bound.

Proposition C.1 *If A is a Sidon set over Λ in a group G of order n and $|A| = m$, then*

$$n \geq |\Lambda^m(H)|.$$

Proposition C.1 provides an upper bound for the size of Sidon sets over Λ in G. The case of equality in Proposition C.1 is of special interest: a Sidon set for which equality holds coincides with perfect spanning sets (see Chapter B).

In the following sections we consider $\sigma_\Lambda(G, H)$ for special coefficient sets Λ.

C.1 Unrestricted sumsets

We first study, for any G and $H \subseteq \mathbb{N}_0$, the quantity

$$\sigma(G,H) = \max\{|A| \mid A \subseteq G, |HA| = |\mathbb{N}_0^{|A|}(H)|\}.$$

If no such set exists, we put $\sigma(G,H) = 0$; for example, we clearly have $\sigma(G,H) = 0$ whenever H contains two elements h_1 and h_2 whose difference is divisible by κ (the exponent of G). Conversely, if all elements of H leave a different remainder when divided by κ, then $\sigma(G,H) \geq 1$, since for any element a of G with order κ, at least the one-element set $\{a\}$ will be a Sidon set for H over $\mathbb{N}_0$.

Proposition C.2 *We have $\sigma(G,H) \geq 1$ if, and only if, the elements of H are pairwise incongruent mod κ. In particular, if $|H| > \kappa$, then $\sigma(G,H) = 0$.*

C.1.1 Fixed number of terms

In this section we investigate, for a given group G and positive integer h so-called *B_h-sets*; that is, sets A whose h-term sums are distinct up to the rearrangement of terms. Thus, A is a B_h-set of size m if, and only if, the equality

$$|hA| = \binom{m+h-1}{h}$$

holds.

We are interested in finding the maximum size of a B_h-set in G—we denote this quantity by $\sigma(G,h)$.

Clearly, every subset of G is a B_1 set, so

$$\sigma(G,1) = n.$$

We can also see that, if A is a B_h set for some positive integer h, then it is also a B_k set for every positive integer $k \leq h$; therefore, $\sigma(G,h)$ is a monotone nonincreasing function of h. When $h \geq \kappa$ in a group of exponent κ, then the only B_h sets in the group are its 1-subsets: Indeed, if a_1 and a_2 are two distinct elements of G, then

$$\kappa a_1 + (h-\kappa)a_2 = ha_2,$$

so no B_h set in the group can contain more than one element. Therefore, by also noting Proposition C.2, we get:

Proposition C.3 *If G is an abelian group with exponent κ and $h \geq \kappa$, then $\sigma(G,h) = 1$.*

According to Proposition C.3, we may limit our investigation to $2 \leq h \leq \kappa - 1$. The case $h = 2$ is worth special mentioning; traditionally, a B_2-set is called a *Sidon set.*

As a consequence of Proposition C.1, we have the following bound.

Proposition C.4 *If A is a B_h set of size m in G, then*

$$n \geq \binom{m+h-1}{h}.$$

From Proposition C.4 we get an upper bound for $\sigma(G,h)$:

Corollary C.5 *The maximum size of a B_h set in an abelian group G of order n satisfies*

$$\sigma(G,h) \leq \left\lfloor \sqrt[h]{h!n} \right\rfloor.$$

With a bit more work, Bravo, Ruiz, and Trujillo improved this as follows:

Theorem C.6 (Bravo, Ruiz, and Trujillo; cf. [43]) *The maximum size of a B_h set in an abelian group G of order n satisfies*

$$\sigma(G,h) \leq \left\lfloor \sqrt[h]{\lfloor h/2 \rfloor ! \lceil h/2 \rceil ! n} \right\rfloor + h - 1.$$

Note that Theorem C.6 is indeed an improvement of Corollary C.5; for example, we get $\sigma(G,2) \leq \lfloor \sqrt{2n} \rfloor$ from Corollary C.5, but $\sigma(G,2) \leq \lfloor \sqrt{n} \rfloor + 1$ from Theorem C.6.

Below we demonstrate these results for $h = 2$ and $h = 3$; in fact, we slightly improve the bound. Note that, if $A = \{a_1, a_2, \dots, a_m\}$ is a Sidon set in G, then the set

$$\{a_i - a_j \mid 1 \leq i \leq m, 1 \leq j \leq m, i \neq j\}$$

must have exactly $m(m-1)$ elements, since all the differences listed above must be distinct. Furthermore, none of these elements are 0. Therefore, $m(m-1) + 1 \leq n$, which yields the following:

Proposition C.7 *For every abelian group G of order n we have*

$$\sigma(G,2) \leq \left\lfloor \frac{\sqrt{4n-3}+1}{2} \right\rfloor.$$

Similarly, one can show, as we do in the proof of Theorem B.8 on page 323, that if A is a B_3 set, then $2A - A$ has cardinality

$$|2A - A| = m(m-1)(m-2)/2 + m(m-1) + m,$$

$A - A$ has cardinality

$$|A - A| = m(m-1) + 1,$$

and that these two sets are disjoint. Therefore,

$$m(m-1)(m-2)/2 + 2m(m-1) + m + 1 \leq n,$$

or

$$m^3 + m^2 \leq 2n - 2,$$

and thus certainly

$$m^3 \leq 2n - 2.$$

This implies:

Theorem C.8 *For every abelian group G of order n we have*

$$\sigma(G,3) \leq \left\lfloor \sqrt[3]{2n-2} \right\rfloor.$$

Let us now turn to constructions of B_h sets. We start with cyclic groups, for which we present three famous results that all guarantee Sidon sets of maximum possible cardinality.

Theorem C.9 (Singer; cf. [187]) *Suppose that q is a prime power and that $n = q^2+q+1$. We then have*

$$\sigma(\mathbb{Z}_n, 2) = q + 1.$$

Theorem C.10 (Bose; cf. [41]) *Suppose that q is a prime power and that $n = q^2 - 1$. We then have*

$$\sigma(\mathbb{Z}_n, 2) = q.$$

Theorem C.11 (Ruzsa; cf. [181]) *Suppose that p is a prime and that $n = p^2 - p$. We then have*

$$\sigma(\mathbb{Z}_n, 2) = p - 1.$$

We can verify that in each of these three results,

$$\sigma(\mathbb{Z}_n, 2) = \left\lfloor \frac{\sqrt{4n-3}+1}{2} \right\rfloor,$$

and thus are best possible. To illuminate this further, we can compute that for positive integers m and n,

$$\left\lfloor \frac{\sqrt{4n-3}+1}{2} \right\rfloor = m$$

holds if, and only if,

$$m^2 - m + 1 \leq n \leq m^2 + m.$$

When $m = q + 1$ for a prime power q, then

$$m^2 - m + 1 = q^2 + q + 1,$$

so Theorem C.9 treats the case of smallest possible n, while if $m = p - 1$ for a prime p, then

$$m^2 + m = p^2 - p,$$

so Theorem C.11 addresses the case of largest possible n. (Theorem C.10 handles an intermediate case.) In particular, note that there is no "waste" in Singer's Theorem C.9, since for his Sidon set A of size $q + 1$, the $(q + 1)q + 1$ elements of $A - A$ cover all of $\mathbb{Z}_n$. Such a set is called a *perfect difference set.*

Other cases of equality in Proposition C.7 also deserve special interest.

Problem C.12 *Find all values of n for which*

$$\sigma(\mathbb{Z}_n, 2) = \left\lfloor \frac{\sqrt{4n-3}+1}{2} \right\rfloor;$$

that is, find all values of m and n with

$$m^2 - m + 1 \leq n \leq m^2 + m,$$

for which a Sidon set of size m exists in $\mathbb{Z}_n$.

For small values of m, we can provide the answer to Problem C.12 as follows. In the table below, for each $m \in \{1, 2, 3, 4, 5, 6\}$, we list the values of n that are in the relevant range, then those values of n for which a Sidon set of size m exists in $\mathbb{Z}_n$ with an example of such a set, and we also list the values of n for which such a Sidon set does not exist.

m	n	yes	set	no
1	1,2	1,2	$\{0\}$	–
2	3,...,6	3,...,6	$\{0,1\}$	–
3	7, ..., 12	7, ..., 12	$\{0,1,3\}$	–
4	13, ..., 20	13, ..., 20	$\{0,1,4,6\}$	–
5	21, ...,30	21; 23, ..., 30	$\{0,2,7,8,11\}$	22
6	31, ..., 42	31; 35, ..., 42	$\{0,1,4,10,12,17\}$	32,33,34

It is easy to verify that the sets given work for each relevant n; the fact that no such sets exist for the four cases listed was verified via the computer program [120].

Of course, we are interested in the value of $\sigma(\mathbb{Z}_n, 2)$ even when it is not of the maximum size given by Proposition C.7:

Problem C.13 *Find $\sigma(\mathbb{Z}_n, 2)$ for all (or at least infinitely many) positive integers n.*

Very little is known about $\sigma(\mathbb{Z}_n, h)$ for $h \geq 3$. As generalizations of Theorems C.9 and C.10, we have the following results:

Theorem C.14 (Bose and Chowla; cf. [42]) *Let $h \geq 2$. Suppose that q is a prime power and that $n = q^h + q^{h-1} + \cdots + 1$. We then have*

$$\sigma(\mathbb{Z}_n, h) \geq q + 1.$$

Theorem C.15 (Bose and Chowla; cf. [42]) *Let $h \geq 2$. Suppose that q is a prime power and that $n = q^h - 1$. We then have*

$$\sigma(\mathbb{Z}_n, h) \geq q.$$

Regarding $\sigma(\mathbb{Z}_n, h)$ in general, we can exhibit a rather obvious lower bound in terms of $\sigma(\mathbb{Z}_{n_0}, h)$ for some n_0 that is much smaller than n. Suppose that A is a set of nonnegative integers, so that A is a B_h set when viewed in $\mathbb{Z}_{n_0}$. Note that the largest element of hA is at most $h(n_0 - 1)$. It is then easy to see that A is also a Sidon set in $\mathbb{Z}_n$ for all $n \geq h(n_0 - 1) + 1$. This yields:

Proposition C.16 *Suppose that n and n_0 are positive integers so that $n \geq hn_0 - h + 1$. We then have $\sigma(\mathbb{Z}_n, h) \geq \sigma(\mathbb{Z}_{n_0}, h)$.*

For example, we can apply Proposition C.16 with Theorem C.11 as follows. By Chebyshev's famous theorem (also known as Bertrand's Postulate), there is always a prime between a positive integer and its double; in particular, for every n, there is a prime p so that

$$\left\lfloor \sqrt{n/8} \right\rfloor < p < 2 \left\lfloor \sqrt{n/8} \right\rfloor.$$

For this prime p, by Theorem C.11 we have $\sigma(\mathbb{Z}_{p^2-p}, 2) \geq p - 1$, and since $n \geq 2(p^2 - p) - 1$, by Proposition C.16 we get $\sigma(\mathbb{Z}_n, 2) \geq p - 1$. Therefore:

Proposition C.17 *For every positive integer n, we have*

$$\sigma(\mathbb{Z}_n, 2) \geq \left\lfloor \sqrt{n/8} \right\rfloor.$$

By using stronger results than Chebyshev's Theorem, stating that primes are denser, one can greatly improve on Proposition C.17.

We seem to be quite far away from the exact answer:

Problem C.18 *Find (better lower bounds for) $\sigma(\mathbb{Z}_n, h)$ for all positive integers n and $h \geq 2$.*

Let us now turn to noncyclic groups; in particular, groups of the form $\mathbb{Z}_k^r$. The situation is quite trivial when $k = 2$, since by Proposition C.3 we have:

Proposition C.19 *For all $r \geq 1$ and $h \geq 2$, we have $\sigma(\mathbb{Z}_2^r, h) = 1$.*

Moving to $k \geq 3$, we have the following exact results:

Theorem C.20 (Babai and Sós; cf. [6]) *Suppose that p is an odd prime and that r is even. We then have*

$$\sigma(\mathbb{Z}_p^r, 2) = p^{r/2}.$$

Theorem C.21 (Cilleruelo; cf. [56]) *Suppose that q is a prime power. We then have*

$$\sigma(\mathbb{Z}_{q-1}^2, 2) = q - 1.$$

By Proposition C.7, we can verify that in both cases the results are maximum possible. We are not aware of other such cases, so we offer:

Problem C.22 *Find all values of k and r for which $\sigma(\mathbb{Z}_k^r, 2)$ equals the maximum value allowed by Proposition C.7.*

For example, when $k = 3$, according to Theorem C.20, the answer to this problem includes all even values of r. Since we have $\sigma(\mathbb{Z}_3, 2) = 2$, as shown by the Sidon set $\{0, 1\}$, and $\sigma(\mathbb{Z}_3^3, 2) = 5$, as demonstrated by

$$\{(0,0,0), (0,0,1), (0,1,0), (1,0,0), (1,1,1)\},$$

the answer also includes $r = 1$ and $r = 3$. It would be nice to know whether $\mathbb{Z}_3^5$ has a Sidon set of size 16.

Note that when r is even, the maximum value allowed by Proposition C.7 equals $k^{r/2}$, so Problem C.22 can be specialized as follows:

Problem C.23 *Find all values of k and even values of r for which*

$$\sigma(\mathbb{Z}_k^r, 2) = k^{r/2};$$

in particular, find all values of k for which

$$\sigma(\mathbb{Z}_k^2, 2) = k.$$

By Theorems C.20 and C.21, for $r = 2$, the answer to Problem C.23 includes all prime values of k, as well as those where $k+1$ is a prime power. The first k for which neither condition applies is $k = 9$. With the computer program [120], we can determine that $\sigma(\mathbb{Z}_9^2, 2) \geq 8$; an example of a Sidon set of size eight is

$$\{(0,0), (0,1), (0,3), (1,0), (1,4), (2,2), (3,7), (5,3)\}.$$

Generalizing Problem C.22, we have:

Problem C.24 *Find all abelian groups G for which $\sigma(G, 2)$ equals the maximum value allowed by Proposition C.7.*

Let us now present some lower bounds for $\sigma(G, 2)$. First, an obvious observation:

Proposition C.25 *For all finite abelian groups G_1 and G_2 and for all positive integers h, we have*

$$\sigma(G_1 \times G_2, h) \geq \sigma(G_1, h).$$

Indeed: if A is a B_h set in G_1, then $A \times \{0\}$ is a B_h set in $G_1 \times G_2$.

As an immediate corollary, from Theorem C.20 we get:

Theorem C.26 (Babai and Sós; cf. [6]) *Suppose that p is an odd prime and that r is odd. We then have*

$$\sigma(\mathbb{Z}_p^r, 2) \geq p^{(r-1)/2}.$$

For a more general lower bound, one can generalize Proposition C.17 as follows:

Theorem C.27 (Babai and Sós; cf. [6]) *Suppose that G is an abelian group of rank r, odd order n, and smallest invariant factor n_1. We then have*

$$\sigma(G, 2) \geq \left\lfloor \sqrt{n_1/8^r} \right\rfloor.$$

It is important to point out that Theorem C.27 is stated for all n in [6], but their concept of a Sidon set there is slightly different from ours and only coincides with ours when n is odd. We should also add that, just like with Proposition C.17, the constant 8 can be largely reduced. In particular, we have:

Corollary C.28 (Babai and Sós; cf. [6]) *For all positive integers r and k, with k odd, we have*

$$\sigma(\mathbb{Z}_k^r, 2) \geq (k/8)^{r/2}.$$

It is worth pointing out that when k is an odd prime, then Corollary C.28 is weaker than Theorem C.20 when r is even, but stronger than Theorem C.26 when r is odd and p is larger than 8^r.

The main conjecture here is as follows:

Conjecture C.29 (Babai; cf. [5]) *There is a positive constant C so that $\sigma(G, 2) \geq C \cdot \sqrt{n}$ holds for all abelian groups G of odd order n.*

Note that, by Proposition C.7, the constant C cannot be more than 1.

Problem C.30 *Prove Conjecture C.29.*

Note that Conjecture C.29 is false for groups of even order; see, for example, Proposition C.19.

Regarding values of $h \geq 3$, we are only aware of the following result, which generalizes Theorem C.20:

Theorem C.31 (Ruiz and Trujillo; cf. [180]) *If p is a prime so that $p > h$ and r_0 is a divisor of r for which $r_0 \geq h$, then*

$$\sigma(\mathbb{Z}_p^r, h) \geq p^{r/r_0}.$$

The assumption that $p > h$ is necessary: see Proposition C.3. We should also note that [180] only addresses the case when $r_0 = h$, but our version immediately follows from that.

The most general question remains largely unsolved:

Problem C.32 *Find $\sigma(G, h)$ for abelian groups G and $h \geq 2$.*

It is also interesting to approach these questions from another viewpoint. Namely, we may fix positive integers m and h, and ask for all groups G that have a B_h set of size m. In other words:

Problem C.33 *Let m and h be given positive integers. Find all finite abelian groups G for which $\sigma(G, h) \geq m$.*

We can answer Problem C.33 easily for $h = 1$, $m = 1$, and $m = 2$ as follows. Since $\sigma(G, 1) = n$, we have $\sigma(G, 1) \geq m$ if, and only if, $n \geq m$. By Proposition C.2, we have $\sigma(G, h) \geq 1$ for all G. Furthermore, we can see that $\sigma(G, h) \geq 2$ if, and only if, the exponent κ of G is at least $h + 1$: The "only if" part follows from Proposition C.3, and for the "if" part, note that the set $\{0, a\}$ is a B_h set for every $a \in G$ of order κ, since if we were to have distinct nonnegative integers λ_1 and λ_2 with $\lambda_1 \leq h$ and $\lambda_2 \leq h$ for which

$$\lambda_1 \cdot 0 + (h - \lambda_1) \cdot a = \lambda_2 \cdot 0 + (h - \lambda_2) \cdot a,$$

then

$$(\lambda_1 - \lambda_2) \cdot a = 0,$$

which is impossible since

$$1 \leq |\lambda_1 - \lambda_2| \leq h < \kappa.$$

Problem C.33 is unsolved for higher values of m and h, and may already be challenging for $m = 3$. For example, we find (via a computer program) that

$$\sigma(\mathbb{Z}_n, h) \geq 3 \text{ holds for } \begin{cases} n \geq 7 & \text{if } h = 2, \\ n \geq 13 & \text{if } h = 3, \\ n = 19, \geq 21 & \text{if } h = 4, \\ n \geq 30 & \text{if } h = 5, \\ n = 37, 39, 40, 41, \geq 43 & \text{if } h = 6. \end{cases}$$

We formulate the following two sub-problems of Problem C.33:

Problem C.34 *Let m and h be given positive integers. Find the smallest positive integer $f(m, h)$ for which $\sigma(\mathbb{Z}_n, h) \geq m$ holds for $n = f(m, h)$.*

Problem C.35 *Let m and h be given positive integers. Find the smallest positive integer $g(m, h)$ for which $\sigma(\mathbb{Z}_n, h) \geq m$ holds for all $n \geq g(m, h)$.*

According to our considerations right above, we have $f(m, 1) = g(m, 1) = m$, $f(1, h) = g(1, h) = 1$, and $f(2, h) = g(2, h) = h + 1$, and as the data above indicate, we have $f(3, 4) = 19$, $g(3, 4) = 21$, and $f(3, 5) = g(3, 5) = 30$.

For $h = 2$, we have the following data:

m	1	2	3	4	5	6	7	8	9	10	11	12	13	14
$f(m, 2)$	1	3	7	13	21	31	48	57	73	91	120	133	168	183

Most of these entries follow directly from our comments on page 134: If $m - 1$ is a prime power, then Theorem C.9 provides the value of $f(m, 2)$. Regarding the remaining three entries: $f(7, 2)$ was found by Graham and Sloane in [100], and $f(11, 2)$ and $f(13, 2)$ by Haanpää, Huima, and Östergård in [107]; note that these three values are given exactly by Theorem C.10.

We can prove the existence and in fact find an upper bound for $g(m, h)$ (and thus for $f(m, h)$), by verifying that for every $m \geq 1$ and $h \geq 2$, the set

$$A = \{1, h, h^2, \ldots, h^{m-1}\}$$

is a B_h set in $\mathbb{Z}_n$ whenever $n \geq h^m$—we will do this on page 329. This yields:

Proposition C.36 *For all positive integers $m \geq 1$ and $h \geq 2$, we have*

$$f(m, h) \leq g(m, h) \leq h^m.$$

As our computations on page 138 show, the bound in Proposition C.36 can be greatly lowered. In particular, note that for $h = 2$, from Proposition C.17 one gets $g(m, 2) \leq 8m^2$ (approximately), substantially better than 2^m.

C.1.2 Limited number of terms

Sidon sets over $\mathbb{N}_0$ for the set $[0,s]$ are called $B_{[0,s]}$*-sets*; that is, $A \subseteq G$ is a $B_{[0,s]}$-set if, and only if, all $\binom{m+s}{m}$ linear combinations of at most s terms of A are distinct (ignoring the order of the terms).

In this section we ought to investigate the quantity $\sigma(G,[0,s])$, denoting the maximum size of a $B_{[0,s]}$-set in G. However, as we now prove, this section is superfluous as it can be reduced to Section C.1.1 above.

Suppose first that A is a $B_{[0,s]}$ set in G for some positive integer s. Then $0A = \{0\}$ and $1A = A$ must be disjoint, so $0 \not\in A$. Let $B = A \cup \{0\}$. Here B must be a B_s set in G, since if two different s-term sums were to be equal, then, after deleting all 0s from these sums (if needed), we get two different sums with at most s terms in each that are equal, contradicting the fact that A is a $B_{[0,s]}$ set in G. Therefore, $\sigma(G,s) \geq \sigma(G,[0,s]) + 1$.

Conversely, suppose that A is a B_s set in G. Note that, for any $g \in G$,

$$A - g = \{a - g \mid a \in A\}$$

is then also a B_s set, since an s-term linear combination of the elements of A is always exactly sg more than the s-term linear combination of the corresponding elements of $A-g$. In particular, for any $a \in A$, the set $B = A - a$ is then a B_s set in G for which $0 \in B$. Let $C = B \setminus \{0\}$. Then C is a $B_{[0,s]}$ set in G, since if two different linear combinations of at most s terms of C were to be equal, then extending each by the necessary terms of 0s we get two different linear combinations of exactly s terms of B that are equal, which is a contradiction. Thus $\sigma(G,[0,s]) \geq \sigma(G,s) - 1$. This yields:

Proposition C.37 *For every group G and all positive integers s we have*

$$\sigma(G,[0,s]) = \sigma(G,s) - 1.$$

C.1.3 Arbitrary number of terms

Here we ought to consider $\sigma(G,\mathbb{N}_0)$, but this quantity (e.g., by Proposition C.2) is clearly 0.

C.2 Unrestricted signed sumsets

Here we study, for any G and $H \subseteq \mathbb{N}_0$, the quantity

$$\sigma_{\pm}(G,H) = \max\{|A| \mid A \subseteq G, |H_{\pm}A| = |\mathbb{Z}^{|A|}(H)|\}.$$

If no such set exists, we put $\sigma_{\pm}(G,H) = 0$.

We clearly have $\sigma_{\pm}(G,H) = 0$ whenever H contains an element $h \in \mathbb{N}$ for which $2h$ is divisible by the exponent κ of G, or distinct elements $h_1, h_2 \in \mathbb{N}_0$ whose sum or difference is divisible by κ. Conversely, if there are no such elements in H, then at least the one-element set $\{a\}$ whose order equals κ will be a Sidon set for H over $\mathbb{Z}$.

Proposition C.38 *We have $\sigma_{\pm}(G,H) \geq 1$ if, and only if, there is no $h \in H$ for which $2h$ is divisible by κ and no distinct elements $h_1, h_2 \in H$ for which $h_1 \pm h_2$ is divisible by κ. In particular, if $|H| > \lceil \kappa/2 \rceil$, then $\sigma_{\pm}(G,H) = 0$.*

C.2.1 Fixed number of terms

In this section we investigate, for a given G and positive integer h, the quantity

$$\sigma_{\pm}(G,h) = \max\{|A| \mid A \subseteq G, |h_{\pm}A| = |\mathbb{Z}^{|A|}(h)|\};$$

that is, the maximum size of a B_h set over $\mathbb{Z}$. Since, according to Section 2.5,

$$|\mathbb{Z}^m(h)| = c(h,m),$$

a B_h set over $\mathbb{Z}$ that has size m has unrestricted signed sumset size

$$c(h,m) = \sum_{i \geq 0} \binom{m}{i}\binom{h-1}{i-1}2^i.$$

The problem of finding $\sigma_{\pm}(G,h)$ is related to $\tau_{\pm}(G,2h)$—see Chapter F.2.1.

Clearly, a subset A of G is a B_1 set over $\mathbb{Z}$ if, and only if, A and $-A$ are disjoint, from which we get:

Proposition C.39 *For any abelian group G of order n we have*

$$\sigma_{\pm}(G,1) = \frac{n-1-|\mathrm{Ord}(G,2)|}{2};$$

in particular,

$$\sigma_{\pm}(\mathbb{Z}_n,1) = \lfloor (n-1)/2 \rfloor.$$

We do not know the value of $\sigma_{\pm}(G,h)$ in general for $h \geq 2$.

It is important to point out that, in contrast to Section C.1.1, a B_h set over $\mathbb{Z}$ is not necessarily a B_k set over $\mathbb{Z}$ for values of $k \leq h$. Suppose, for example, that n is a positive integer that is divisible by 4, and consider the set $A = \{n/4\}$ in $\mathbb{Z}_n$. We can then see that A is a B_h set over $\mathbb{Z}$ in $\mathbb{Z}_n$ if, and only if, h is odd: Indeed,

$$h \cdot n/4 = -h \cdot n/4$$

in $\mathbb{Z}_n$ if, and only if, h is even. In particular,

$$\sigma_{\pm}(\mathbb{Z}_4,h) = \begin{cases} 1 & \text{if } h \text{ is odd,} \\ 0 & \text{if } h \text{ is even.} \end{cases}$$

As a consequence of Proposition C.1, we have the following bound.

Proposition C.40 *If A is a B_h set over $\mathbb{Z}$ of size m in G, then*

$$n \geq c(h,m).$$

From Proposition C.40 we get an upper bound for $\sigma_{\pm}(G,h)$; in particular, since $c(2,m) = 2m^2$, we have:

Corollary C.41 *For any abelian group G of order n we have*

$$\sigma_{\pm}(G,2) \leq \left\lfloor \sqrt{n/2} \right\rfloor.$$

It would be particularly interesting to classify situations with equality in Corollary C.41:

Problem C.42 *For each $n \in \mathbb{N}$, find all groups G of order n for which*

$$\sigma_{\pm}(G, 2) = \left\lfloor \sqrt{n/2} \right\rfloor .$$

The following problems are wide open.

Problem C.43 *Find $\sigma_{\pm}(\mathbb{Z}_n, h)$ for all positive integers n and $h \geq 2$.*

Problem C.44 *Find $\sigma_{\pm}(G, h)$ for all finite abelian groups G and integers $h \geq 2$.*

It is also interesting to approach these questions from another viewpoint. Namely, we may fix positive integers m and h, and ask for all groups G that have a B_h set over $\mathbb{Z}$ of size m. In other words:

Problem C.45 *Let m and h be given positive integers. Find all finite abelian groups G for which $\sigma_{\pm}(G, h) \geq m$.*

We can answer Problem C.45 for $m = 1$ and for $h = 1$ easily; by Propositions C.38 and Proposition C.39, we get the following:

Proposition C.46 *Suppose that G is a finite abelian group with exponent κ; as usual, let $\mathrm{Ord}(G, 2)$ denote the set of involutions (elements of order 2) in G.*

1. *For a given positive integer h, we have $\sigma_{\pm}(G, h) \geq 1$ if, and only if, $2h$ is not divisible by κ.*

2. *For a given positive integer m, we have $\sigma_{\pm}(G, 1) \geq m$ if, and only if,*

$$|\mathrm{Ord}(G, 2)| \leq n - 2m - 1.$$

The answer to Problem C.45 is not known when $h \geq 2$ and $m \geq 2$.

We formulate the following two sub-problems of Problem C.45:

Problem C.47 *Let m and h be given positive integers. Find the smallest positive integer $f_{\pm}(m, h)$ for which $\sigma_{\pm}(\mathbb{Z}_n, h) \geq m$ holds for $n = f_{\pm}(m, h)$.*

Problem C.48 *Let m and h be given positive integers. Find the smallest positive integer $g_{\pm}(m, h)$ for which $\sigma_{\pm}(\mathbb{Z}_n, h) \geq m$ holds for all $n \geq g_{\pm}(m, h)$.*

It is easy to see that Proposition C.46 implies the following:

Proposition C.49 *With $f_{\pm}(m, h)$ and $g_{\pm}(m, h)$ defined above,*

1. *$f_{\pm}(1, h)$ equals the smallest integer that is not a divisor of $2h$, and $g_{\pm}(1, h) = 2h + 1$; and*

2. *$f_{\pm}(m, 1) = g_{\pm}(m, 1) = 2m + 1$.*

To verify the second statement, note that $\mathrm{Ord}(\mathbb{Z}_n, 2)$ is empty when n is odd, and contains exactly one element when n is even.

We can prove the existence and in fact find an upper bound for $g_{\pm}(m, h)$ (and thus for $f_{\pm}(m, h)$), by verifying that for every $m \geq 1$ and $h \geq 2$, the set

$$A = \{1, 2h, (2h)^2, \ldots, (2h)^{m-1}\}$$

is a B_h set in $\mathbb{Z}_n$ whenever $n \geq 2h^m + 1$—we will do this on page 330. This yields:

Proposition C.50 *For all positive integers m and h, we have*

$$f_{\pm}(m,h) \leq g_{\pm}(m,h) \leq (2h)^m + 1.$$

Observe that, by Proposition C.49, for $m = 1$, equality holds in Proposition C.50.

By Proposition C.50, we have

$$g_{\pm}(2,h) \leq 4h^2 + 1.$$

However, Day in [61] conjectured that the set $A = \{a_1, h\}$, with $a_1 = 2$ if h is odd and $a_1 = 1$ if h is even provides a signed sumset of size $c(h,2) = 4h$ when $n > 2h^2$. Indeed, we can prove the following.

Proposition C.51 *For all positive integers h we have*

$$g_{\pm}(2,h) \leq 2h^2 + 1.$$

For a proof of Proposition C.51, see page 331. Note that, by Proposition C.40, we have

$$g_{\pm}(2,h) \geq 4h;$$

therefore, almost certainly, one can even improve the bound $2h^2 + 1$.

Problem C.52 *Find a better upper bound for $g_{\pm}(2,h)$.*

For $m = 3$, from Propositions C.40 and C.50 we have

$$4h^2 + 2 \leq g_{\pm}(3,h) \leq 8h^3 + 1.$$

Day in [61] conjectured that the set $A = \{a_1, h, a_3\}$, with $a_1 = 2$ and $a_3 = h^2 + h + 4$ if h is odd and $a_1 = 1$ and $a_3 = h^2 + h + 1$ if h is even provides a signed sumset of size $c(h,3)$ when $n > 2ha_3$, from which we can conjecture the following.

Conjecture C.53 (Day; cf. [61]) *For all positive integers h we have*

$$g_{\pm}(3,h) \leq 2h^3 + 2h^2 + 8h + 1.$$

Problem C.54 *Prove (or disprove) Conjecture C.53, or—even better—find an improved upper bound for $g_{\pm}(3,h)$.*

C.2.2 Limited number of terms

Here we ought to study, for a given G and nonnegative integer s, the quantity

$$\sigma_{\pm}(G,[0,s]) = \max\{|A| \mid A \subseteq G, |[0,s]_{\pm}A| = |\mathbb{Z}^{|A|}([0,s])|\},$$

that is, the size of the largest subset A of G for which all signed sums of at most s terms of A are distinct (ignoring the order of the terms). However, as we proved on page 70, this quantity agrees with the maximum size of a subset A of G for which no signed sum of at most $2s$ terms of A, other than the trivial one with zero terms, is equal to 0—in Section F.2.2, we denote this latter quantity by $\tau_{\pm}(G,[1,2s])$. Therefore:

Proposition C.55 *For every finite abelian group G and every positive integer s, we have*

$$\sigma_{\pm}(G,[0,s]) = \tau_{\pm}(G,[1,2s]).$$

In addition, for $s = 0$, we trivially have

$$\sigma_{\pm}(G,[0,0]) = n;$$

thus, we reduced the question of finding $\sigma_{\pm}(G,[0,s])$ to Section F.2.2.

C.2.3 Arbitrary number of terms

Here we ought to consider $\sigma_{\pm}(G, \mathbb{N}_0)$, but this quantity (e.g., by Proposition C.38) is clearly 0.

C.3 Restricted sumsets

In this section we study, for any G and $H \subseteq \mathbb{N}_0$, the quantity

$$\sigma\hat{}(G, H) = \max\{|A| \mid A \subseteq G, |H\hat{}A| = |\hat{\mathbb{N}}_0^{|A|}(H)|\};$$

that is, the size of the largest subset of G whose restricted sumset (for a given H) has the same size as the corresponding index set. If no such set exists, we put $\sigma\hat{}(G, H) = 0$.

C.3.1 Fixed number of terms

The analogue of a B_h set for restricted addition is called a weak B_h set. More precisely, we call a subset A of G (with $|A| = m$) a *weak* B_h *set* if

$$|h\hat{}A| = |\hat{\mathbb{N}}_0^m(h)| = \binom{m}{h}.$$

The maximum size of a weak B_h set in G is denoted by $\sigma\hat{}(G, h)$; if no such subset exists, we write $\sigma\hat{}(G, h) = 0$.

Note that, if A is a B_h set, then it is also a weak B_h set, and therefore we have

$$\sigma\hat{}(G, h) \geq \sigma(G, h).$$

We trivially have $\sigma\hat{}(G, 1) = n$, since every subset of G is a weak B_h set. We do not know the value of $\sigma\hat{}(G, h)$ in general for $h \geq 2$.

As a consequence of Proposition C.1, we have the following bound.

Proposition C.56 *If A is a weak B_h set of size m in G, then*

$$n \geq \binom{m}{h}.$$

From Proposition C.56 we get an upper bound for $\sigma\hat{}(G, h)$:

Corollary C.57 *For any abelian group G of order n we have*

$$\sigma\hat{}(G, h) \leq \left\lfloor \sqrt[h]{h!n} \right\rfloor + h - 1.$$

A weak B_h set for $h = 2$ is called a weak Sidon set. Weak Sidon sets were introduced and studied by Ruzsa in [181], though the same concept under the name "well spread set" was investigated both earlier (cf. [136]) and later (cf. [127], [170]). For weak Sidon sets a better upper bound is known, in terms of the number of order-two elements of G:

Theorem C.58 (Haanpää and Östergård; cf. [108]) *For any abelian group G of order n we have*

$$\sigma\hat{}(G, 2) \leq \left\lfloor \frac{\sqrt{4n + 4|\mathrm{Ord}(G, 2)| + 5} + 3}{2} \right\rfloor.$$

In particular, since in a cyclic group $|\mathrm{Ord}(G,2)| \leq 1$, from Theorem C.58 we get the previously published result:

Corollary C.59 (Haanpää, Huima, and Östergård; cf. [107]) *For any positive integer n we have*

$$\sigma\hat{}(\mathbb{Z}_n, 2) \leq \left\lfloor \frac{\sqrt{4n+9}+3}{2} \right\rfloor.$$

As for a lower bound, we have the result of Babai and Sós (which, unlike with Theorem C.27 in Section C.1.1, applies to both even and odd values of n):

Theorem C.60 (Babai and Sós; cf. [6]) *Suppose that G is an abelian group of rank r, odd order n, and smallest invariant factor n_1. We then have*

$$\sigma\hat{}(G, 2) \geq \left\lfloor \sqrt{n_1/8^r} \right\rfloor.$$

(We note that it is possible to improve the constant 8.)
The following problems are wide open.

Problem C.61 *Find $\sigma\hat{}(\mathbb{Z}_n, h)$ for positive integers n and $h \geq 2$.*

Problem C.62 *Find $\sigma\hat{}(G, h)$ for finite abelian groups G and integers $h \geq 2$.*

It is also interesting to approach these questions from another viewpoint. Namely, we may fix positive integers m and h, and ask for all groups G that have a weak B_h set of size m. In other words:

Problem C.63 *Let m and h be given positive integers. Find all finite abelian groups G for which $\sigma\hat{}(G, h) \geq m$.*

We formulate the following two sub-problems of Problem C.63:

Problem C.64 *Let m and h be given positive integers. Find the smallest positive integer $f\hat{}(m, h)$ for which $\sigma\hat{}(\mathbb{Z}_n, h) \geq m$ holds for $n = f\hat{}(m, h)$.*

Problem C.65 *Let m and h be given positive integers. Find the smallest positive integer $g\hat{}(m, h)$ for which $\sigma\hat{}(\mathbb{Z}_n, h) \geq m$ holds for all $n \geq g\hat{}(m, h)$.*

Problems C.64 and C.65 are trivial when $m < h$, and we have

$$f\hat{}(h, h) = g\hat{}(h, h) = h.$$

Furthermore, the answer is clear for $h = 1$: since $\sigma\hat{}(G, 1) = n$, we have

$$f\hat{}(m, 1) = g\hat{}(m, 1) = m.$$

The general answers to Problems C.64 and C.65 are not known. For $h = 2$, Haanpää, Huima, and Östergård computed $f\hat{}(m, 2)$ for $m \leq 15$ (see [107]; some of these values had been determined earlier by Graham and Sloane in [100]), and Maturo and Yager-Elorriaga computed $g\hat{}(m, 2)$ for $m \leq 9$ (see [154]):

m	2	3	4	5	6	7	8	9	10	11	12	13	14	15
$f\hat{}(m, 2)$	2	3	6	11	19	28	40	56	72	96	114	147	178	183
$g\hat{}(m, 2)$	2	3	6	11	19	28	42	56						

As we can see from the table, we have $f\hat{}(m,2) = g\hat{}(m,2)$ for $m \leq 9$, with one exception: $\mathbb{Z}_n$ contains a weak Sidon set of size 8 for $n = 40$ and for every $n \geq 42$, but not for $n = 41$.

We can easily find a general upper bound for both $f\hat{}(m,h)$ and $g\hat{}(m,h)$ (and thus prove their existence). Note that, for every $h \in \mathbb{N}$, the set

$$\{1, 2, 2^2, \ldots, 2^{m-1}\}$$

is a weak B_h set in $\mathbb{Z}_n$ when n is at least 2^m. Therefore:

Proposition C.66 *For all positive integers m and h we have*

$$f\hat{}(m,h) \leq g\hat{}(m,h) \leq 2^m.$$

Undoubtedly, this bound—which does not even depend on h—can be greatly reduced.

A bit more ambitiously than Problem C.64, but less ambitiously than Problem C.63, we may pose the following:

Problem C.67 *Let m and h be given positive integers. Find the smallest positive integer $F\hat{}(m,h)$, for which there exists a group G of order $F\hat{}(m,h)$ such that $\sigma\hat{}(G,h) = m$.*

For $h = 2$, Haanpää and Östergård computed $F\hat{}(m,2)$ for $m \leq 15$ (see [108]):

m	2	3	4	5	6	7	8	9	10	11	12	13	14	15
$F\hat{}(m,2)$	2	3	6	11	16	24	40	52	72	96	114	147	178	183

When comparing values of $F\hat{}(m,2)$ and $f\hat{}(m,2)$ above, we see that they agree for all $m \leq 15$, except as follows:

- $F\hat{}(6,2) = f\hat{}(6,2) - 3$, as $\mathbb{Z}_2^2 \times \mathbb{Z}_4$ and $\mathbb{Z}_2^4$ both have weak Sidon sets of size 6;
- $F\hat{}(7,2) = f\hat{}(7,2) - 4$, as $\mathbb{Z}_2^3 \times \mathbb{Z}_3$ has a weak Sidon sets of size 7; and
- $F\hat{}(9,2) = f\hat{}(9,2) - 4$, as $\mathbb{Z}_2^2 \times \mathbb{Z}_{13}$ has a weak Sidon sets of size 9.

C.3.2 Limited number of terms

We say that an m-subset A of G is a *weak $B_{[0,s]}$-set* if, and only if, the equality

$$|[0,s]\hat{}A| = |\hat{\mathbb{N}}_0^m([0,s])| = \sum_{h=0}^{s} \binom{m}{h}$$

holds. We are interested in finding the maximum size of a weak $B_{[0,s]}$-set in G—we denote this quantity by $\sigma\hat{}(G,[0,s])$.

Clearly, every subset of $G \setminus \{0\}$ is a weak $B_{[0,1]}$ set, so

$$\sigma\hat{}(G,[0,1]) = n - 1.$$

We do not know the value of $\sigma\hat{}(G,[0,s])$ in general for $s \geq 2$.

As a consequence of Proposition C.1, we have the following bound.

Proposition C.68 *If A is a weak $B_{[0,s]}$ set of size m in G, then*

$$n \geq \sum_{h=0}^{s} \binom{m}{h}.$$

Proposition C.68 yields an upper bound for $\sigma\hat{}(G, [0, s])$; for example, for $s = 2$ we get:

Corollary C.69 *The maximum size of a $B_{[0,2]}$ set in an abelian group G of order n satisfies*

$$\sigma\hat{}(G, [0, 2]) \leq \left\lfloor \frac{\sqrt{8n-7}-1}{2} \right\rfloor.$$

The following problems are wide open.

Problem C.70 *Find $\sigma\hat{}(\mathbb{Z}_n, [0, s])$ for positive integers n and $s \geq 2$.*

Problem C.71 *Find $\sigma\hat{}(G, [0, s])$ for finite abelian groups G and integers $s \geq 2$.*

It is also interesting to approach these questions from another viewpoint. Namely, we may fix positive integers m and s, and ask for all groups G that have a weak $B_{[0,s]}$ set of size m. In other words:

Problem C.72 *Let m and s be given positive integers. Find all finite abelian groups G for which $\sigma\hat{}(G, [0, s]) \geq m$.*

We formulate the following two sub-problems of Problem C.72:

Problem C.73 *Let m and s be given positive integers. Find the smallest positive integer $f\hat{}(m, [0, s])$ for which $\sigma\hat{}(\mathbb{Z}_n, [0, s]) \geq m$ holds for $n = f\hat{}(m, [0, s])$.*

Problem C.74 *Let m and s be given positive integers. Find the smallest positive integer $g\hat{}(m, [0, s])$ for which $\sigma\hat{}(\mathbb{Z}_n, [0, s]) \geq m$ holds for all $n \geq g\hat{}(m, [0, s])$.*

We can easily find a general upper bound for both $f\hat{}(m, [0, s])$ and $g\hat{}(m, [0, s])$ (and thus prove their existence). Note that, for every $s \in \mathbb{N}$, the set

$$\{1, 2, 2^2, \ldots, 2^{m-1}\}$$

is a weak $B_{[0,s]}$ set in $\mathbb{Z}_n$ when n is at least 2^m. Therefore:

Proposition C.75 *For all positive integers m and s we have*

$$g\hat{}(m, [0, s]) \leq 2^m.$$

As Soma in [190] noticed, one can improve on the bound in Proposition C.75 slightly as follows. For a fixed positive integer s, consider the sequence

$$\mathbf{a}(s) = (a_1(s), a_2(s), \ldots),$$

defined as

$$a_i(s) = \begin{cases} 2^{i-1} & \text{if } 1 \leq i \leq s, \\ 1 + \sum_{j=i-s}^{i-1} a_j(s) & \text{if } i > s. \end{cases}$$

For example,

$$\mathbf{a}(2) = (1, 2, 4, 7, 12, 20, 33, 54, \ldots);$$

the i-th term we recognize as the $(i + 1)$-st Fibonacci number minus 1. Clearly, if $n \geq a_{m+1}(s) - 1$, then

$$A = \{a_1(s), a_2(s), \ldots, a_m(s)\}$$

is a weak $B_{[0,s]}$ set in $\mathbb{Z}_n$. While the bound on n is hard to compute explicitly in general, for $s = 2$ we get $F_{m+2} - 2$; that is:

Proposition C.76 (Soma; cf. [190]) *For all positive integers m we have*

$$g\hat{}(m,[0,2]) \leq \left[\frac{1}{\sqrt{5}}\left(\frac{1+\sqrt{5}}{2}\right)^{m+3}\right] - 2.$$

(As usual, $[x]$ denotes the closest integer to the real number x.) This bound is still exponential in m but its base is smaller than 2, thus it improves Proposition C.75. The measure of improvement is difficult to state for $s > 2$ and becomes less pronounced as s gets closer to m.

C.3.3 Arbitrary number of terms

Here we should consider, for a given group G,

$$\sigma\hat{}(G,\mathbb{N}_0) = \max\{m \mid A \subseteq G,\ |A| = m,\ |\Sigma A| = |\hat{\mathbb{N}}_0^m(\mathbb{N}_0)| = 2^m\};$$

that is, the maximum size of a subset of G for which all restricted sums are distinct.

However, as we now show, this subsection is redundant in that the following holds:

Proposition C.77 *For any subset A of G we have $|\Sigma A| = 2^m$ if, and only if, $0 \notin \cup_{h=1}^{\infty} h\hat{\pm}A$.*

To see this, simply note that two restricted sums are equal if, and only if, their difference is zero; in particular, no two "different" restricted sums being equal is equivalent to no "nontrivial" signed sum equaling zero.

As a consequence of Proposition C.77, we have

$$\sigma\hat{}(G,\mathbb{N}_0) = \tau\hat{\pm}(G,\mathbb{N}),$$

so it suffices to study this quantity in Section F.4.3.

C.4 Restricted signed sumsets

C.4.1 Fixed number of terms

C.4.2 Limited number of terms

C.4.3 Arbitrary number of terms

Here we consider, for a given group G,

$$\sigma\hat{\pm}(G,\mathbb{N}_0) = \max\{m \mid A \subseteq G,\ |A| = m,\ |\Sigma_{\pm}A| = |\hat{\mathbb{Z}}^m(\mathbb{N}_0)| = 3^m\}.$$

The case when G is cyclic is easy. Suppose that $G = \mathbb{Z}_n$, and let

$$m = \lfloor \log_3 n \rfloor;$$

note that we then have

$$3^m \leq n < 3^{m+1}.$$

Consider the set

$$A = \{1, 3, \ldots, 3^{m-1}\}$$

in $\mathbb{Z}_n$. Observe that $|A| = m$. We can easily see (see Section A.4.3) that

$$\Sigma_{\pm}A = \left\{-\frac{3^m-1}{2}, \ldots, -1, 0, 1, \ldots, \frac{3^m-1}{2}\right\},$$

so

$$|\Sigma_{\pm}A| = 3^m;$$

therefore,

$$\sigma\hat{_\pm}(\mathbb{Z}_n, \mathbb{N}_0) \geq m.$$

To show that we cannot do better, suppose, indirectly, that $A' \subseteq \mathbb{Z}_n$, $|A'| = m+1$, and

$$|\Sigma_{\pm}A'| = 3^{m+1}.$$

But this implies that

$$n \geq 3^{m+1},$$

a contradiction. Thus we have proved the following.

Proposition C.78 *For all positive integers n we have*

$$\sigma\hat{_\pm}(\mathbb{Z}_n, \mathbb{N}_0) = \lfloor \log_3 n \rfloor.$$

So, the case of cyclic groups has been settled, which leaves us with the following problem.

Problem C.79 *Find the value of $\sigma\hat{_\pm}(G, \mathbb{N}_0)$ for noncyclic groups G.*

Chapter D

Minimum sumset size

In this chapter we attempt to answer the following question: Given a finite abelian group G and a positive integer m, how small can a sumset of an m-subset of G be? More precisely, our objective is to determine, for any G, m, Λ, and H, the quantity

$$\rho_\Lambda(G, m, H) = \min\{|H_\Lambda A| \mid A \subseteq G, |A| = m\}.$$

In the following sections we consider $\rho_\Lambda(G, m, H)$ for special $\Lambda \subseteq \mathbb{Z}$ and $H \subseteq \mathbb{N}_0$.

D.1 Unrestricted sumsets

Our goal in this section is to investigate the quantity

$$\rho(G, m, H) = \min\{|HA| \mid A \subseteq G, |A| = m\}$$

where HA is the union of all h-fold sumsets hA for $h \in H$. We consider three special cases: when H consists of a single nonnegative integer h, when H consists of all nonnegative integers up to some value s, and when H is the entire set of nonnegative integers.

D.1.1 Fixed number of terms

Here we consider

$$\rho(G, m, h) = \min\{|hA| \mid A \subseteq G, |A| = m\},$$

that is, the minimum size of an h-fold sumset of an m-element subset of G.

Clearly, for all A we have $0A = \{0\}$ and $1A = A$, so

$$\rho(G, m, 0) = 1$$

and

$$\rho(G, m, 1) = m.$$

It is also easy to see that for all $A = \{a_1, \ldots, a_m\}$ and $h \geq 1$ we have

$$\{(h-1)a_1 + a_i \mid i = 1, 2, \ldots, m\} \subseteq hA,$$

and thus we have the following obvious bound.

Proposition D.1 *For all $h \geq 1$ we have $\rho(G, m, h) \geq m$.*

As we have said above, equality always holds when $h = 1$; in Proposition D.6 below, we classify all other cases for equality in Proposition D.1. But first we provide a general construction that, as it turns out, provides the minimum possible size of an h-fold sumset that an m-element subset of G can have.

How can one find m-subsets A in a group G that have small h-fold sumsets hA? Two ideas come to mind. First, observe that if A is a subset of a subgroup H of G, then hA will be a subset of H as well; a bit more generally, if A is a subset of any coset $g+H$ of H, then hA will be a subset of the coset $hg+H$. For example, with $G = \mathbb{Z}_{15}$, $m = 4$, and $h = 2$, we may choose A to be any 4-subset of $H = \{0, 3, 6, 9, 12\}$; we then have $2A \subseteq H$ and thus can conclude that $\rho(\mathbb{Z}_{15}, 4, 2) \leq 5$. (More generally, A may be a 4-subset of any coset of H.)

Our second idea is based on the observation that, when A is an arithmetic progression

$$A = \{a, a+g, a+2g, \ldots, a+(m-1)g\}$$

for some $a, g \in G$, then many of the h-fold sums coincide; in particular, we have

$$hA = \{ha, ha+g, ha+2g, \ldots, ha+(hm-h)g\}.$$

For example, with $G = \mathbb{Z}_{15}$, $m = 4$, and $h = 2$, we may choose A to be the set $\{0, 1, 2, 3\}$, in which case $2A = \{0, 1, 2, \ldots, 6\}$ and thus $\rho(\mathbb{Z}_{15}, 4, 2) \leq 7$. While this result is worse than that of the one obtained above, in other instances the second construction may be better; for example, with $G = \mathbb{Z}_{15}$, $m = 7$, and $h = 2$, not having a subgroup H of G of size at least 7 other than G itself, the first construction gives $\rho(\mathbb{Z}_{15}, 7, 2) \leq 15$, while the second construction yields the better bound $\rho(\mathbb{Z}_{15}, 7, 2) \leq 13$. (We will soon see that, in fact, $\rho(\mathbb{Z}_{15}, 4, 2) = 5$ and $\rho(\mathbb{Z}_{15}, 7, 2) = 13$.)

The general construction we are about to present is based on the combination of these two ideas: we choose a subgroup H of G, and then select an m-subset A so that its elements are in as few cosets of H as possible; furthermore, we want these cosets to form an arithmetic progression.

More explicitly, let us discuss this for the cyclic group $G = \mathbb{Z}_n$. Fixing a divisor d of n, we consider the (unique) subgroup H of order d of $\mathbb{Z}_n$,

$$H = \left\{ j \cdot \frac{n}{d} \mid j = 0, 1, 2, \ldots, d-1 \right\}.$$

With $|A| = m$ and $|H| = d$, the number of cosets of H that we need is $\lceil \frac{m}{d} \rceil$. (Note that $m \leq n$ assures that $\lceil \frac{m}{d} \rceil \leq \frac{n}{d}$ and thus A has m distinct elements.) Let k be the positive remainder of m when divided by d; that is, write $m = cd + k$ with integers c and k satisfying $1 \leq k \leq d$. We note, in passing, that c and k can be computed as

$$c = \left\lceil \frac{m}{d} \right\rceil - 1$$

and

$$k = m - cd = m - \left(\left\lceil \frac{m}{d} \right\rceil - 1 \right) \cdot d,$$

respectively. Now we choose our cosets to be $i + H$ with $i = 0, 1, 2, \ldots, c$, and set

$$A_d(n, m) = \bigcup_{i=0}^{c-1} (i + H) \cup \left\{ c + j \cdot \frac{n}{d} \mid j = 0, 1, 2, \ldots, k-1 \right\}.$$

Then $A_d(n, m)$ has size m. Note that, when $c = 0$ (that is, when $m \leq d$), then

$$\bigcup_{i=0}^{c-1} (i + H) = \emptyset,$$

so A lies entirely within a single coset and forms the arithmetic progression

$$\left\{j \cdot \frac{n}{d} \mid j = 0, 1, 2, \ldots, m-1\right\}.$$

On the other hand, $k \geq 1$, so

$$\left\{c + j \cdot \frac{n}{d} \mid j = 0, 1, 2, \ldots, k-1\right\}$$

is never the empty set.

It is easy to see that we have

$$hA_d(n,m) = \bigcup_{i=0}^{hc-1} (i+H) \cup \left\{hc + j \cdot \frac{n}{d} \mid j = 0, 1, 2, \ldots, h(k-1)\right\},$$

and thus

$$\begin{aligned} |hA_d(n,m)| &= \min\{n,\ hcd + \min\{d,\ h(k-1)+1\}\} \\ &= \min\{n,\ (hc+1)d,\ hcd + h(k-1)+1\} \\ &= \min\{n,\ \left(h\left\lceil \frac{m}{d} \right\rceil - h + 1\right) \cdot d,\ hm - h + 1\}. \end{aligned}$$

Recalling our notation

$$f_d = f_d(m,h) = \left(h\left\lceil \frac{m}{d} \right\rceil - h + 1\right) \cdot d$$

from page 83, we get the following:

Proposition D.2 *For integers n, m, h, a given divisor d of n, and $A_d(n,m)$ defined as above, we have*

$$|hA_d(n,m)| = \min\{n,\ f_d,\ hm - h + 1\}.$$

Recalling from page 83 that

$$u(n,m,h) = \min\{f_d \mid d \in D(n)\},$$

and noting that $f_n = n$ and $f_1 = hm - h + 1$, we can summarize our findings to say that

$$\rho(\mathbb{Z}_n, m, h) \leq \min\{|hA_d(n,m)| \mid d \in D(n)\} = u(n,m,h).$$

It is somewhat surprising that we get the same bound by taking a potentially larger set that is easier to work with. Namely, completing the part of $A_d(n,m)$ that falls into the last coset, we get

$$\overline{A}_d(n,m) = \bigcup_{i=0}^{c} (i+H),$$

a set that is the union of $c + 1 = \left\lceil \frac{m}{d} \right\rceil$ cosets and thus is of size

$$|\overline{A}_d(n,m)| = \left\lceil \frac{m}{d} \right\rceil \cdot d \geq m.$$

We now find that

$$|h\overline{A}_d(n,m)| = \min\{n,\ (hc+1)d\}$$

$$= \min\{n,\ f_d\},$$

and therefore

$$\rho(\mathbb{Z}_n, m, h) \leq \min\{|h\overline{A}_d(n,m)| \mid d \in D(n)\} = u(n,m,h).$$

Next, we ask whether it is possible for the h-fold sumset of an m-subset of $\mathbb{Z}_n$ to have size less than $u(n,m,h)$. The question of whether one can improve on the bound

$$\rho(G,m,h) \leq u(n,m,h)$$

or not for $\mathbb{Z}_n$ or other abelian groups has a long history. Cauchy's result from 1813 in [52] implies that for $h = 2$ and for cyclic groups of prime order we have equality above; since for a prime p we have

$$u(p,m,2) = \min\{p, 2m-1\}$$

(see Proposition 4.14), we thus have

$$\rho(\mathbb{Z}_p, m, 2) = \min\{p, 2m-1\}.$$

In 1935, Davenport in [59] rediscovered Cauchy's result, which is now known as the *Cauchy–Davenport Theorem.* (See page 334 for the general statement of the theorem. Davenport was unaware of Cauchy's result until twelve years later; see [60].) Finally, after various partial results by several researchers, a sequence of papers at the beginning of the twenty-first century, including [73] by Eliahou and Kervaire; [172] by Plagne; [76] by Eliahou, Kervaire, and Plagne; and [173] by Plagne established that, indeed, equality holds for all finite abelian groups and all h:

Theorem D.3 (Plagne; cf. [173]) *For any finite abelian group G of order n we have*

$$\rho(G,m,h) = u(n,m,h).$$

It may be worthwhile to state the following more explicit consequence of Theorem D.3 in light of Proposition 4.16.

Corollary D.4 *Let G be any finite abelian group of order $n \geq m$, and let p be the smallest prime divisor of n. We then have*

$$\rho(G,m,h) \geq \min\{p, hm-h+1\},$$

with equality if, and only if, $m \leq p$. In particular,

$$\rho(\mathbb{Z}_p, m, h) = \min\{p, hm-h+1\}$$

for all $m \leq p$.

Note that Corollary D.4 is a generalization of the Cauchy–Davenport Theorem.

According to Theorem D.3, the question of finding the minimum size $\rho(G,m,h)$ of the h-fold sumset of an m-subset of G is solved. We then may ask for a classification of all subsets A of G for which $|hA| = \rho(G,m,h)$. Since, as we have seen, we have

$$|0A| = \rho(G,m,0) = 1$$

and

$$|1A| = \rho(G,m,1) = m$$

for all G and A (with $|A| = m \leq n$), we may assume that $h \geq 2$.

Problem D.5 *For each G, m, and $h \geq 2$, find a characterization of all m-subsets of G with an h-fold sumset of size $\rho(G, m, h)$.*

For example, let us consider $G = \mathbb{Z}_{15}$, $m = 6$, and $h = 2$, for which we have $\rho(\mathbb{Z}_{15}, 6, 2) = 9$ (see page 83). As we can verify, a 6-subset A of G has a 2-fold sumset of size 9 if, and only if, it is of the form

$$A = (a_1 + H) \cup (a_2 + H)$$

where $H = \{0, 5, 10\}$ is the subgroup of G of order 3. Thus, in this situation, the only subsets with minimum sumset size are those constructed above.

We get a more mixed picture when we consider $G = \mathbb{Z}_{15}$, $m = 7$, and $h = 2$. This time, $\rho(\mathbb{Z}_{15}, 7, 2) = 13$, and we can determine that 7-subsets of G with a 2-fold sumset of size 13 come in a variety of forms: as an arithmetic progression

$$\{a, a + g, a + 2g, \ldots, a + 6g\}$$

for some a and g of G; as

$$(a_1 + H) \cup \{a_2, a_3\}$$

with H being the subgroup of G of order 5 and for certain $a_1, a_2, a_3 \in G$; or in the form

$$(a_1 + H) \cup (a_2 + H) \cup \{a_3\},$$

where H is the subgroup of G of order 3, and $a_1, a_2, a_3 \in G$. For example,

$$\{0, 1, 2, 3, 4, 5, 6\},$$

$$\{0, 3, 6, 9, 12\} \cup \{1, 4\},$$

and

$$\{0, 5, 10\} \cup \{1, 6, 11\} \cup \{2\}$$

all yield minimum sumset size. As these examples indicate, Problem D.5 may be quite difficult in general.

We are able, however, to say more about the cases when $u(n, m, h)$ achieves its extreme values. Recall that, by Proposition 4.15, we have

$$m \leq u(n, m, h) \leq \min\{n, hm - h + 1\},$$

with $u(n, m, h) = m$ if, and only if, $h = 1$ or n is divisible by m.

In the case when $u(n, m, h) = m$ and $h \geq 2$, we must have $m \in D(n)$, and therefore a subgroup of G of order m exists. Taking A to be a coset of this subgroup, hA will also be a coset of the same subgroup, and thus $|hA| = m$. It turns out that the converse of this holds as well, as we have the following.

Proposition D.6 *Let A be an m-subset of G and $h \geq 1$. Then $|hA| \geq m$ with equality if, and only if, $h = 1$ or A is a coset of a subgroup of G.*

For the proof of Proposition D.6, see page 333. In light of Proposition D.6, Problem D.5 is solved in the case when $u(n, m, h) = m$.

Let us now turn to the upper bounds of $u(n, m, h)$ in Proposition 4.15. The case when $u(n, m, h) = n$ translates to the situation when every m-subset A of G has $hA = G$; sets like these are called *h-fold bases* (cf. Chapter B), and the minimum value of m for which every m-subset of G is an h-fold basis of G is called the *h-critical number of G*. We allocate a separate chapter to critical numbers; in particular, we will study the h-critical number of groups in Section E.1.1.

The case when $u(n, m, h) = hm - h + 1 < n$ is also quite interesting, and we offer this question as a special case of Problem D.5.

Problem D.7 *Find all G, m, and $h \geq 2$ for which $\rho(G, m, h) = hm - h + 1 < n$, and for each such instance, find a characterization of all m-subsets of G with an h-fold sumset of size $hm - h + 1$.*

Problem D.7, while only a special case of Problem D.5, is still quite elusive in general, as the example of $G = \mathbb{Z}_{15}$, $m = 7$, and $h = 2$ above indicates. We do have a complete answer, however, when G is of prime order.

Theorem D.8 *Let p be a prime, $h \geq 2$, and suppose that $hm-h+1 < p$. Then an m-subset A of $\mathbb{Z}_p$ satisfies $|hA| = hm - h + 1$ if, and only if, A is an arithmetic progression; that is,*

$$A = \{a, a+g, a+2g, \dots, a+(m-1)g\}$$

for some $a, g \in \mathbb{Z}_p$.

Theorem D.8 can be reduced to Vosper's Theorem (cf. [197]); see page 333. As we pointed out above, the "only if" part of Theorem D.8 doesn't hold when $h = 1$, and it is also false when $hm - h + 1 = p$. For example, the set $A = \{0, 1, 2, 4\}$ is not an arithmetic progression in $\mathbb{Z}_7$ (something that can be easily verified), but $|2A| = 7 = 2 \cdot 4 - 2 + 1$.

Let us see what we can say when n is not prime and the lower bound of Corollary D.4 is achieved so that

$$\rho(G, m, h) = \min\{p, hm - h + 1\},$$

where p is the smallest prime divisor of n. According to Corollary D.4, we then have $p \geq m$. We examine three cases.

When $m \leq p < hm - h + 1$, we can find m-subsets A of G for which

$$|hA| = p = \min\{p, hm - h + 1\},$$

as follows. Let H be any subgroup of G with $|H| = p$. Then for any $g \in G$, the coset $g + H$ contains p elements, and its h-fold sumset is another coset of H. Thus, any m-subset A of $g + H$ has $|hA| = p$. It turns out that there are no others:

Theorem D.9 *Let p be the smallest prime divisor of n, A be an m-subset of G, and assume that $m \leq p < hm - h + 1$. Then $|hA| = p$ if, and only if, A is contained in a coset of some subgroup H of G with $|H| = p$.*

As Grynkiewicz pointed out in [104], Theorem D.9 follows easily from Kneser's Theorem [133]; we explain this on page 334.

Assume now that $p > hm - h + 1$. This time, we can find m-subsets A of G for which

$$|hA| = hm - h + 1 = \min\{p, hm - h + 1\},$$

as follows. Let $g \in G$ be of order p. If A is the arithmetic progression

$$A = \{a, a+g, a+2g, \dots, a+(m-1)g\},$$

then we clearly have

$$hA = \{ha, ha+g, ha+2g, \dots, ha+h(m-1)g\};$$

since $p > h(m-1) + 1$, these $hm - h + 1$ elements are all distinct.

As a generalization of Theorem D.8 above, we can prove that the converse holds as well:

Theorem D.10 *Let $h \geq 2$, p be the smallest prime divisor of n, A be an m-subset of G, and assume that $p > hm - h + 1$. Then $|hA| = hm - h + 1$ if, and only if, A is an arithmetic progression.*

The proof of Theorem D.10 follows from Kemperman's famous results in [129]—see page 335.

Finally, the third case, when $p = hm - h + 1$: this case seems more complicated. Not only do we have the option of arithmetic progressions of length m and cosets of a subgroup of order p, but there are other possibilities as well. For example, with $p = 7$, $m = 4$, and $h = 2$ (as is the case, for example, with $\rho(\mathbb{Z}_{49}, 4, 2) = 7$) the subset

$$A = \{0, a, (n-a)/2, (n+a)/2\}$$

works as well, as we get

$$2A = \{0, a, 2a, (n-a)/2, (n+a)/2, (n+3a)/2, n-a\}$$

(since $p = 7$ implies that n must be odd, we need to assume that a is odd). The following problem seems particularly intriguing:

Problem D.11 *Suppose that n, mm, and h are positive integers, p is the smallest prime divisor of n, and $p = hm - h + 1$. Classify all m-subsets A of G for which $|hA| = p$.*

D.1.2 Limited number of terms

Here we ought to consider, for a given group G, positive integer m (with $m \leq n = |G|$), and nonnegative integer s,

$$\rho(G, m, [0, s]) = \min\{|[0, s]A| \mid A \subseteq G, |A| = m\},$$

that is, the minimum size of $\cup_{h=0}^{s} hA$ for an m-element subset A of G.

However, we have the following result.

Proposition D.12 *For any group G, positive integer $m \leq n$, and nonnegative integer s we have*

$$\rho(G, m, [0, s]) = \rho(G, m, s).$$

We can prove Proposition D.12 as follows. Suppose first that A is a subset of G of size m and that it has a minimum-size $[0, s]$-fold sumset:

$$|[0, s]A| = \rho(G, m, [0, s]).$$

Clearly, $sA \subseteq [0, s]A$, so

$$\rho(G, m, [0, s]) = |[0, s]A| \geq |sA| \geq \rho(G, m, s).$$

For the other direction, choose a subset A of G of size m for which

$$|sA| = \rho(G, m, s).$$

By Proposition 3.2, we may assume that $0 \in A$, so $[0, s]A = sA$, which implies that

$$\rho(G, m, [0, s]) \leq |[0, s]A| = |sA| = \rho(G, m, s).$$

Proposition D.12 makes this subsection superfluous.

D.1.3 Arbitrary number of terms

Here we consider, for a given group G and positive integer m (with $m \leq n = |G|$) the quantity

$$\rho(G, m, \mathbb{N}_0) = \min\{|\langle A\rangle| \mid A \subseteq G, |A| = m\}.$$

Recall that $\langle A\rangle$ is the subgroup of G generated by A; this immediately gives the following result:

Proposition D.13 *For any group G and positive integer $m \leq n$, we have*

$$\rho(G, m, \mathbb{N}_0) = \min\{d \in D(n) \mid d \geq m\}.$$

D.2 Unrestricted signed sumsets

Our goal in this section is to investigate the quantity

$$\rho_{\pm}(G, m, H) = \min\{|H_{\pm}A| \mid A \subseteq G, |A| = m\}$$

where $H_{\pm}A$ is the union of all h-fold signed sumsets $h_{\pm}A$ for $h \in H$. We consider three special cases: when H consists of a single nonnegative integer h, when H consists of all nonnegative integers up to some value s, and when H is the entire set of nonnegative integers.

D.2.1 Fixed number of terms

Here we consider

$$\rho_{\pm}(G, m, h) = \min\{|h_{\pm}A| \mid A \subseteq G, |A| = m\},$$

that is, the minimum size of an h-fold signed sumset of an m-element subset of G.

It is easy to see that

$$\rho_{\pm}(G, 1, h) = 1,$$

$$\rho_{\pm}(G, m, 0) = 1,$$

and

$$\rho_{\pm}(G, m, 1) = m.$$

(To see the last equality, it suffices to verify that one can always find a *symmetric* subset of size m in G, that is, an m-subset A of G for which $A = -A$.) Therefore, for the rest of this subsection, we assume that $m \geq 2$ and $h \geq 2$.

Perhaps surprisingly, we find that, while the h-fold signed sumset of a given set is generally much larger than its sumset, $\rho_{\pm}(G, m, h)$ often agrees with $\rho(G, m, h)$; in particular, this is always the case when G is cyclic:

Theorem D.14 (Bajnok and Matzke; cf. [22]) *For all positive integers n, m, and h, we have*

$$\rho_{\pm}(\mathbb{Z}_n, m, h) = \rho(\mathbb{Z}_n, m, h).$$

The situation seems considerably more complicated for noncyclic groups: in contrast to $\rho(G, m, h)$, the value of $\rho_{\pm}(G, m, h)$ depends on the structure of G rather than just the order n of G. We do not have the answer to the following general problem:

Problem D.15 *Find the value of $\rho_{\pm}(G, m, h)$ for all noncyclic groups G.*

We do, however, have some tight bounds. Observe that by Theorem D.3, we have the lower bound

$$\rho_{\pm}(G,m,h) \geq u(n,m,h) = \min\{f_d(m,h) \ : \ d \in D(n)\}.$$

In [22], we proved that with a certain subset $D(G,m)$ of $D(n)$, we have

$$\rho_{\pm}(G,m,h) \leq u_{\pm}(G,m,h) = \min\{f_d(m,h) \ : \ d \in D(G,m)\};$$

here $D(G,m)$ is defined in terms of the *type* $(n_1,\dots,n_r)$ of G, that is, via integers $n_1,\dots,n_r$ such that $n_1 \geq 2$, n_i is a divisor of n_{i+1} for each $i \in \{1,\dots,r-1\}$, and for which G is isomorphic to the invariant product

$$\mathbb{Z}_{n_1} \times \cdots \times \mathbb{Z}_{n_r}.$$

Namely, we proved the following result:

Theorem D.16 (Bajnok and Matzke; cf. [22]) *The minimum size of the h-fold signed sumset of an m-subset of a group G of type* $(n_1,\dots,n_r)$ *satisfies*

$$\rho_{\pm}(G,m,h) \leq u_{\pm}(G,m,h),$$

where

$$u_{\pm}(G,m,h) = \min\{f_d(m,h) \ : \ d \in D(G,m)\}$$

with

$$D(G,m) = \{d \in D(n) \ : \ d = d_1 \cdots d_r, d_1 \in D(n_1),\dots,d_r \in D(n_r), dn_r \geq d_r m\}.$$

Observe that, for cyclic groups of order n, $D(G,m)$ is simply $D(n)$.

We are aware of only one type of scenario where $u_{\pm}(G,m,h)$ does not yield the actual value of $\rho_{\pm}(G,m,h)$; it only occurs when $h=2$.

As a prime example, consider the group $\mathbb{Z}_p^2$ for an odd prime p, let $m=(p^2-1)/2$, and $h=2$. We find that

$$D(\mathbb{Z}_p^2,(p^2-1)/2) = \{p,p^2\};$$

and we have

$$f_p = f_{p^2} = p^2,$$

hence

$$u_{\pm}(\mathbb{Z}_p^2,(p^2-1)/2,2) = p^2.$$

However, observe that each nonzero element of $\mathbb{Z}_p^2$ has order $p \geq 3$, thus we can partition $\mathbb{Z}_p^2$ as

$$\{0\} \cup A \cup (-A)$$

for some subset A of $\mathbb{Z}_p^2$ of size $(p^2-1)/2$. Since clearly $0 \not\in 2_{\pm}A$, we have

$$\rho_{\pm}(\mathbb{Z}_p^2,(p^2-1)/2,2) \leq p^2-1.$$

A bit more generally, if d is an odd element of $D(n)$ so that $d \geq 2m+1$, then the same argument yields

$$\rho_{\pm}(G,m,2) \leq d-1,$$

and therefore we have the following:

Proposition D.17 (Bajnok and Matzke; cf. [22]) *Suppose that G is an abelian group of order n and type $(n_1, \ldots, n_r)$. Let $m \leq n$, and let d_m be the smallest odd element of $D(n)$ that is at least $2m+1$; if no such element exists, set $d_m = \infty$. We then have*

$$\rho_{\pm}(G, m, 2) \leq \min\{u_{\pm}(G, m, 2), d_m - 1\}.$$

We make the following conjecture:

Conjecture D.18 (Bajnok and Matzke; cf. [22]) *Suppose that G is an abelian group of order n and type $(n_1, \ldots, n_r)$.*

If $h \geq 3$, then

$$\rho_{\pm}(G, m, h) = u_{\pm}(G, m, h).$$

If each odd divisor of n is less than $2m$, then

$$\rho_{\pm}(G, m, 2) = u_{\pm}(G, m, 2).$$

If there are odd divisors of n greater than $2m$, let d_m be the smallest one. We then have

$$\rho_{\pm}(G, m, 2) = \min\{u_{\pm}(G, m, 2), d_m - 1\}.$$

Problem D.19 *Prove or disprove Conjecture D.18.*

We are able to say more about minimum sumset size in elementary abelian groups; in particular, we wish to study $\rho_{\pm}(\mathbb{Z}_p^r, m, h)$, where p denotes a positive prime and r is a positive integer. By Theorem D.14, we assume that $r \geq 2$, and, since obviously

$$\rho_{\pm}(\mathbb{Z}_2^r, m, h) = \rho(\mathbb{Z}_2^r, m, h)$$

for all m, h, and r, we will also assume that $p \geq 3$.

Let us first exhibit a sufficient condition for $\rho_{\pm}(\mathbb{Z}_p^r, m, h)$ to equal $\rho(\mathbb{Z}_p^r, m, h)$. When $p \leq h$, our result is easy to state:

Theorem D.20 (Bajnok and Matzke; cf. [23]) *If $p \leq h$, then for all values of $1 \leq m \leq p^r$ we have*

$$\rho_{\pm}(\mathbb{Z}_p^r, m, h) = \rho(\mathbb{Z}_p^r, m, h).$$

The case $h \leq p-1$ is more complicated and delicate. In order to state our results, we will need to introduce some notations. Suppose that $m \geq 2$ is a given positive integer. First, we let k be the maximal integer for which

$$p^k + \delta \leq hm - h + 1,$$

where δ equals 0 or 1, depending on whether $p-1$ is divisible by h or not. Second, we let c be the maximal integer for which

$$(hc+1) \cdot p^k + \delta \leq hm - h + 1.$$

Note that k and c are nonnegative integers and $c \leq p-1$, since for $c \geq p$ we would have

$$(hc+1) \cdot p^k \geq p^{k+1} + \delta > hm - h + 1.$$

It is also worth noting that

$$f_1(m, h) = hm - h + 1.$$

Our sufficient condition can now be stated as follows:

Theorem D.21 (Bajnok and Matzke; cf. [23]) *Suppose that $2 \leq h \leq p-1$, and let k and c be the unique nonnegative integers defined above. If*

$$m \leq (c+1) \cdot p^k,$$

then

$$\rho_{\pm}(\mathbb{Z}_p^r, m, h) = \rho(\mathbb{Z}_p^r, m, h).$$

In fact, we believe that this condition is also necessary:

Conjecture D.22 (Bajnok and Matzke; cf. [23]) *The converse of Theorem D.21 is true as well; that is, if $2 \leq h \leq p-1$, k and c are the unique nonnegative integers defined above, and*

$$m > (c+1) \cdot p^k,$$

then

$$\rho_{\pm}(\mathbb{Z}_p^r, m, h) > \rho(\mathbb{Z}_p^r, m, h).$$

Problem D.23 *Prove or disprove Conjecture D.22.*

We are able to prove that Conjecture D.22 holds in the case of $\rho_{\pm}(\mathbb{Z}_p^2, m, 2)$:

Theorem D.24 (Bajnok and Matzke; cf. [23]) *Let p be an odd prime and $m \leq p^2$ be a positive integer. Then*

$$\rho_{\pm}(\mathbb{Z}_p^2, m, 2) = \rho(\mathbb{Z}_p^2, m, 2),$$

if, and only if, one of the following holds:

- *$m \leq p$,*
- *$m \geq (p^2+1)/2$, or*
- *there is a positive integer $c \leq (p-1)/2$ for which*

$$c \cdot p + (p+1)/2 \leq m \leq (c+1) \cdot p.$$

Theorem D.24 does not tell us the value of $\rho_{\pm}(\mathbb{Z}_p^2, m, 2)$ when it is more than $\rho(\mathbb{Z}_p^2, m, 2)$; for these cases, we have the following recent result of Lee:

Theorem D.25 (Lee; cf. [139]) *Let p be an odd prime.*

1. *If $m = c \cdot p + v$ with $1 \leq c \leq (p-3)/2$ and $1 \leq v \leq (p-1)/2$, then*

$$\rho_{\pm}(\mathbb{Z}_p^2, m, 2) = (2c+1)p.$$

2. *If $m = c \cdot p + v$ with $c = (p-1)/2$ and $1 \leq v \leq (p-1)/2$, then*

$$\rho_{\pm}(\mathbb{Z}_p^2, m, 2) = p^2 - 1.$$

Combining Theorems D.24 and D.25, we get:

Corollary D.26 (Bajnok and Matzke; cf. [23] and Lee; cf. [139]) *Let us write m as $m = cp + v$ with*

$$0 \leq c \leq p-1 \quad and \quad 1 \leq v \leq p.$$

We then have:

c	v	$\rho(\mathbb{Z}_p^2, m, 2)$		$\rho_\pm(\mathbb{Z}_p^2, m, 2)$		$u_\pm(\mathbb{Z}_p^2, m, 2)$
0	$v \leq (p-1)/2$	$2m-1$	$=$	$2m-1$	$=$	$2m-1$
	$v \geq (p+1)/2$	p	$=$	p	$=$	p
$1 \leq c \leq (p-3)/2$	$v \leq (p-1)/2$	$2m-1$	$<$	$(2c+1)p$	$=$	$(2c+1)p$
	$v \geq (p+1)/2$	$(2c+1)p$	$=$	$(2c+1)p$	$=$	$(2c+1)p$
$c = (p-1)/2$	$v \leq (p-1)/2$	$2m-1$	$<$	p^2-1	$<$	p^2
	$v \geq (p+1)/2$	p^2	$=$	p^2	$=$	p^2
$c \geq (p+1)/2$	*any* v	p^2	$=$	p^2	$=$	p^2

We can also see that we have settled Conjecture D.18 for the group $\mathbb{Z}_p^2$. Indeed, from Corollary D.26, we see that

$$\rho_\pm(\mathbb{Z}_p^2, m, 2) = u_\pm(\mathbb{Z}_p^2, m, 2)$$

in all cases, except when $m = cp + v$ with $c = (p-1)/2$ and $1 \leq v \leq (p-1)/2$, this case coincides exactly with

$$d_m - 1 < u_\pm(\mathbb{Z}_p^2, m, 2)$$

for

$$d_m = \min\{d \in D(n) \mid d \text{ odd and } d > 2m\}.$$

Therefore:

Corollary D.27 (Bajnok and Matzke; cf. [23] and Lee; cf. [139]) *Let p be an odd prime. Conjecture D.18 holds for elementary abelian p-groups of rank two.*

According to Theorem D.24, for a given p, there are exactly $(p-1)^2/4$ values of m for which $\rho_\pm(\mathbb{Z}_p^2, m, 2)$ and $\rho(\mathbb{Z}_p^2, m, 2)$ disagree—fewer than $1/4$ of all possible values. We have not been able to find any groups where this proportion is higher than $1/4$, so make the following conjecture:

Conjecture D.28 *For any abelian group G of order n, we have fewer than $n/4$ values of m for which $\rho_\pm(G, m, 2)$ and $\rho(G, m, 2)$ disagree.*

Problem D.29 *Prove (or disprove) Conjecture D.28.*

We do not know the generalization of Theorem D.24 for higher rank:

Problem D.30 *Generalize Theorem D.24 for rank $r \geq 3$.*

We also have the following "inverse type" result from [22] regarding subsets that achieve $\rho_\pm(G, m, h)$. Given a group G and a positive integer $m \leq |G|$, we define a certain collection $\mathcal{A}(G, m)$ of m-subsets of G. We let

- $\text{Sym}(G, m)$ be the collection of *symmetric* m-subsets of G, that is, m-subsets A of G for which $A = -A$;

- $\mathrm{Nsym}(G,m)$ be the collection of *near-symmetric* m-subsets of G, that is, m-subsets A of G that are not symmetric, but for which $A \setminus \{a\}$ is symmetric for some $a \in A$;

- $\mathrm{Asym}(G,m)$ be the collection of *asymmetric* m-subsets of G, that is, m-subsets A of G for which $A \cap (-A) = \emptyset$.

We then let

$$\mathcal{A}(G,m) = \mathrm{Sym}(G,m) \cup \mathrm{Nsym}(G,m) \cup \mathrm{Asym}(G,m).$$

In other words, $\mathcal{A}(G,m)$ consists of those m-subsets of G that have exactly m, $m-1$, or 0 elements whose inverse is also in the set.

Theorem D.31 (Bajnok and Matzke; cf. [22]) *For every G, m, and h, we have*

$$\rho_\pm(G,m,h) = \min\{|h_\pm A| \ : \ A \in \mathcal{A}(G,m)\}.$$

We should add that each of the three types of sets are essential as can be seen by examples (cf. [22]). However, we do not know the answers to the following problems:

Problem D.32 *Classify each situation where the minimum signed sumset size is achieved by symmetric sets.*

Problem D.33 *Classify each situation where the minimum signed sumset size is achieved by near-symmetric sets.*

Problem D.34 *Classify each situation where the minimum signed sumset size is achieved by asymmetric sets.*

Problem D.35 *Classify each situation where the minimum signed sumset size is achieved by sets that are not in $\mathcal{A}(G,m)$.*

D.2.2 Limited number of terms

Here we consider, for a given group G, positive integer m (with $m \leq n = |G|$), and nonnegative integer s,

$$\rho_\pm(G,m,[0,s]) = \min\{|[0,s]_\pm A| \mid A \subseteq G, |A| = m\},$$

that is, the minimum size of $[0,s]_\pm A$ for an m-element subset A of G.

We note that we don't have a version of Proposition 3.2 for signed sumsets, so we are not able to reduce this entire section to Section D.2.1. (However, one may be able to apply similar techniques.)

It is easy to see that, for every m, we have

$$\rho_\pm(G,m,[0,0]) = 1,$$

and for every s, we have

$$\rho_\pm(G,1,[0,s]) = 1.$$

Furthermore, we can evaluate $\rho_\pm(G,m,[0,1])$ as follows. Note that

$$[0,1]_\pm A = A \cup (-A) \cup \{0\},$$

so $|[0,1]_\pm A| \geq m$ with equality if, and only if, A is symmetric (that is, $A = -A$) and $0 \in A$. Let us see the conditions that allow for such a set A. We can partition G into the pairwise disjoint union of four (potentially empty) parts: $\{0\}$, $\mathrm{Ord}(G,2)$, K, and $-K$. Therefore, if $\mathrm{Ord}(G,2) \neq \emptyset$, we can take A to be the set containing 0, together with the right number of pairs of the form $\pm k$ with $k \in K$, and some elements from $\mathrm{Ord}(G,2)$. Likewise, if $\mathrm{Ord}(G,2) = \emptyset$ but m is odd, we can let A be 0 together with $(m-1)/2$ elements of K together with their inverses.

This leaves only the case when $\mathrm{Ord}(G,2) = \emptyset$ (that is, n is odd), and m is even. It is easy to find sets A with $|[0,1]_{\pm}A| = m+1$, and we can see that that is the best we can do. Indeed, we will have either $0 \not\in A$ (but $0 \in [0,1]_{\pm}A$), or an element $a \in A$ for which $-a \not\in A$ (but $-a \in [0,1]_{\pm}A$).

Summarizing our results thus far, we have the following.

Proposition D.36 *Let G be an abelian group G of order n, and let m and s positive integers.*

- *For all m we have*
$$\rho_{\pm}(G,m,[0,0]) = 1.$$
- *For all s we have*
$$\rho_{\pm}(G,1,[0,s]) = 1.$$
- *For all m we have*
$$\rho_{\pm}(G,m,[0,1]) = \begin{cases} m & \text{if } n \text{ is even or } m \text{ is odd,} \\ m+1 & \text{if } n \text{ is odd and } m \text{ is even.} \end{cases}$$

For values of $s \geq 2$ and $m \geq 2$, we have no exact values for $\rho_{\pm}(G,m,[0,s])$ in general.

However, Matzke in [155] provided the following upper bound for $\rho_{\pm}(G,m,[0,s])$ for the case when G is cyclic. First, we introduce some notations. For a positive integer k we let $v(k)$ denote the highest power of 2 that is a divisor of k. For given positive integers n and m we then define
$$D_1(n) = \{d \in D(n) \mid v(n) \geq v(d\lceil m/d \rceil)\}$$
and
$$D_2(n) = \{d \in D(n) \mid v(n) < v(d\lceil m/d \rceil)\}.$$
Finally, we set
$$u_{\pm}(n,m,[0,s]) = \min\{\min\{f_d(m,s) \mid d \in D_1(n)\},\ \min\{f_d(m+d,s) \mid d \in D_2(n)\}\}.$$
(Note that for $d \in D_2(n)$ we have $d\lceil m/d \rceil < n$ and thus $m+d \leq n$.)

Theorem D.37 (Matzke; cf. [155]) *With the notations just introduced, we have*
$$\rho_{\pm}(\mathbb{Z}_n,m,[0,s]) \leq u_{\pm}(n,m,[0,s]).$$

Furthermore, Matzke believes that equality holds in Theorem D.37:

Conjecture D.38 (Matzke; cf. [155]) *With the notations introduced above, we have*
$$\rho_{\pm}(\mathbb{Z}_n,m,[0,s]) = u_{\pm}(n,m,[0,s]).$$

Problem D.39 *Prove or disprove Conjecture D.38.*

As Grynkiewicz pointed out (cf. [104]), we can prove Conjecture D.38 for prime values of n:

Theorem D.40 *For odd prime values of p we have*

$$\rho_{\pm}(\mathbb{Z}_p, m, [0,s]) = u_{\pm}(p, m, [0,s]) = \min\{p, 2s\lfloor m/2 \rfloor + 1\}.$$

We present the short proof on page 335.

We do not know much about the value of $\rho_{\pm}(G, m, [0,s])$ for noncyclic groups:

Problem D.41 *Find the value of (or, at least, find good bounds for) $\rho_{\pm}(G, m, [0,s])$ for noncyclic groups G and integers m and s.*

D.2.3 Arbitrary number of terms

This subsection is identical to Subsection D.1.3.

D.3 Restricted sumsets

Our goal in this section is to investigate the quantity

$$\rho\hat{\ }(G, m, H) = \min\{|H\hat{\ }A| \mid A \subseteq G, |A| = m\}$$

where $H\hat{\ }A$ is the union of all h-fold restricted sumsets $h\hat{\ }A$ for $h \in H$. We consider three special cases: when H consists of a single nonnegative integer h, when H consists of all nonnegative integers up to some value s, and when H is the entire set of nonnegative integers.

D.3.1 Fixed number of terms

Here we consider

$$\rho\hat{\ }(G, m, h) = \min\{|h\hat{\ }A| \mid A \subseteq G, |A| = m\},$$

that is, the minimum size of an h-fold restricted sumset of an m-element subset of G.

Note that, when $h > m$, we obviously have $h\hat{\ }A = \emptyset$ for every m-subset A of G, thus we may assume that $h \leq m$. Clearly, for all A we have $0\hat{\ }A = \{0\}$ and $1\hat{\ }A = A$, so

$$\rho\hat{\ }(G, m, 0) = 1$$

and

$$\rho\hat{\ }(G, m, 1) = m.$$

Furthermore, as we saw in Section A.3.1, for all m, h, and m-subsets A of G we have

$$|(m-h)\hat{\ }A| = |h\hat{\ }A|,$$

and thus

$$\rho\hat{\ }(G, m, m-h) = \rho\hat{\ }(G, m, h).$$

Therefore, it suffices to study cases when

$$2 \leq h \leq \left\lfloor \frac{m}{2} \right\rfloor.$$

By these considerations, we see that the only case with $m \leq 4$ in which $\rho\hat{\ }(G, m, h)$ is not immediate is $m = 4$ and $h = 2$, for which we prove the following result.

Proposition D.42 *Let G be an abelian group with* $\mathrm{Ord}(G,2)$ *as the set of its elements of order 2. We have*

$$\rho\hat{}(G,4,2) = \begin{cases} 3 & \text{if } |\mathrm{Ord}(G,2)| \geq 2, \\ 4 & \text{if } |\mathrm{Ord}(G,2)| = 1, \\ 5 & \text{if } |\mathrm{Ord}(G,2)| = 0. \end{cases}$$

The proof of Proposition D.42 can be found on page 336; the proposition on page 336 also provides a complete characterization of 4-subsets A of G with $|2\hat{}A|$ attaining all possible values. We can rephrase the conditions in Proposition D.42 by using the invariant factorization of G. Let

$$G \cong \mathbb{Z}_{n_1} \times \cdots \times \mathbb{Z}_{n_r}$$

with $2 \leq n_1$, $n_1|n_2|\cdots|n_r$. We then have

$$\rho\hat{}(G,4,2) = \begin{cases} 3 & \text{if } r \geq 2 \text{ and } n_{r-1} \text{ is even}, \\ 4 & \text{if } r \geq 2 \text{ and } n_{r-1} \text{ is odd but } n_r \text{ is even, or } r = 1 \text{ and } n_r \text{ is even}, \\ 5 & \text{if } n_r \text{ is odd}. \end{cases}$$

(Note that n_r is even if, and only if, $n = |G|$ is even.)

As Proposition D.42 demonstrates, the value of $\rho\hat{}(G,4,h)$ may be less than 4. (This is in contrast to $\rho(G,m,h) \geq m$; see Proposition D.1.) However, we can easily prove that

$$\rho\hat{}(G,5,2) \geq 5,$$

as follows. Let $A = \{a_1, \dots, a_5\}$ (with $|A| = 5$); we then have

$$2\hat{}A = \{a_i + a_j \mid 1 \leq i < j \leq 5\}.$$

It is easy to see that

$$B = \{a_1 + a_i \mid i = 2,3,4,5\}$$

is a 4-subset of $2\hat{}A$, thus $|2\hat{}A| \geq 4$; furthermore, for $2\hat{}A$ to have size 4, we must have (among other things) $a_2 + a_3 \in B$ and $a_4 + a_5 \in B$. Since

$$a_2 + a_3 \not\in \{a_1 + a_2, a_1 + a_3\},$$

we have, w.l.o.g.,

$$a_2 + a_3 = a_1 + a_4;$$

similarly,

$$a_4 + a_5 = a_1 + a_2.$$

Adding these two equations and cancelling yields

$$a_3 + a_5 = 2a_1,$$

and, therefore, $2a_1 \in B$, but that cannot happen. Therefore, $|2\hat{}A| \geq 5$.

In fact, Carrick evaluated $\rho\hat{}(G,5,2)$ for every group G:

Theorem D.43 (Carrick; cf. [51]) *For an abelian group G of order n we have*

$$\rho\hat{}(G,5,2) = \begin{cases} 5 & \text{if } n \text{ is divisible by } 5, \\ 6 & \text{if } n \text{ is not divisible by } 5 \text{ but is divisible by } 6, \\ 7 & \text{otherwise.} \end{cases}$$

The evaluation of $\rho\hat{}(G,m,2)$ gets more difficult as m increases. We still offer:

Problem D.44 *Evaluate $\rho\hat{}(G,6,2)$ for every finite abelian group G.*

The values of $\rho\hat{}(G,m,h)$ are largely unknown in general; we first attempt to evaluate them in the case when G is cyclic. The general question regarding $\rho\hat{}(\mathbb{Z}_n,m,h)$ remains open:

Problem D.45 *Find the exact value of $\rho\hat{}(\mathbb{Z}_n,m,h)$ for all n, m, and h.*

Below we summarize what we know about $\rho\hat{}(\mathbb{Z}_n,m,h)$.

We start by wondering what the minimum possible value of $\rho\hat{}(\mathbb{Z}_n,m,h)$ is. Suppose that $1 \leq h \leq m-1$. Let $A = \{a_1,\ldots,a_m\}$, but assume that the elements (as integers) are between 0 and $n-1$ and that they are in increasing order. Now consider the sets

$$A_1 = \{a_1 + a_2 + \cdots + a_{h-1} + a_i \mid i = h, h+1, \ldots, m\}$$

and

$$A_2 = \{a_1 + a_2 + \cdots + a_h - a_j + a_m \mid j = 1, 2, \ldots, h-1\}.$$

Note that $A_1 \subseteq h\hat{}A$ and (since $h \leq m-1$) $A_2 \subseteq h\hat{}A$. Clearly, $|A_1|+|A_2| = m$; furthermore, A_1 and A_2 are disjoint, since (as integers), we have

$$a_1+a_2+\cdots+a_h < \cdots < a_1+\cdots+a_{h-1}+a_m < a_1+\cdots+a_{h-2}+a_h+a_m < \cdots < a_2+\cdots+a_h+a_m,$$

and the smallest and largest sums differ by $a_m - a_1 < n$, and thus the m elements are distinct in $\mathbb{Z}_n$. Therefore, we have the following.

Proposition D.46 *For all $1 \leq h \leq m-1$ we have $\rho\hat{}(\mathbb{Z}_n,m,h) \geq m$.*

Clearly, equality holds for $h = 1$ and $h = m-1$. Another obvious example of $\rho\hat{}(\mathbb{Z}_n,m,h) = m$ is the case when m is a divisor of n: the (unique) subgroup A of $\mathbb{Z}_n$ of order m has $|h\hat{}A| = |A| = m$; more generally, A can be any coset of this subgroup. As it turns out (see below), there are no other cases where equality holds for the cyclic group.

We now turn to the question of finding upper bounds for $\rho\hat{}(\mathbb{Z}_n,m,h)$. An obvious upper bound is that

$$\rho\hat{}(\mathbb{Z}_n,m,h) \leq \rho(\mathbb{Z}_n,m,h),$$

where, according to Theorem D.3,

$$\rho(\mathbb{Z}_n,m,h) = u(n,m,h) = \min\left\{\left(h\cdot\left\lceil\frac{m}{d}\right\rceil - h + 1\right)\cdot d \mid d \in D(n)\right\}.$$

Recall that we provided, for each $d \in D(n)$, an m-subset $A_d(n,m)$ of $\mathbb{Z}_n$ whose h-fold sumset has size

$$|hA_d(n,m)| = \min\{n, f_d, hm - h + 1\},$$

where

$$f_d = \left(h\cdot\left\lceil\frac{m}{d}\right\rceil - h + 1\right)\cdot d$$

(see Proposition D.2). To get an upper bound for $\rho\hat{}(\mathbb{Z}_n,m,h)$, we now compute the size of the restricted h-fold sumset of $A_d(n,m)$.

We recall the construction from Section D.1.1 as follows. For a divisor d of n, consider the (unique) subgroup H of order d of $\mathbb{Z}_n$, namely,

$$H = \left\{j\cdot\frac{n}{d} \mid j = 0, 1, 2, \ldots, d-1\right\},$$

and then let $A_d(n,m)$ be a certain subset of the "first" $\lceil \frac{m}{d} \rceil$ cosets of H. Namely, we set

$$A_d(n,m) = \bigcup_{i=0}^{c-1}(i+H) \cup \left\{ \left\lceil \frac{m}{d} \right\rceil - 1 + j \cdot \frac{n}{d} \mid j = 0,1,2,\dots,k-1 \right\}$$

where

$$c = \left\lceil \frac{m}{d} \right\rceil - 1$$

and

$$k = m - dc$$

assures that $A = A_d(n,m)$ has size m. We also see that $1 \leq k \leq d$ and thus we take at least 1 but at most d elements of the coset $c + H$; on the other hand, when $c = 0$ (that is, when $m \leq d$), we see that

$$\bigcup_{i=0}^{c-1}(i+H) = \emptyset,$$

so A lies entirely within a single coset and forms the arithmetic progression

$$\left\{ j \cdot \frac{n}{d} \mid j = 0,1,2,\dots,m-1 \right\}.$$

Recall also the function $u\hat{}(n,m,h)$ from Section III. For that, we introduced the notations $h = qd + r$ with

$$q = \left\lceil \frac{h}{d} \right\rceil - 1$$

and

$$r = h - dq,$$

then set

$$f\hat{d}(n,m,h) = \begin{cases} \min\{n, f_d, hm - h^2 + 1\} & \text{if } h \leq \min\{k, d-1\}, \\ \min\{n, hm - h^2 + 1 - \delta_d\} & \text{otherwise;} \end{cases}$$

where δ_d is a "correction term" defined as

$$\delta_d(n,m,h) = \begin{cases} (k-r)r - (d-1) & \text{if } r < k, \\ (d-r)(r-k) - (d-1) & \text{if } k < r < d, \\ d-1 & \text{if } k = r = d, \\ 0 & \text{otherwise.} \end{cases}$$

We can then prove the following result.

Theorem D.47 (Bajnok; cf. [15]) *Suppose that $1 \leq h < m \leq n$. We then have*

$$|h\hat{}A_d(n,m)| = f\hat{d}(n,m,h).$$

The proof of Proposition D.47 can be found on page 337.

Obviously, $|h\hat{}A_d(n,m)|$ provides an upper bound for $\rho\hat{}(\mathbb{Z}_n,m,h)$ for every $d \in D(n)$. In Section III we also defined

$$u\hat{}(n,m,h) = \min\{f\hat{d}(n,m,h) \mid d \in D(n)\},$$

with which Theorem D.47 implies:

Corollary D.48 *Suppose that* $1 \leq h < m \leq n$. *For the cyclic group of order* n *we have*

$$\rho\hat{}(\mathbb{Z}_n, m, h) \leq u\hat{}(n, m, h).$$

Manandhar in [148] has computed the value of $\rho\hat{}(\mathbb{Z}_n, m, h)$ for all $n \leq 25$ and $m \leq 12$; Malec in [147] extended this search to the range $n \leq 40$ and all m. They found that in all cases, $\rho\hat{}(\mathbb{Z}_n, m, h)$ agrees with $u\hat{}(n, m, h)$, with the following exceptions:

n	m	h	$u\hat{}(n, m, h)$	$\rho\hat{}(\mathbb{Z}_n, m, h)$
10	6	3	10	9
12	7	2	11	10
15	8	3	15	14
20	6	3	10	9
20	11	2	19	18
24	13	2	23	22
28	15	2	27	26
35	12	5	35	34
40	11	2	19	18

(Only those with $h \leq \lfloor \frac{m}{2} \rfloor$ are listed.)

From these data it seems that $u\hat{}(n, m, h)$ is a remarkably good upper bound for $\rho\hat{}(\mathbb{Z}_n, m, h)$: indeed, $u\hat{}(n, m, h)$ agrees with $\rho\hat{}(\mathbb{Z}_n, m, h)$ in the overwhelming majority (over 99%) of cases, and when it does not, it differs by only 1. In light of this we pose the following:

Problem D.49 *Classify all situations when*

$$\rho\hat{}(\mathbb{Z}_n, m, h) < u\hat{}(n, m, h);$$

in particular, how much smaller than $u\hat{}(n, m, h)$ *can* $\rho\hat{}(\mathbb{Z}_n, m, h)$ *be?*

As we just mentioned, in all cases that we are aware of, we have $\rho\hat{}(\mathbb{Z}_n, m, h) = u\hat{}(n, m, h)$ or $\rho\hat{}(\mathbb{Z}_n, m, h) = u\hat{}(n, m, h) - 1$.

Let us mention three—as it turns out, quite predicative—examples of the case when $\rho\hat{}(\mathbb{Z}_n, m, h) = u\hat{}(n, m, h) - 1$.

- $u\hat{}(12, 7, 2) = 11$, but $\rho\hat{}(\mathbb{Z}_{12}, 7, 2) = 10$ as shown by the set

$$C_1 = \{0, 4\} \cup \{1, 5, 9\} \cup \{6, 10\}.$$

- $u\hat{}(10, 6, 3) = 10$, but $\rho\hat{}(\mathbb{Z}_{10}, 6, 3) = 9$ as shown by the set

$$C_2 = \{0, 2, 4, 6\} \cup \{7, 9\}.$$

- $u\hat{}(15, 8, 3) = 15$, but $\rho\hat{}(\mathbb{Z}_{15}, 8, 3) = 14$ as shown by the set

$$C_3 = \{0, 3, 6, 9\} \cup \{10, 13, 1, 4\}.$$

Needless to say, we chose to represent our sets in the particular formats above for a reason. In fact, all known situations where $\rho\hat{}(\mathbb{Z}_n, m, h) < u\hat{}(n, m, h)$ can be understood by a particular modification of our sets $A_d(n, m)$, as we now describe.

Observe that the m elements of $A_d(n,m)$ are within $\lceil m/d \rceil$ cosets of the order d subgroup H of $\mathbb{Z}_n$, and at most one of these cosets does not lie entirely in $A_d(n,m)$. We now consider the situation when the m elements are still within $\lceil m/d \rceil$ cosets of H, but exactly two of these cosets don't lie entirely in our set. In order to do so, we write m in the form

$$m = k_1 + (c-1)d + k_2$$

for some positive integers c, k_1, and k_2; we assume that $k_1 < d$, $k_2 < d$, but $k_1 + k_2 > d$. We then are considering m-subsets B of $\mathbb{Z}_n$ of the form

$$B = B_d(n,m;k_1,k_2,g,j_0) = B' \cup \bigcup_{i=1}^{c-1} (ig + H) \cup B'',$$

where H is the subgroup of $\mathbb{Z}_n$ with order d, g is an element of $\mathbb{Z}_n$, B' is a proper subset of H given by

$$B' = \left\{ j \cdot \frac{n}{d} \ : \ j = 0,1,\ldots,k_1 - 1 \right\},$$

and B'' is a proper subset of $cg + H$ of the form

$$B'' = \left\{ cg + (j_0 + j) \cdot \frac{n}{d} \ : \ j = 0,1,\ldots,k_2 - 1 \right\}$$

for some integer j_0 with $0 \leq j_0 \leq d-1$.

It turns out that our set B (under some additional assumptions to be made precise in [15]) has the potential to have a restricted h-fold sumset of size less than $u\hat{\ }(n,m,h)$ in only three cases:

- $h = 2$, $m-1$ is not a power of 2, and n is divisible by $2m-2$;
- $h = 3$, $m = 6$, and n is divisible by 10; or
- h is odd, $m+2$ is divisible by $h+2$, and n is divisible by $hm - h^2$.

In particular, we have the following.

Theorem D.50 (Bajnok; cf. [15]) *Let $B_d(n,m;k_1,k_2,g,j_0)$ be the m-subset of $\mathbb{Z}_n$ defined above, and let h be a positive integer with $h \leq m-1$.*

- *If $h = 2$, $m-1$ is not a power of 2, n is divisible by $2m-2$, and d is an odd divisor of $m-1$ with $d > 1$, then*

  $$B_d\left(n, m; \tfrac{d+1}{2}, \tfrac{d+1}{2}, \tfrac{n}{2m-2}, \tfrac{d-1}{2}\right)$$

  has a restricted 2-fold sumset of size $2m-4$.

- *If $h = 3$, $m = 6$, and n is divisible by 10, then*

  $$B_5\left(n, 6; 4, 2, \tfrac{n}{10}, 3\right)$$

  has a restricted 3-fold sumset of size $3m - 9 = 9$.

- *If h is odd, $m+2$ is divisible by $h+2$, and n is divisible by $hm-h^2$, then*

  $$B_{h+2}\left(n, m; h+1, h+1, \tfrac{n}{hm-h^2}, \tfrac{h+3}{2}\right)$$

  has a restricted h-fold sumset of size $hm - h^2 - 1$.

Our examples above demonstrate the three cases of Theorem D.50 in order: we have

$$C_1 = B_3(12, 7; 2, 2, 1, 1),$$

$$C_2 = B_5(10, 6; 4, 2, 1, 3),$$

and

$$C_3 = B_5(15, 8; 4, 4, 1, 3).$$

The proof of Theorem D.50 is an easy exercise; it also follows from [15] where we verify that, in a certain sense that we make precise, there are no other such sets.

While we seem far away from knowing the value of $\rho\hat{}(\mathbb{Z}_n, m, h)$ in general, we do have the answer in the case when n is prime. Recall that, by Proposition 4.25, for a prime p we have

$$u\hat{}(p, m, h) = \min\{p, hm - h^2 + 1\},$$

thus Corollary D.48, when n equals a prime number p, simplifies to

$$\rho\hat{}(\mathbb{Z}_p, m, h) \leq \min\{p, hm - h^2 + 1\}.$$

The conjecture that equality holds here has been known since the 1960s as the *Erdős–Heilbronn Conjecture* (not mentioned in [80] but in [79]). Three decades later, Dias Da Silva and Hamidoune [66] succeeded in proving the Erdős–Heilbronn Conjecture and thus we have

Theorem D.51 (Dias Da Silva and Hamidoune) *For a prime p and integers $1 \leq h \leq m \leq p$ we have*

$$\rho\hat{}(\mathbb{Z}_p, m, h) = \min\{p, hm - h^2 + 1\}.$$

(The result was reestablished and extended, using different methods, by Alon, Nathanson, and Ruzsa; see [3], [4], and [161].)

For composite values of n, the evaluation of $\rho\hat{}(\mathbb{Z}_n, m, h)$ seems considerably more difficult; we attempt to summarize here what is known for $h = 2$ and $h = 3$.

For $h = 2$, recall Proposition 4.22:

$$u\hat{}(n, m, 2) = \begin{cases} \min\{u(n, m, 2), 2m - 4\} & \text{if } n \text{ and } m \text{ are both even,} \\ \min\{u(n, m, 2), 2m - 3\} & \text{otherwise.} \end{cases}$$

We can then use Theorem D.50 to determine the values of n and m that allow for an improvement over $u\hat{}(n, m, 2)$. Only one case applies: when $m - 1$ is not a power of 2 and n is divisible by $2m - 2$, in which a special set B exists with $|2\hat{}B| = 2m - 4$. Therefore, using Theorem D.3 as well, we have the following result.

Corollary D.52 *For all positive integers n and m with $3 \leq m \leq n$ we have*

$$\rho\hat{}(\mathbb{Z}_n, m, 2) \leq \begin{cases} \min\{\rho(\mathbb{Z}_n, m, 2), 2m - 4\} & \text{if } 2|n \text{ and } 2|m, \text{ or} \\ & (2m-2)|n \text{ and } \nexists k \in \mathbb{N}, m = 2^k + 1; \\ \min\{\rho(\mathbb{Z}_n, m, 2), 2m - 3\} & \text{otherwise.} \end{cases}$$

We have performed a computer search for all m-subsets of $\mathbb{Z}_n$ with $3 \leq m \leq n \leq 40$, and in each case we found that equality holds in Corollary D.52.

Conjecture D.53 *For all n and m, we have equality in Corollary D.52.*

Problem D.54 *Prove or disprove Conjecture D.53.*

We can carry out a similar analysis for the case of $h = 3$, by first recalling from Proposition 4.23:

$$u\hat{}(n,m,3) = \begin{cases} \min\{u(n,m,3), 3m-3-\gcd(n,m-1)\} & \text{if } \gcd(n,m-1) \geq 8; \\ \min\{u(n,m,3), 3m-10\} & \text{if } \gcd(n,m-1) = 7, \text{ or} \\ & \gcd(n,m-1) \leq 5,\ 3|n, \text{ and } 3|m; \\ \min\{u(n,m,3), 3m-9\} & \text{if } \gcd(n,m-1) = 6; \\ \min\{u(n,m,3), 3m-8\} & \text{otherwise.} \end{cases}$$

We then examine Theorem D.50 for the case $h = 3$ to see if we can do better. We find two such instances: when n is divisible by 10 and $m = 6$, and when n is divisible by $3m - 9$ and $m - 3$ is divisible by 5. Observe that, in the latter case, we may assume that m is even, since otherwise $d_0 = (3m - 9)/2 \in D(n)$ and thus

$$u\hat{}(n,m,3) \leq u(n,m,3) \leq f_{d_0} = d_0 \leq 3m - 10.$$

Therefore, we have the following.

Corollary D.55 *Let n and m be positive integers, $4 \leq m \leq n$, and set $d_0 = \gcd(n, m-1)$. We then have*

$$\rho\hat{}(\mathbb{Z}_n,m,3) \leq \begin{cases} \min\{u(n,m,3), 3m-3-d_0\} & \textit{if } d_0 \geq 8; \\ \min\{u(n,m,3), 3m-10\} & \textit{if } d_0 = 7, \textit{ or} \\ & d_0 \leq 5,\ 3|n, \textit{ and } 3|m, \textit{ or} \\ & d_0 \leq 5,\ (3m-9)|n, \textit{ and } 5|(m-3); \\ \min\{u(n,m,3), 3m-9\} & \textit{if } d_0 = 6, \textit{ or} \\ & m = 6 \textit{ and } 10|n \textit{ but } 3\nmid n; \\ \min\{u(n,m,3), 3m-8\} & \textit{otherwise.} \end{cases}$$

Motivated by computational data mentioned above, we have:

Conjecture D.56 *For all n and m, we have equality in Corollary D.55.*

Problem D.57 *Prove or disprove Conjecture D.56.*

Few other results or even conjectures are known for the exact value of $\rho\hat{}(\mathbb{Z}_n, m, h)$ or (especially) for $\rho\hat{}(G, m, h)$ in general. One such result is the following:

Proposition D.58 *Let G be an abelian group of order n. We have $\rho\hat{}(G, m, 2) = n$ if, and only if,*

$$m \geq \frac{n + |\mathrm{Ord}(G,2)| + 3}{2}.$$

In particular,

- *$\rho\hat{}(G, n, 2) = n$ if, and only if, G is not isomorphic to the elementary abelian 2-group; and*

- $\rho\hat{}(\mathbb{Z}_n, m, 2) = n$ *if, and only if,* $m \geq \lfloor n/2 \rfloor + 2$.

Note that $n + |\mathrm{Ord}(G, 2)| + 3$ is always even. We should also mention that Proposition D.58 appeared several times in the literature: cf. [179] by Roth and Lempel; [53] (for the "if" part only) and [54] by Chiaselotti; and [16] by Bajnok.

According to Proposition D.58, the largest value of m for which $\rho\hat{}(G, m, 2)$ is less than n equals

$$m_0 = \frac{n + |\mathrm{Ord}(G, 2)| + 1}{2}.$$

It is then an interesting question to find $\rho\hat{}(G, m_0, 2)$. By Proposition D.58,

$$\rho\hat{}(G, m_0, 2) \leq n - 1.$$

We prove the following:

Proposition D.59 *For*

$$m_0 = \frac{n + |\mathrm{Ord}(G, 2)| + 1}{2}$$

we have

$$\rho\hat{}(G, m_0, 2) \leq \begin{cases} n - 1 & \textit{if } \exists k \in \mathbb{N}, n = 2^k; \\ n - 2 & \textit{otherwise.} \end{cases}$$

In particular,

$$\rho\hat{}\left(\mathbb{Z}_n, \left\lfloor \frac{n}{2} \right\rfloor + 1, 2\right) \leq \begin{cases} n - 1 & \textit{if } \exists k \in \mathbb{N}, n = 2^k; \\ n - 2 & \textit{otherwise.} \end{cases}$$

As we mentioned, the first case of our result holds by Proposition D.58; the proof of the second case can be found on page 344.

In 2002, Gallardo, Grekos, et al. proved that, when G is cyclic, Proposition D.59 holds with equality:

Theorem D.60 (Gallardo, Grekos, et al.; cf. [89]) *For every positive integer* $n \geq 2$ *we have*

$$\rho\hat{}\left(\mathbb{Z}_n, \left\lfloor \frac{n}{2} \right\rfloor + 1, 2\right) = \begin{cases} n - 1 & \textit{if } \exists k \in \mathbb{N}, n = 2^k; \\ n - 2 & \textit{otherwise.} \end{cases}$$

Note that Conjecture D.53, once established, would generalize Theorem D.60.

We do not know if Proposition D.59 holds with equality when G is not cyclic:

Problem D.61 *Decide whether equality holds in Proposition D.59 when G is noncyclic.*

We can generalize Problem D.61 as follows. In Chapter E, we define and investigate the *critical numbers* of groups; in particular, in Section E.3.1 we discuss the *restricted h-critical number* $\chi\hat{}(G, h)$ of G, defined as the smallest positive integer m (if it exists) for which $\rho\hat{}(G, m, h) = n$ or, equivalently, the smallest positive integer m for which $h\hat{}A = G$ holds for every m-subset A of G. For example, according to Proposition D.58, for every group G we have

$$\chi\hat{}(G, 2) = \frac{n + |\mathrm{Ord}(G, 2)| + 3}{2}.$$

The value of $\chi\hat{}(G, h)$ is not known in general—in Section E.3.1 we summarize what we know.

With this notation, we can restate Theorem D.60 in the form

$$\rho\hat{}(\mathbb{Z}_n, \chi\hat{}(\mathbb{Z}_n, 2) - 1, 2) = \begin{cases} n-1 & \text{if } \exists k \in \mathbb{N}, n = 2^k; \\ n-2 & \text{otherwise.} \end{cases}$$

We can then ask for the following:

Problem D.62 *For each G and h, find $\rho\hat{}(G, \chi\hat{}(G,h) - 1, h)$.*

As an example, for $G = \mathbb{Z}_{15}$ we have the following values:

h	$\chi\hat{}(G,h)$	$\rho\hat{}(G, \chi\hat{}(G,h) - 1, h)$
1	15	14
2	9	13
3	9	14
4	8	13
5	9	14
6	9	13
$7-13$	$h+2$	$h+1$
14	15	1

The most general problem, of course, is:

Problem D.63 *Find the exact value of $\rho\hat{}(G, m, h)$ for all groups G and positive integers m and h.*

We should point out that, while $\rho(G, m, h)$ depends only on the order n of G and not on the structure of G, this is definitely not the case for $\rho\hat{}(G, m, h)$ (see, for example, Proposition D.42 or further results below).

Very few general results are known for $\rho\hat{}(G, m, h)$ when G is not cyclic; indeed, most exact values thus far have been for $\rho\hat{}(\mathbb{Z}_p^r, m, 2)$ when p is prime.

The case $p = 2$ is easy. Consider any m-subset $A = \{a_1, \ldots, a_m\}$ of G. We then have

$$2\hat{}A = \{a_i + a_j \mid 1 \leq i < j \leq m\}.$$

Since

$$2A = \{a_i + a_j \mid 1 \leq i \leq j \leq m\} = 2\hat{}A \cup \{0\}$$

and $0 \notin 2\hat{}A$, we have the following result.

Proposition D.64 *For all positive integers r and m, we have*

$$\rho\hat{}(\mathbb{Z}_2^r, m, 2) = \rho(\mathbb{Z}_2^r, m, 2) - 1 = u(2^r, m, 2) - 1.$$

The corresponding answer for $h \geq 3$ is not known.

Problem D.65 *Find $\rho\hat{}(\mathbb{Z}_2^r, m, h)$ for $h \geq 3$.*

For $p \geq 3$ we do not have a complete answer even for $h = 2$. Here is what we do know:

Theorem D.66 (Eliahou and Kervaire; cf. [73]) *Suppose that p is an odd prime, and r and m are positive integers with $2 \leq m \leq p^r$.*

1. *The following statements are equivalent:*

(a) $\binom{2m-2}{m-1}$ *is divisible by* p.

(b) *The base* p *representation of* $m-1$ *contains a digit that is at least* $\frac{p+1}{2}$.

(c) $\rho(\mathbb{Z}_p^r, m, 2) = u(p^r, m, 2) < 2m-1$.

(d) $\rho\hat{}(\mathbb{Z}_p^r, m, 2) = u(p^r, m, 2)$.

2. *The following statements are equivalent:*

(a) $\binom{2m-2}{m-1}$ *is not divisible by* p.

(b) *Each digit in the base* p *representation of* $m-1$ *is at most* $\frac{p-1}{2}$.

(c) $\rho(\mathbb{Z}_p^r, m, 2) = u(p^r, m, 2) = 2m-1$.

(d) $\rho\hat{}(\mathbb{Z}_p^r, m, 2) < u(p^r, m, 2)$ *and* $\rho\hat{}(\mathbb{Z}_p^r, m, 2) \in \{2m-3, 2m-2\}$.

The fact that statements (a), (b), (c) are equivalent—which, obviously, only need to be proved in 1 or 2—can be established by some elementary considerations (recall that $u(n, m, 2) \leq f_1(m, 2) = 2m-1$ holds for all n and m). Observe that the only unsettled case in Theorem D.66 occurs in case 2: we don't know whether the value of $\rho\hat{}(\mathbb{Z}_p^r, m, 2)$ equals $2m-2$ or $2m-3$. In [74] the authors write "The problem of deciding this alternative is unsolved and looks amazingly difficult in general." They have the following additional partial results.

Theorem D.67 (Eliahou and Kervaire; cf. [73, 74, 75]) *Keeping the notations of Theorem D.66, suppose that we are in case 2.*

1. *If* $m-1$ *is not divisible by* p, *then* $\rho\hat{}(\mathbb{Z}_p^r, m, 2) = 2m-3$.

2. *For* $m = p+1$ *we have* $\rho\hat{}(\mathbb{Z}_p^r, m, 2) = 2m-3$ *when* $p=3$ *and* $\rho\hat{}(\mathbb{Z}_p^r, m, 2) = 2m-2$ *when* $p \geq 5$.

3. *Let* $m = 9k+4$ *for some nonnegative integer* k; *we then have* $\rho\hat{}(\mathbb{Z}_3^r, m, 2) = 2m-3$.

4. *Let* $m = 9 \cdot 3^k + 1$ *for some nonnegative integer* k; *we then have* $\rho\hat{}(\mathbb{Z}_3^r, m, 2) = 2m-2$.

We should point out that in the second statement of Theorem D.67 the case $p=3$ follows immediately from Proposition D.42; this simple fact, in turn, implies the third statement of the theorem. After Theorems D.66 and D.67 we see that $\rho\hat{}(\mathbb{Z}_3^r, m, 2)$ has been settled for all cases except when $m \equiv 1 \bmod 27$ or when $m \equiv 10 \bmod 27$. For $p \geq 5$, the simplest unsolved case seems to be when $m = 2p+1$.

Problem D.68 *Find the value of* $\rho\hat{}(\mathbb{Z}_3^r, m, 2)$ *for each* $r \geq 2$ *and* $m \equiv 1$ *mod* 27 *or* $m \equiv 10$ *mod* 27.

Problem D.69 *Find the value of* $\rho\hat{}(\mathbb{Z}_p^r, 2p+1, 2)$ *for each* $r \geq 2$ *and each prime* $p \geq 5$.

Furthermore:

Problem D.70 *Find the value of* $\rho\hat{}(\mathbb{Z}_p^r, m, h)$ *for each prime* $p \geq 3$, $r \geq 2$, $m \geq 5$, *and* $h \geq 3$.

We should also mention an "inverse" result of the same authors:

Theorem D.71 (Eliahou and Kervaire; cf. [74]) *Suppose that $p \geq 5$ is prime, $A \subset \mathbb{Z}_p^r$ with $|A| = p+1$, and that $|2\hat{\ }A| = \rho\hat{\ }(\mathbb{Z}_p^r, p+1, 2) = 2p$. Then A is of the form*

$$A = (a_1 + H) \cup \{a_2\}$$

for some $a_1, a_2 \in G$ and $H \leq G$ with $|H| = p$.

It is easy to verify that the set A above has a restricted 2-fold sumset of size $2p$; according to Theorem D.71, all sets with minimum sumset size are of this form.

Turning now to general abelian groups, we have the following (unsurprising) result for the case of $m = n$:

Theorem D.72 *For positive integers r, we have $2\hat{\ }\mathbb{Z}_2^r = \mathbb{Z}_2^r \setminus \{0\}$, but for all other finite abelian groups G and $1 \leq h \leq \lfloor n/2 \rfloor$, we have $h\hat{\ }G = G$. Consequently,*

$$\rho\hat{\ }(G, n, h) = \begin{cases} n-1 & \text{if } G \cong \mathbb{Z}_2^r \text{ and } h = 2; \\ n & \text{otherwise.} \end{cases}$$

The proof of Theorem D.72 can be found on page 346 (for $h = 2$, see Proposition D.58 as well).

With exact results for $\rho\hat{\ }(G, m, h)$ few and far between, we turn to lower and upper bounds.

We start with some lower bounds on $\rho\hat{\ }(G, m, h)$. Repeating an argument from earlier, we see that for all $A = \{a_1, \ldots, a_m\} \subseteq G$ and $1 \leq h \leq m$, we have

$$\{a_1 + a_2 + \cdots + a_{h-1} + a_i \mid i = h, h+1, \ldots, m\} \subseteq h\hat{\ }A,$$

and these $m - h + 1$ elements are distinct, thus $|h\hat{\ }A| \geq m - h + 1$. The dual inequality (switching h and $m - h$) gives $|h\hat{\ }A| \geq h + 1$, and therefore we have the following obvious lower bound.

Proposition D.73 *For all $1 \leq h \leq m - 1$ we have*

$$\rho\hat{\ }(G, m, h) \geq \max\{m - h + 1, h + 1\};$$

consequently, for $1 \leq h \leq \left\lfloor \frac{m}{2} \right\rfloor$ we have

$$\rho\hat{\ }(G, m, h) \geq m - h + 1.$$

Naturally, we would like to know about the cases where equality occurs in Proposition D.73—this has been very recently answered by Girard, Griffiths, and Hamidoune in [99]. For $h = 1$, the answer, of course, is obvious: equality holds for all G and m. For $h = 2$, we have already seen that equality holds if $m = 4$ and G has more than one element of order 2. More generally, if A is any coset $g + H$ of some subgroup $H = \{0, h_2, \ldots, h_m\}$ of $G_2 = \text{Ord}(G, 2) \cup \{0\}$, then, since $h_i + h_j \in H \setminus \{0\}$ for any $2 \leq i < j \leq m$, we have

$$2\hat{\ }A = \{2g + h_2, \ldots, 2g + h_m\},$$

and thus $|2\hat{\ }A| = m - 1$. It turns out that these are the only cases when equality holds in Proposition D.73:

Theorem D.74 (Girard, Griffiths, and Hamidoune; cf. [99]) *Suppose that A is an m-subset of G and that $1 \leq h \leq \left\lfloor \frac{m}{2} \right\rfloor$. Then*

$$|h\hat{\ }A| = m - h + 1$$

if, and only if, $h = 1$ (and A arbitrary) or $h = 2$ and A is a coset of some subgroup of $\text{Ord}(G, 2) \cup \{0\}$.

As a consequence of Theorem D.74, we get that

$$\rho\hat{}(G,m,h) \geq m-1,$$

with equality if, and only if, $h \in \{2, m-2\}$ and $\mathrm{Ord}(G,2) \cup \{0\}$ possesses a subgroup of order m (and thus $m = 2^k$ for some $2 \leq k \leq e$ where e is the number of even orders in the invariant factorization of G).

As it turns out, we also have a full characterization of cases where

$$\rho\hat{}(G,m,h) = m.$$

This equality clearly holds for $h = 1$, $h = m-1$, or when m is a divisor of n (in which case G has a subgroup of order m).

Let us demonstrate another example. Let $H = \{0, h_2, \ldots, h_m, h_{m+1}\}$ be a subgroup of $G_2 = \mathrm{Ord}(G,2) \cup \{0\}$ in a (noncyclic) group G, and let A be a coset of H in G with one element removed:

$$A = \{g, g+h_2, \ldots, g+h_m\}.$$

As above, we see that

$$2\hat{}A = \{2g+h_2, \ldots, 2g+h_m\},$$

and thus $|2\hat{}A| = m$. The question of lower bounds for $\rho\hat{}(G,m,h)$ is settled for general G by the following results.

Theorem D.75 (Girard, Griffiths, and Hamidoune; cf. [99]) *Suppose that $m \geq 5$ and $1 \leq h \leq \lfloor \frac{m}{2} \rfloor$, and let A be an m-subset of G that is not a coset of some subgroup of* $\mathrm{Ord}(G,2) \cup \{0\}$. *Then $|h\hat{}A| \geq m$. Furthermore, $|h\hat{}A| = m$ if, and only if, (at least) one of the following holds:*

(i) $h = 1$ (and A arbitrary),

(ii) A is a coset of some subgroup of G, or

(iii) $h = 2$ and A is a coset of some subgroup of $\mathrm{Ord}(G,2) \cup \{0\}$ *minus one element.*

Note that the assumption that $m \geq 5$ is permissible as we have the corresponding results for $m \leq 4$ in Proposition D.42 above.

It is worthwhile to compare Theorem D.75 to Proposition D.6. We state the following corollary explicitly.

Corollary D.76 *Suppose that $m \geq 5$ and $1 \leq h \leq m-1$. Let e be the number of even orders in the invariant factorization of G. We then have*

$$\rho\hat{}(G,m,h) \begin{cases} = m-1 & \text{if } h \in \{2, m-2\} \text{ and } m = 2^k \text{ for some } 2 \leq k \leq e; \\ = m & \text{if } h \in \{1, m-1\}, \text{ or} \\ & m | n, \text{ or} \\ & h \in \{2, m-2\} \text{ and } m = 2^k - 1 \text{ for some } 2 \leq k \leq e; \\ \geq m+1 & \text{otherwise.} \end{cases}$$

We can attempt to find a different type of lower bound for $\rho\hat{}(G,m,h)$ by considering Proposition 4.31. We make the following conjecture:

Conjecture D.77 *Let G be a group of order n, p be the smallest prime divisor of n, and assume that $1 \leq h < m \leq n$. We then have*

$$\rho\hat{}(G,m,h) \geq \min\{p, hm - h^2 + 1\},$$

with equality if, and only if, $p \geq m$.

Note that Conjecture D.77 is a generalization of Theorem D.51.
Károlyi succeeded in providing a proof for the case $h = 2$:

Theorem D.78 (Károlyi; cf. [124, 125]) *For any abelian group G of order n, we have*

$$\rho\hat{}(G,m,2) \geq \min\{p, 2m - 3\},$$

where p is the smallest prime divisor of n. Equality may occur if, and only if, $p \geq m$.

A very difficult problem is the following.

Problem D.79 *Prove Conjecture D.77 for $h \geq 3$.*

Let us attempt to characterize situations where the lower bound of Conjecture D.77 is achieved and

$$\rho\hat{}(G,m,h) = \min\{p, hm - h^2 + 1\},$$

where p is the smallest prime divisor of n. We examine four cases.

First, let us assume that $p < m$, in which case $\min\{p, hm - h^2 + 1\} = p$ for all h. If p is odd, then n is odd, and thus, by Theorem D.75, we have $|h\hat{}A| \geq m$, so we cannot have $|h\hat{}A| = p$. If $p = 2 < m$, then $|h\hat{}A| = p$ cannot happen either, since by Proposition D.73 this would imply that

$$2 \geq \max\{m - h + 1, h + 1\},$$

which cannot hold for any m and h with $2 < m$. So this case yields no examples, and we, in fact, verified the "only if" part of the last claim in Conjecture D.77; we are about to prove the "if" part as well.

Suppose now that $m \leq p < hm - h^2 + 1$. Let H be any subgroup of G with $|H| = p$, and let A be any m-subset of H. By Theorem D.51,

$$|h\hat{}A| = \min\{p, hm - h^2 + 1\} = p.$$

Therefore, any m-subset of H and, more generally, any m-subset of any coset of H, is an example of a set for which the lower bound of Conjecture D.77 is achieved. We then question whether there are any other such sets:

Conjecture D.80 *Let p be the smallest prime divisor of n, A be an m-subset of G, and assume that $m \leq p < hm - h^2 + 1$. Then $|h\hat{}A| = p$ if, and only if, A is contained in a coset of some subgroup H of G with $|H| = p$.*

Problem D.81 *Prove (or disprove) Conjecture D.80.*

Assume now that $p > hm - h^2 + 1$. This time, we can find m-subsets A of G for which

$$|h\hat{}A| = hm - h^2 + 1 = \min\{p, hm - h^2 + 1\},$$

as follows. First of all, any m-subset A of G will do, if $h = 1$ or $h = m - 1$. Second, by our proposition on page 336, we see that when $h = 2$ and $m = 4$, 4-subsets of G of the form

$$A = \{a, a + d_1, a + d_2, a + d_1 + d_2\}$$

work for any $a, d_1, d_2 \in G$. (Note that, since $p > hm - h^2 + 1 = 5$ in this case, n is odd, and thus G has no elements of order 2. Therefore, $|2\hat{}A| = 5$.)

We can find further examples by using arithmetic progressions. Let $g \in G$ be of order p. If A is the arithmetic progression

$$A = \{a, a+g, a+2g, \ldots, a+(m-1)g\},$$

then, as we have seen,

$$h\hat{}A = \left\{ha + \frac{h^2-h}{2}g, \ldots, ha + \left(h(m-1) - \frac{h^2-h}{2}\right)g\right\};$$

since $p > hm - h^2 + 1$, these $hm - h^2 + 1$ elements are all distinct. We believe that there are no other examples.

Conjecture D.82 *Let p be the smallest prime divisor of n, A be an m-subset of G, and assume that $p > hm - h^2 + 1$. Then $|h\hat{}A| = hm - h^2 + 1$ if, and only if, (at least) one of the following holds:*

(i) $h = 1$ (and A arbitrary);

(ii) $h = m - 1$ (and A arbitrary);

(iii) $h = 2$, $m = 4$, and A is of the form $A = \{a, a+d_1, a+d_2, a+d_1+d_2\}$ (for arbitrary $a, d_1, d_2 \in G$); or

(iv) A is an arithmetic progression (of size m).

By a result of Károlyi, we have:

Theorem D.83 (Károlyi; cf. [126]) *Conjecture D.82 holds for $h = 2$.*

A difficult problem is the following.

Problem D.84 *Prove (or disprove) Conjecture D.82 for $h \geq 3$.*

Finally, suppose that $p = hm - h^2 + 1$. This case (mirroring the analogous situation for unrestricted sumsets) is more complicated. For example, with $p = 7$, $m = 5$, and $h = 2$ (as is the case, for example, with $\rho\hat{}(\mathbb{Z}_{49}, 5, 2) = 7$) we see that the set

$$A = \{0, a, 2a, (n+a)/2, (n+3a)/2\}$$

provides an example, as we have

$$2\hat{}A = \{a, 2a, 3a, (n+a)/2, (n+3a)/2, (n+5a)/2, (n+7a)/2\}.$$

We have the following intriguing problem:

Problem D.85 *Suppose that n, m, and h are positive integers, p is the smallest prime divisor of n, and $p = hm - h^2 + 1$. Classify all m-subsets A of G for which $\rho\hat{}(G, m, h) = p$.*

Of course, these last types of lower bounds on $\rho\hat{}(G, m, h)$ are not meaningful when p (the smallest prime divisor of $|G|$) is small. For example, when a subset of G is within a subgroup of small size (or within a coset of that subgroup), then its restricted sumset—indeed, any sumset—will not be larger than that subgroup (a coset of that subgroup). Therefore, when there is a divisor d of n with $d \geq m$, then $\rho\hat{}(G, m, h)$ cannot be more than d. Thus, to form more meaningful lower bounds, one may wonder what happens if we assume that no such divisor d exists. The following theorem provides one answer to this question.

Theorem D.86 (Hamidoune, Lladó, and Serra; cf. [113]) *Suppose that G is an abelian group that is either cyclic or is of odd order, and let m be an integer with $m \geq 33$ and $m \geq 21$ in these two cases, respectively. Furthermore, suppose that the only divisor of n that is greater than or equal to m is n itself. Then*

$$\rho\hat{\ }(G, m, 2) \geq \min\{n, 3m/2\}.$$

We can prove that Conjecture D.53, with the additional assumption that the only divisor of n that is greater than or equal to m is n itself, implies Theorem D.86 when G is cyclic, and we only need to know that $m \geq 8$. (The claim is false for $m = 7$ as example C_1 on page 167 demonstrates.) Indeed, we will show that

$$\min\{u(n, m, 2), 2m - 4\} \geq \min\{n, 3m/2\}.$$

Note that $m \geq 8$ implies that $2m - 4 \geq 3m/2$; therefore, our claim clearly holds when

$$u(n, m, 2) \geq \min\{n, 3m/2\}.$$

We can show that, in fact, if the only divisor of n that is greater than or equal to m is n itself, then

$$u(n, m, h) \geq \min\{n, (h + 1)m/2\}$$

holds for all positive integers h.

To see this, let $d \in D(n)$ be such that

$$u(n, m, h) = f_d(m, h) = \left(h \left\lceil \frac{m}{d} \right\rceil - h + 1\right) d.$$

Note that when $d < m \leq 2d$, we have

$$\left(h \left\lceil \frac{m}{d} \right\rceil - h + 1\right) d = (h + 1)d \geq (h + 1)m/2,$$

and in the case when $m > 2d$, we have

$$\left(h \left\lceil \frac{m}{d} \right\rceil - h + 1\right) d \geq \left(h \frac{m}{d} - h + 1\right) d = hm - (h - 1)d > hm - (h - 1)m/2 = (h + 1)m/2.$$

The remaining case to consider is when $m \leq d$, but then, by assumption, $d = n$, for which we have $f_n = n$, and thus our claim is established.

To see that Theorem D.86 is not true for an arbitrary finite abelian group, consider $G = \mathbb{Z}_2^r \times \mathbb{Z}_3$, and let $A = \mathbb{Z}_2^r \times \{0, 1\}$. Note that $n = 3 \cdot 2^r$, $m = 2 \cdot 2^r$, so the only divisor of n that is greater than or equal to m is n itself. We can see that

$$2\hat{\ }A = G \setminus \{(0, 0), (0, 2)\},$$

thus

$$|2\hat{\ }A| = n - 2 = 3m/2 - 2.$$

This example prompts us to wonder the following:

Problem D.87 *Is there a constant C (independent of n and m) for which the inequality*

$$\rho\hat{\ }(G, m, 2) \geq \min\{n, 3m/2\} - C$$

holds for every abelian group G and for every n and m with the property that the only divisor of n that is greater than or equal to m is n itself?

As our example above demonstrates, if there is such a constant, it must be at least 2.

A similar argument shows that, for $h = 3$, Conjecture D.56 implies that, if $m \geq 8$ and the only divisor of n that is greater than or equal to m is n itself, then

$$\rho\hat{}(\mathbb{Z}_n, m, 3) \geq \min\{n, 2m-2\}.$$

Indeed, according to Conjecture D.56, we have

$$\rho\hat{}(\mathbb{Z}_n, m, 3) \geq \min\{u(n, m, 3), 3m-10, 3m-3-\gcd(n, m-1)\}.$$

The fact that our assumptions imply that

$$u(n, m, 3) \geq \min\{n, 2m\} \geq \min\{n, 2m-2\}$$

was shown above, and for $m \geq 8$ we obviously have $3m - 10 \geq 2m - 2$ and $3m - 3 - \gcd(n, m-1) \geq 2m - 2$.

Therefore, we pose the following:

Problem D.88 *Is there a constant C (independent of n and m) for which the inequality*

$$\rho\hat{}(G, m, 3) \geq \min\{n, 2m\} - C$$

holds for every abelian group G and for every n and m with the property that the only divisor of n that is greater than or equal to m is n itself?

Since $\rho\hat{}(\mathbb{Z}_{15}, 6, 3) = u\hat{}(15, 6, 3) = 8$, C would have to be at least 4.

More generally:

Problem D.89 *Let h be a given positive integer. Is there a constant $C(h)$ (independent of n and m but dependent on h) for which the inequality*

$$\rho\hat{}(G, m, h) \geq \min\{n, (h+1)m/2\} - C(h)$$

holds for every abelian group G and for every n and m with the property that the only divisor of n that is greater than or equal to m is n itself?

We can propose some other lower bounds for $\rho\hat{}(G, m, h)$ (at least for $h = 2$) in terms of $\rho(G, m, h)$. One such conjecture is the following:

Conjecture D.90 (Plagne; cf. [174]) *For any finite abelian group G and positive integer $m \geq 2$ we have*

$$\rho\hat{}(G, m, 2) \geq \rho(G, m, 2) - 2.$$

If true, Conjecture D.90 limits the value of $\rho\hat{}(G, m, 2)$ to one of only three possibilities: $\rho(G, m, 2)$, $\rho(G, m, 2) - 1$, or $\rho(G, m, 2) - 2$. We can verify that each exact result mentioned in this section satisfies this conjecture.

Problem D.91 *Prove or disprove Conjecture D.90.*

We should point out that Conjecture D.53 implies Conjecture D.90 for cyclic groups. Indeed, we have

$$\rho(\mathbb{Z}_n, m, 2) \leq f_1 = 2m - 1;$$

furthermore, when n is even, we have

$$\rho(\mathbb{Z}_n, m, 2) \leq f_2 = 2m - 2,$$

and when n is divisible by $2m - 2$, we find that

$$\rho(\mathbb{Z}_n, m, 2) \leq f_{2m-2} = 2m - 2.$$

Another conjecture of this type is the following:

Conjecture D.92 (Lev; cf. [140]) *For any G and m, we have*

$$\rho\hat{}(G,m,2) \geq \min\{\rho(G,m,2), 2m-3-|\mathrm{Ord}(G,2)|\}.$$

Problem D.93 *Prove or disprove Conjecture D.92.*

Lev proved the somewhat weaker result that

$$\rho\hat{}(G,m,2) \geq \min\{\rho(G,m,h), \theta m-3-|\mathrm{Ord}(G,2)|\}$$

where $\theta = \frac{1+\sqrt{5}}{2} \approx 1.6$ is the golden ratio. For the case when G is cyclic, $\mathrm{Ord}(G,2)$ can have at most one element (namely, $\frac{n}{2}$ if n is even), and thus Conjecture D.53 implies Conjecture D.92 for cyclic groups.

We can further illuminate Conjectures D.90 and D.92 by the following example of a set A for which we have both

$$|2\hat{}A| = |2A| - 2$$

and

$$|2\hat{}A| = 2|A| - 3 - |\mathrm{Ord}(G,2)|.$$

Let G be any group of even order that has at least one element of order more than 2 (and thus G is not isomorphic to $\mathbb{Z}_2^r$).

Recall that, by the Fundamental Theorem of Finite Abelian Groups, we have

$$G \cong \mathbb{Z}_2^{\alpha_1} \times \mathbb{Z}_4^{\alpha_2} \times \cdots \times \mathbb{Z}_{2^k}^{\alpha_k} \times H,$$

where $k \in \mathbb{N}$ (since n is even), $\alpha_i \in \mathbb{N}_0$ for each $i = 1, 2, \ldots, k$ with at least one being positive, and H is a group of odd order (perhaps $|H| = 1$). One can also easily verify that

$$G_2 = \mathrm{Ord}(G,2) \cup \{0\} = \{g \in G \mid 2g = 0\}$$

is a subgroup of G. Let a be an element of G of order at least 3, and set

$$A = G_2 \cup \{a + g_2 \mid g_2 \in G_2\}.$$

We can then check that

$$2A = G_2 \cup \{a + g_2 \mid g_2 \in G_2\} \cup \{2a + g_2 \mid g_2 \in G_2\}$$

and

$$2\hat{}A = 2A \setminus \{0, 2a\}.$$

(Note that the sum of two distinct elements of order 2 in G is again an element of order 2, and the sum of 0 and an element of order 2 is, of course, an element of order 2.) Therefore, we have

$$|2\hat{}A| = |2A| - 2 = 2|A| - 3 - |\mathrm{Ord}(G,2)|.$$

By Proposition 4.28, for every $h \geq 3$ and for every positive real number C, one can find positive integers n and m for which

$$\rho\hat{}(\mathbb{Z}_n, m, h) \leq u\hat{}(n,m,h) < u(n,m,h) - C = \rho(\mathbb{Z}_n, m, h) - C.$$

Therefore, for $h \geq 3$, one cannot expect a claim similar to Conjecture D.90. However, we offer:

Problem D.94 *Generalize Conjecture D.92 for $h \geq 3$.*

Let us now turn to upper bounds for $\rho\hat{}(G, m, h)$. Of course, most trivially, we have

$$\rho\hat{}(G, m, h) \leq \rho(G, m, h).$$

More meaningful bounds were given by Plagne for $h = 2$:

Theorem D.95 (Plagne; cf. [174]) *For any G and m, we have*

$$\rho\hat{}(G, m, 2) \leq \min\{\rho(G, m, 2), 2m - 2\}.$$

Furthermore, if n has a prime divisor p that does not divide $m - 1$, then

$$\rho\hat{}(G, m, 2) \leq \min\{\rho(G, m, 2), 2m - 3\}.$$

We are not aware of similar bounds for $h \geq 3$ and thus we pose:

Problem D.96 *Find a function $f(h)$ (as "small" as possible) for which*

$$\rho\hat{}(G, m, h) \leq \min\{\rho(G, m, h), hm - h^2 + 1 + f(h)\}$$

holds for any G, m, and h.

Recall that, by Proposition 4.29 we have

$$u\hat{}(n, m, h) \leq \min\{u(n, m, h), hm - h^2 + 1\}$$

for any n, m, and h.

A particularly intriguing question is the following:

Problem D.97 *Find a set analogous to $A_d(n, m)$ (described earlier) that yields a good upper bound for $\rho\hat{}(G, m, h)$ when G has rank 2 (or more).*

Before closing this section, we define two new quantities related to restricted addition. For positive integers m and h, we let

$$\rho\hat{}(m, h)_{\min} = \min\{\rho\hat{}(G, m, h) \mid |G| \geq m\},$$

and

$$\rho\hat{}(m, h)_{\max} = \max\{\rho\hat{}(G, m, h) \mid |G| \geq m\}.$$

(The condition that the order of G is at least m is necessary to avoid trivialities.)

Clearly,

$$\rho\hat{}(m, 1)_{\min} = \rho\hat{}(m, 1)_{\max} = m,$$

and, as before, we may assume $2 \leq h \leq \lfloor m/2 \rfloor$. By these considerations, the first non-trivial example is $(m, h) = (4, 2)$: According to Proposition D.42, we have $\rho\hat{}(4, 2)_{\min} = 3$ and $\rho\hat{}(4, 2)_{\max} = 5$. Thus we may assume that $m \geq 5$.

The value of $\rho\hat{}(m, h)_{\min}$ is already known for all (m, h): By Corollary D.76, we have:

Theorem D.98 *Suppose that m and h are positive integers with $m \geq 5$ and $h \leq \lfloor m/2 \rfloor$. Then*

$$\rho\hat{}(m, h)_{\min} = \begin{cases} m - 1 & \text{if } h = 2 \text{ and } m = 2^k \text{ for some } k \in \mathbb{N}; \\ m & \text{otherwise.} \end{cases}$$

The value of $\rho\hat{}(m,h)_{\max}$ is not known in general but, from previous results, we can find lower and upper bounds. For a lower bound, recall that by Theorem D.51, when p is a prime with

$$p \geq hm - h^2 + 1,$$

then

$$\rho\hat{}(\mathbb{Z}_p, m, h) = hm - h^2 + 1.$$

This yields the lower bound $hm - h^2 + 1$.

For an upper bound, recall that by Theorem D.3, we have

$$\rho(G, m, h) = u(n, m, h),$$

and if

$$n \geq hm - h + 1,$$

then by Proposition 4.15,

$$u(n, m, h) \leq hm - h + 1,$$

which then is an upper bound for $\rho\hat{}(m,h)_{\max}$. In summary:

Theorem D.99 *Suppose that m and h are positive integers with $m \geq 5$ and $h \leq \lfloor m/2 \rfloor$. Then*

$$hm - h^2 + 1 \leq \rho\hat{}(m,h)_{\max} \leq hm - h + 1.$$

For $h = 2$, we know a bit more: by Theorem D.95, we see that

$$\rho\hat{}(m,2)_{\max} \leq 2m - 2.$$

Therefore:

Theorem D.100 *For all positive integers m, $\rho\hat{}(m,2)_{\max}$ is either $2m-3$ or $2m-2$.*

We offer the following intriguing problem:

Problem D.101 *Find the value of $\rho\hat{}(m,2)_{\max}$ for all positive integers m.*

We note that, by Theorem D.67, we have

$$\rho\hat{}(m,2)_{\max} = 2m - 2$$

when $m - 1 \geq 5$ is a prime, and also when $m = 9 \cdot 3^k + 1$ for some $k \in \mathbb{N}_0$. As a consequence of Theorem D.43, for $m = 5$ we have:

Theorem D.102 (Carrick; cf. [51]) *We have $\rho\hat{}(5,2)_{\max} = 7$.*

We know very little about $\rho\hat{}(m,h)_{\max}$ for $h \geq 3$:

Problem D.103 *Find the value of $\rho\hat{}(m,h)_{\max}$ for all positive integers m and $3 \leq h \leq \lfloor m/2 \rfloor$.*

D.3.2 Limited number of terms

Here we consider, for a given group G, positive integer m (with $m \leq n = |G|$), and nonnegative integer s,

$$\rho\hat{}(G, m, [0, s]) = \min\{|[0, s]\hat{}A| \mid A \subseteq G, |A| = m\},$$

that is, the minimum size of $\cup_{h=0}^{s} h\hat{}A$ for an m-element subset A of G. Clearly, we may restrict our attention to $0 \leq s \leq m$; in fact, when $s \geq m$,

$$\rho\hat{}(G, m, [0, s]) = \rho\hat{}(G, m, \mathbb{N}_0)$$

(cf. Subsection D.3.3).

Note that, although we always have

$$\rho(G, m, [0, s]) = \rho(G, m, s)$$

(cf. Section D.1.2), the quantities $\rho\hat{}(G, m, [0, s])$ and $\rho\hat{}(G, m, s)$ are not necessarily equal. (Of course, the inequality

$$\rho\hat{}(G, m, [0, s]) \geq \rho\hat{}(G, m, s)$$

holds.)

Problem D.104 *Find $\rho\hat{}(G, m, [0, s])$ for any group G, positive integer $m \leq n$, and nonnegative integer s.*

Some authors have investigated the minimum size of $[0, s]\hat{}A$ among all m-subsets of G that possess some additional properties. One such pursuit is the attempt to find

$$\rho_A\hat{}(G, m, [0, s]) = \min\{|[0, s]\hat{}A| \mid A \subseteq G, |A| = m, A \cap -A = \emptyset\},$$

that is, the minimum size of $[0, s]\hat{}A$ among *asymmetric* m-subsets of G. Since no element of $\mathrm{Ord}(G, 2) \cup \{0\}$ may be in an asymmetric set in G, for an asymmetric set of size m to exist, we need to assume that

$$m \leq \frac{n - |\mathrm{Ord}(G, 2)| - 1}{2}.$$

We can get a lower bound in the case of the cyclic group $\mathbb{Z}_n$ by considering the set

$$A = \{1, 2, \ldots, m\};$$

this results in

$$[0, s]\hat{}A = \{0, 1, 2, \ldots, ms - (s^2 - s)/2\}$$

and thus we get:

Proposition D.105 *For positive integers n, m, and s with $m \leq \lfloor (n-1)/2 \rfloor$ and $s \leq m$, we have*

$$\rho_A\hat{}(\mathbb{Z}_n, m, [0, s]) \leq \min\{n, ms - (s^2 - s)/2 + 1\}.$$

As was recently proved by Balandraud in [27], equality occurs in Proposition D.105 when n is prime:

Theorem D.106 (Balandraud; cf. [27]) *If p is prime and $m \leq \lfloor (p-1)/2 \rfloor$, we have*

$$\rho_A\hat{}(\mathbb{Z}_p, m, [0, s]) = \min\{p, ms - (s^2 - s)/2 + 1\}.$$

The value of $\rho_A\hat{}(G, m, [0, s])$ is not known for groups with composite order.

Problem D.107 *Find $\rho_A\hat{}(\mathbb{Z}_n, m, [0, s])$ for composite values of n.*

Problem D.108 *Find $\rho_A\hat{}(G, m, [0, s])$ for noncyclic groups G.*

D.3.3 Arbitrary number of terms

In this subsection we discuss what we know about

$$\rho\hat{}(G, m, \mathbb{N}_0) = \min\{|\Sigma A| \mid A \subseteq G, |A| = m\},$$

that is, the minimum size of

$$\Sigma A = \cup_{h=0}^{m} h\hat{}A$$

among all m-subsets A of G.

Suppose that G is cyclic. We can develop an upper bound for $\rho\hat{}(\mathbb{Z}_n, m, \mathbb{N}_0)$ as follows. Let d be an arbitrary positive divisor of n; we write m as

$$m = cd + k$$

where $1 \leq k \leq d$ and thus

$$c = \lceil m/d \rceil - 1.$$

We also let H be the subgroup of $\mathbb{Z}_n$ of size d:

$$H = \{j \cdot n/d \mid j = 0, 1, \ldots, d-1\}.$$

We separate two cases depending on the parity of c.

When c is odd, we let

$$A = \bigcup_{i=-(c-1)/2}^{(c-1)/2} (i + H) \bigcup \{(c+1)/2 + j \cdot n/d \mid j = 0, 1, \ldots, k-1\}.$$

Then $|A| = m$. We also see that

$$\Sigma A = \bigcup_{i=-\frac{c^2-1}{8}\cdot d}^{\frac{c^2-1}{8}\cdot d + \frac{c+1}{2}\cdot k} (i + H),$$

and thus

$$\begin{aligned} |\Sigma A| &= \min\left\{n, \left(\frac{c^2-1}{4} \cdot d + \frac{c+1}{2} \cdot k + 1\right) \cdot d\right\} \\ &= \min\left\{n, \left(\frac{c^2-1}{4} \cdot d + \frac{c+1}{2} \cdot (m - cd) + 1\right) \cdot d\right\} \\ &= \min\left\{n, \left(\frac{c+1}{2} \cdot m - \frac{(c+1)^2}{4} \cdot d + 1\right) \cdot d\right\}. \end{aligned}$$

The case when c is even is similar; this time we let

$$A = \bigcup_{i=-c/2}^{c/2-1} (i + H) \bigcup \{c/2 + j \cdot n/d \mid j = 0, 1, \ldots, k-1\}.$$

Again $|A| = m$, and we have

$$\Sigma A = \bigcup_{i=-\frac{c^2+2c}{8}\cdot d}^{\frac{c^2-2c}{8}\cdot d + \frac{c}{2}\cdot k} (i + H),$$

and thus

$$\begin{aligned}|\Sigma A| &= \min\left\{n, \left(\frac{c^2}{4}\cdot d + \frac{c}{2}\cdot k + 1\right)\cdot d\right\} \\ &= \min\left\{n, \left(\frac{c^2}{4}\cdot d + \frac{c}{2}\cdot (m-cd) + 1\right)\cdot d\right\} \\ &= \min\left\{n, \left(\frac{c}{2}\cdot m - \frac{c^2}{4}\cdot d + 1\right)\cdot d\right\}.\end{aligned}$$

We can combine our two cases and say that

$$\begin{aligned}|\Sigma A| &= \min\left\{n, \left(\left\lceil\frac{c}{2}\right\rceil\cdot m - \left\lceil\frac{c}{2}\right\rceil^2\cdot d + 1\right)\cdot d\right\} \\ &= \min\left\{n, \left(\left\lceil\frac{\lceil m/d-1\rceil}{2}\right\rceil\cdot m - \left\lceil\frac{\lceil m/d-1\rceil}{2}\right\rceil^2\cdot d + 1\right)\cdot d\right\} \\ &= \min\left\{n, \left(\left\lceil\frac{m/d-1}{2}\right\rceil\cdot m - \left\lceil\frac{m/d-1}{2}\right\rceil^2\cdot d + 1\right)\cdot d\right\}.\end{aligned}$$

Observe also that for $d=n$, we have $\lceil (m/n-1)/2\rceil = 0$, so

$$\left(\left\lceil\frac{m/n-1}{2}\right\rceil\cdot m - \left\lceil\frac{m/n-1}{2}\right\rceil^2\cdot n + 1\right)\cdot n = n.$$

This provides the following upper bound:

Proposition D.109 *With $D(n)$ denoting the set of positive divisors of n, we have*

$$\rho\hat{\ }(\mathbb{Z}_n, m, \mathbb{N}_0) \leq \min\left\{\left(\left\lceil\frac{m/d-1}{2}\right\rceil\cdot m - \left\lceil\frac{m/d-1}{2}\right\rceil^2\cdot d + 1\right)\cdot d \mid d\in D(n)\right\}.$$

We risk the following:

Conjecture D.110 *For all positive integers n and m with $m \leq n$, we have equality in Proposition D.109.*

Problem D.111 *Prove or disprove Conjecture D.110.*

Evaluating our expression in Proposition D.109 for $d=1$ yields

$$\lceil (m-1)/2\rceil\cdot m - \lceil (m-1)/2\rceil^2 + 1 = \lfloor m^2/4\rfloor + 1,$$

so we get the following:

Corollary D.112 *For every positive integer n we have*

$$\rho\hat{\ }(\mathbb{Z}_n, m, \mathbb{N}_0) \leq \min\left\{n, \lfloor m^2/4\rfloor + 1\right\}.$$

We can, in fact, prove that, for cyclic groups of prime order p, equality holds in Corollary D.112, since clearly

$$\rho\hat{\ }(\mathbb{Z}_p, m, \mathbb{N}_0) \geq \rho\hat{\ }(\mathbb{Z}_p, m, \lfloor m/2\rfloor),$$

and, by Theorem D.51, we have

$$\rho\hat{\ }(\mathbb{Z}_p, m, \lfloor m/2\rfloor) = \min\left\{p, \lfloor m/2\rfloor\cdot m - \lfloor m/2\rfloor^2 + 1\right\} = \min\left\{p, \lfloor m^2/4\rfloor + 1\right\}.$$

Proposition D.113 *For all positive primes p and positive integers $m \leq p$,*

$$\rho\hat{}(\mathbb{Z}_p, m, \mathbb{N}_0) = \min\left\{p, \lfloor m^2/4 \rfloor + 1\right\}.$$

We do not know the value of $\rho\hat{}(G, m, \mathbb{N}_0)$ for noncyclic groups:

Problem D.114 *Find the value of $\rho\hat{}(G, m, \mathbb{N}_0)$ for all positive integers $m \leq n$ and noncyclic groups G.*

Some authors have investigated the minimum size of ΣA among all m-subsets of G that possess some additional properties. One such pursuit is the attempt to find

$$\rho_A\hat{}(G, m, \mathbb{N}_0) = \min\{|\Sigma A| \mid A \subseteq G, |A| = m, A \cap -A = \emptyset\},$$

that is, the minimum size of ΣA among *asymmetric* m-subsets of G. Since no element of $\mathrm{Ord}(G,2) \cup \{0\}$ may be in an asymmetric set in G, for an asymmetric set of size m to exist, we need to assume that

$$m \leq \frac{n - |\mathrm{Ord}(G,2)| - 1}{2}.$$

We can get a lower bound in the case of the cyclic group $\mathbb{Z}_n$ by considering the set

$$A = \{1, 2, \ldots, m\};$$

this results in

$$\Sigma A = \{0, 1, 2, \ldots, m(m+1)/2\}$$

and thus we get:

Proposition D.115 *For positive integers n and m with $m \leq \lfloor (n-1)/2 \rfloor$, we have*

$$\rho_A\hat{}(\mathbb{Z}_n, m, \mathbb{N}_0) \leq \min\{n, (m^2 + m + 2)/2\}.$$

As a consequence of Balandraud's Theorem D.106 with $s = m$ (and as proved by him in [25]), equality occurs in Proposition D.115 when n is prime:

Theorem D.116 (Balandraud; cf. [25], [26]) *If p is prime and $m \leq \lfloor (p-1)/2 \rfloor$, we have*

$$\rho_A\hat{}(\mathbb{Z}_p, m, \mathbb{N}_0) = \min\{p, (m^2 + m + 2)/2\}.$$

The value of $\rho_A\hat{}(G, m, \mathbb{N}_0)$ is not known for groups with composite order.

Problem D.117 *Find $\rho_A\hat{}(\mathbb{Z}_n, m, \mathbb{N}_0)$ for composite values of n.*

Problem D.118 *Find $\rho_A\hat{}(G, m, \mathbb{N}_0)$ for noncyclic groups G.*

As a variation, we may consider

$$\rho_A\hat{}(G, m, \mathbb{N}) = \min\{|\Sigma^* A| \mid A \subseteq G, |A| = m, A \cap -A = \emptyset\},$$

that is, the minimum size of

$$\Sigma^* A = \cup_{h=1}^{\infty} h\hat{}A$$

among asymmetric m-subsets of G.

The subset

$$A = \{1, 2, \ldots, m\}$$

of $\mathbb{Z}_n$ provides a lower bound again:

Proposition D.119 *For positive integers n and m with $m \leq \lfloor (n-1)/2 \rfloor$, we have*

$$\rho_A\hat{}(\mathbb{Z}_n, m, \mathbb{N}_0) \leq \min\{n, (m^2+m)/2\}.$$

And we have:

Theorem D.120 (Balandraud; cf. [25], [26]) *If p is prime and $m \leq \lfloor (p-1)/2 \rfloor$, we have*

$$\rho_A\hat{}(\mathbb{Z}_p, m, \mathbb{N}) = \min\{p, (m^2+m)/2\}.$$

Problem D.121 *Find $\rho_A\hat{}(\mathbb{Z}_n, m, \mathbb{N})$ for composite values of n.*

Problem D.122 *Find $\rho_A\hat{}(G, m, \mathbb{N})$ for noncyclic groups G.*

Another related quantity of interest is

$$\widehat{\rho_*}\hat{}(G, m, \mathbb{N}_0) = \min\{|\Sigma A| \mid A \subseteq G \setminus \{0\}, |A| = m, \langle A \rangle = G\},$$

that is, the minimum size of ΣA among those m-subsets A of G that do not contain zero (hence the $_*$) but generate the entire group (hence the $\widehat{\ }$).

A relevant result is the following:

Theorem D.123 (Hamidoune; cf. [112]) *Let $S \subseteq G \setminus \{0\}$ be such that $|S| \geq 3$ and $\langle S \rangle = G$; furthermore, if $|S| = 3$, then suppose that S is not of the form $\{\pm a, 2a\}$ for any $a \in G$, and that if $|S| = 4$, then S is not of the form $\{\pm a, \pm 2a\}$ for any $a \in G$. Then*

$$|\Sigma S| \geq \min\{n-1, 2|S|\}.$$

Combining Theorem D.123 with the fact that if $n \geq 10$, then $\chi\hat{}(G^*, \mathbb{N}) \leq n/2$ (see Section E.3.3), we get the following:

Corollary D.124 *Suppose that $n \geq 10$, and let $S \subseteq G \setminus \{0\}$ be such that $|S| \geq 5$ and $\langle S \rangle = G$. Then*

$$|\Sigma S| \geq \min\{n, 2|S|\}.$$

Indeed, if $|S| \leq (n-1)/2$, then the result follows from Theorem D.123; if $|S| \geq n/2$, then it is due to the fact that $\chi\hat{}(G^*, \mathbb{N}) \leq n/2$. (We mention that Corollary D.124 was stated in [112] only for cyclic groups G; its proof there, however, relied on Lemma 3.3 in [112], which is actually false for all groups of even order; cf. [17].)

Furthermore, as it was noted by Hamidoune in [112], if n is divisible by 3, and

$$S = (H \setminus \{0\}) \cup \{s\}$$

for an index 3 subgroup H of G and $s \in G \setminus H$, then $|\Sigma S| = 2n/3$. Consequently,

Corollary D.125 *If n is divisible by 3 and $n \geq 15$, then*

$$\widehat{\rho_*}\hat{}(G, n/3, \mathbb{N}_0) = 2n/3.$$

Problem D.126 *Evaluate $\widehat{\rho_*}\hat{}(G, m, \mathbb{N}_0)$ for other G and m.*

Yet another variation, introduced by Eggleton and Erdős in [72], is to find the following quantity:

$$\rho_Z\hat{}(G, m, \mathbb{N}) = \min\{|\Sigma^* A| \mid A \subseteq G, |A| = m, 0 \notin \Sigma^* A\},$$

that is, the minimum size of $\Sigma^* A$ among all weakly zero-sum-free m-subsets A of G; if no such set A exists, we set $\rho_Z\hat{}(G, m, \mathbb{N}) = \infty$.

For $m = 1$ the answer is obvious, since $|\Sigma^* A| = 1$ for all 1-subsets of G. The case of $m = 2$ is not much harder: if $A = \{a, b\}$ with $a \neq b$, $a \neq 0$, and $b \neq 0$, then $\Sigma^* A = \{a, b, a+b\}$ has size 3; we just need to make sure that $a + b \neq 0$. We can summarize:

Proposition D.127 *We have:*

$$\rho\hat{z}(G,1,\mathbb{N}) = \begin{cases} \infty & \text{if } n = 1, \\ 1 & \text{if } n \geq 2; \end{cases}$$

and

$$\rho\hat{z}(G,2,\mathbb{N}) = \begin{cases} \infty & \text{if } n \leq 3, \\ 3 & \text{if } n \geq 4. \end{cases}$$

For $m = 3$, we have the following result:

Proposition D.128 *We have:*

$$\rho\hat{z}(G,3,\mathbb{N}) = \begin{cases} \infty & \text{if } n \leq 5; \\ 5 & \text{if } n \geq 6,\ n \text{ is even, and } G \not\cong \mathbb{Z}_2^r; \\ 6 & \text{if } n \geq 7 \text{ and } n \text{ is odd;} \\ 7 & \text{if } n \geq 8 \text{ and } G \cong \mathbb{Z}_2^r. \end{cases}$$

The proof of Proposition D.128 is on page 349.

Problem D.129 *Evaluate $\rho\hat{z}(G,4,\mathbb{N})$, $\rho\hat{z}(G,5,\mathbb{N})$, etc. for all abelian groups G.*

A considerably easier, but still largely unknown, special case is the evaluation of $\rho\hat{z}(G,m,\mathbb{N})$ for cyclic groups G; we review what is currently known.

From Section F.3.3, we have that the set $A = \{1,2,\ldots,m\}$ is weakly zero-sum-free in $\mathbb{Z}_n$ when $n \geq (m^2+m+2)/2$; since

$$\Sigma^* A = \{1,2,\ldots,(m^2+m)/2\},$$

this implies:

Proposition D.130 *If $n \geq (m^2+m+2)/2$, then*

$$\rho\hat{z}(\mathbb{Z}_n,m,\mathbb{N}) \leq (m^2+m)/2.$$

The two constructions of Selfridge mentioned on page 261 yield slightly better bounds for certain values of n. The first, based on the fact that

$$A = \{1,-2,3,4,\ldots,m\}$$

is weakly zero-sum-free in $\mathbb{Z}_n$ when $n \geq (m^2+m-2)/2$ and $n \geq 6$, has an advantage over Proposition D.130 only when $n = (m^2+m-2)/2$ or $n = (m^2+m)/2$, since we have

$$\Sigma^* A = \{-1,-2,1,2,\ldots,(m^2+m-4)/2\},$$

which has the same size as the set yielding that theorem when $n \geq (m^2+m+2)/2$. However, for $n = (m^2+m-2)/2$,

$$-2 = (m^2+m-6)/2$$

and

$$-1 = (m^2+m-4)/2;$$

and for $n = (m^2+m)/2$,

$$-2 = (m^2+m-4)/2.$$

We thus get that if A is the m-subset of the cyclic group of order $(m^2+m-2)/2$ described above, then A is weakly zero-sum-free with

$$|\Sigma^* A| = (m^2+m-4)/2;$$

and if A is the m-subset of the cyclic group of order $(m^2+m)/2$ described above, then A is weakly zero-sum-free with

$$|\Sigma^* A| = (m^2+m-2)/2.$$

We can take this further by noting that whenever A is weakly zero-sum-free in $\mathbb{Z}_n$, then $d \cdot A$ is weakly zero-sum-free in $\mathbb{Z}_{dn}$; this holds for every $d \geq 1$ and neither the size of the subset nor the size of its sumset changes. We thus arrive at the following result:

Proposition D.131 *If $n \geq 6$ and n is divisible by $(m^2+m-2)/2$, then*

$$\rho_Z\hat{}(\mathbb{Z}_n, m, \mathbb{N}) \leq (m^2+m-4)/2,$$

and if $n \geq 6$ and n is divisible by $(m^2+m)/2$, then

$$\rho_Z\hat{}(\mathbb{Z}_n, m, \mathbb{N}) \leq (m^2+m-2)/2.$$

The second construction assumes that n is even, and considers the set

$$A = \{1, 2, \dots, \lfloor (m-1)/2 \rfloor\} \cup \{n/2, n/2+1, n/2+2, \dots, n/2 + \lfloor m/2 \rfloor\},$$

for which

$$\Sigma^* A = \{1, 2, \dots, \lfloor m^2/4 \rfloor\} \cup \{n/2, n/2+1, n/2+2, \dots, n/2 + \lfloor m^2/4 \rfloor\}.$$

Therefore, if $n/2 \geq \lfloor m^2/4 \rfloor + 1$, that is, if $n \geq \lfloor m^2/2 \rfloor + 2$, then A is weakly zero-sum-free. We get:

Proposition D.132 *If n is even and $n \geq \lfloor m^2/2 \rfloor + 2$, then*

$$\rho_Z\hat{}(\mathbb{Z}_n, m, \mathbb{N}) \leq \lfloor m^2/2 \rfloor + 1.$$

Let us see now what our propositions tell us for small values of m. The cases of $m \leq 3$ follow from Propositions D.127 and D.128. For $m = 4$,

(i) Proposition D.130 yields that if $n \geq 11$, then $\rho_Z\hat{}(\mathbb{Z}_n, m, \mathbb{N}) \leq 10$;

(ii) Proposition D.131 yields that if $9|n$, then $\rho_Z\hat{}(\mathbb{Z}_n, m, \mathbb{N}) \leq 8$;

(iii) Proposition D.131 also yields that if $10|n$, then $\rho_Z\hat{}(\mathbb{Z}_n, m, \mathbb{N}) \leq 9$; and

(iv) Proposition D.132 yields that if $n \geq 10$ and $2|n$, then $\rho_Z\hat{}(\mathbb{Z}_n, m, \mathbb{N}) \leq 9$.

Note that statement (iii) follows from (iv) and thus is unnecessary. In [33], Bhowmik, Halupczok, and Schlage-Puchta presented one additional construction: if $n \geq 12$ and $3|n$, then the set

$$A = \{1, n/3, n/3+1, 2n/3+1\}$$

is weakly zero-sum-free in $\mathbb{Z}_n$, and $\Sigma^* A$ has size 9. Furthermore, relying on a computer program, they proved that one can never do better:

Theorem D.133 (Bhowmik, Halupczok, and Schlage-Puchta; cf. [33]) *We have:*

$$\rho_Z\hat{}(\mathbb{Z}_n, 4, \mathbb{N}) = \begin{cases} \infty & \text{if } n \leq 8; \\ 8 & \text{if } 9|n; \\ 9 & \text{if } n \geq 10 \text{ and } 9 \not| n \text{ but } (2|n \text{ or } 3|n); \\ 10 & \text{otherwise.} \end{cases}$$

For $m = 5$ and $m = 6$, the same authors proved that our Propositions D.130, D.131, and D.132 provide the right values:

Theorem D.134 (Bhowmik, Halupczok, and Schlage-Puchta; cf. [33]) *We have:*

$$\rho_Z\hat{}(\mathbb{Z}_n, 5, \mathbb{N}) = \begin{cases} \infty & \text{if } n \leq 13; \\ 13 & \text{if } n \geq 14 \text{ and } 2|n; \\ 14 & \text{if } 15|n; \\ 15 & \text{otherwise;} \end{cases}$$

and

$$\rho_Z\hat{}(\mathbb{Z}_n, 6, \mathbb{N}) = \begin{cases} \infty & \text{if } n \leq 19; \\ 19 & \text{if } n \geq 20 \text{ and } 2|n; \\ 20 & \text{if } 21|n; \\ 21 & \text{otherwise.} \end{cases}$$

The same authors also exhibited the precise (but quite a bit more complicated) formula for $\rho_Z\hat{}(\mathbb{Z}_n, 7, \mathbb{N})$ (see [33]).

Problem D.135 *Evaluate* $\rho_Z\hat{}(\mathbb{Z}_n, 8, \mathbb{N})$, $\rho_Z\hat{}(\mathbb{Z}_n, 9, \mathbb{N})$, *etc. for all values of* n.

There is more known about $\rho_Z\hat{}(\mathbb{Z}_n, m, \mathbb{N})$ for prime values of n. First, recall that, according to Balandraud's result from [25] (cf. Theorem F.124), if

$$1 + 2 + \cdots + m \geq p,$$

then $\mathbb{Z}_p$ has no weakly zero-sum-free subsets of size m. We can restate this as follows:

Theorem D.136 (Balandraud; cf. [25], [26]) *If p is prime for which $p \leq (m^2 + m)/2$, then $\rho_Z\hat{}(\mathbb{Z}_p, m, \mathbb{N}) = \infty$.*

On the other hand, from Proposition D.130 we get that if $p \geq (m^2 + m + 2)/2$, then

$$\rho_Z\hat{}(\mathbb{Z}_p, m, \mathbb{N}) \leq (m^2 + m)/2.$$

Olson proved that when p is large, equality holds:

Theorem D.137 (Olson; cf. Theorem 2 in [166]) *Let p be a prime for which $p \geq m^2 + m - 1$ when m is even, and $p \geq m^2 + (3m - 5)/2$ when m is odd. Then*

$$\rho_Z\hat{}(\mathbb{Z}_p, m, \mathbb{N}) = (m^2 + m)/2.$$

This raises the following:

Problem D.138 *Decide whether*

$$\rho_Z\hat{}(\mathbb{Z}_p, m, \mathbb{N}) = (m^2 + m)/2$$

holds for all primes p with $p \geq (m^2 + m + 2)/2$.

According to Theorem D.137, only finitely many primes need to be considered to answer Problem D.138 for a given value of m. We should observe that the answer to Problem D.138 is affirmative for $m = 1$ and $m = 2$ (by Proposition D.127), for $m = 3$ (by Proposition D.128), and for $m \in \{4,5,6,7\}$ (by [33]).

Rather than finding $\rho_Z\hat{}(G, m, \mathbb{N})$ for all G and m, Eggleton and Erdős in [72] proposed the potentially easier problem of evaluating $f(m)$, defined as the minimum possible value of $\rho_Z\hat{}(G, m, \mathbb{N})$ for any group G. According to Propositions D.127 and D.128, we have $f(1) = 1$, $f(2) = 3$, and $f(3) = 5$. We also have:

- $f(4) = 8$ (Eggleton and Erdős; cf. [72]);
- $f(5) = 13$ (Gao, et al.; cf. [95]);
- $f(6) = 19$ (Gao, et al.; cf. [95]);
- $f(7) = 24$ (Yuan and Zeng; cf. [203]).

The proof of $f(6) = 19$ is long with a very large number of cases, and the proof of $f(7) = 24$ relies on a computer program.

Problem D.139 *Evaluate (perhaps relying on a computer program) $f(8)$, $f(9)$, etc.*

We can also observe that for $m \leq 7$, $f(m)$ agrees with the smallest possible value of $\rho_Z\hat{}(G, m, \mathbb{N})$ for cyclic groups G, and Eggleton and Erdős believed that this is always the case:

Conjecture D.140 (Eggleton and Erdős; cf. [72]) *For every $m \in \mathbb{N}$ there is an $n \in \mathbb{N}$ for which $f(m) = \rho_Z\hat{}(\mathbb{Z}_n, m, \mathbb{N})$.*

Problem D.141 *Prove Conjecture D.140.*

We should mention that by Proposition D.132, we get:

Corollary D.142 *For every $m \in \mathbb{N}$, we have $f(m) \leq \lfloor m^2/2 \rfloor + 1$.*

From our stated values above, we have equality in Corollary D.142 for $m \in \{1,2,3,5,6\}$ but not for $m \in \{4,7\}$.

A particularly intriguing problem is the following:

Problem D.143 *Find infinitely many values of m for which $f(m) \leq \lfloor m^2/2 \rfloor$ or prove that this is not possible.*

We also have a lower bound for $f(m)$:

Theorem D.144 (Olson; cf. [167]) *For every $m \in \mathbb{N}$, we have $f(m) \geq \lceil m^2/9 \rceil$.*

Problem D.145 *Find a real number $c > 1/9$ so that $f(m) \geq \lceil c \cdot m^2 \rceil$ holds for all (but finitely many) $m \in \mathbb{N}$.*

D.4 Restricted signed sumsets

D.4.1 Fixed number of terms

D.4.2 Limited number of terms

D.4.3 Arbitrary number of terms

Chapter E

The critical number

Recall that in Chapter D we investigated, for given Λ and H, the minimum sumset size of an m-subset of G:

$$\rho_\Lambda(G, m, H) = \min\{|H_\Lambda A| \mid A \subseteq G, |A| = m\}.$$

As a special case, here we are interested in the minimum value of m for which

$$\rho_\Lambda(G, m, H) = n;$$

that is, the minimum value of m for which every m-subset of G spans all of G. This value, if it exists, is called the (Λ, H)-*critical number* of G and is denoted by $\chi_\Lambda(G, H)$.

In the following sections we consider $\chi_\Lambda(G, H)$ for special $\Lambda \subseteq \mathbb{Z}$ and $H \subseteq \mathbb{N}_0$.

E.1 Unrestricted sumsets

Our goal in this section is to investigate $\chi(G, H)$, the minimum value of m for which

$$HA = G$$

holds for every m-subset of G. (Recall that HA is the union of all h-fold sumsets hA for $h \in H$.) Since $0A = \{0\}$ for every subset A of G but $hG = G$ for every positive integer h, we see that $\chi(G, H)$ does not exist for $n \geq 2$ when $H = \{0\}$, but $\chi(G, H)$ does exist and is at most n when H contains at least one positive integer.

We consider three special cases: when H consists of a single nonnegative integer h, when H consists of all nonnegative integers up to some value s, and when H is the entire set of nonnegative integers.

E.1.1 Fixed number of terms

Here we ought to consider, for fixed G and positive integer h, the quantity $\chi(G, h)$, that is, the minimum value of m for which the h-fold sumset of every m-element subset of G is G itself. However, according to Theorem D.3, the h-critical number of a group of order n is the minimum value of m for which $u(n, m, h) = n$, and this value was determined by Theorem 4.17 to be $v_1(n, h) + 1$ where

$$v_1(n, h) = \max\left\{\left(\left\lfloor \frac{d-2}{h} \right\rfloor + 1\right) \cdot \frac{n}{d} \mid d \in D(n)\right\}.$$

Thus we have:

Theorem E.1 *For all finite abelian groups G of order n and all positive integers h we have*

$$\chi(G,h) = v_1(n,h) + 1.$$

Having found the value of $\chi(G,h)$, we are now interested in the inverse problem of classifying all m-subsets A of G with

$$m = \chi(G,h) - 1 = v_1(n,h)$$

for which $hA \neq G$. The problem being trivial for $h = 1$, we let $h \geq 2$.

We consider $h = 2$ first. (As we explain below, the case of $h = 2$ seems more complicated than the case of $h \geq 3$.)

When n is even, we have $v_1(n,2) = n/2$. Recall that, by Theorem D.3, we have

$$\rho(G,m,h) = u(n,m,h) = \min\{f_d(m,h) \mid d \in D(n)\},$$

where $D(n)$ is the set of positive divisors of n and

$$f_d(m,h) = (h \cdot \lceil m/d \rceil - h + 1) \cdot d.$$

In particular, when $f_d(n/2,2) < n$ for some $d \in D(n)$, then we are guaranteed to find subsets A_d of G with $|A_d| = n/2$ and $|2A_d| < n$.

One can easily determine that

$$(2 \cdot \lceil n/(2d) \rceil - 1) \cdot d = \begin{cases} n-d & \text{if } n/d \text{ is even,} \\ n & \text{if } n/d \text{ is odd.} \end{cases}$$

When n and n/d are both even for some $d \in D(n)$, we can, in fact, find explicit subsets A_d of G of size $n/2$ whose two-fold sumset has size $n-d$; we will explain this here for the case when G is cyclic. Recall the set $A_d(n,m)$ from page 150. In particular, for $m = n/2$ and when n/d is even, we have

$$A_d(n,n/2) = \cup_{i=0}^{n/(2d)-1}(i+H),$$

where H is the subgroup of $\mathbb{Z}_n$ with order d. We see that $|A_d(n,n/2)| = n/2$, and

$$2A_d(n,n/2) = \cup_{i=0}^{n/d-2}(i+H),$$

so $|2A_d(n,n/2)| = n-d$. We thus have:

Proposition E.2 *Suppose n is even and that it has a divisor d for which n/d is also even. Then $\mathbb{Z}_n$ has a subset A of size $n/2$ for which $2A$ has size $n-d$.*

For example, $\mathbb{Z}_{20}$ has subsets $A_d(20,10)$ for $d \in \{1,2,5,10\}$, each of size ten, so that $2A_d(20,10)$ has size $20-d$. Using the computer program [120], we checked that for $d \in \{2,5,10\}$, there are essentially (ignoring equivalences) no other 10-subsets A with $2A \neq \mathbb{Z}_{20}$ besides $A_d(20,10)$. However, there are many 10-subsets whose 2-fold sumset has size 19 other than the (arithmetic progression) $A_1(20,10)$ constructed above: for example,

$$\{0,1,2,3,4,5,6,7\} \cup C$$

with $C = \{8,10\}$, $C = \{9,11\}$, $C = \{16,17\}$, etc.

We pose the following questions, in increasing order of difficulty:

Problem E.3 *For each even value of n, classify all subsets of $\mathbb{Z}_n$ of size $n/2$ whose two-fold sumset is not $\mathbb{Z}_n$.*

Problem E.4 *For each abelian group G of even order n, classify all subsets of size $n/2$ whose two-fold sumset is not G.*

Problem E.5 *For each abelian group G of odd order n, classify all subsets of size $(n-1)/2$ whose two-fold sumset is not G.*

Let us now turn to the case of $h \geq 3$; we again assume that n is even, in which case $v_1(n,h) = n/2$ by Corollary 4.5. For divisors $d \in D(n)$, we can compute $f_d(n/2,h)$ as follows. We see that $f_n(n/2,h) = n$, $f_{n/2}(n/2,h) = n/2$, and (in the case when n is divisible by 3) $f_{n/3}(n/2,h) = (h+1) \cdot n/3 > n$. For any other d we get

$$f_d = (h \cdot \lceil n/(2d) \rceil - h + 1) \cdot d \geq (h \cdot n/(2d) - h + 1) \cdot d \geq h \cdot n/2 - (h-1) \cdot n/4 \geq n.$$

So $f_d < n$ holds only for $d = n/2$, in which case $f_d = n/2$. This suggests that, if A is a subset of G of size $n/2$ for which $hA \neq G$, then $A = H$ or $A = G \setminus H$ where $H \leq G$ has order $n/2$ (and in both cases $|hA| = n/2$).

Conjecture E.6 *Suppose that G is an abelian group of even order n, h is an integer with $h \geq 3$, and A is a subset of G with $|A| = n/2$ and $|hA| < n$. Then G has a subgroup H of order $n/2$ for which $A = H$ or $A = G \setminus H$.*

We know that this claim holds for cyclic groups:

Theorem E.7 (Navarro; cf. [162]) *Conjecture E.6 holds when G is cyclic.*

We have the following open questions:

Problem E.8 *Prove Conjecture E.6 for noncyclic groups G.*

Problem E.9 *For each n and h with n odd and $h \geq 3$, characterize all subsets A of $\mathbb{Z}_n$ of size $v_1(n,h)$ for which $hA \neq \mathbb{Z}_n$.*

Problem E.10 *For each G of odd order n and for each $h \geq 3$, characterize all subsets A of G of size $v_1(n,h)$ for which $hA \neq G$.*

We have the following result of Lev that not only answers Problem E.10 in a special case, but accomplishes more:

Theorem E.11 (Lev; cf. [144]) *Suppose that A is an m-subset of $\mathbb{Z}_5^r$ with*

$$(3 \cdot 5^{r-1} - 1)/2 \leq m \leq 2 \cdot 5^{r-1} = v_1(5^r, 3)$$

for which $3A \neq \mathbb{Z}_5^r$. Then A is contained in a union of two cosets of a subgroup of index 5.

Note that if A is contained in a union of two cosets of a subgroup of index 5, then indeed $3A \neq \mathbb{Z}_5^r$; furthermore, Lev constructed an example in [144] that shows that the lower bound on m is tight.

We should mention that, as an analogue of $\widehat{\chi}(G,[0,s])$ discussed in Section E.1.2 below, one may define the variation where only generating subsets of G are considered:

$$\widehat{\chi}(G,h) = \min\{m \mid A \subseteq G, \langle A \rangle = G, |A| \geq m \Rightarrow hA = G\}.$$

We have the following—somewhat surprising—result:

Theorem E.12 (Bajnok; cf. [18]) *For all G and h, we have*

$$\hat{\chi}(G,h)=\chi(G,h)=v_1(n,h)+1.$$

It is likely an interesting question to find inverse results as well:

Problem E.13 *For each G and h, classify all generating subsets A of G for which $|A|=\hat{\chi}(G,h)-1$ and $hA\neq G$.*

E.1.2 Limited number of terms

For given groups G and positive integer s, here we consider $\chi(G,[0,s])$, that is, the minimum value of m for which the $[0,s]$-fold sumset of every m-subset of G is G itself. By Proposition D.12 and Theorem E.1, we have:

Theorem E.14 *For all finite abelian groups G of order n and all positive integers s we have*

$$\chi(G,[0,s])=v_1(n,s)+1.$$

While

$$\chi(G,[0,s])=\min\{m \mid A\subseteq G, |A|\geq m \Rightarrow [0,s]A=G\}$$

has thus been evaluated for each G and s, there is a variation that has been considered in the literature for which much less is known. Namely, Klopsch and Lev in [132] have investigated the quantity $\hat{\chi}(G,[0,s])$: the minimum value of m for which the $[0,s]$-fold sumset of every m-subset of G that generates G is G itself, that is,

$$\hat{\chi}(G,[0,s])=\min\{m \mid A\subseteq G, \langle A\rangle=G, |A|\geq m \Rightarrow [0,s]A=G\}.$$

It turns out that the following relative of the arithmetic function $v_1(n,s)$ of page 76 will be useful: we define

$$\hat{v}(n,s)=\max\left\{\left(\left\lfloor\frac{d-2}{s}\right\rfloor+1\right)\cdot\frac{n}{d} \mid d\in D(n), d\geq s+2\right\};$$

we adhere to the convention that the maximum element of the empty set equals zero, and thus if $n\leq s+1$, we have $\hat{v}(n,s)=0$. (Note also that we omitted the index 1 as unnecessary here.)

From the paper [132] of Klopsch and Lev we are able to deduce the following upper bound for $\hat{\chi}(G,[0,s])$:

Theorem E.15 *For every G and s we have*

$$\hat{\chi}(G,[0,s])\leq\hat{v}(n,s)+1.$$

Since this result was not stated in [132], we provide a proof—see page 349.

We can easily see that equality holds in Theorem E.15, when G is cyclic. This is obvious when $n\leq s+1$, since then

$$\hat{\chi}(\mathbb{Z}_n,[0,s])=\hat{v}(n,s)+1=1.$$

Assume now that $n\geq s+2$. Recall from Section D.1.1 (see page 150) that, for $1\leq m\leq n$ and a divisor d of n, we defined the set $A_d(n,m)$. Suppose now that $d\in D(n)$, $d\geq s+2$ (possible since $n\geq s+2$), and

$$m=\hat{v}(n,s)=\left(\left\lfloor\frac{d-2}{s}\right\rfloor+1\right)\cdot\frac{n}{d}.$$

Then, with H as the order n/d subgroup of $\mathbb{Z}_n$ and $c = \lfloor (d-2)/s \rfloor$, we get

$$A_{n/d}(n,m) = \cup_{i=0}^{c}(i+H).$$

Now $A_{n/d}(n,m)$ has size m, and

$$[0,s]A_{n/d}(n,m) = \cup_{i=0}^{sc}(i+H)$$

has size

$$(sc+1)\cdot\frac{n}{d} = \left(s\left\lfloor\frac{d-2}{s}\right\rfloor+1\right)\cdot\frac{n}{d} \leq (d-1)\cdot\frac{n}{d} < n,$$

so $[0,s]A \neq \mathbb{Z}_n$. Furthermore, since $d \geq s+2$, we have $c \geq 1$ and thus $1 \in A$, which implies that A generates $\mathbb{Z}_n$. This yields:

Proposition E.16 *For all positive integers n and s we have*

$$\widehat{\chi}(\mathbb{Z}_n,[0,s]) \geq \widehat{v}(n,s)+1.$$

Combining Proposition E.16 with Theorem E.15, we get:

Theorem E.17 (Klopsch and Lev; cf. [132]) *For all positive integers n and s we have*

$$\widehat{\chi}(\mathbb{Z}_n,[0,s]) = \widehat{v}(n,s)+1.$$

Considerably less is known about $\widehat{\chi}(G,[0,s])$ for noncyclic G; in fact, in contrast to $\chi(G,[0,s])$, which only depends on the order n of G (see Theorem E.14 above), $\widehat{\chi}(G,[0,s])$ is generally greatly dependent on the structure of G itself.

We have the following general lower bound:

Proposition E.18 (Bajnok; cf. [18]) *Let G be an abelian group of order n, and let H be a subgroup of G of index $d>1$ for which G/H is of type $(d_1,\dots,d_t)$. For each $i=1,\dots,t$, let c_i be a positive integer with $c_i \leq d_i - 1$, and suppose that*

$$\Sigma_{i=1}^{t}\lceil (d_i-1)/c_i \rceil \geq s+1.$$

Then we have

$$\widehat{\chi}(G,[0,s]) \geq \left(1+\Sigma_{i=1}^{t}c_i\right)\cdot n/d+1.$$

As an application, we consider $\mathbb{Z}_2^r$, the elementary abelian 2-group of rank r. Trivially, when $r \leq s$, we have

$$\widehat{\chi}(\mathbb{Z}_2^r,[0,s]) = 1,$$

so assume that $s+1 \leq r$, and let t be an integer with

$$s+1 \leq t \leq r.$$

Then choosing $H = \mathbb{Z}_2^t$ and $c_i = 1$ for all $i \in \{1,\dots,t\}$, Proposition E.18 implies that

$$\widehat{\chi}(\mathbb{Z}_2^r,[0,s]) \geq (t+1)\cdot 2^{r-t}+1;$$

in particular, we have

$$\widehat{\chi}(\mathbb{Z}_2^r,[0,s]) \geq (s+2)\cdot 2^{r-s-1}+1.$$

It turns out that equality holds:

Theorem E.19 (Lev; cf. [142]) *Let r and s be positive integers, $s \geq 2$. If $r \leq s$, then $\widehat{\chi}(\mathbb{Z}_2^r, [0,s]) = 1$; otherwise we have*

$$\widehat{\chi}(\mathbb{Z}_2^r, [0,s]) = (s+2) \cdot 2^{r-s-1} + 1.$$

We thus find that

$$\widehat{\chi}(\mathbb{Z}_2^r, [0,2]) = \widehat{\chi}(\mathbb{Z}_{2^r}, [0,2])$$

for all $r \geq 3$, but

$$\widehat{\chi}(\mathbb{Z}_2^r, [0,s]) < \widehat{\chi}(\mathbb{Z}_{2^r}, [0,s])$$

for all $r \geq s+1 \geq 4$.

Problem E.20 *Find other applications of Proposition E.18.*

The following additional results are known:

Theorem E.21 (Klopsch and Lev; cf. [132]) *Suppose that G is a finite abelian group of order $n \geq 2$, rank r, and invariant factorization $\mathbb{Z}_{n_1} \times \cdots \times \mathbb{Z}_{n_r}$; we set*

$$D = n_1 + \cdots + n_r - r.$$

(D is called the positive diameter *of G.)*

1. *If $G \not\cong \mathbb{Z}_2$, then $\widehat{\chi}(G, [0,1]) = n$.*
2. *If $G \not\cong \mathbb{Z}_2, \mathbb{Z}_2^2$, then $\widehat{\chi}(G, [0,2]) = \lfloor n/2 \rfloor + 1$.*
3. *If $G \not\cong \mathbb{Z}_2^r$, then*

$$\widehat{\chi}(G,[0,3]) = \begin{cases} \left(1+\frac{1}{d}\right)\cdot\frac{n}{3}+1 & \text{if } G \text{ has a subgroup whose order is congruent to 2 mod 3 and which is not isomorphic to an elementary abelian 2-group, and } d \text{ is the minimum size of such a subgroup;} \\ \left\lfloor \frac{n}{3} \right\rfloor + 1 & \text{otherwise.} \end{cases}$$

4. *If $G \not\cong \mathbb{Z}_2$, then $\widehat{\chi}(G, [0, D-1]) = r+2$.*
5. *If $s \geq D$, then $\widehat{\chi}(G, [0,s]) = 1$.*

We can observe that, according to Theorem E.21, for $s \in \{1,2\}$ we have

$$\widehat{\chi}(G,[0,s]) = \chi(G,[0,s]) = v_1(n,s)+1.$$

To assess the case of $s = 3$, recall from page 78 that

$$v_1(n,3) = \begin{cases} \left(1+\frac{1}{p}\right)\frac{n}{3} & \text{if } n \text{ has prime divisors congruent to 2 mod 3, and } p \text{ is the smallest such divisor,} \\ \left\lfloor \frac{n}{3} \right\rfloor & \text{otherwise.} \end{cases}$$

Similarly, we get

$$\widehat{v}_1(n,3) = \begin{cases} \left(1+\frac{1}{d}\right)\frac{n}{3} & \text{if } n \text{ has divisors congruent to 2 mod 3 that are greater than 2, and } d \text{ is the smallest such divisor,} \\ \left\lfloor \frac{n}{3} \right\rfloor & \text{otherwise.} \end{cases}$$

(Note that d need not be prime.) Therefore, we see that

$$\widehat{\chi}(G,[0,3]) \leq \widehat{v}_1(n,3)+1 \leq \chi(G,[0,3]) = v_1(n,3)+1.$$

While equality holds throughout for cyclic groups, this may not be the case for noncyclic groups; for example, for $G = \mathbb{Z}_2^2 \times \mathbb{Z}_6$, we get $\widehat{\chi}(G,[0,3]) = 9$, $\widehat{v}_1(n,3)+1 = 10$, and $\chi(G,[0,3]) = v_1(n,3)+1 = 13$.

Combining all results above, we see that $\widehat{\chi}(G,[0,s])$ has been determined for all G and s, except for noncyclic groups of exponent more than two and for $4 \leq s \leq D-2$.

Problem E.22 *Find $\widehat{\chi}(G,[0,s])$ for every noncyclic group G and every $4 \leq s \leq D-2$.*

Since the general problem is probably difficult, we offer the following special cases:

Problem E.23 *Find $\widehat{\chi}(G,[0,4])$ for every noncyclic group G.*

Problem E.24 *Find $\widehat{\chi}(G,[0,D-2])$ for every noncyclic group G.*

Problem E.25 *Find $\widehat{\chi}(\mathbb{Z}_k^2,[0,s])$ for all $k \geq 4$ and $s \geq 4$.*

Problem E.26 *Find $\widehat{\chi}(G,[0,s])$ for every $s \geq 4$ and every noncyclic group G of odd order.*

While exact values of $\widehat{\chi}(G,[0,s])$ might be difficult to get in general, we can find a tight upper bound for it. Recall that by Theorem E.15, we have

$$\widehat{\chi}(G,[0,s]) \leq \widehat{v}(n,s)+1.$$

Observe that, when $d \in D(n)$ for which $d \geq s+2$, then

$$\begin{aligned}
\left(\left\lfloor \frac{d-2}{s} \right\rfloor + 1\right) \cdot \frac{n}{d} &\leq \left(\frac{d-2}{s}+1\right) \cdot \frac{n}{d} \\
&= \left(\frac{s-2}{d}+1\right) \cdot \frac{n}{s} \\
&\leq \left(\frac{s-2}{s+2}+1\right) \cdot \frac{n}{s} \\
&= \frac{2n}{s+2}.
\end{aligned}$$

Therefore, we have

$$\widehat{\chi}(G,[0,s]) \leq \frac{2n}{s+2}+1,$$

with equality if, and only if, n is divisible by $s+2$.

We have the following extension of this result:

Theorem E.27 (Klopsch and Lev; cf. [132]) *For every G abelian group of order n and integer s we have*

$$\widehat{\chi}(G,[0,s]) \leq \frac{2n}{s+2}+1;$$

furthermore, when $s \geq 3$, equality holds if, and only if, there is a subgroup H of order $n/(s+2)$ in G for which G/H is cyclic.

Indeed, when H is a subgroup of order $n/(s+2)$ in G for which G/H is cyclic with $g+H$ as a generator, then the set

$$A = H \cup (g+H)$$

has size $|A| = 2n/(s+2)$, and we have $\langle A \rangle = G$, but

$$[0,s]A = \cup_{i=0}^{s}(ig+H) \neq G.$$

Alternately, as did Margotta in [152], instead of $\widehat{\chi}(G,[0,s])$, we may study the quantity $\widehat{s}(G,m)$: the minimum value of s for which the $[0,s]$-fold sumset of every m-subset of G that generates G is G itself, that is,

$$\widehat{s}(G,m) = \min\{s \mid A \subseteq G, \langle A \rangle = G, |A| \geq m \Rightarrow [0,s]A = G\}.$$

While, in theory, it suffices to study only one of $\widehat{s}(G,m)$ or $\widehat{\chi}(G,[0,s])$, it may be possible to gain different results via the two different perspectives.

To establish a lower bound for $\widehat{s}(\mathbb{Z}_n,m)$ for all $n \geq m \geq 2$, we can observe that for the subset

$$A = \{0,1,2,\ldots,m-1\}$$

of $\mathbb{Z}_n$ we get

$$[0,s]A = \{0,1,2,\ldots,s(m-1)\},$$

which immediately implies the following:

Proposition E.28 *For every positive integer n and m with $n \geq m \geq 2$, we have*

$$\widehat{s}(\mathbb{Z}_n,m) \geq \left\lfloor \frac{n+m-3}{m-1} \right\rfloor.$$

It turns out that for prime values of n, equality holds in Proposition E.28: Indeed, by Corollary D.4 (the generalization of the Cauchy–Davenport Inequality), with

$$s = \left\lfloor \frac{p+m-3}{m-1} \right\rfloor,$$

for all m-subsets A of $\mathbb{Z}_p$ we get

$$\begin{aligned}
|[0,s]A| &\geq |sA| \\
&\geq \min\{p, sm-s+1\} \\
&= \min\left\{p, \left\lfloor \frac{p+m-3}{m-1} \right\rfloor \cdot (m-1)+1\right\} \\
&\geq \min\left\{p, \frac{p-1}{m-1} \cdot (m-1)+1\right\} \\
&= p.
\end{aligned}$$

Therefore:

Proposition E.29 *For every prime p and positive integer m with $p \geq m \geq 2$, we have*

$$\widehat{s}(\mathbb{Z}_p,m) = \left\lfloor \frac{p+m-3}{m-1} \right\rfloor.$$

While determining $\widehat{s}(G,m)$ was easy when the order of G is prime, this seems not to be the case when G has subgroups of many different sizes. As a case in point, observe that for every divisor d of n, by Proposition E.16 above we have

$$\widehat{\chi}(\mathbb{Z}_n,[0,n/d-2]) \geq \widehat{v}(n,n/d-2)+1 \geq \left(\left\lfloor \frac{n/d-2}{n/d-2} \right\rfloor +1\right)\cdot \frac{n}{n/d}+1 = 2d+1.$$

Therefore, if $d \geq m/2$, then there must be an m-subset A of $\mathbb{Z}_n$ that generates $\mathbb{Z}_n$ but for which

$$[0,n/d-2]A \neq \mathbb{Z}_n;$$

this yields the following lower bound:

Proposition E.30 *Suppose that n and m are positive integers, and $d \in D(n)$ with $d \geq m/2$. Then*

$$\widehat{s}(\mathbb{Z}_n,m) \geq n/d-1.$$

It turns out that for $m \leq 5$ our two lower bounds above actually determine $\widehat{s}(\mathbb{Z}_n,m)$; it can be shown that (for $n \geq m$):

$$\widehat{s}(\mathbb{Z}_n,2) = n-1;$$

$$\widehat{s}(\mathbb{Z}_n,3) = \lfloor n/2 \rfloor;$$

$$\widehat{s}(\mathbb{Z}_n,4) = \begin{cases} n/2-1 & \text{if } n \text{ is even;} \\ \lfloor (n+1)/3 \rfloor & \text{if } n \text{ is odd;} \end{cases}$$

$$\widehat{s}(\mathbb{Z}_n,5) = \begin{cases} n/3-1 & \text{if } n \text{ is divisible by 3;} \\ \lfloor (n+2)/4 \rfloor & \text{if } n \text{ is not divisible by 3.} \end{cases}$$

(Margotta in [152] conjectured the first three formulae.) However, as m increases, the result becomes less transparent; for example, for $n \geq 7$ we get

$$\widehat{s}(\mathbb{Z}_n,6) = \begin{cases} n/3-1 & \text{if } n \text{ is divisible by 3;} \\ \lfloor n/4 \rfloor & n \text{ is even but not divisible by 3;} \\ \lfloor (n+3)/5 \rfloor & \text{otherwise.} \end{cases}$$

We pose the following (potentially difficult) problems:

Problem E.31 *Find a concise formula for $\widehat{s}(\mathbb{Z}_n,m)$ for all $m \leq n$.*

Problem E.32 *Evaluate, or find bounds for $\widehat{s}(G,m)$ for arbitrary m and noncyclic group G.*

E.1.3 Arbitrary number of terms

Here we consider, for a given group G, the quantity

$$\chi(G,\mathbb{N}_0) = \min\{m \mid A \subseteq G, |A| \geq m \Rightarrow \langle A \rangle = G\},$$

where $\langle A \rangle$ is the subgroup of G generated by A. By Proposition D.13, we have:

Proposition E.33 *Let G be any abelian group of order $n \geq 2$, and let p be the smallest prime divisor of n. Then*

$$\chi(G, \mathbb{N}_0) = n/p + 1.$$

We also note that the variation

$$\widehat{\chi}(G, \mathbb{N}_0) = \min\{m \mid A \subseteq G, \langle A \rangle = G, |A| \geq m \Rightarrow \langle A \rangle = G\}$$

that would correspond to the analogous $\widehat{\chi}(G, [0, s])$ of Subsection E.1.2 is trivial and thus of no interest.

E.2 Unrestricted signed sumsets

Our goal in this section is to investigate $\chi_\pm(G, H)$, the minimum value of m for which

$$H_\pm A = G$$

holds for every m-subset of G. (Recall that $H_\pm A$ is the union of all h-fold signed sumsets $h_\pm A$ for $h \in H$.) Since $0_\pm A = \{0\}$ for every subset A of G but $h_\pm G = G$ for every positive integer h, we see that $\chi_\pm(G, H)$ does not exist for $n \geq 2$ when $H = \{0\}$, but $\chi_\pm(G, H)$ does exist and is at most n when H contains at least one positive integer.

We consider three special cases: when H consists of a single nonnegative integer h, when H consists of all nonnegative integers up to some value s, and when H is the entire set of nonnegative integers.

E.2.1 Fixed number of terms

Our goal here is to find, for a given group G and positive integer h, the quantity

$$\chi_\pm(G, h) = \min\{m \mid A \subseteq G, |A| \geq m \Rightarrow h_\pm A = G\}.$$

It is easy to see that

$$\chi_\pm(G, 1) = n$$

for each group G: indeed, $1_\pm G = G$, but $1_\pm(G \setminus \{0\}) = G \setminus \{0\}$.

We can also evaluate $\chi_\pm(G, 2)$. First, observe that

$$\chi_\pm(G, 2) \leq \chi(G, 2) = \lfloor n/2 \rfloor + 1.$$

Clearly, if n is even, then for a subgroup H of order $n/2$ we have $2_\pm H = H$, so $\chi_\pm(G, 2)$ cannot be $n/2$ or less. When n is odd, G can be partitioned as

$$G = \{0\} \cup K \cup (-K);$$

here $0 \not\in 2_\pm K$, so $\chi_\pm(G, 2)$ cannot be $(n-1)/2$ or less. In summary, we have:

Proposition E.34 *For all groups G of order n, we have*

$$\chi_\pm(G, 1) = n$$

and

$$\chi_\pm(G, 2) = \lfloor n/2 \rfloor + 1.$$

Furthermore, as an immediate consequence of Theorems D.14 and E.1, we get:

Theorem E.35 *For all n and h we have*

$$\chi_\pm(\mathbb{Z}_n, h) = v_1(n, h) + 1.$$

This leaves us with the following problem:

Problem E.36 *Evaluate $\chi_\pm(G, h)$ for noncyclic groups G and integers $h \geq 3$.*

E.2.2 Limited number of terms

For given groups G and positive integer s, here we consider $\chi_{\pm}(G,[0,s])$, that is, the minimum value of m for which the $[0,s]$-fold signed sumset of every m-subset of G is G itself.

By Proposition D.36, we have:

Proposition E.37 *For all finite abelian groups G of order $n \geq 3$, we have*

$$\chi_{\pm}(G,[0,1]) = \begin{cases} n-1 & \text{if } n \text{ is odd,} \\ n & \text{if } n \text{ is even.} \end{cases}$$

For $s \geq 2$, we do not know the value of $\chi_{\pm}(G,[0,s])$ in general, but we have the following obvious upper bound:

Proposition E.38 *For every G and s we have*

$$\chi_{\pm}(G,[0,s]) \leq \chi_{\pm}(G,s).$$

We can also establish a lower bound in the case when G is cyclic as follows. Let d be any positive divisor of n, and let H be a subgroup of order n/d in G. Consider the set

$$A = \bigcup_{i=-\lfloor (d-2)/(2s) \rfloor}^{\lfloor (d-2)/(2s) \rfloor} (i+H).$$

We then see that A has size

$$\left(2 \cdot \left\lfloor \frac{d-2}{2s} \right\rfloor + 1\right) \cdot \frac{n}{d}.$$

Furthermore,

$$[0,s]_{\pm}A = \bigcup_{i=-s\lfloor (d-2)/(2s) \rfloor}^{s\lfloor (d-2)/(2s) \rfloor} (i+H),$$

so $[0,s]_{\pm}A$ has size

$$\left(2s \cdot \left\lfloor \frac{d-2}{2s} \right\rfloor + 1\right) \cdot \frac{n}{d} \leq (d-1) \cdot \frac{n}{d} < n,$$

and thus $[0,s]_{\pm}A \neq \mathbb{Z}_n$. Recalling the function

$$v_{\pm}(n,h) = \max\left\{\left(2 \cdot \left\lfloor \frac{d-2}{2h} \right\rfloor + 1\right) \cdot \frac{n}{d} \mid d \in D(n)\right\}$$

from page 81, we get:

Proposition E.39 *For all positive integers n and s we have*

$$\chi_{\pm}(\mathbb{Z}_n,[0,s]) \geq v_{\pm}(n,s)+1.$$

Combining Propositions E.35, E.38, and E.39, we get:

Proposition E.40 *For all positive integers n and s we have*

$$v_{\pm}(n,s)+1 \leq \chi_{\pm}(\mathbb{Z}_n,[0,s]) \leq v_1(n,s)+1.$$

We believe that the lower bound is exact:

Conjecture E.41 *For all positive integers n and s we have*

$$\chi_{\pm}(\mathbb{Z}_n, [0,s]) = v_{\pm}(n,s) + 1.$$

Problem E.42 *Prove (or disprove) Conjecture E.41.*

From Propositions 4.9 and E.37, we see that Conjecture E.41 holds for $s = 1$. By Proposition E.40, Conjecture E.41 also holds whenever $v_{\pm}(n,s) = v_1(n,s)$. In particular, from Propositions 4.5 and 4.12 we get:

Proposition E.43 *When n is even and $s \geq 2$, we have*

$$\chi_{\pm}(\mathbb{Z}_n, [0,s]) = v_{\pm}(n,s) + 1 = v_1(n,s) + 1 = n/2 + 1.$$

For the case when n is odd and divisible by 3, by Proposition 4.12, Conjecture E.41 becomes:

Conjecture E.44 *When n is odd and divisible by 3 and $s \geq 3$, we have*

$$\chi_{\pm}(\mathbb{Z}_n, [0,s]) = v_{\pm}(n,s) + 1 = n/3 + 1.$$

As a modest step toward Conjecture E.41, we offer:

Problem E.45 *Prove (or disprove) Conjecture E.44.*

We can also prove Conjecture E.41 for groups of prime order p. Recall that, by Theorem D.40, we have

$$\rho_{\pm}(\mathbb{Z}_p, m, [0,s]) = \min\{p, 2s\lfloor m/2 \rfloor + 1\}.$$

Since for

$$m \geq 2\lfloor (p-2)/(2s) \rfloor + 2$$

we have

$$2s\lfloor m/2 \rfloor + 1 \geq 2s\lfloor (p-2)/(2s) \rfloor + 2s + 1 \geq 2s\left(\frac{p-2-(2s-1)}{2s}\right) + 2s + 1 = p,$$

but for

$$m \leq 2\lfloor (p-2)/(2s) \rfloor + 1$$

we have

$$2s\lfloor m/2 \rfloor + 1 \leq 2s\lfloor (p-2)/(2s) \rfloor + 1 \leq p - 1,$$

we get

$$\chi_{\pm}(\mathbb{Z}_p, [0,s]) = 2\lfloor (p-2)/(2s) \rfloor + 2.$$

Therefore, recalling Proposition 4.8, this yields:

Theorem E.46 *For every prime p and positive integer s, we have*

$$\chi_{\pm}(\mathbb{Z}_p, [0,s]) = v_{\pm}(p,s) + 1 = 2\lfloor (p-2)/(2s) \rfloor + 2.$$

We know little about $\chi_{\pm}(G, [0,s])$ for noncyclic groups:

Problem E.47 *Find $\chi_{\pm}(G, [0,2])$ for all noncyclic groups G.*

Problem E.48 *Find $\chi_{\pm}(G, [0,s])$ for all noncyclic groups G and positive integers $s \geq 3$.*

Analogous to $\widehat{\chi}(G,[0,s])$ of Subsection E.1.2 above, Klopsch and Lev in [131] investigated the quantity $\widehat{\chi}_{\pm}(G,[0,s])$: the minimum value of m for which the $[0,s]$-fold signed sumset of every m-subset of G that generates G is G itself, that is,

$$\widehat{\chi}_{\pm}(G,[0,s]) = \min\{m \mid A \subseteq G, \langle A \rangle = G, |A| \geq m \Rightarrow [0,s]_{\pm}A = G\}.$$

It turns out that the following relative of the function $v_1(n,s)$ of page 76 will be useful: if $n \geq 2s+2$, we define

$$\widehat{v}_{\pm}(n,s) = \max\left\{\left(2\left\lfloor \frac{d-2}{2s} \right\rfloor + 1\right) \cdot \frac{n}{d} \mid d \in D(n), d \geq 2s+2\right\};$$

we adhere to the convention that the maximum element of the empty set equals zero, and thus if $n \leq 2s+1$, we have $\widehat{v}(n,s)=0$. (Note also that we omitted the index 1 as unnecessary here.)

We can easily see that for all n and s we have

$$\widehat{\chi}_{\pm}(\mathbb{Z}_n,[0,s]) \geq \widehat{v}_{\pm}(n,s)+1.$$

This is trivial if $s \geq \lfloor n/2 \rfloor$, so assume $s \leq \lfloor n/2 \rfloor - 1$. Suppose that $d \in D(n)$, $d \geq 2s+2$, and

$$m = \widehat{v}_{\pm}(n,s) = \left(2\left\lfloor \frac{d-2}{2s} \right\rfloor + 1\right) \cdot \frac{n}{d}.$$

Then, with H as the order n/d subgroup of $\mathbb{Z}_n$ and $c = \lfloor (d-2)/(2s) \rfloor$, we define

$$A = \cup_{i=-c}^{c}(i+H).$$

Now A has size m, and

$$[0,s]_{\pm}A = \cup_{i=-sc}^{sc}(i+H)$$

has size

$$(2sc+1) \cdot \frac{n}{d} = \left(2s\left\lfloor \frac{d-2}{2s} \right\rfloor + 1\right) \cdot \frac{n}{d} \leq (d-1) \cdot \frac{n}{d} < n,$$

so $[0,s]_{\pm}A \neq \mathbb{Z}_n$. Furthermore, since $d \geq 2s+2$, we have $c \geq 1$ and thus $1 \in A$, which implies that A generates $\mathbb{Z}_n$. This yields:

Proposition E.49 *For positive integers n and s, we have*

$$\widehat{\chi}_{\pm}(\mathbb{Z}_n,[0,s]) \geq \widehat{v}_{\pm}(n,s)+1.$$

As it turns out, we have equality in Proposition E.49:

Theorem E.50 (Klopsch and Lev; cf. [131]) *Let n and s be positive integers; $n \geq 2$. We have*

$$\widehat{\chi}_{\pm}(\mathbb{Z}_n,[0,s]) = \widehat{v}_{\pm}(n,s)+1.$$

As a special case, we have:

Corollary E.51 *For every prime p and positive integer s, we have*

$$\widehat{\chi}_{\pm}(\mathbb{Z}_p,[0,s]) = \begin{cases} 2\lfloor (p-2)/(2s) \rfloor + 2 & \textit{if} \quad s \leq (p-3)/2; \\ 1 & \textit{if} \quad s \geq (p-1)/2. \end{cases}$$

Let us turn to $\widehat{\chi}_{\pm}(G,[0,s])$ for noncyclic G. First, we recall that, by Theorem E.19, we have

$$\widehat{\chi}_{\pm}(\mathbb{Z}_2^r,[0,s]) = \widehat{\chi}(\mathbb{Z}_2^r,[0,s]) = (s+2)\cdot 2^{r-s-1}+1.$$

The following additional results are known:

Theorem E.52 (Klopsch and Lev; cf. [131]) *Suppose that G is a finite abelian group of order $n \geq 2$, rank r, and invariant factorization $\mathbb{Z}_{n_1} \times \cdots \times \mathbb{Z}_{n_r}$; we set*

$$D_{\pm} = \lfloor n_1/2 \rfloor + \cdots + \lfloor n_r/2 \rfloor.$$

($D_{\pm}$ is called the diameter *of G.)*

1. *If $D_{\pm} \geq 2$, then*

$$\widehat{\chi}_{\pm}(G,[0,1]) = \begin{cases} n-1 & \text{if } n \text{ is odd,} \\ n & \text{if } n \text{ is even.} \end{cases}$$

2. *If $D_{\pm} \geq 3$, then*

$$\widehat{\chi}_{\pm}(G,[0,2]) = \begin{cases} \frac{n_r-1}{2}\cdot\frac{n}{n_r} & \text{if } n_r \equiv 1 \ (4), \\ \frac{n_r-1}{2}\cdot\frac{n}{n_r}+1 & \text{if } n_r \equiv 3 \ (4), \\ \frac{n}{2} & \text{if } G \cong \mathbb{Z}_{2^k} \text{ for some } k \in \mathbb{N}, \\ \frac{n}{2}+1 & \text{otherwise.} \end{cases}$$

3. *If $D_{\pm} \geq 2$, then*

$$\widehat{\chi}_{\pm}(G,[0,D_{\pm}-1]) = 2r+2-n(\mathbb{Z}_2)+2\cdot\lfloor n(\mathbb{Z}_3)/2\rfloor,$$

where $n(\mathbb{Z}_2)$ and $n(\mathbb{Z}_3)$ are the number of $\mathbb{Z}_2$ and $\mathbb{Z}_3$ factors in the invariant factorization of G, respectively.

4. *If $s \geq D_{\pm}$, then $\widehat{\chi}_{\pm}(G,[0,s]) = 1$.*

Thus we see that $\widehat{\chi}_{\pm}(G,[0,s])$ has been determined for all G and s, except for noncyclic groups of exponent more than two and for $3 \leq s \leq D_{\pm}-2$.

Problem E.53 *Find $\widehat{\chi}_{\pm}(G,[0,s])$ for every noncyclic group G and every $3 \leq s \leq D_{\pm}-2$.*

Since the general problem is probably difficult, we offer the following special cases:

Problem E.54 *Find $\widehat{\chi}_{\pm}(G,[0,3])$ for every noncyclic group G.*

Problem E.55 *Find $\widehat{\chi}_{\pm}(G,[0,D_{\pm}-2])$ for every noncyclic group G.*

While exact values of $\widehat{\chi}_{\pm}(G,[0,s])$ might be difficult to get in general, there is a tight upper bound for it. Recall that by Proposition E.49, for all n and s we have

$$\widehat{\chi}_{\pm}(\mathbb{Z}_n,[0,s]) \geq \widehat{v}_{\pm}(n,s)+1;$$

in the case when n happens to be divisible by $2s+2$, we can further see from our formula for $\widehat{v}_{\pm}(n,s)$ that

$$\widehat{\chi}_{\pm}(\mathbb{Z}_n,[0,s]) \geq \frac{3n}{2s+2}+1.$$

As Klopsch and Lev proved in [131], this is as good a bound as one can get:

Theorem E.56 (Klopsch and Lev; cf. [131]) *For every G abelian group of order n and integer $s \geq 3$ we have*

$$\hat{\chi}_{\pm}(G, [0, s]) \leq \frac{3n}{2s+2} + 1;$$

furthermore, equality holds if, and only if, there is a subgroup H of order $n/(2s+2)$ in G for which G/H is cyclic.

E.2.3 Arbitrary number of terms

This subsection is identical to Subsection E.1.3.

E.3 Restricted sumsets

Our goal in this section is to investigate $\chi\hat{}(G, H)$, the minimum value of m for which

$$H\hat{}A = G$$

holds for every m-subset of G. (Recall that $H\hat{}A$ is the union of all restricted h-fold sumsets $h\hat{}A$ for $h \in H$.) In contrast to $\chi(G, H)$, which exists for every group G when H contains at least one positive element, there are some less trivial situations for which $\chi\hat{}(G, H)$ does not exist.

In the subsections below, we consider three special cases: when H consists of a single nonnegative integer h, when H consists of all nonnegative integers up to some value s, and when H is the entire set of nonnegative integers.

E.3.1 Fixed number of terms

Analogous to the h-critical number of a group G, we define the *restricted h-critical number of G* to be the minimum value of m for which every m-subset A of G has $h\hat{}A = G$; this quantity, if it exists, is denoted by $\chi\hat{}(G, h)$.

Let us make some initial observations. First, note that $0\hat{}A = \{0\}$ for every subset A of G, thus $\chi\hat{}(G, 0)$ only exists when $n = 1$ (in which case it obviously equals 1). Second, since $1\hat{}A = A$ for every $A \subseteq G$, the restricted 1-critical number of any G is clearly just n. Third, if $h > n$, then $h\hat{}G = \emptyset$, and when $h = n$, then $h\hat{}G$ consists of exactly one element. Furthermore, for $h = n - 1$, we have $h\hat{}G = G$: to see this, note that

$$(n-1)\hat{}G = \{-g + \Sigma_{g \in G} g \mid g \in G\} = G,$$

so $\chi\hat{}(G, n-1)$ exists for all G and is at most n. Since we have $|(n-1)\hat{}A| = 1$ for every subset A of G that has size $n - 1$, we also see that $\chi\hat{}(G, n-1) = n$.

We summarize our findings as follows:

Proposition E.57 *Let G be an abelian group of order n.*

1. *If $n = 1$, then $\chi\hat{}(G, 0) = 1$; if $n \geq 2$, then $\chi\hat{}(G, 0)$ does not exist.*
2. *We have $\chi\hat{}(G, 1) = n$.*
3. *We have $\chi\hat{}(G, n-1) = n$.*
4. *If $n = 1$, then $\chi\hat{}(G, n) = 1$; if $n \geq 2$, then $\chi\hat{}(G, n)$ does not exist.*
5. *If $h > n$, then $\chi\hat{}(G, h)$ does not exist.*

By Proposition E.57, it suffices to investigate $\chi\hat{}(G,h)$ for $2 \leq h \leq n-2$. The question then arises: when does $\chi\hat{}(G,h)$ exist? This question is clearly equivalent to deciding when $h\hat{}G = G$ holds, for which the answer is provided by Theorem D.72. Therefore:

Theorem E.58 *The restricted h-critical number $\chi\hat{}(G,h)$ of an abelian group G of order n exists for all G and $1 \leq h \leq n-1$, except for the elementary abelian 2-group for $h = 2$ or $h = n-2$.*

For $h = 2$ we recall Proposition D.58, which can be rephrased as follows:

Proposition E.59 *Suppose that G is of order $n \geq 3$ and is not isomorphic to the elementary abelian 2-group. Then*

$$\chi\hat{}(G,2) = \frac{n + |\mathrm{Ord}(G,2)| + 3}{2}.$$

In particular,

$$\chi\hat{}(\mathbb{Z}_n,2) = \lfloor n/2 \rfloor + 2.$$

For $h \geq 3$, we know the value of $\chi\hat{}(G,h)$ for all h when n is even:

Theorem E.60 (Roth and Lempel; cf. [179]) *Suppose that G is an abelian group of even order $n \geq 12$.*

If $G \in \{\mathbb{Z}_2^r, \mathbb{Z}_2^{r-1} \times \mathbb{Z}_4\}$ and $h \in \{3, n/2-2\}$, then $\chi\hat{}(G,h) = n/2+2$.

In all other cases:

$$\chi\hat{}(G,h) = \begin{cases} n/2+1 & \text{if } 3 \leq h \leq n/2-2; \\ h+3 & \text{if } n/2-1 \leq h \leq (n+|\mathrm{Ord}(G,2)|-3)/2; \\ h+2 & \text{if } (n+|\mathrm{Ord}(G,2)|-1)/2 \leq h \leq n-2. \end{cases}$$

Observe that the assumption that $n \geq 12$ is necessary: we have $\chi\hat{}(\mathbb{Z}_{10},3) = 7$ as shown by the subset $A = \{1,2,4,6,8,9\}$ for which $0 \notin 3\hat{}A$. We also note that Theorem E.60 was proved independently for cyclic groups by Bajnok; cf. [16].

Let us now turn to the case of odd values of n, for which we know much less.

First, we find a lower bound as follows. Assume that A is an $(h+1)$-subset of G. Then

$$|h\hat{}A| = h+1 \leq n-1.$$

Therefore:

Proposition E.61 *For all abelian groups G of order n and positive integers $h \leq n-2$ we have $\chi\hat{}(G,h) \geq h+2$.*

Next, we show that, in fact, equality holds in Proposition E.61 for every group G of odd order and for all

$$(n-1)/2 \leq h \leq n-2.$$

Let A be an $(h+2)$-subset of G. Then, by symmetry, $|h\hat{}A| = |2\hat{}A|$; since

$$|A| = h+2 \geq (n+3)/2,$$

by Proposition E.59 we have

$$|h\hat{}A| = n.$$

This proves the following:

Proposition E.62 *Let G be an abelian group of odd order n, and suppose that h is a positive integer with*

$$(n-1)/2 \leq h \leq n-2.$$

Then $\chi\hat{}(G,h) = h+2$.

This leaves us with the following problem:

Problem E.63 *For each abelian group G of odd order n and each h with*

$$3 \leq h \leq (n-3)/2,$$

find the restricted h-critical number of G.

We now summarize what we know about cyclic groups. First, a lower bound. Consider the set

$$A = \{1, 2, \dots, \lfloor (n-2)/h \rfloor + h\}$$

in $\mathbb{Z}_n$. We can easily see that

$$h\hat{}A = \{h(h+1)/2, h(h+1)/2+1, \dots, h\lfloor (n-2)/h \rfloor + h(h+1)/2\};$$

in particular,

$$h(h+1)/2 - 1 \not\in h\hat{}A.$$

Therefore:

Proposition E.64 *For all positive integers n and h with $h \leq n-1$ we have*

$$\chi\hat{}(\mathbb{Z}_n, h) \geq \lfloor (n-2)/h \rfloor + h + 1.$$

For cyclic groups of prime order, Theorem D.51 implies that equality holds in Proposition E.64:

Theorem E.65 *For any positive integer h and prime p with $h \leq p-1$ we have*

$$\chi\hat{}(\mathbb{Z}_p, h) = \lfloor (p-2)/h \rfloor + h + 1.$$

For cyclic groups, this leaves us with the following question:

Problem E.66 *Find the restricted h-critical number of $\mathbb{Z}_n$ for each odd composite value of n and for* $3 \leq h \leq (n-3)/2$.

As an example, we mention that for $n = 15$ we find the following values:

$$\chi\hat{}(\mathbb{Z}_{15}, h) = \begin{cases} 15 & h=1 \\ 9 & h=2,3 \\ 8 & h=4 \\ 9 & h=5,6 \\ h+2 & h=7,8,\dots,13 \\ 15 & h=14. \end{cases}$$

The answers for $h \leq 2$ and for $h \geq 7$ follow from our results above; the rest were determined by a computer program (of course, $\chi\hat{}(\mathbb{Z}_{15}, h)$ does not exist for $h=0$ or $h \geq 15$). As these values indicate, Problem E.66 may be challenging in general.

Relying on Corollary D.55, we have additional results for $h = 3$:

Proposition E.67 (Bajnok; cf. [16]) *Let n be an arbitrary integer with $n \geq 16$.*

1. *If n has prime divisors congruent to 2 mod 3 and p is the smallest such divisor, then*

$$\chi\hat{}(\mathbb{Z}_n, 3) \geq \begin{cases} \left(1+\frac{1}{p}\right)\frac{n}{3}+2 & \text{if } n = 3p; \\ \left(1+\frac{1}{p}\right)\frac{n}{3}+1 & \text{otherwise.} \end{cases}$$

2. *If n has no prime divisors congruent to 2 mod 3, then*

$$\chi\hat{}(\mathbb{Z}_n, 3) \geq \begin{cases} \left\lfloor\frac{n}{3}\right\rfloor+4 & \text{if } n \text{ is divisible by } 9; \\ \left\lfloor\frac{n}{3}\right\rfloor+3 & \text{otherwise.} \end{cases}$$

Observe that the case when n is even follows from Theorem E.60, since

$$\left(1+\frac{1}{2}\right)\frac{n}{3}+1 = \frac{n}{2}+1;$$

and the case when n is prime follows from Theorem E.65 since

$$\left\lfloor\frac{p-2}{3}\right\rfloor+3+1 = \begin{cases} \left(1+\frac{1}{p}\right)\frac{p}{3}+3 & \text{if } p \equiv 2 \text{ mod } 3; \\ \left\lfloor\frac{p}{3}\right\rfloor+3 & \text{otherwise.} \end{cases}$$

We make the following conjecture:

Conjecture E.68 *For all values of $n \geq 16$, equality holds in Proposition E.67.*

We have verified that Conjecture E.68 holds for all values of $n \leq 50$, and by Theorems E.65 and E.60, it holds when n is prime or even. As additional support, we have the following:

Theorem E.69 (Bajnok; cf. [16]) *Conjecture D.56 implies Conjecture E.68.*

It is worth mentioning the following special case of Conjecture E.68:

Conjecture E.70 *If $n \geq 31$ is an odd integer, then*

$$\chi\hat{}(\mathbb{Z}_n, 3) \leq \tfrac{2}{5}n + 1.$$

(The additive constant could be adjusted to include odd integers less than 31.) This conjecture was made by Gallardo, Grekos, et al. in [89], and (for large n) proved by Lev via the following more general result:

Theorem E.71 (Lev; cf. [141]) *Let G be an abelian group of order n with*

$$n \geq 312 \cdot |\mathrm{Ord}(G, 2)| + 1235.$$

Then for any subset A of G, at least one of the following possibilities holds:

- *$|A| \leq \frac{5}{13}n$;*
- *A is contained in a coset of an index-two subgroup of G;*
- *A is contained in a union of two cosets of an index-five subgroup of G; or*

- $3\hat{}A = G$.

So, in particular, if n is odd, is at least 1235, and a subset A of G has size more than $2n/5$, then the last possibility must hold, so we get:

Corollary E.72 (Lev; cf. [141]) *If $n \geq 1235$ is an odd integer, then*

$$\chi\hat{}(\mathbb{Z}_n, 3) \leq \tfrac{2}{5}n + 1.$$

The bound on n in Corollary E.72 can hopefully be reduced to the one in Conjecture E.70:

Problem E.73 *Prove that*

$$\chi\hat{}(\mathbb{Z}_n, 3) \leq \tfrac{2}{5}n + 1$$

holds for odd integer values of n between 31 and 1235 (inclusive).

As another special case of Conjecture E.68, we have:

Conjecture E.74 *If $n \geq 83$ is odd and not divisible by five, then*

$$\chi\hat{}(\mathbb{Z}_n, 3) \leq \tfrac{4}{11}n + 1.$$

Theorem E.71 does not quite yield Conjecture E.74: while a careful read of [141] enables us to reduce the coefficient $5/13$ to $(3-\sqrt{5})/2$ (at least for large enough n), this is still higher than $4/11$. Hence we pose:

Problem E.75 *Prove Conjecture E.74.*

Combining Theorem E.1 with Conjecture E.68, we claim that, when $n \geq 11$, we have

$$\chi(\mathbb{Z}_n, 3) \leq \chi\hat{}(\mathbb{Z}_n, 3) \leq \chi(\mathbb{Z}_n, 3) + 3.$$

Before closing this subsection, we should mention that, unlike in Subsection E.3.3 below, there is no point in considering the quantity

$$\chi\hat{}(G^*, h) = \min\{m \mid A \subseteq G \setminus \{0\}, |A| \geq m \Rightarrow h\hat{}A = G\}.$$

(The study of critical numbers originated with the paper [80] of Erdős and Heilbronn, where they studied only subsets of $G \setminus \{0\}$.) Indeed, we have the following easy result:

Proposition E.76 *Let G be a finite abelian group of order $n \geq 6$, and let h be an integer with $2 \leq h \leq n-2$. Suppose that $\chi\hat{}(G, h)$ exists (that is, if $h \in \{2, n-2\}$ then G is not an elementary abelian 2-group; cf. Proposition E.58). Then*

$$\chi\hat{}(G^*, h) = \chi\hat{}(G, h).$$

The short proof can be found on page 350.

E.3.2 Limited number of terms

For a finite abelian group G and a nonnegative integer s, we define the *restricted* $[0,s]$-*critical number of* G to be the minimum value of m for which every m-subset A of G has $[0,s]\hat{}A = G$; this quantity, if it exists, is denoted by $\chi\hat{}(G,[0,s])$.

Let us make some initial observations. First, note that $0\hat{}A = \{0\}$ for every subset A of G, thus $\chi\hat{}(G,[0,0])$ only exists when $n = 1$ (in which case it obviously equals 1). Second, since $1\hat{}A = A$ for every $A \subseteq G$, the restricted $[0,1]$-critical number of any G is clearly just n. Third, since for each $s \geq 1$,

$$G = 1\hat{}G \subseteq [0,s]\hat{}G,$$

$\chi\hat{}(G,[0,s])$ exists (and is at most n) for all G and $s \geq 1$.

We summarize our observations as follows:

Proposition E.77 *Let G be an abelian group of order n, and let s be a nonnegative integer.*

1. *If $n = 1$, then $\chi\hat{}(G,[0,0]) = 1$; if $n \geq 2$, then $\chi\hat{}(G,[0,0])$ does not exist.*
2. *For all $s \geq 1$, $\chi\hat{}(G,[0,s])$ exists and is at most n.*
3. *We have $\chi\hat{}(G,[0,1]) = n$.*

Furthermore, $\chi\hat{}(G,s)$ is clearly an upper bound for $\chi\hat{}(G,[0,s])$, and $\chi(G,[0,s])$ is a lower bound for it, so by Theorem E.14, we have:

Proposition E.78 *For all G and $s \geq 1$,*

$$v_1(n,s) + 1 = \chi(G,[0,s]) \leq \chi\hat{}(G,[0,s]) \leq \chi\hat{}(G,s).$$

For $s = 2$ we see that by Proposition E.59, we have

$$\chi\hat{}(\mathbb{Z}_n,[0,2]) \leq \chi\hat{}(\mathbb{Z}_n,2) = \lfloor n/2 \rfloor + 2;$$

we can show that equality holds by finding a subset A of $\mathbb{Z}_n$ for which $|A| = \lfloor n/2 \rfloor + 1$ but $[0,2]\hat{}A \neq \mathbb{Z}_n$. Indeed, we see that when n is odd and

$$A = \{0,1,\dots,(n-1)/2\},$$

then $|A| = \lfloor n/2 \rfloor + 1 = (n+1)/2$ and $n-1 \not\in [0,2]\hat{}A$; when n is divisible by 4, then with

$$A = \{0,1,\dots,n/4\} \cup \{n/2+1, n/2+2, \dots, 3n/4\},$$

$|A| = n/2+1$ and $n/2 \not\in [0,2]\hat{}A$; and when $n-2$ is divisible by 4, then with

$$A = \{0,1,\dots,(n-2)/4\} \cup \{n/2, n/2+1, \dots, (3n-2)/4\},$$

$|A| = n/2+1$ and $n/2-1 \not\in [0,2]\hat{}A$. Therefore:

Proposition E.79 *For all integers $n \geq 3$,*

$$\chi\hat{}(\mathbb{Z}_n,[0,2]) = \lfloor n/2 \rfloor + 2.$$

(We mention that Lemma 3.3 in [112] says that if $A \subseteq \mathbb{Z}_n \setminus \{0\}$ and $|A| \geq n/2$, then $[1,2]\hat{}A = \mathbb{Z}_n$, but, as we have just seen, this is always false when n is even—see [17] for more information.)

Recall that, by Theorem E.60, for even values of $n \geq 12$ we have

$$\chi\hat{}(\mathbb{Z}_n, 3) = n/2 + 1.$$

But for even n, we must have

$$\chi\hat{}(\mathbb{Z}_n, [0, s]) \geq n/2 + 1$$

for all positive integers s, so for $s \geq 3$ we have

$$n/2 + 1 \leq \chi\hat{}(\mathbb{Z}_n, [0, s]) \leq \chi\hat{}(\mathbb{Z}_n, [0, 3]) \leq \chi\hat{}(\mathbb{Z}_n, 3) = n/2 + 1.$$

Therefore:

Theorem E.80 *For all even values of $n \geq 12$ and every $s \geq 3$, we have*

$$\chi\hat{}(\mathbb{Z}_n, [0, s]) = n/2 + 1.$$

This leaves us with the following open question:

Problem E.81 *Find $\chi\hat{}(\mathbb{Z}_n, [0, s])$ for all $s \geq 3$ and for odd values of n.*

We can get a lower bound for $\chi\hat{}(\mathbb{Z}_n, [0, s])$ as follows. Suppose that $n \geq s^2 - s + 2$, and consider

$$A = \{0, 1, \ldots, \lfloor (n-2)/s) \rfloor\} \cup \{n - s + 1, n - s + 2, \ldots, n - 1\};$$

we may consider this set as the interval

$$A = \{-(s-1), -(s-2), \ldots, -1, 0, 1, \ldots, \lfloor (n-2)/s) \rfloor\}.$$

For the size of A we have

$$|A| = \lfloor (n-2)/s) \rfloor + s$$

(note that this value is less than n).

By the assumption that $n \geq s^2 - s + 2$, we have

$$\lfloor (n-2)/s) \rfloor \geq s - 1,$$

and thus for the $[0, s]$-fold restricted sumset of A we get

$$[0, s]\hat{}A = \{-s(s-1)/2, -s(s-1)/2 + 1, \ldots, s\lfloor (n-2)/s) \rfloor - s(s-1)/2\},$$

which we can rewrite as

$$[0, s]\hat{}A = \{0, 1, \ldots, s\lfloor (n-2)/s) \rfloor - s(s-1)/2\} \cup \{n - s(s-1)/2, n - s(s-1)/2 + 1, \ldots, n-1\}.$$

Here

$$0 \leq s\lfloor (n-2)/s) \rfloor - s(s-1)/2 < n - s(s-1)/2 - 1 < n - s(s-1)/2 < n;$$

so

$$n - s(s-1)/2 - 1 \not\in [0, s]\hat{}A.$$

We just proved the following:

Proposition E.82 *For all positive integers s and n with $n \geq s^2 - s + 2$, we have*

$$\chi\hat{}(\mathbb{Z}_n, [0, s]) \geq \lfloor (n-2)/s) \rfloor + s + 1.$$

We should mention that Proposition E.82 is not tight in that the condition $n \geq s^2 - s + 2$ is not necessary for the conclusion to hold, and the bound does not always give the value of $\chi\hat{}(\mathbb{Z}_n, [0, s])$. So Problem E.81 is very much still open.

We know little about noncyclic groups:

Problem E.83 *Find $\chi\hat{}(G, [0, 2])$ for all noncyclic groups G.*

Problem E.84 *Find $\chi\hat{}(G, [0, s])$ for all noncyclic groups G and for all $s \geq 3$.*

E.3.3 Arbitrary number of terms

In this subsection we determine the *restricted critical number* of G, which we define as

$$\chi\hat{}(G,\mathbb{N}_0) = \min\{m \mid A \subseteq G, |A| \geq m \Rightarrow \Sigma A = G\}$$

where, for $A = \{a_1, \dots, a_m\} \subseteq G$,

$$\Sigma A = \cup_{h=0}^{\infty} h\hat{}A = \{\lambda_1 a_1 + \cdots + \lambda_m a_m \mid \lambda_1, \dots, \lambda_m \in \{0,1\}\}.$$

Before doing so, we introduce three variations:

$$\chi\hat{}(G^*,\mathbb{N}_0) = \min\{m \mid A \subseteq G \setminus \{0\}, |A| \geq m \Rightarrow \Sigma A = G\},$$

$$\chi\hat{}(G,\mathbb{N}) = \min\{m \mid A \subseteq G, |A| \geq m \Rightarrow \Sigma^* A = G\},$$

$$\chi\hat{}(G^*,\mathbb{N}) = \min\{m \mid A \subseteq G \setminus \{0\}, |A| \geq m \Rightarrow \Sigma^* A = G\},$$

where

$$\Sigma^* A = \cup_{h=1}^{\infty} h\hat{}A = \{\lambda_1 a_1 + \cdots + \lambda_m a_m \mid \lambda_1, \dots, \lambda_m \in \{0,1\}, \lambda_1 + \cdots + \lambda_m \geq 1\}.$$

We can determine if these four quantities are well-defined as follows. Since

$$G = 1\hat{}G \subseteq \Sigma^* G \subseteq \Sigma G,$$

we see that $\chi\hat{}(G,\mathbb{N}_0)$ and $\chi\hat{}(G,\mathbb{N})$ are well-defined (and are at most n) for any group G. Similarly,

$$G = \{0\} \cup 1\hat{}(G \setminus \{0\}) \subseteq \Sigma(G \setminus \{0\}),$$

so $\chi\hat{}(G^*,\mathbb{N}_0)$ is also well-defined (and at most $n-1$) for any group G of order $n \geq 2$. The same way, we see that $\chi\hat{}(G^*,\mathbb{N})$ is well-defined (and is at most $n-1$) if, and only if, $0 \in \Sigma^*(G \setminus \{0\})$. This is clearly the case if G has an element of order three or more (the element and its inverse are distinct and add to zero); if G is an elementary abelian 2-group of rank two or more, then, for example, $e_1 = 1000\ldots$, $e_2 = 0100\ldots$, and $e_1 + e_2$ are three distinct elements that add to zero. That leaves us with the group of order two, but $\chi\hat{}(\mathbb{Z}_2^*,\mathbb{N})$ cannot exist. In summary:

Proposition E.85 *The quantities $\chi\hat{}(G,\mathbb{N}_0)$ and $\chi\hat{}(G,\mathbb{N})$ are well-defined for every group G; $\chi\hat{}(G^*,\mathbb{N}_0)$ is well-defined for every group G of order at least two; and $\chi\hat{}(G^*,\mathbb{N})$ is well-defined for every group G of order at least three.*

The four quantities are strongly related—see Theorem E.98 below. The last quantity, $\chi\hat{}(G^*,\mathbb{N})$, is the one that has been studied most; it is in fact the one that has been coined the *critical number* of G. Thus we begin our investigation with $\chi\hat{}(G^*,\mathbb{N})$.

First, following a construction of Erdős and Heilbronn in [80] that was improved by Griggs in [102], we show that

$$\chi\hat{}(\mathbb{Z}_n^*,\mathbb{N}) \geq \lfloor 2\sqrt{n-2} \rfloor.$$

Assume that $n \geq 3$ (as noted above, $\chi\hat{}(\mathbb{Z}_2^*,\mathbb{N})$ does not exist). Letting $k = \lfloor 2\sqrt{n-2} \rfloor$, we set

$$A = \begin{cases} \{\pm 1, \pm 2, \dots, \pm (k-1)/2\} & \text{if } k \text{ is odd;} \\ \{\pm 1, \pm 2, \dots, \pm (k-2)/2, k/2\} & \text{if } k \text{ is even.} \end{cases}$$

Then $|A| = k-1$, and $\Sigma^* A$ is an interval consisting of all integers between $-(k^2-1)/8$ and $(k^2-1)/8$ (inclusive) when k is odd, and between $-(k^2-2k)/8$ and $(k^2+2k)/8$ (inclusive) when k is even. This yields

$$|\Sigma^* A| = \begin{cases} \frac{k^2-1}{8} + 1 + \frac{k^2-1}{8} = \frac{k^2+3}{4} & \text{if } k \text{ is odd;} \\ \\ \frac{k^2-2k}{8} + 1 + \frac{k^2+2k}{8} = \frac{k^2+4}{4} & \text{if } k \text{ is even.} \end{cases}$$

Therefore,

$$|\Sigma^* A| \leq \frac{4(n-2)+4}{4} = n-1,$$

and we get:

Proposition E.86 (Cf. [80], [102]) *With $n \geq 3$, we have*

$$\chi\hat{\ }(\mathbb{Z}_n^*, \mathbb{N}) \geq \lfloor 2\sqrt{n-2} \rfloor.$$

It turns out that, when n is prime, the lower bound of Proposition E.86 is sharp. We will need the following inequality:

Lemma E.87 *For an odd integer $n \geq 3$, $k = \lfloor 2\sqrt{n-2} \rfloor$, and $h = \lfloor (k+1)/2 \rfloor$ we have*

$$\lfloor (n-2)/h \rfloor + h \leq k.$$

The short and easy proof of Lemma E.87 is on page 351.

We now show how Theorem E.65 (via Lemma E.87) implies that, for any odd prime p,

$$\chi\hat{\ }(\mathbb{Z}_p^*, \mathbb{N}) \leq \lfloor 2\sqrt{p-2} \rfloor.$$

We follow the proof of Dias Da Silva and Hamidoune in [66].

The claim is obvious for $p = 3$, so we assume that $p \geq 5$. Consider any subset A of $\mathbb{Z}_p \setminus \{0\}$ of size $k = \lfloor 2\sqrt{p-2} \rfloor$. Then $B = A \cup \{0\}$ has size $k+1$, so by Lemma E.87, for $h = \lfloor (k+1)/2 \rfloor$, we have

$$|B| \geq \lfloor (p-2)/h \rfloor + h + 1.$$

Note also that $h \leq p-1$. Therefore, Theorem E.65 implies that $h\hat{\ }B = \mathbb{Z}_p$. But

$$h\hat{\ }B = h\hat{\ }(A \cup \{0\}) = h\hat{\ }A \cup (h-1)\hat{\ }A,$$

so (since $p \geq 5$ implies that $h \geq 2$) $\Sigma^* A = \mathbb{Z}_p$, proving our claim.

Combining this upper bound with the lower bound of Proposition E.86, we get:

Theorem E.88 (Dias Da Silva and Hamidoune; cf. [66] and Griggs; cf. [102]) *If p is an odd prime, then*

$$\chi\hat{\ }(\mathbb{Z}_p^*, \mathbb{N}) = \lfloor 2\sqrt{p-2} \rfloor.$$

Let us now turn to groups of composite order. We can easily find a lower bound for $\chi\hat{\ }(G^*, \mathbb{N})$ as follows.

Following Diderrich's construction in [67], we let p denote the smallest prime divisor of n, and consider the set

$$A = (H \setminus \{0\}) \cup (g + K)$$

where H is a subgroup of G with index p, K is a subset of H of size $p-2$, and g is any element of $G \setminus H$. (This is possible as n being composite implies that $p - 2 \leq n/p$.) Since $(p-1) \cdot g$ (and, in fact, every element of $(p-1) \cdot g + H$) is outside of $\Sigma^* A$, we get the following lower bound:

Proposition E.89 (Cf. [67]) *Let n be a composite integer with smallest prime divisor p. Then for every abelian group G of order n, we have*

$$\chi\hat{}(G^*, \mathbb{N}) \geq n/p + p - 2.$$

It took about 35 years after Diderrich's lower bound to determine the value of $\chi\hat{}(G^*, \mathbb{N})$ for all groups G. As it turns out, in most cases, the lower bound above is sharp. In particular, we have the following results:

Theorem E.90 (Diderrich and Mann; cf. [68]) *If G is of even order $n \geq 4$, then*

$$\chi\hat{}(G^*, \mathbb{N}) = \begin{cases} n/2+1 & \text{if } G \cong \mathbb{Z}_4, \mathbb{Z}_6, \mathbb{Z}_8, \mathbb{Z}_2^2, \text{ or } \mathbb{Z}_2 \times \mathbb{Z}_4; \\ n/2 & \text{otherwise.} \end{cases}$$

Theorem E.91 (Mann and Wou; cf. [149]) *Let p be an odd prime. We have*

$$\chi\hat{}((\mathbb{Z}_p^2)^*, \mathbb{N}) = \begin{cases} 2p-1 & \text{if } p = 3; \\ 2p-2 & \text{otherwise.} \end{cases}$$

Theorem E.92 (Gao and Hamidoune; cf. [93]) *Suppose that G is an abelian group of odd order n. Let p be the smallest prime divisor of n. If n/p is a composite number, then*

$$\chi\hat{}(G^*, \mathbb{N}) = n/p + p - 2.$$

These results leave us with the case of cyclic groups whose order n is the product of two (not necessarily different) odd primes. When the two primes are far from one another, we have the following:

Theorem E.93 (Diderrich; cf. [67]) *Suppose that p and q are odd primes. If $q \geq 2p+1$, then*

$$\chi\hat{}(\mathbb{Z}_{pq}^*, \mathbb{N}) = p + q - 2.$$

In the same paper, Diderrich also proved that

$$\chi\hat{}(\mathbb{Z}_{pq}^*, \mathbb{N}) \leq p + q - 1$$

holds for all odd primes p and q, and thus, by Proposition E.89, the critical number of $\mathbb{Z}_{pq}$ is either $p + q - 2$ or $p + q - 1$. We can observe that, by Proposition E.86, we also have

$$\chi\hat{}(\mathbb{Z}_{pq}^*, \mathbb{N}) \geq \lfloor 2\sqrt{pq-2} \rfloor.$$

We claim that for odd integers p and q with $3 \leq p \leq q$, we have

$$\lfloor 2\sqrt{pq-2} \rfloor \leq p + q - 1,$$

with equality if, and only if,

$$q \leq p + \lfloor 2\sqrt{p-2} \rfloor + 1.$$

Indeed, the first claim follows from the fact that

$$2\sqrt{pq-2} < p + q$$

is equivalent to

$$(q-p)^2 + 8 > 0.$$

To prove the second claim, we can square and rearrange the inequality

$$2\sqrt{pq-2} \geq p+q-1$$

to get

$$(q-p-1)^2 \leq 4p-8,$$

from which the claim follows.

Therefore, if p and q are odd integers with

$$3 \leq p \leq q \leq p + \lfloor 2\sqrt{p-2} \rfloor + 1,$$

then

$$\chi\hat{}(\mathbb{Z}_{pq}^*, \mathbb{N}) \geq p+q-1.$$

Consequently, we have the following result:

Theorem E.94 (Cf. Diderrich; cf. [67] and Griggs; cf. [102]) *Suppose that p and q are odd primes. If*

$$p \leq q \leq p + \lfloor 2\sqrt{p-2} \rfloor + 1,$$

then

$$\chi\hat{}(\mathbb{Z}_{pq}^*, \mathbb{N}) = p+q-1 = \lfloor 2\sqrt{pq-2} \rfloor.$$

(Griggs in [102] provided a more constructive proof for the lower bound than our argument above: he showed that, given the condition for p and q in Theorem E.94, the set

$$A = \{\pm 1, \pm 2, \ldots, \pm (p+q-2)/2\}$$

does not generate the elements $\pm(pq-1)/2$ in $\mathbb{Z}_{pq}$.)

Finally, nearly a half century after Erdős and Heilbronn posed the original problem of finding the critical number of an abelian group, the remaining case was decided in [85] (see also [86] for a correction):

Theorem E.95 (Freeze, Gao, and Geroldinger; cf. [85], [86]) *Suppose that p and q are odd primes. If*

$$p + \lfloor 2\sqrt{p-2} \rfloor + 1 < q < 2p+1,$$

then

$$\chi\hat{}(\mathbb{Z}_{pq}^*, \mathbb{N}) = p+q-2.$$

(We note that

$$p + \lfloor 2\sqrt{p-2} \rfloor + 1 < 2p+1$$

holds for all odd primes p.)

We can summarize these results as follows:

Theorem E.96 *Suppose that $n \geq 3$ is an integer, let p be the smallest prime divisor of n, and set $k = \lfloor 2\sqrt{p-2} \rfloor$. Then*

$$\chi\hat{}(G^*, \mathbb{N}) = \begin{cases} k & \text{if } n = p; \\ n/p+p-1 & \text{if } G \cong \mathbb{Z}_4, \mathbb{Z}_6, \mathbb{Z}_8, \mathbb{Z}_2 \times \mathbb{Z}_4, \text{ or } \mathbb{Z}_3^2, \\ & \text{or } G \text{ is cyclic, } n/p \text{ is prime, and } 3 \leq p \leq n/p \leq p+k+1; \\ n/p+p-2 & \text{otherwise.} \end{cases}$$

As we noted above, for odd integers p and n/p with

$$3 \leq p \leq n/p \leq p + \lfloor 2\sqrt{p-2} \rfloor + 1,$$

we have

$$n/p + p - 1 = \lfloor 2\sqrt{n-2} \rfloor.$$

Thus for $n \geq 10$ this allows for the second line in Theorem E.96 to be combined with the first:

Corollary E.97 *Suppose that $n \geq 10$, and let p be the smallest prime divisor of n. Then*

$$\chi\hat{}(G^*, \mathbb{N}) = \begin{cases} \lfloor 2\sqrt{n-2} \rfloor & \text{if } G \text{ is cyclic of order } n = p \text{ or } n = pq \text{ where} \\ & q \text{ is prime and } 3 \leq p \leq q \leq p + \lfloor 2\sqrt{p-2} \rfloor + 1, \\ n/p + p - 2 & \text{otherwise.} \end{cases}$$

With $\chi\hat{}(G^*, \mathbb{N})$ thus determined for any finite abelian group G, let us now turn to our other three quantities: $\chi\hat{}(G^*, \mathbb{N}_0)$, $\chi\hat{}(G, \mathbb{N}_0)$, and $\chi\hat{}(G, \mathbb{N})$.

First, we note the obvious facts that

$$\chi\hat{}(G^*, \mathbb{N}_0) \leq \chi\hat{}(G^*, \mathbb{N})$$

and

$$\chi\hat{}(G, \mathbb{N}_0) \leq \chi\hat{}(G, \mathbb{N}).$$

Next, we show that

$$\chi\hat{}(G, \mathbb{N}) \leq \chi\hat{}(G^*, \mathbb{N}) + 1$$

and

$$\chi\hat{}(G, \mathbb{N}_0) \leq \chi\hat{}(G^*, \mathbb{N}_0) + 1.$$

Indeed, if A is a subset of G of size $\chi\hat{}(G, \mathbb{N}) - 1$ so that $\Sigma^* A \neq G$, then $A \setminus \{0\}$ is a subset of $G \setminus \{0\}$ of size at least $\chi\hat{}(G, \mathbb{N}) - 2$ and $\Sigma^*(A \setminus \{0\}) \neq G$. This implies our first inequality; the second can be shown similarly.

We now prove that for every group G of order at least ten, we have

$$\chi\hat{}(G, \mathbb{N}_0) \geq \chi\hat{}(G^*, \mathbb{N}) + 1.$$

(We here ignore the exceptional cases of Theorems E.90 and E.91 that may occur when $n \leq 9$.)

Our strategy is to point to a(n already-mentioned) subset A of G for which (i) $0 \notin A$, (ii) $0 \in \Sigma^* A$, (iii) $\Sigma^* A \neq G$, and (iv) $|A| = \chi\hat{}(G^*, \mathbb{N}) - 1$. Then, by (i) and (iv), $B = A \cup \{0\}$ has size $|B| = \chi\hat{}(G^*, \mathbb{N})$; and by (ii) and (iii), $\Sigma B = \Sigma^* A \neq G$. Therefore,

$$\chi\hat{}(G, \mathbb{N}_0) \geq |B| + 1 = \chi\hat{}(G^*, \mathbb{N}) + 1$$

follows.

To find a set A in G satisfying properties (i)–(iv), recall that, when G is cyclic of order n, the set A we exhibited on page 214 has size

$$|A| = k - 1 = \lfloor 2\sqrt{n-2} \rfloor - 1,$$

and satisfies (i), (ii), and (iii) above; furthermore, when n is prime or a product of odd primes p and q with

$$p \leq q \leq p + k + 1,$$

then, by Theorem E.97, $|A| = \chi\hat{}(G^*, \mathbb{N}) - 1$. Additionally, with p denoting the smallest prime divisor of the composite number n, the set leading to Proposition E.89 above also satisfies properties (i), (ii), and (iii); and by Theorem E.97, in all remaining cases (that is, when G is not cyclic, or n is even, or n/p is composite, or when n/p equals a prime q that is greater than $p+k+1$), its size $n/p+p-3$ equals $\chi\hat{}(G^*, \mathbb{N}) - 1$ as well. Therefore,

$$\chi\hat{}(G, \mathbb{N}_0) \geq \chi\hat{}(G^*, \mathbb{N}) + 1,$$

as claimed.

The combination of our five inequalities enables us to evaluate each of $\chi\hat{}(G^*, \mathbb{N}_0)$, $\chi\hat{}(G, \mathbb{N}_0)$, and $\chi\hat{}(G, \mathbb{N})$ in terms of the already-determined value of $\chi\hat{}(G^*, \mathbb{N})$:

Theorem E.98 *For any abelian group G of order at least ten we have*

$$\chi\hat{}(G^*, \mathbb{N}_0) = \chi\hat{}(G^*, \mathbb{N}) \lessdot \chi\hat{}(G, \mathbb{N}_0) = \chi\hat{}(G, \mathbb{N}),$$

where $x \lessdot y$ means that x is exactly one less than y.

In particular, for $\chi\hat{}(G, \mathbb{N}_0)$ we have:

Corollary E.99 *Suppose that $n \geq 10$ is an integer, and let p be the smallest prime divisor of n. Then*

$$\chi\hat{}(G, \mathbb{N}_0) = \begin{cases} \lfloor 2\sqrt{n-2} \rfloor + 1 & \text{if } G \text{ is cyclic of order } n = p \text{ or } n = pq \text{ where} \\ & q \text{ is prime and } 3 \leq p \leq q \leq p + \lfloor 2\sqrt{p-2} \rfloor + 1, \\ n/p + p - 1 & \text{otherwise.} \end{cases}$$

Now that we have determined the restricted critical number of all finite groups, we move on to the inverse problem of classifying the extremal sets. Given that we have four different versions for the critical number, we have four corresponding inverse problems:

P1 Classify all subsets A of G with size $\chi\hat{}(G, \mathbb{N}_0) - 1$ for which $\Sigma A \neq G$. (We shall refer to these sets as P1-sets.)

P2 Classify all subsets A of $G \setminus \{0\}$ with size $\chi\hat{}(G^*, \mathbb{N}_0) - 1$ for which $\Sigma A \neq G$. (We shall refer to these sets as P2-sets.)

P3 Classify all subsets A of G with size $\chi\hat{}(G, \mathbb{N}) - 1$ for which $\Sigma^* A \neq G$. (We shall refer to these sets as P3-sets.)

P4 Classify all subsets A of $G \setminus \{0\}$ with size $\chi\hat{}(G^*, \mathbb{N}) - 1$ for which $\Sigma^* A \neq G$. (We shall refer to these sets as P4-sets.)

We can prove that (for $n \geq 10$) the first three problems are equivalent and that the fourth problem incorporates the first three:

Theorem E.100 *Let G be an abelian group of order at least ten. The following are equivalent:*

- *A is a P1-set in G;*
- *$0 \in A$ and $A \setminus \{0\}$ is a P2-set in G; and*
- *A is a P3-set in G.*

Furthermore, each of the above implies that $0 \in A$ *and* $A \setminus \{0\}$ *is a P4-set in* G.

The short proof is on page 351. We believe that P4 is also equivalent to the other three problems:

Problem E.101 *Prove that for groups* G *of order ten or more, every P4-set in* G *is also a P2-set in* G.

Note that having a P4-set in G that is not a P2-set would mean that there is a subset A of $G \setminus \{0\}$ of size $\chi\hat{}(G^*, \mathbb{N}) - 1$ for which $\Sigma^* A = G \setminus \{0\}$.

Assuming that the four problems are indeed equivalent, we focus on P1:

Problem E.102 *For each finite abelian group* G*, classify all subsets* A *of size* $\chi\hat{}(G, \mathbb{N}_0) - 1$ *for which* $\Sigma A \neq G$.

Problem E.102 is quite extensive; we discuss next what we already know.

Let us first consider the case when G is cyclic of order n. It is helpful to view the elements of $\mathbb{Z}_n$ as integers in the interval $(-n/2,\ n/2]$. Suppose that $A \subseteq \mathbb{Z}_n$, and let

$$A = \{a_1, \dots, a_t, -a_{t+1}, \dots, -a_m\}$$

with $a_1, \dots, a_t$ nonnegative and $-a_{t+1}, \dots, -a_m$ negative (of course, we may have $t = 0$ or $t = m$). The *norm* of A, denoted by $||A||$, is the sum $a_1 + \cdots + a_m$; for example, for the set

$$A = \{0, 2, 5, 8\} = \{0, 2, 5, -2\}$$

in $\mathbb{Z}_{10}$, we have

$$||A|| = 0 + 2 + 5 + 2 = 9.$$

We will show that if A is a subset of $\mathbb{Z}_n$ with norm $||A|| \leq n-2$, then $\Sigma A \neq \mathbb{Z}_n$; in fact, we can easily see that, using our notations from above,

$$a_1 + \cdots + a_t + 1 \not\in \Sigma A.$$

Indeed, this follows right away from our assumption, since it is equivalent to

$$a_1 + \cdots + a_t + 1 < n - (a_{t+1} + \cdots + a_m).$$

Thus we have:

Proposition E.103 *Let* $A \subseteq \mathbb{Z}_n$. *If* $||A|| \leq n - 2$*, then* $\Sigma A \neq \mathbb{Z}_n$.

Note also that if $\Sigma A \neq \mathbb{Z}_n$ for some $A \subseteq \mathbb{Z}_n$, then for any $b \in \mathbb{Z}_n$, $\Sigma(b \cdot A) \neq \mathbb{Z}_n$ as well, where

$$b \cdot A = \{b \cdot a \mid a \in A\}$$

is a *dilate* of A; indeed, the size of $\Sigma(b \cdot A)$ cannot be more than the size of ΣA.

We have the following question:

Problem E.104 *Find all odd prime values* p *for which whenever* $\mathbb{Z}_p$ *contains a subset* A *of size*

$$|A| = \chi\hat{}(\mathbb{Z}_p, \mathbb{N}_0) - 1 = \lfloor 2\sqrt{p-2} \rfloor$$

so that $\Sigma A \neq \mathbb{Z}_p$*, then there is an element* $b \in \{1, \dots, p-1\}$ *for which* $||b \cdot A|| \leq p - 2$.

We have verified that all primes less than 30 satisfy the requirements of Problem E.104, except for $p = 17$: the subset

$$A = \{0, 1, 3, 4, 5, 12, 14\}$$

of $\mathbb{Z}_{17}$ (for example) has size $|A| = \chi\hat{}(\mathbb{Z}_{17}, \mathbb{N}_0) - 1 = 7$ and norm 21, does not generate the element 11, and has no dilate with a norm less than 21.

Related to Problem E.104, we mention two results:

Theorem E.105 (Nguyen, Szemerédi, and Vu; cf. [163]) *There is a positive constant C, so that whenever A is a subset of $\mathbb{Z}_p$ for an odd prime p, if $|A| \geq 1.99\sqrt{p}$ and $\Sigma A \neq \mathbb{Z}_p$, then there is an element $b \in \{1, \ldots, p-1\}$ for which*

$$||b \cdot A|| \leq p + C\sqrt{p}.$$

Theorem E.106 (Nguyen and Vu; cf. [164]) *There is a positive constant c, so that whenever A is a subset of $\mathbb{Z}_p$ for an odd prime p, if $\Sigma A \neq \mathbb{Z}_p$, then there is a subset A' of A of size at most $cp^{6/13} \log p$ and an element $b \in \{1, \ldots, p-1\}$, for which*

$$||b \cdot (A \setminus A')|| < p.$$

Theorem E.105 says that, for large values of p, if A has size near $\chi\hat{}(\mathbb{Z}_p, \mathbb{N}_0) - 1$ but $\Sigma A \neq \mathbb{Z}_p$, then A has a dilate with norm not much larger than p; Theorem E.106 tells us that, again for large values of p, if $\Sigma A \neq \mathbb{Z}_p$, then after removing a relatively small subset from A, the remaining part has a dilate with norm less than p.

The inverse problem for groups of prime order is thus not fully settled:

Problem E.107 *For each odd prime p, classify all subsets A of $\mathbb{Z}_p$ of size $\chi\hat{}(\mathbb{Z}_p, \mathbb{N}_0) - 1$ for which $\Sigma A \neq \mathbb{Z}_p$.*

Moving on to groups of composite order n, let us first consider the case when n is even. According to Corollary E.99, for $n \geq 10$ we have

$$\chi\hat{}(G, \mathbb{N}_0) = n/2 + 1.$$

Clearly, if A is a subgroup of G of order $n/2$, then $\Sigma A \neq G$; it turns out that (for groups of order at least 16) the converse of this is true as well:

Theorem E.108 *Suppose that G is an abelian group of order $n \geq 16$ and that n is even. Let A be a subset of G with $|A| = n/2$. Then $\Sigma A \neq G$ if, and only if, A is a subgroup of G.*

We will show how results from the paper [94] by Gao, Hamidoune, Lladó, and Serra imply Theorem E.108; see page 352. We should point out that the lower bound of 16 on n cannot be reduced: in $\mathbb{Z}_{14}$, for example, the set

$$\{0, \pm 1, \pm 2, \pm 3\}$$

is not a subgroup, yet it does not generate 7.

Suppose now that n is odd, divisible by 3, and $n/3$ is composite. We then have the following result:

Theorem E.109 *Suppose that G is an abelian group of order n where n is odd, divisible by 3, $n/3$ is composite, and $n \neq 27, 45$. Let A be a subset of G with $|A| = n/3 + 1$. Then $\Sigma A \neq G$ if, and only if, there is a subgroup H of G of size $n/3$ and an element $a \in A$ so that*

$$A = H \cup \{a\}.$$

Like for Theorem E.108, our proof for Theorem E.109 will come from a careful reexamination of a corresponding result by Gao, Hamidoune, Lladó, and Serra in [94]; see page 353. We can settle the case of $n = 27$: The claim is false in $\mathbb{Z}_{27}$ as seen by the set

$$\{0, \pm 1, \pm 2, \pm 3, \pm 4, 5\},$$

but true in the other two groups of order 27, as verified by the computer program [120]. The case of $n = 45$ is still open and poses the following:

Problem E.110 *Decide if the claim of Theorem E.109 above is true for the two groups of order 45.*

In a similar fashion:

Theorem E.111 *Suppose that G is an abelian group of order n where n is odd, has smallest prime divisor $p \geq 5$, n/p is composite, and $n \neq 125$. Let A be a subset of G with $|A| = n/p + p - 2$. Then $\Sigma A \neq G$ if, and only if,*

$$A = H \cup A_1 \cup A_2,$$

where H is a subgroup of G of size n/p, and there is an element $a \in A$ so that $A_1 \subset a + H$ and $A_2 \subset -a + H$.

This result is essentially Theorem 4.1 in [94] by Gao, Hamidoune, Lladó, and Serra. Note that their assumption that $n \geq 7p^2 + 7p$ can be reduced to $n \geq 7p^2$ in our case, which, considering that $p \geq 5$, excludes only $n = 125$.

Problem E.112 *Decide if the claim of Theorem E.111 above is true for the three groups of order 125.*

This leaves us with the case when n is the product of two odd primes. One such case was resolved by Qu, Wang, Wang, and Guo:

Theorem E.113 (Qu, Wang, Wang, and Guo; cf. [176]) *Suppose that p and q are odd primes so that $q \geq 2p + 3$. Let A be a subset of $\mathbb{Z}_{pq}$ of size $p + q - 2$ for which $\Sigma A \neq \mathbb{Z}_{pq}$. Then*

$$A = H \cup A_1 \cup A_2,$$

where H is a subgroup of G of size q, and there is an element $a \in A$ so that $A_1 \subset a + H$ and $A_2 \subset -a + H$.

(We should caution that the paper [176] contains some inaccuracies: In Theorem A, the critical numbers of $\mathbb{Z}_2^2$ and cyclic groups of order p^2 with prime p are given incorrectly; the other main result in the paper, regarding groups of even order, had been done previously (cf. Theorem E.90); and Example 4.1 is the wrong example of the point that the authors make.)

We separate the remaining cases of Problem E.102 into three parts:

Problem E.114 *For each prime $p \geq 5$, classify all subsets A of size $2p - 2$ of $\mathbb{Z}_p^2$ for which $\Sigma A \neq \mathbb{Z}_p^2$.*

Problem E.115 *For all primes p and q with*

$$3 \leq p \leq q \leq p + \lfloor 2\sqrt{p-2} \rfloor + 1,$$

classify all subsets A of size $p + q - 1$ of $\mathbb{Z}_{pq}$ for which $\Sigma A \neq \mathbb{Z}_{pq}$.

Problem E.116 *For all primes p and q with*

$$p + \lfloor 2\sqrt{p-2} \rfloor + 2 \leq q \leq 2p + 2,$$

classify all subsets A of size $p+q-2$ of $\mathbb{Z}_{pq}$ for which $\Sigma A \neq \mathbb{Z}_{pq}$.

Recall that our examples of sets in Problems E.114 and E.116 were generated by cosets of subgroups, but for Problem E.115, they came from sets with small norm.

E.4 Restricted signed sumsets

E.4.1 Fixed number of terms

E.4.2 Limited number of terms

E.4.3 Arbitrary number of terms

Chapter F

Zero-sum-free sets

Recall that for a given finite abelian group G, m-subset $A = \{a_1, \ldots, a_m\}$ of G, $\Lambda \subseteq \mathbb{Z}$, and $H \subseteq \mathbb{N}_0$, we defined the sumset of A corresponding to Λ and H as

$$H_\Lambda A = \{\lambda_1 a_1 + \cdots + \lambda_m a_m \mid (\lambda_1, \ldots, \lambda_m) \in \Lambda^m(H)\}$$

where the index set $\Lambda^m(H)$ is defined as

$$\Lambda^m(H) = \{(\lambda_1, \ldots, \lambda_m) \in \Lambda^m \mid |\lambda_1| + \cdots + |\lambda_m| \in H\}.$$

In this chapter we investigate the maximum possible size of a zero-sum-free set over Λ in a given finite abelian group G. Namely, our objective is to determine, for any G, $\Lambda \subseteq \mathbb{Z}$, and $H \subseteq \mathbb{N}_0$ the quantity

$$\tau_\Lambda(G, H) = \max\{|A| \mid A \subseteq G, 0 \not\in H_\Lambda A\}.$$

If no zero-sum-free set exists, we put $\tau_\Lambda(G, H) = 0$.

Since, by definition, for every subset A of size $\chi_\Lambda(G, H)$ or more $H_\Lambda A$ equals the entire group, we have the following:

Proposition F.1 *Let $\Lambda \subseteq \mathbb{Z}$ and $H \subseteq \mathbb{N}_0$. If $\chi_\Lambda(G, H)$ exists, then*

$$\tau_\Lambda(G, H) \leq \chi_\Lambda(G, H) - 1.$$

In the following sections we consider $\tau_\Lambda(G, H)$ for special coefficient sets Λ.

F.1 Unrestricted sumsets

Our goal in this section is to investigate the maximum possible size of a zero-H-sum-free set, that is, the quantity

$$\tau(G, H) = \max\{|A| \mid A \subseteq G, 0 \not\in HA\}.$$

Clearly, we always have $\tau(G, H) = 0$ when $0 \in H$; in fact, $\tau(G, H) = 0$ whenever H contains any multiple of the exponent of G. However, when H contains no multiples of the exponent κ, then $\tau(G, H) \geq 1$: for any $a \in G$ with order κ, at least for the one-element set $A = \{a\}$ we have $0 \not\in HA$.

It is often useful to consider G of the form $G_1 \times G_2$. (We may do so even when G is cyclic if its order has at least two different prime divisors.) It is not hard to see that, if $A_1 \subseteq G_1$ is zero-H-sum-free in G_1, then

$$A = \{(a, g) \mid a \in A_1, g \in G_2\}$$

is zero-H-sum-free in G. Indeed, if hA were to contain $(0, 0)$ (the zero-element of $G = G_1 \times G_2$) for some $h \in H$, then we would have h (not necessarily distinct) elements of A adding to $(0, 0)$; this would then mean that h (not necessarily distinct) elements of A_1 would add to 0 in G_1, a contradiction. Thus, we have the following.

Proposition F.2 *For all finite abelian groups G_1 and G_2 and for all $H \subseteq \mathbb{N}_0$ we have*

$$\tau(G_1 \times G_2, H) \geq \tau(G_1, H) \cdot |G_2|.$$

Below we consider two special cases: when H consists of a single positive integer h and when H consists of all positive integers up to some value t. The cases when $H = \mathbb{N}_0$ or $H = \mathbb{N}$, as we just mentioned, yield no zero-sum-free sets.

F.1.1 Fixed number of terms

In this section we investigate, for a given group G and positive integer h, the quantity

$$\tau(G, h) = \max\{|A| \mid A \subseteq G, 0 \not\in hA\},$$

that is, the maximum size of a zero-h-sum-free subset of G.

For $h = 1$, we see that a subset of G is zero-1-sum-free if, and only if, it does not contain 0, hence the unique maximal zero-1-sum-free set in G is $G \setminus \{0\}$.

We can also easily determine the value of $\tau(G, 2)$. First, note that a zero-2-sum-free set A cannot contain any element of $\{0\} \cup \text{Ord}(G, 2)$ (the elements of order at most 2), and neither can it contain any element with its negative; to get a maximum zero-2-sum-free set in G, take exactly one of each element or its negative in $G \setminus \text{Ord}(G, 2) \setminus \{0\}$. To summarize:

Proposition F.3 *We have*

$$\tau(G, 1) = n - 1$$

and

$$\tau(G, 2) = \frac{n - |\text{Ord}(G, 2)| - 1}{2};$$

in particular,

$$\tau(\mathbb{Z}_n, 2) = \lfloor (n-1)/2 \rfloor.$$

Suppose now that $h = 3$. For the cyclic group $\mathbb{Z}_n$, we can find explicit zero-3-sum-free sets as follows. For every n, the positive integers that are less than $n/3$ form a zero-3-sum-free set; that is, the set

$$A = \{1, 2, \ldots, \lfloor (n-1)/3 \rfloor\}$$

is zero-3-sum-free in $\mathbb{Z}_n$, since

$$3A = \{3, 4, \ldots, 3 \cdot \lfloor (n-1)/3 \rfloor\}$$

does not contain 0.

We can do better in some cases. For example, when n is even, we may take the larger set

$$\{1, 3, 5, \ldots, n-1\},$$

which is zero-3-sum-free since no three odd numbers can add to a number divisible by n when n is even. More generally, suppose that n has a prime divisor p which is congruent to 2 mod 3. It is not hard to see that the set

$$\{(p+1)/3 + i + pj \mid i = 0, 1, \ldots, (p-2)/3, \; j = 0, 1, \ldots, n/p - 1\}$$

is zero-3-sum-free. Indeed, if

$$k = (p+1) + (i_1 + i_2 + i_3) + p(j_1 + j_2 + j_3) = 0$$

in $\mathbb{Z}_n$ for some

$$i_1, i_2, i_3 \in \{0, 1, \ldots, (p-2)/3\}$$

and

$$j_1, j_2, j_3 \in \{0, 1, \ldots, n/p - 1\},$$

then the integer k is divisible by n and thus by p. Therefore, $1 + (i_1 + i_2 + i_3)$ would have to be divisible by p, but this is not possible as

$$1 \leq 1 + (i_1 + i_2 + i_3) \leq p - 1.$$

Recalling the function $v_g(n, h)$ from page 76, we see that we have established the following.

Proposition F.4 *We have*

$$\tau(\mathbb{Z}_n, 3) \geq v_3(n, 3) = \begin{cases} \left(1 + \frac{1}{p}\right) \frac{n}{3} & \text{if } n \text{ has prime divisors congruent to 2 mod 3,} \\ & \text{and } p \text{ is the smallest such divisor;} \\ \left\lfloor \frac{n-1}{3} \right\rfloor & \text{otherwise.} \end{cases}$$

We believe that equality holds in Proposition F.4, but no proof of this has been discovered. More generally, we make the following conjecture.

Conjecture F.5 *For all positive integers h and n, the maximum size of a zero-h-sum-free set in the cyclic group $\mathbb{Z}_n$ is given by*

$$\tau(\mathbb{Z}_n, h) = v_h(n, h).$$

We have already seen that Conjecture F.5 holds when $h = 1$ or $h = 2$, and we have the following result.

Theorem F.6 *For all positive integers h and n, we have*

$$v_h(n, h) \leq \tau(\mathbb{Z}_n, h) \leq v_1(n, h).$$

The upper bound here follows directly from Proposition F.1 and Theorem E.1; the proof of the lower bound can be found on page 354.

The lower and upper bounds in Theorem F.6 often agree—in which case the value of $\tau(\mathbb{Z}_n, h)$ is determined. For example, for $h = 3$, we have $v_3(n, 3) = v_1(n, 3)$ if, and only if, n has at least one prime divisor which is congruent to 2 mod 3 or if all its prime divisors are congruent to 1 mod 3. So, for $h = 3$, the first few values of n for which $v_3(n, 3) \neq v_1(n, 3)$ are

$$n = 3, 9, 21, 27, 39, 57, 63, \ldots.$$

A potentially difficult problem is the following.

Problem F.7 *Prove or disprove Conjecture F.5; in particular, settle the case for $h = 3$.*

It is worth mentioning that, using Theorem F.6, we can determine the size of the largest zero-h-sum-free set in cyclic groups of prime order. According to Proposition 4.1, for a prime p and an arbitrary positive integer h we have

$$v_1(p,h) = \left\lfloor \frac{p-2}{h} \right\rfloor + 1$$

and

$$v_h(p,h) = \begin{cases} 0 & \text{if } p|h, \\ \lfloor \frac{p-2}{h} \rfloor + 1 & \text{otherwise.} \end{cases}$$

Therefore, if h is not divisible by p, then $v_1(p,h) = v_h(p,h)$, and by Theorem F.6 this yields $\tau(\mathbb{Z}_p, h)$. On the other hand, if h is divisible by p, then clearly no (nonempty) subset of $\mathbb{Z}_p$ is zero-h-sum-free since for every element a of the group we have $ha = 0$. In summary, we have the following.

Theorem F.8 *The size of the largest zero-h-sum-free set in cyclic group of prime order p is*

$$\tau(\mathbb{Z}_p, h) = v_h(p,h) = \begin{cases} 0 & \textit{if } p|h, \\ \lfloor \frac{p-2}{h} \rfloor + 1 & \textit{otherwise.} \end{cases}$$

At this point we know considerably less about the value of $\tau(G,h)$ for $h \geq 3$ when G is not cyclic. An immediate consequence of Proposition F.2 and Theorem F.6 is the following.

Corollary F.9 *Suppose that G is an abelian group of order n and exponent κ. Then, for all positive integers h we have*

$$v_h(\kappa, h) \cdot \frac{n}{\kappa} \leq \tau(G,h) \leq v_1(n,h).$$

When the two bounds above coincide, then of course we have the exact value of $\tau(G,h)$. Two such instances are worth mentioning: one coming from Corollary 4.5, and the other from our formulas for $v_1(n,3)$ and $v_3(n,3)$ on page 78:

Corollary F.10 *If n is even and $h \geq 3$ is odd, then*

$$\tau(G,h) = v_1(n,h) = n/2.$$

Corollary F.11 *If n is divisible by a prime p with $p \equiv 2 \bmod 3$ and p is the smallest such prime, then*

$$\tau(G,3) = v_1(n,3) = \left(1 + \frac{1}{p}\right) \frac{n}{3}.$$

We also know the value of $\tau(G,h)$ when h is relatively prime to n:

Theorem F.12 *If h is relatively prime to n, then*

$$\tau(G,h) = \chi(G,n) - 1 = v_1(n,h).$$

We verify Theorem F.12 following an idea of Lemma 2.1 in the paper [179] of Roth and Lempel. Note that it suffices to find a set A in G of size $\chi(G,n)-1$ for which $0 \notin hA$. Since h and n are relatively prime, we have a positive integer h' for which $hh' \equiv 1 \mod n$. Now let B be a subset of G of size $\chi(G,n)-1$ for which $hB \neq G$, choose an element $g \in G \setminus hB$, and set $A = -h'g + B$. As

$$hA = -hh'g + hB = -g + hB,$$

we indeed find that $0 \notin hA$.

Next, for a positive prime p and $r \geq 1$, we consider $\mathbb{Z}_p^r$, the elementary abelian p-group of rank r. Let $h \geq 2$. When h is divisible by p, then, as we have mentioned above, $\tau(\mathbb{Z}_p^r, h) = 0$. If h is not divisible by p, then p cannot be divisible by h either (since that would imply $p = h$ and thus $p|h$). If p leaves a remainder of at least 2 mod h, then, by Corollary F.9 and by Propositions 4.1 and 4.2,

$$\tau(\mathbb{Z}_p^r, h) = v_1(p^r, h) = p^{r-1} \cdot (1 + \lfloor p/h \rfloor).$$

This leaves one case: when $p \equiv 1 \mod h$, which can be treated using Finn's construction in [81] that we explain (in a generalized form) next.

Suppose that G_i is a finite abelian group for $i = 1, \ldots, r$; A_i is a zero-$[1,h]$-sum-free set in G_i for $i = 1, \ldots, r-1$; and A_r is a zero-h-sum-free set in G_r. (We say that A is zero-$[1,h]$-sum-free in G if $0 \notin \cup_{j=1}^h jA$—see Section F.1.2.) Now set

$$G = G_1 \times \cdots \times G_r$$

and

$$A = (A_1 \times G_2 \times \cdots \times G_r) \cup (\{0\} \times A_2 \times G_3 \times \cdots \times G_r) \cup \cdots \cup (\{0\} \times \cdots \times \{0\} \times A_r).$$

It is easy to verify that A is zero-h-sum-free in G, and therefore we get the following result:

Proposition F.13 (Finn; cf. [81]) *For all abelian groups $G_1, \ldots, G_r$ we have*

$$\tau(G_1 \times \cdots \times G_r, h) \geq \sum_{i=1}^{r} \tau(G_i, [1,h]) \cdot \Pi_{j=i+1}^r |G_j|.$$

(Of course, $\Pi_{j=r}^r |G_j| = |G_r|$ and $\Pi_{j=r+1}^r |G_j| = 1$; $\tau(G,[1,h])$ denotes the maximum size of a zero-$[1,h]$-sum-free set in G.)

We can apply Proposition F.13 to the case when each G_i is cyclic by observing that the set

$$\{1, 2, \ldots, \lfloor (n-1)/h \rfloor\}$$

is zero-$[1,h]$-free (and thus also zero-h-sum-free) in $\mathbb{Z}_n$. In particular, when p is a prime with $p \equiv 1 \mod h$, then

$$\tau(\mathbb{Z}_p, [1,h]) \geq (p-1)/h,$$

so in this case

$$\tau(\mathbb{Z}_p^r, h) \geq \frac{p-1}{h} \cdot (p^{r-1} + \cdots + p + 1) = \frac{p^r - 1}{h} = v_1(p^r, h).$$

Therefore, we have the following:

Theorem F.14 (Finn; cf. [81]) *Let p be a positive prime and $r \geq 1$.*
If h is divisible by p, then $\tau(\mathbb{Z}_p^r, h) = 0$.
If h is not divisible by p, then

$$\tau(\mathbb{Z}_p^r, h) = v_1(p^r, h) = \begin{cases} (p^r - 1)/h & \text{if } p \equiv 1 \bmod h; \\ p^{r-1} \cdot (1 + \lfloor p/h \rfloor) & \text{if } p \not\equiv 1 \bmod h. \end{cases}$$

Note that Theorem F.14 is a generalization of Theorem F.8.

The general problem of finding $\tau(G, h)$ for noncyclic groups G is quite intriguing, even for $h = 3$.

Problem F.15 *Determine the value of $\tau(G, h)$ for noncyclic G.*

F.1.2 Limited number of terms

Here we investigate, for a given group G and positive integer t, the quantity

$$\tau(G, [1, t]) = \max\{|A| \mid A \subseteq G, 0 \not\in [1, t]A\},$$

that is, the maximum size of a zero-$[1, t]$-sum-free subset of G.

For $t = 1$ and $t = 2$ our considerations of Section F.1.1 above yield the answers here as well:

Proposition F.16 *We have*

$$\tau(G, [1, 1]) = n - 1$$

and

$$\tau(G, [1, 2]) = \frac{n - |\mathrm{Ord}(G, 2)| - 1}{2};$$

in particular,

$$\tau(\mathbb{Z}_n, [1, 2]) = \lfloor (n - 1)/2 \rfloor.$$

The following general upper bound is easy to establish (see Corollary 2.3 in [1]):

Theorem F.17 (Alon; cf. [1]) *In any group of order n we have*

$$\tau(G, [1, t]) \leq \lfloor (n - 1)/t \rfloor.$$

Note that the set

$$\{1, 2, \ldots, \lfloor (n - 1)/t \rfloor\}$$

is zero-$[1, t]$-free in $\mathbb{Z}_n$, hence we get:

Corollary F.18 *For all positive integers t and n, we have*

$$\tau(\mathbb{Z}_n, [1, t]) = \lfloor (n - 1)/t \rfloor.$$

The case of noncyclic groups remains open:

Problem F.19 *Find the value of $\tau(G, [1, t])$ for noncyclic groups G for $t \geq 3$.*

F.1.3 Arbitrary number of terms

Here we ought to consider

$$\tau(G,H) = \max\{|A| \mid A \subseteq G, 0 \not\in HA\}$$

for the case when H is the set of all nonnegative or all positive integers. However, as we have already mentioned, we have $\tau(G,H) = 0$ whenever H contains a multiple of the exponent of the group (including 0). Thus, there are no zero-sum-free sets when the addition of an arbitrary number of terms is allowed.

F.2 Unrestricted signed sumsets

Our goal in this section is to investigate the maximum possible size of a zero-sum-free set over the set of all integers, that is, the quantity

$$\tau_\pm(G,H) = \max\{|A| \mid A \subseteq G, 0 \not\in H_\pm A\}.$$

Clearly, we have $\tau_\pm(G,H) = 0$ whenever H contains a multiple of the exponent of the group (including 0). However, when H contains no multiples of the exponent κ, then $\tau_\pm(G,H) \geq 1$: for any $a \in G$ with order κ, at least the one-element set $\{a\}$ will be zero-sum-free for H.

It is important to note that Proposition F.2 does not carry through to zero-sum-free sets over the set of all integers. For example, the subset $A_1 = \{1\}$ of $\mathbb{Z}_{10}$ is clearly zero-4-sum-free over the integers, since $0 \not\in 4_\pm A$, but $A_1 \times \mathbb{Z}_{10}$ is not zero-4-sum-free over the integers in $\mathbb{Z}_{10}^2$, since (for example)

$$(1,1) + (1,6) - (1,3) - (1,4) = (0,0).$$

We consider two special cases: when H consists of a single positive integer h and when H consists of all positive integers up to some value t. The cases when $H = \mathbb{N}_0$ or $H = \mathbb{N}$ are trivial as we then have $\tau_\pm(G,H) = 0$ since $0 \in H_\pm A$ for any nonempty subset A of G.

F.2.1 Fixed number of terms

In this section we investigate, for a given group G and positive integer h, the quantity

$$\tau_\pm(G,h) = \max\{|A| \mid A \subseteq G, 0 \not\in h_\pm A\},$$

that is, the maximum size of a zero-h-sum-free set over $\mathbb{Z}$.

Since both for $h = 1$ and for $h = 2$, a subset of G is zero-h-sum-free over $\mathbb{Z}$ if, and only if, it is zero-h-sum-free (over $\mathbb{N}_0$), Proposition F.3 implies:

Proposition F.20 *We have*

$$\tau_\pm(G,1) = n - 1$$

and

$$\tau_\pm(G,2) = \frac{n - |\mathrm{Ord}(G,2)| - 1}{2};$$

in particular,

$$\tau_\pm(\mathbb{Z}_n,2) = \lfloor (n-1)/2 \rfloor.$$

Now consider the case $h = 3$. For a subset A of G, we have $0 \not\in 3_{\pm}A$ if, and only if, A is both zero-3-sum-free and sum-free, that is, $0 \not\in 3A$ and $0 \not\in 2A - A$. Let us consider $G = \mathbb{Z}_n$. In Section F.1.1, we showed that

$$\tau(\mathbb{Z}_n, 3) \geq v_3(n,3) = \begin{cases} \left(1+\frac{1}{p}\right)\frac{n}{3} & \text{if } n \text{ has prime divisors congruent to 2 mod 3,} \\ & \text{and } p \text{ is the smallest such divisor,} \\ \left\lfloor \frac{n-1}{3} \right\rfloor & \text{otherwise.} \end{cases}$$

Here we show that $\tau_{\pm}(\mathbb{Z}_n, 3) \geq v_3(n,3)$ as well. As before, when n has a divisor p which is congruent to 2 mod 3, we take

$$A = \{(p+1)/3 + i + pj \mid i = 0, 1, \ldots, (p-2)/3, \; j = 0, 1, \ldots, n/p - 1\}.$$

We have already seen that A is zero-3-sum-free, it is also easy to see that it is sum-free as well. Indeed, if

$$k = (p+1)/3 + (i_1 + i_2 - i_3) + p(j_1 + j_2 - j_3) = 0$$

in $\mathbb{Z}_n$ for some

$$i_1, i_2, i_3 \in \{0, 1, \ldots, (p-2)/3\}$$

and

$$j_1, j_2, j_3 \in \{0, 1, \ldots, n/p - 1\},$$

then the integer k is divisible by n and thus by p. Therefore, $(p+1)/3 + (i_1 + i_2 - i_3)$ would have to be divisible by p, but this is not possible as

$$1 \leq (p+1)/3 + (i_1 + i_2 - i_3) \leq p - 1.$$

Now suppose that n has no such divisor—in this case, n itself cannot be congruent to 2 mod 3. Then n is either divisible by 3, in which case the set

$$\left\{\frac{n}{3} + 1, \frac{n}{3} + 2, \ldots, 2 \cdot \frac{n}{3} - 1\right\}$$

works, or it is congruent to 1 mod 3, in which case we can take

$$\left\{\frac{n-1}{3} + 1, \frac{n-1}{3} + 2, \ldots, 2 \cdot \frac{n-1}{3}\right\}.$$

Both of these sets have size $\left\lfloor \frac{n-1}{3} \right\rfloor$, completing the proof of the inequality above.

Therefore, we have the following.

Proposition F.21 *For all positive integers n, we have*

$$\tau_{\pm}(\mathbb{Z}_n, 3) \geq v_3(n,3) = \begin{cases} \left(1+\frac{1}{p}\right)\frac{n}{3} & \textit{if } n \textit{ has prime divisors congruent to 2 mod 3,} \\ & \textit{and } p \textit{ is the smallest such divisor,} \\ \left\lfloor \frac{n-1}{3} \right\rfloor & \textit{otherwise.} \end{cases}$$

Our conjecture is that, in Proposition F.21, the lower bound gives the actual value:

Conjecture F.22 *For all positive integers n, we have*

$$\tau_{\pm}(\mathbb{Z}_n, 3) = v_3(n,3).$$

We pose the following problem.

Problem F.23 *Prove Conjecture F.22.*

We should mention that Conjecture F.22 obviously holds when $v_3(n,3) = v_1(n,3)$, as is the case, for example, if n is even (more generally, has a prime divisor congruent to 2 mod 3) or is prime.

Let us now turn to noncyclic groups. As we noted on page 231, Proposition F.2 does not carry through to zero-sum-free sets over the integers.

However, if $h \geq 3$ is odd, and for each odd integer $k \leq h$, A_1 is zero-k-sum-free over the integers in G_1, then we still see that $A_1 \times G_2$ is zero-h-sum-free over the integers in $G_1 \times G_2$: Indeed, if we were to have

$$(a_1, g_1) + \cdots + (a_i, g_i) - (a_{i+1}, g_{i+1}) - \cdots - (a_h, g_h) = (0,0)$$

with $a_i \in A_1$ and $g_i \in G_2$ for some $i = 1, 2, \ldots, h$ (ignoring subtractions when $i = h$), then, looking at the first components and cancelling identical terms that are both added and subtracted (if any), we arrive at a signed sum of k elements of A_1 for some odd integer $k \leq h$, a contradiction. In fact, assuming that A_1 is zero-1-sum-free is not necessary as that is implied by it being zero-h-sum-free for h. This results in the following:

Proposition F.24 *Suppose that G_1 and G_2 are finite abelian groups, $h \geq 3$ is odd, and that A_1 is zero-k-sum-free over the integers in G_1 for each $k \in \{3, 5, \ldots, h\}$. Then $A_1 \times G_2$ is zero-h-sum-free over the integers in $G_1 \times G_2$.*

(A special case of this result was discovered by Matys in [156].)

As a consequence of Propositions F.9, F.21, and F.24, we get:

Corollary F.25 *If G is an abelian group of order n and exponent κ, then*

$$v_3(\kappa, 3) \cdot \frac{n}{\kappa} \leq \tau_\pm(G,3) \leq v_1(n,3).$$

We do not have a conjecture for $\tau_\pm(G,3)$ when G is not cyclic.

Problem F.26 *Evaluate $\tau_\pm(G,3)$ for noncyclic groups G.*

Turning to the case of $h = 4$, we see a radical difference: as we are about to see, we cannot expect a formula for $\tau_\pm(\mathbb{Z}_n, 4)$ that is a linear function of n. Indeed, we have the following results.

Proposition F.27 *Let $A \subseteq G$, and suppose that $h \in \mathbb{N}$. If A is a B_h set over $\mathbb{Z}$, then A is also a zero-$2h$-sum-free set over $\mathbb{Z}$, and therefore*

$$\sigma_\pm(G,h) \leq \tau_\pm(G,2h).$$

Proposition F.28 *Let $A \subseteq G$, and suppose that h is a positive integer that is divisible by 4. If A is a zero-h-sum-free set over $\mathbb{Z}$, then A is also a B_2 set over $\mathbb{Z}$, and therefore*

$$\tau_\pm(G,h) \leq \sigma_\pm(G,2).$$

For the proofs of Propositions F.27 and F.28, see pages 355 and 356, respectively.

By these results:

Corollary F.29 *A set is zero-4-sum-free set over $\mathbb{Z}$ in G if, and only if, it is a B_2 set over $\mathbb{Z}$ in G.*

Therefore, by Proposition C.7, we must have

$$\tau_{\pm}(G,4) = \sigma_{\pm}(G,2) \leq \sigma(G,2) \leq \lfloor \sqrt{n} \rfloor + 1.$$

We do not know $\tau_{\pm}(\mathbb{Z}_n, 4)$.

Problem F.30 *Find the value of $\tau_{\pm}(\mathbb{Z}_n, 4)$ for all n.*

A bit more modestly:

Problem F.31 *Find a positive constant c and an explicit zero-4-sum-free set over $\mathbb{Z}$ in $\mathbb{Z}_n$ which has at least $c \cdot \sqrt{n}$ elements for every large enough n.*

The following result comes short of the demands of Problem F.31: it provides a set of (asymptotically) smaller size, and the set is not given explicitly.

Proposition F.32 *For each h there exists a positive constant c_h for which*

$$\tau_{\pm}(\mathbb{Z}_n, h) > c_h \cdot n^{1/(h-1)}.$$

In particular, if

$$n > \frac{4}{3}m^3 + \frac{8}{3}m,$$

then there is a zero-4-sum-free set over $\mathbb{Z}$ in $\mathbb{Z}_n$ that has size m.

The proof of Proposition F.32 is on page 357. (The result for $h = 4$ was found by Phillips in [169].)

We also offer the following problem.

Problem F.33 *Find the value of $\tau_{\pm}(\mathbb{Z}_n, h)$ for even values of $h \geq 6$.*

The situation is very different for odd values of h. We have already discussed the case $h = 3$ above; here we see what we can say about the general case when $h \geq 5$ is odd. As Matys in [156] pointed out, we cannot hope for a result similar to Proposition F.25: we see that $v_5(9,5) = 3$, but $\tau_{\pm}(\mathbb{Z}_9, 5) = 2$. (We can verify this last assertion as follows. Suppose that A is a zero-5-sum-free set over the integers in $\mathbb{Z}_9$. Clearly, $0 \not\in A$. Also, if $a \in A$, then $4a \not\in A$; since the sequence $(1, 4, 16, 64)$ becomes $(1, 4, 7, 1)$ mod 9, at most one of 1, 4, or 7 can be in A. Similarly, at most one of 2, 8, or 5 can be in A. Furthermore, neither 3 nor 6 can be in A together with any of 1, 4, 7, 2, 8, or 5: for example, if $3 \in A$ and $a \in A$ for some $a \equiv 1$ mod 3, then

$$3 + 3 + a + a + a = 3(a + 2) \equiv 0 \text{ mod } 9,$$

and if $3 \in A$ and $a \in A$ for some $a \equiv 2$ mod 3, then

$$-3 - 3 + a + a + a = 3(a - 2) \equiv 0 \text{ mod } 9.$$

This proves that $\tau_{\pm}(\mathbb{Z}_9, 5) \leq 2$; the set $\{4, 5\}$ shows that equality holds.)

Here is what we know about $\tau_{\pm}(G, h)$ for odd $h \geq 5$ when G is cyclic. As Matys in [156] observed, when n is even, then the odd elements of $\mathbb{Z}_n$ form a zero-h-sum-free set over the integers, hence $\tau_{\pm}(\mathbb{Z}_n, h) \geq n/2$ then. Since $\tau(\mathbb{Z}_n, h)$, and thus $v_1(n, h)$ provides an upper bound and, by Corollary 4.5, $v_1(n, h) = n/2$ in this case, we get:

Proposition F.34 (Matys; cf. [156]) *If n is even and $h \geq 3$ is odd, then $\tau_{\pm}(\mathbb{Z}_n, h) = n/2$.*

When n and h are both odd, we can find a linear lower bound for $\tau_{\pm}(\mathbb{Z}_n, h)$ as follows. Let A be the set of integers (viewed as elements of $G = \mathbb{Z}_n$) that are strictly between $\frac{h-1}{2}\frac{n}{h}$ and $\frac{h+1}{2}\frac{n}{h}$. We can then prove that $0 \not\in h_{\pm}A$, and evaluating the size of A results in the following proposition.

Proposition F.35 (Matys; cf. [156]) *For all odd positive integers n and h,*

$$\tau_{\pm}(\mathbb{Z}_n, h) \geq 2\left\lfloor \frac{n+h-2}{2h} \right\rfloor.$$

The proof of Proposition F.35 (different from the one given by Matys in [156]) is on page 357.

The following problem seems rather intriguing:

Problem F.36 *Find $\tau_{\pm}(\mathbb{Z}_n, h)$ for odd values of $h \geq 5$.*

As always, we are also interested in noncyclic groups.

Problem F.37 *Find the value of $\tau_{\pm}(G, h)$ for noncyclic groups and for $h \geq 4$.*

F.2.2 Limited number of terms

A subset A of G for which

$$0 \not\in [1,t]_{\pm}A = \cup_{h=1}^{t} h_{\pm}A$$

for some positive integer t is called a *t-independent set* in G. Here we investigate

$$\tau_{\pm}(G, [1,t]) = \max\{|A| \mid A \subseteq G, 0 \not\in [1,t]_{\pm}A\},$$

that is, the maximum size of a t-independent set in G.

As we pointed out above, with κ denoting the exponent of the group, adding any element of the group to itself κ times results in 0. Therefore, if $t \geq \kappa$, then no element of G can belong to a t-independent set and we have $\tau_{\pm}(G, [1,t]) = 0$. However, if $t < \kappa$, then $\tau_{\pm}(G, [1,t]) \geq 1$: at the least the one-element set $\{a\}$, where a is any element of order κ, will be t-independent.

The cases of $t = 1$ and $t = 2$ are easy to handle. Clearly, a set A is 1-independent if, and only if, $0 \not\in A$. Regarding $t = 2$, first note that a 2-independent set cannot contain any element of $\{0\} \cup \mathrm{Ord}(G,2)$ (the elements of order at most 2); to get a maximum 2-independent set in G, take exactly one of each element or its negative in $G \setminus \mathrm{Ord}(G,2) \setminus \{0\}$. In summary:

Proposition F.38 *For all groups G we have*

$$\tau_{\pm}(G, [1,1]) = n - 1$$

and

$$\tau_{\pm}(G, [1,2]) = \frac{n - |\mathrm{Ord}(G,2)| - 1}{2};$$

in particular,

$$\tau_{\pm}(\mathbb{Z}_n, [1,2]) = \lfloor (n-1)/2 \rfloor.$$

Let us now consider $t = 3$. Before we proceed, it is helpful to state that a subset A is 3-independent in G if, and only if, none of the equations

$$x = 0,\ x + y = 0,\ x + y + z = 0, \text{and } x + y = z$$

have a solution in A.

We first find explicit 3-independent sets in the cyclic group $\mathbb{Z}_n$ as follows. For every n, the odd integers that are less than $n/3$ form a 3-independent set; that is, the set

$$\{2i+1 \mid i=0,1,\dots,\lfloor n/6 \rfloor -1\}$$

is 3-independent in $\mathbb{Z}_n$. Furthermore, if n is even, we can go up to (but not including) $n/2$ as then the sum of two odd integers cannot equal n; so, when n is even, the set

$$\{2i+1 \mid i=0,1,\dots,\lfloor n/4 \rfloor -1\}$$

is 3-independent in $\mathbb{Z}_n$.

We can do better in one special case when n is odd; namely, when n has a prime divisor p that is congruent to 5 mod 6, one can show that the set

$$\{pi_1+2i_2+1 \mid i_1=0,1,\dots,n/p-1, \ i_2=0,1,\dots,(p-5)/6\}$$

is 3-independent. For example, when $n=25$, we may take $p=5$, with which we get

$$\{5i_1+2i_2+1 \mid i_1=0,1,\dots,4, \ i_2=0\} = \{1,6,11,16,21\}.$$

This set is 3-independent in $G=\mathbb{Z}_{25}$. Note that, when determining the independence number of a subset, an element and its negative play the same role, thus the set $\{1,6,11,16,21\}$ above is essentially the same as the set $\{1,4,6,9,11\}$ of the example on page 70.

In summary, we have

$$\tau_\pm(\mathbb{Z}_n,[1,3]) \geq \begin{cases} \left\lfloor \frac{n}{4} \right\rfloor & \text{if } n \text{ is even,} \\ \left(1+\frac{1}{p}\right)\frac{n}{6} & \text{if } n \text{ is odd, has prime divisors congruent to 5 mod 6,} \\ & \text{and } p \text{ is the smallest such divisor,} \\ \left\lfloor \frac{n}{6} \right\rfloor & \text{otherwise.} \end{cases}$$

Let us now turn to noncyclic groups. First, an observation. Suppose that G is of the form $G_1 \times G_2$. (Note that every group can be written in the form $G_1 \times G_2$ with $G_2 = \mathbb{Z}_\kappa$ where κ is the exponent of G.) It is not hard to see that, if $A_2 \subseteq G_2$ is 3-independent in G_2, then

$$A = \{(g,a) \mid g \in G_1, a \in A_2\}$$

is 3-independent in G, so we have the following:

Proposition F.39 *Let G_1 and G_2 be finite abelian groups, $G=G_1 \times G_2$, and suppose that $A_2 \subseteq G_2$ is 3-independent in G_2. Then*

$$A = \{(g,a) \mid g \in G_1, a \in A_2\}$$

is 3-independent in G; in particular,

$$\tau_\pm(G_1 \times G_2,[1,3]) \geq |G_1| \cdot \tau_\pm(G_2,[1,3]).$$

(This proposition does not hold for t-independent sets for $t \geq 4$.)

Combining Proposition F.39 with our results for cyclic groups above, we see that, for a group of exponent κ we get:

$$\tau_\pm(G,[1,3]) \geq \begin{cases} \frac{n}{4} & \text{if } \kappa \text{ is divisible by 4,} \\ \frac{n}{\kappa}\cdot\frac{\kappa-2}{4} & \text{if } \kappa \text{ is even but not divisible by 4,} \\ \left(1+\frac{1}{p}\right)\frac{n}{6} & \text{if } \kappa \text{ is odd, has prime divisors congruent to 5 mod 6,} \\ & \text{and } p \text{ is the smallest such divisor,} \\ \left\lfloor \frac{\kappa}{6} \right\rfloor \cdot \frac{n}{\kappa} & \text{otherwise.} \end{cases}$$

Next, we show that we can do slightly better when κ is even but not divisible by 4. Let $G = G_1 \times \mathbb{Z}_\kappa$. Suppose that A_1 is a 2-independent set in G_1, and A_2 is the 3-independent set

$$A_2 = \{2i+1 \mid i = 0, 1, \ldots, (\kappa-6)/4\}$$

in $\mathbb{Z}_\kappa$. We can then verify that the set

$$A = \{(g,a) \mid g \in G_1, a \in A_2\} \cup \{(a, \kappa/2) \mid a \in A_1\}$$

is 3-independent in G. (When checking that the equations $x+y+z=0$ and $x+y=z$ have no solutions in A, note that κ is even, but $\kappa/2$ and all elements of A_2 are odd.)

Suppose further that A_1 is of maximum size; that is,

$$|A_1| = \frac{|G_1| - |\mathrm{Ord}(G_1,2)| - 1}{2};$$

since

$$|\mathrm{Ord}(G_1,2)| = \frac{|\mathrm{Ord}(G,2)| - 1}{2},$$

we have

$$|A_1| = \frac{n}{2\kappa} - \frac{|\mathrm{Ord}(G,2)| + 1}{4},$$

with which

$$|A| = \frac{\kappa-2}{4}\cdot\frac{n}{\kappa} + \frac{n}{2\kappa} - \frac{|\mathrm{Ord}(G,2)|+1}{4} = \frac{n - |\mathrm{Ord}(G,2)| - 1}{4}.$$

Therefore, we have

$$\tau_\pm(G,[1,3]) \geq \begin{cases} \frac{n}{4} & \text{if } \kappa \text{ is divisible by 4,} \\ \frac{n-|\mathrm{Ord}(G,2)|-1}{4} & \text{if } \kappa \text{ is even but not divisible by 4,} \\ \left(1+\frac{1}{p}\right)\frac{n}{6} & \text{if } \kappa \text{ is odd, has prime divisors congruent to 5 mod 6,} \\ & \text{and } p \text{ is the smallest such divisor,} \\ \left\lfloor \frac{\kappa}{6} \right\rfloor \cdot \frac{n}{\kappa} & \text{otherwise.} \end{cases}$$

In [24], Bajnok and Ruzsa showed that, in the first three cases, equality holds; for the last case, they only proved that $\tau_\pm(G,[1,3])$ cannot be more than $\left\lfloor \frac{n}{6} \right\rfloor$. Namely, we have the following result.

Theorem F.40 (Bajnok and Ruzsa; cf. [24]) *As usual, let κ be the exponent of G. We have*

$$\tau_{\pm}(G,[1,3]) = \begin{cases} \frac{n}{4} & \text{if } \kappa \text{ is divisible by } 4, \\ \frac{n-|\mathrm{Ord}(G,2)|-1}{4} & \text{if } \kappa \text{ is even but not divisible by } 4, \\ \left(1+\frac{1}{p}\right)\frac{n}{6} & \text{if } \kappa \text{ is odd, has prime divisors congruent to 5 mod 6,} \\ & \text{and } p \text{ is the smallest such divisor;} \end{cases}$$

furthermore, if κ (iff n) is odd and has no prime divisors congruent to 5 mod 6, then

$$\left\lfloor \frac{\kappa}{6} \right\rfloor \frac{n}{\kappa} \leq \tau_{\pm}(G,[1,3]) \leq \frac{n}{6}.$$

As a consequence, we see that Theorem F.40 has settled the problem of finding the maximum size of a 3-independent set in cyclic groups:

Theorem F.41 (Bajnok and Ruzsa; cf. [24]) *For the cyclic group $G = \mathbb{Z}_n$ we have*

$$\tau_{\pm}(\mathbb{Z}_n,[1,3]) = \begin{cases} \left\lfloor \frac{n}{4} \right\rfloor & \text{if } n \text{ is even,} \\ \left(1+\frac{1}{p}\right)\frac{n}{6} & \text{if } n \text{ is odd, has prime divisors congruent to 5 mod 6,} \\ & \text{and } p \text{ is the smallest such divisor,} \\ \left\lfloor \frac{n}{6} \right\rfloor & \text{otherwise.} \end{cases}$$

As a corollary to Theorem F.41, we get the following.

$$\tau_{\pm}(\mathbb{Z}_n,[1,3]) = \begin{cases} 1 & \text{if } n = \mathbf{4}, 5, 6, 7, 9; \\ 2 & \text{if } n = \mathbf{8}, 10, 11, 13; \\ 3 & \text{if } n = \mathbf{12}, 14, 15, 17, 19, 21; \\ 4 & \text{if } n = \mathbf{16}, 18, 23, 27; \\ 5 & \text{if } n = \mathbf{20}, 22, 25, 29, 31; \\ 6 & \text{if } n = \mathbf{24}, 26, 33, 37, 39. \end{cases}$$

(Entries in boldface mark tight 3-independent sets, as we will explain shortly.)

Since the publication of [24], Green and Ruzsa succeeded in determining the maximum size of a sum-free set in any G:

$$\mu(G,\{1,2\}) = v_1(\kappa,3) \cdot \frac{n}{\kappa}$$

(cf. Theorem G.18). We can use this result to determine $\tau_{\pm}(G,[1,3])$ when κ is only divisible by primes congruent to 1 mod 6. Indeed, since a 3-independent set A must be asymmetric: $A \cap -A$ must be empty. Furthermore, the set $A \cup -A$ has to be sum-free, so we get

$$2|A| = |A \cup -A| \leq v_1(\kappa,3) \cdot \frac{n}{\kappa} = \left\lfloor \frac{\kappa}{3} \right\rfloor \cdot \frac{n}{\kappa} = \frac{\kappa-1}{3} \cdot \frac{n}{\kappa},$$

and thus

$$|A| \leq \frac{\kappa-1}{6} \cdot \frac{n}{\kappa}$$

(note that $(\kappa-1)/3$ is an even integer). Since in our case

$$\left\lfloor \frac{\kappa}{6} \right\rfloor \frac{n}{\kappa} = \frac{\kappa-1}{6} \cdot \frac{n}{\kappa}$$

as well, from the last case of Theorem F.40 we get the following:

Corollary F.42 *Suppose that G has exponent κ and that every prime divisor of κ is congruent to 1 mod 6. Then*

$$\tau_{\pm}(G,[1,3]) = \frac{\kappa-1}{6} \cdot \frac{n}{\kappa}.$$

This leaves only one case open:

Problem F.43 *Suppose that G is not cyclic and that its exponent is the product of a positive integer power of 3 and perhaps some primes that are congruent to 1 mod 6. Find $\tau_{\pm}(G,[1,3])$.*

For example, by Theorem F.40 we see that $\tau_{\pm}(\mathbb{Z}_9^2,[1,3])$ is at least 9 and at most 13; according to Laza (see [138]), we have $\tau_{\pm}(\mathbb{Z}_9^2,[1,3]) \geq 11$ as the set

$$A = \{(0,1),(1,0),(1,3),(1,6),(2,1),(2,4),(2,7),(4,1),(4,4),(4,7),(6,1)\}$$

is 3-independent in $\mathbb{Z}_9^2$. (It is worth noting that $[1,3]_{\pm}A$ contains all nonzero elements of the group.)

For $t \geq 4$, exact results seem more difficult. With the help of a computer, Laza (see [138]) generated the following values:

$$\tau_{\pm}(\mathbb{Z}_n,[1,4]) = \begin{cases} 1 & \text{if } n = \mathbf{5}, 6, \dots, 12; \\ 2 & \text{if } n = \mathbf{13}, 14, \dots, 26; \\ 3 & \text{if } n = 27, 28, \dots, 45, \text{ and } n = 47; \\ 4 & \text{if } n = 46,\ n = 48, 49, \dots, 68, \text{ and } n = 72, 73; \\ 5 & \text{if } n = 69, 70, 71, \text{ and } n = 74, 75, \dots, 102; \end{cases}$$

$$\tau_{\pm}(\mathbb{Z}_n,[1,5]) = \begin{cases} 1 & \text{if } n = \mathbf{6}, 7, \dots, 17, \text{ and } n = 19, 20; \\ 2 & \text{if } n = \mathbf{18},\ n = 21, 22, \dots, 37,\ n = 39, 40, 41,\ n = 43, 44, 45, 47; \\ 3 & \text{if } n = \mathbf{38}, 42, 46,\ n = 48, 49, \dots, 69, \\ & \text{n=71, 72, 73, 75, 76, 77, 79, 81, 83, 85, 87;} \end{cases}$$

$$\tau_{\pm}(\mathbb{Z}_n,[1,6]) = \begin{cases} 1 & \text{if } n = \mathbf{7}, 8, 9, \dots, 24; \\ 2 & \text{if } n = \mathbf{25}, 26, 27, \dots, 69; \\ 3 & \text{if } n = 70, 71, \dots, 151, \text{ and } n = 153, 154, 155, 158, 159, 160. \end{cases}$$

(Values marked in boldface will be discussed shortly.)

The following two problems seem difficult, but even partial answers (those for certain special values of n) would be very interesting:

Problem F.44 *Find $\tau_{\pm}(\mathbb{Z}_n,[1,4])$ (at least for some infinite family of n values).*

Problem F.45 *Find $\tau_{\pm}(\mathbb{Z}_n,[1,5])$ (at least for some infinite family of n values).*

It is also interesting to investigate t-independent sets from the opposite viewpoint: given a nonnegative integer m and positive integer t, what are the possible groups G for which $\tau_{\pm}(G,[1,t]) = m$?

The answer for $m = 0$ is clear and has already been discussed. For $m = 1$ we can see from Laza's work [138] that, in the cyclic group $\mathbb{Z}_n$, we have

$$\tau_{\pm}(\mathbb{Z}_n,[1,t])=1 \Leftrightarrow \begin{cases} n=3-4 & \text{if } t=2; \\ n=4-7,9 & \text{if } t=3; \\ n=5-12 & \text{if } t=4; \\ n=6-17,19-20 & \text{if } t=5; \\ n=7-24 & \text{if } t=6; \\ n=8-31,33,35 & \text{if } t=7; \\ n=9-40 & \text{if } t=8; \\ n=10-49,51-53 & \text{if } t=9; \\ n=11-60 & \text{if } t=10. \end{cases}$$

We can prove the following:

Proposition F.46 *As usual, let G be an abelian group of order n and exponent κ. Let $t \geq 2$ be an integer.*

1. *We have $\tau_{\pm}(G,[1,t])=0$ if, and only if, $\kappa \leq t$.*
2. *Set $b_t = \lfloor t^2/2 \rfloor + t$.*
   (a) *Suppose that t is even. Then $\tau_{\pm}(\mathbb{Z}_n,[1,t]) = 1$ if, and only if, $t+1 \leq n \leq b_t$. In particular, the set $\{1\}$ is t-independent in $\mathbb{Z}_n$ for $n \geq t+1$, and the set $\{t/2, t/2+1\}$ is t-independent in $\mathbb{Z}_n$ for $n \geq b_t+1$.*
   (b) *Suppose that t is odd.*
      i. *If $t+1 \leq n \leq b_t$, then $\tau_{\pm}(\mathbb{Z}_n,[1,t]) = 1$; in particular, the set $\{1\}$ is t-independent in $\mathbb{Z}_n$.*
      ii. **(Miller; cf. [157])** *If $\tau_{\pm}(\mathbb{Z}_n,[1,t]) = 1$, then $t+1 \leq n \leq b_t + (t+1)/2$; in particular, the set $\{(t+1)/2, (t+3)/2\}$ is t-independent in $\mathbb{Z}_n$ for $n \geq b_t + (t+3)/2$.*
      iii. *If $n = b_t + 1$, then $\tau_{\pm}(\mathbb{Z}_n,[1,t]) = 2$; in particular, the set $\{1,t\}$ is t-independent in $\mathbb{Z}_n$.*
      iv. **(Miller; cf. [157])** *If t is congruent to 3 mod 4, n is even, and $n \geq b_t + 1$, then $\tau_{\pm}(\mathbb{Z}_n,[1,t]) \geq 2$; in particular, the set $\{(t-1)/2, (t+3)/2\}$ is t-independent in $\mathbb{Z}_n$.*
      v. **(Miller; cf. [157])** *If t is congruent to 1 mod 4, n is congruent to 2 mod 4, and $n \geq b_t+1$, then $\tau_{\pm}(\mathbb{Z}_n,[1,t]) \geq 2$; in particular, the set $\{(t-3)/2, (t+5)/2\}$ is t-independent in $\mathbb{Z}_n$.*

Some of these statements we have already seen: the set $\{1\}$ is clearly t-independent in $\mathbb{Z}_n$ for $n \geq t+1$, and Proposition F.52 implies that

- if there is a t-independent set of size 1 in $\mathbb{Z}_n$, then $n \geq t+1$, and
- if there is a t-independent set of size 2 in $\mathbb{Z}_n$, then $n \geq \lfloor t^2/2 \rfloor + t + 1$.

To complete the proof, we need to verify that the given two-element sets are indeed t-independent in their corresponding groups—for that, see page 358. We can observe that, when $t \equiv 1$ mod 4, then $b_t + 1 \equiv 2$ mod 4, so statements (iv) and (v) together imply that if $n = b_t + 1$, then $\tau_{\pm}(\mathbb{Z}_n,[1,t]) = 2$—cf. statement (iii).

We should point out that Proposition F.46 does not completely determine all n for which $\tau_{\pm}(\mathbb{Z}_n,[1,t]) = 1$, although for $t \leq 10$ it rules out all n values other than those listed above. For example, for $t=9$, $n \leq 9$ and $n \geq 55$ are ruled out by statement 2 (b) (ii), $n=50$ is ruled out by both 2 (b) (iii) and (v), and $n=54$ is ruled out by 2 (b) (v).

We offer the following open problems.

Problem F.47 *Examine the remaining cases of statement 2 (b) of Proposition F.46 to determine, for each odd value of t, all values of n for which* $\tau_{\pm}(\mathbb{Z}_n, [1,t]) = 1$.

Problem F.48 *For each even value of t, find all values of n for which* $\tau_{\pm}(\mathbb{Z}_n, [1,t]) = 2$.

Problem F.49 *For each odd value of t, find all values of n for which* $\tau_{\pm}(\mathbb{Z}_n, [1,t]) = 2$.

The general problem for cyclic groups is as follows:

Problem F.50 *For each value of t and each* $m \geq 3$, *find all values of n for which* $\tau_{\pm}(\mathbb{Z}_n, [1,t]) = m$.

For noncyclic groups, we offer:

Problem F.51 *For each value of t, find all noncyclic groups for which* $\tau_{\pm}(G, [1,t]) = 1$.

With exact values few and far between, we are also interested in some good bounds for the value of $\tau_{\pm}(G, [1,t])$. We can derive an upper bound for $\tau_{\pm}(G, [1,t])$ as follows. Since the case $t = 1$ is trivial, we assume that $t \geq 2$.

When t is even, we already pointed out in Section C.2.2 that $A = \{a_1, \dots, a_m\}$ (with $|A| = m$) being t-independent is equivalent to

$$|[0, \tfrac{t}{2}]_{\pm}A| = |\mathbb{Z}^m([0, \tfrac{t}{2}])|$$

where

$$\mathbb{Z}^m([0, \tfrac{t}{2}])) = \{(\lambda_1, \dots, \lambda_m) \in \mathbb{Z}^m \mid |\lambda_1| + \cdots + |\lambda_m| \leq \tfrac{t}{2}\}.$$

Therefore, for even values of t, if a set A of size m is t-independent in G, then

$$|\mathbb{Z}^m([0, \tfrac{t}{2}])| \leq n$$

must hold. Recalling from Section 2.5 that

$$|\mathbb{Z}^m([0, \tfrac{t}{2}])| = a(m, t/2) = \sum_{i \geq 0} \binom{m}{i} \binom{t/2}{i} 2^i,$$

we thus must have

$$a(m, t/2) = \sum_{i \geq 0} \binom{m}{i} \binom{t/2}{i} 2^i \leq n.$$

If t is odd, we similarly must have

$$|[0, \tfrac{t-1}{2}]_{\pm}A| = |\mathbb{Z}^m([0, \tfrac{t-1}{2}])|;$$

in addition, signed sums

$$\lambda_1 a_1 + \cdots + \lambda_m a_m$$

corresponding to the index set

$$\mathbb{Z}^m(\tfrac{t+1}{2})_{1+} = \{(\lambda_1, \dots, \lambda_m) \in \mathbb{Z}^m \mid \lambda_1 \geq 1, \lambda_1 + |\lambda_2| + \cdots + |\lambda_m| = \tfrac{t+1}{2}\}$$

must be pairwise distinct and distinct from $[0, \tfrac{t}{2}]_{\pm}A$ as well. Therefore, for odd values of t, if a set A of size m is t-independent in G, then

$$|\mathbb{Z}^m([0, \tfrac{t-1}{2}])| + |\mathbb{Z}^m(\tfrac{t+1}{2})_{1+}| \leq n$$

must hold.

Of course, there is nothing special about using λ_1 in the argument above—we could have replaced λ_1 by any one of the indices. It is important to note that, while our condition above for the case when t is even is equivalent for A to be t-independent, for odd t we only have a necessary condition: we could, in theory, still have, say, $\frac{t+1}{2} \cdot a_2$ equal one of the elements of $[0, \frac{t-1}{2}]_\pm A$, preventing A from being t-independent.

Recalling from Section 2.5 that

$$|\mathbb{Z}^m(\tfrac{t+1}{2})_{1+}| = a(m-1,(t-1)/2) = \sum_{i\geq 0} \binom{m-1}{i}\binom{(t-1)/2}{i}2^i,$$

we must have

$$a(m,(t-1)/2) + a(m-1,(t-1)/2) = \sum_{i\geq 0}\left[\binom{m}{i} + \binom{m-1}{i}\right]\binom{(t-1)/2}{i}2^i \leq n.$$

Note also that by Proposition 2.1, we can rewrite the left-hand side as

$$a(m,(t-1)/2) + a(m-1,(t-1)/2) = c(m,(t+1)/2) = \sum_{i\geq 0}\binom{m-1}{i-1}\binom{(t+1)/2}{i}2^i.$$

In summary, we have the following:

Proposition F.52 *Suppose that $t \geq 2$ and that A is a t-independent set of size m in a group G of order n. Then*

$$n \geq \begin{cases} a(m,t/2) = \sum_{i\geq 0}\binom{m}{i}\binom{t/2}{i}2^i & \text{if } t \text{ is even,} \\ c(m,(t+1)/2) = \sum_{i\geq 0}\binom{m-1}{i-1}\binom{(t+1)/2}{i}2^i & \text{if } t \text{ is odd.} \end{cases}$$

Proposition F.52 gives a lower bound for $\tau_\pm(G,[1,t])$. For example, we have

$$\begin{aligned} \tau_\pm(G,[1,2]) &\leq \left\lfloor \frac{n-1}{2} \right\rfloor, \\ \tau_\pm(G,[1,3]) &\leq \left\lfloor \frac{n}{4} \right\rfloor, \\ \tau_\pm(G,[1,4]) &\leq \left\lfloor \frac{\sqrt{2n-1}-1}{2} \right\rfloor, \\ \tau_\pm(G,[1,5]) &\leq \left\lfloor \frac{\sqrt{n-2}}{2} \right\rfloor; \end{aligned}$$

the bound is less explicit for $t \geq 6$, but we certainly have

$$\tau_\pm(G,[1,t]) \leq \left\lfloor \frac{1}{2}\sqrt[s]{s!\cdot n} + s \right\rfloor,$$

where $s = \lfloor t/2 \rfloor$. In particular:

Corollary F.53 *For every $t \geq 2$ there is a positive constant C_t so that*

$$\tau_\pm(G,[1,t]) \leq C_t \cdot \sqrt[s]{n},$$

where $s = \lfloor t/2 \rfloor$.

For a lower bound, Bajnok and Ruzsa proved the following:

Theorem F.54 (Bajnok and Ruzsa; cf. [24]) *For every $t \geq 2$ there is a positive constant c_t so that*

$$\tau_{\pm}(\mathbb{Z}_n, [1,t]) \geq c_t \cdot \sqrt[s]{n},$$

where $s = \lfloor t/2 \rfloor$.

Letting

$$c \cdot g_1(n) \lesssim f(n) \lesssim C \cdot g_2(n)$$

mean that for every real number $\epsilon > 0$, if n is large enough (depending on ϵ), then

$$(c - \epsilon) \cdot g_1(n) \leq f(n) \leq (C + \epsilon) \cdot g_2(n),$$

we can combine Corollary F.53 and Theorem F.54 to write

$$c_t \cdot \sqrt[s]{n} \lesssim \tau_{\pm}(\mathbb{Z}_n, [1,t]) \lesssim C_t \cdot \sqrt[s]{n}.$$

What can we say about the values of c_t and C_t? From Proposition F.38 we see that we may take $c_2 = C_2 = 1/2$, and therefore

$$\lim \frac{\tau_{\pm}(\mathbb{Z}_n, [1,2])}{n} = 1/2.$$

From Theorem F.41, we can have $c_3 = 1/6$ and $C_3 = 1/4$; we also see that both of these are tight as $\tau_{\pm}(\mathbb{Z}_n, [1,3])$ approaches both $n/6$ and $n/4$ infinitely often and thus

$$\lim \frac{\tau_{\pm}(\mathbb{Z}_n, [1,3])}{n}$$

does not exist.

For $t = 4$ and $t = 5$ we have:

Theorem F.55 (Bajnok and Ruzsa; cf. [24]) *We have*

$$1/\sqrt{8} \cdot \sqrt{n} \lesssim \tau_{\pm}(\mathbb{Z}_n, [1,4]) \lesssim 1/\sqrt{2} \cdot \sqrt{n}$$

and

$$1/\sqrt{15} \cdot \sqrt{n} \lesssim \tau_{\pm}(\mathbb{Z}_n, [1,5]) \lesssim 1/\sqrt{2} \cdot \sqrt{n}.$$

The following problems seem intriguing:

Problem F.56 *Find (if possible) a value higher than $1/\sqrt{8}$ for c_4 and one lower than $1/\sqrt{2}$ for C_4.*

Problem F.57 *Find (if possible) a value higher than $1/\sqrt{15}$ for c_5 and one lower than $1/\sqrt{2}$ for C_5.*

In [24] we find the following conjectures:

Conjecture F.58 (Bajnok and Ruzsa; cf. [24]) *We have*

$$\lim \frac{\tau_{\pm}(\mathbb{Z}_n, [1,4])}{\sqrt{n}} = 1/\sqrt{3},$$

but

$$\lim \frac{\tau_{\pm}(\mathbb{Z}_n, [1,5])}{\sqrt{n}}$$

does not exist.

Note that this conjecture, if true, would imply that one can take $c_4 = C_4$ but not $c_5 = C_5$.

While we see that for a given $t \geq 3$, $\tau_{\pm}(\mathbb{Z}_n, [1,t])$ is not a monotone function of n, we may find that the subsequence of even or odd n values is monotonic:

Problem F.59 *Decide whether for each even value of t (but at least for $t = 4$), the sequence $\tau_{\pm}(\mathbb{Z}_n, [1,t])$ is monotone for odd values of n.*

Problem F.60 *Decide whether for each odd value of t (but at least for $t = 5$), the sequence $\tau_{\pm}(\mathbb{Z}_n, [1,t])$ is monotone for even values of n.*

We are also interested in 4-independent and 5-independent sets in noncyclic groups:

Problem F.61 *Determine the value of $\tau_{\pm}(G, [1,4])$ for noncyclic groups G.*

Problem F.62 *Determine the value of $\tau_{\pm}(G, [1,5])$ for noncyclic groups G.*

We close this section with a particularly intriguing question: when can the size of t-independent sets achieve the maximum possible value allowed by the upper bound of Proposition F.52? (We mention that for $t = 1$ the answer is trivial: the set $G \setminus \{0\}$ is the maximum size 1-independent set in G.) For $t \geq 2$, we are thus interested in classifying all *tight t-independent sets*, that is, t-independent sets of size m with $n = b(m,t)$ where

$$b(m,t) = \begin{cases} a(m,t/2) = \sum_{i \geq 0} \binom{m}{i}\binom{t/2}{i} 2^i & \text{if t is even,} \\ c(m,(t+1)/2) = \sum_{i \geq 0} \binom{m-1}{i-1}\binom{(t+1)/2}{i} 2^i & \text{if t is odd.} \end{cases}$$

The values of $b(m,t)$ can be tabulated for small values of m and t:

b(m,t)	t=2	t=3	t=4	t=5	t=6	t=7	t=8
m=1	**3**	**4**	**5**	**6**	**7**	**8**	**9**
m=2	**5**	**8**	**13**	**18**	**25**	**32**	**41**
m=3	**7**	**12**	25	**38**	63	88	129
m=4	**9**	**16**	41	66	129	192	321
m=5	**11**	**20**	61	102	231	360	681
m=6	**13**	**24**	85	146	377	608	1289

Cases where there exists a group of size $b(m,t)$ with a known tight t-independent set are marked with boldface. The following proposition exhibits tight t-independent sets for all known parameters.

Proposition F.63 *Let m and t be positive integers, $t \geq 2$, and let G be an abelian group of order n and exponent κ.*

1. *If $n = 2m+1$, then $G \setminus \{0\}$ can be partitioned into parts K and $-K$, and both K and $-K$ are tight 2-independent sets in G. For example, the set $\{1, 2, \ldots, m\}$ is a tight 2-independent set in $\mathbb{Z}_n$.*

2. *If $n = 4m$ and κ is divisible by 4, then the set*

$$\{(g, 2i+1) \mid g \in G_1, i = 0, 1, \ldots, \kappa/4 - 1\}$$

is a tight 3-independent set in $G = G_1 \times \mathbb{Z}_\kappa$. For example, the set $\{1, 3, \ldots, 2m-1\}$ is a tight 3-independent set in $\mathbb{Z}_n$.

3. *If $n = t + 1$, then the set $\{1\}$ is a perfect t-independent set in $\mathbb{Z}_n$.*

4. *Let $n = \lfloor t^2/2 \rfloor + t + 1$.*

   - *If t is even, then the set $\{t/2, t/2 + 1\}$ is a tight t-independent set in $\mathbb{Z}_n$.*
   - *If t is odd, then the set $\{1, t\}$ is a tight t-independent set in $\mathbb{Z}_n$.*
   - *The set $\{(1, 1), (1, 3)\}$ is a tight 5-independent set in $\mathbb{Z}_3 \times \mathbb{Z}_6$.*

5. *The set $\{1, 7, 11\}$ is a tight 5-independent set in $\mathbb{Z}_{38}$.*

Each part of Proposition F.63, other than the two specific examples (which can be easily verified by hand or on a computer), is either obvious or follows from previously discussed results. Note that the sets given in Proposition F.63 are not unique.

We could not find tight independent sets for $t \geq 4$ and $m \geq 3$ for any n other than the seemingly sporadic example listed last. It might be an interesting problem to find and classify all tight independent sets.

Problem F.64 *For each positive integer n, find all tight t-independent sets in $\mathbb{Z}_n$ of size 2.*

Problem F.65 *For any given noncyclic group G, find all tight t-independent sets in G of size 2.*

Problem F.66 *Find tight t-independent sets in $G = \mathbb{Z}_n$ of size m for some values $t \geq 4$ and $m \geq 3$ other than for $t = 5$ and $n = 38$, or prove that such tight independent sets do not exist.*

Problem F.67 *Find tight t-independent sets in noncyclic groups of size m for values $t \geq 4$ and $m \geq 2$ (cf. the example of $\mathbb{Z}_3 \times \mathbb{Z}_6$ above).*

Finally, a more general problem.

Problem F.68 *For each given value of n, find all values of m and t for which the group $\mathbb{Z}_n$ has a t-independent set of size m.*

Proposition F.52 puts a necessary condition on n, m, and t, but that inequality is not necessarily sufficient.

F.2.3 Arbitrary number of terms

Here we ought to consider

$$\tau_{\pm}(G, H) = \max\{|A| \mid A \subseteq G, 0 \not\in H_{\pm}A\}$$

for the case when H is the set of all nonnegative or all positive integers. However, as we have already mentioned, we have $\tau_{\pm}(G, H) = 0$ whenever H contains a multiple of the exponent of the group (including 0). Thus, there are no "infinitely independent" sets in a finite group.

F.3 Restricted sumsets

In this section we investigate the maximum possible size of a weakly zero-H-sum-free set, that is, the quantity

$$\tau\hat{}(G,H) = \max\{|A| \mid A \subseteq G, 0 \not\in H\hat{}A\}$$

(if there is no subset A for which $0 \not\in H\hat{}A$, we let $\tau\hat{}(G,H) = 0$). Clearly, we always have $\tau\hat{}(G,H) = 0$ when $0 \in H$; however, when $0 \not\in H$ and $n \geq 2$, then $\tau\hat{}(G,H) \geq 1$: for any $a \in G \setminus \{0\}$, for the one-element set $A = \{a\}$ we obviously have $0 \not\in H\hat{}A$.

It is important to note that Proposition F.2 does not carry through for $\tau\hat{}(G,H)$: for example, $\{1\}$ is trivially weakly zero-$\mathbb{N}$-sum-free in $\mathbb{Z}_5$, but $\{1\} \times \mathbb{Z}_5$ is not weakly zero-$\mathbb{N}$-sum-free in $\mathbb{Z}_5^2$ since

$$(1,0) + (1,1) + (1,2) + (1,3) + (1,4) = (0,0).$$

We consider three special cases: when H consists of a single positive integer h, when H consists of all positive integers up to some value t, and when $H = \mathbb{N}$. As noted above, we have $\tau\hat{}(G,H) = 0$ whenever $0 \in H$, so the cases when $H = [0,t]$ or $H = \mathbb{N}_0$ yield no weakly zero-sum-free sets.

F.3.1 Fixed number of terms

The analogue of a zero-h-sum-free set for restricted addition is called a weak or weakly zero-h-sum-free set. In particular, in this section we investigate, for a given group G and positive integer h, the quantity

$$\tau\hat{}(G,h) = \max\{|A| \mid A \subseteq G, 0 \not\in h\hat{}A\},$$

that is, the maximum size of a weak zero-h-sum-free subset of G.

Since we trivially have $\tau\hat{}(G,h) = n$ for every $h \geq n+1$, we assume below that $h \leq n$.

We start with the following obvious bounds:

Proposition F.69 *When $\chi\hat{}(G,h)$ exists, then we have*

$$\tau(G,h) \leq \tau\hat{}(G,h) \leq \chi\hat{}(G,h) - 1.$$

As before, we can easily verify the following:

Proposition F.70 *In a group of order n, we have*

$$\tau\hat{}(G,1) = n-1$$

and

$$\tau\hat{}(G,2) = \frac{n + |\text{Ord}(G,2)| + 1}{2};$$

as a special case, for the cyclic group of order n we have

$$\tau\hat{}(\mathbb{Z}_n,2) = \left\lfloor \frac{n+2}{2} \right\rfloor.$$

Next, we consider the other end of the spectrum: $h = n$. As Blyler in ([40]) observed, in $\mathbb{Z}_n$ we have

$$0 + 1 + \cdots + (n-1) = \frac{(n-1)\cdot n}{2} \begin{cases} \neq 0 & \text{if } n \text{ even;} \\ = 0 & \text{if } n \text{ odd,} \end{cases}$$

and thus

$$\tau\hat{}(\mathbb{Z}_n, n) = \begin{cases} n & \text{if } n \text{ even;} \\ n-1 & \text{if } n \text{ odd.} \end{cases}$$

More generally, suppose that G has exponent κ and $G \cong G_1 \times \mathbb{Z}_\kappa$. The elements of $G_1 \times \mathbb{Z}_\kappa$ sum to

$$\sum_{g \in G_1} \left(\kappa \cdot g, \sum_{i=0}^{\kappa-1} i \right) = \sum_{g \in G_1} \left(0, \sum_{i=0}^{\kappa-1} i \right) = \left(0, |G_1| \cdot \sum_{i=0}^{\kappa-1} i \right) = \left(0, |G_1| \cdot \frac{(\kappa-1)\cdot\kappa}{2} \right).$$

Here the second component is zero if, and only if, $|G_1| = n/\kappa$ is even or if κ is odd. Therefore:

Proposition F.71 *Suppose that G has order n and exponent κ. Then*

$$\tau\hat{}(G, n) = \begin{cases} n & \text{if } \kappa \text{ is even and } n/\kappa \text{ is odd;} \\ n-1 & \text{otherwise.} \end{cases}$$

We can also evaluate $\tau\hat{}(G, n-1)$. Let $g \in G$ denote the sum of all n elements of G. Then $G \setminus \{g\}$ is not a weak zero-$(n-1)$-sum-free set, since its elements add to zero, so $\tau\hat{}(G, n-1) \leq n-1$. On the other hand, for every $g' \in G \setminus \{g\}$, the set $G \setminus \{g'\}$ is a weak zero-$(n-1)$-sum-free set, since its elements add to $g - g' \neq 0$, so $\tau\hat{}(G, n-1) \geq n-1$. Thus:

Proposition F.72 *For every group G of order n we have $\tau\hat{}(G, n-1) = n-1$.*

Furthermore, Bajnok and Edwards determined the value of $\tau\hat{}(G, h)$ for every G in a wide range of h values:

Theorem F.73 (Bajnok and Edwards; cf. [21]) *For every abelian group G of order n and all*

$$\frac{n + |\mathrm{Ord}(G,2)| - 1}{2} \leq h \leq n-2,$$

we have $\tau\hat{}(G, h) = h + 1$, with the following exceptions:

- *$\tau\hat{}(G, n-3) = n-3$ when $\kappa = 3$ (that is, G is isomorphic to an elementary abelian 3-group);*
- *$\tau\hat{}(G, n-2) = n-2$ when $|\mathrm{Ord}(G,2)| = 1$ and $\kappa \equiv 2 \bmod 4$.*

Note that Theorem F.73 does not apply to elementary abelian 2-groups; for these groups we have the value of $\tau\hat{}(G, h)$ for all $h \geq n/2 - 1$:

Theorem F.74 (Bajnok and Edwards; cf. [21]) *For every positive integer r we have*

$$\tau\hat{}(\mathbb{Z}_2^r, h) = \begin{cases} h+2 & \text{if } 2^{r-1} - 1 \leq h \leq 2^r - 5 \text{ or } h = 2^r - 3; \\ h & \text{if } h = 2^r - 4. \end{cases}$$

For the elementary abelian 2-group, we also have the value $\tau\hat{}(\mathbb{Z}_2^r, 3)$. We readily have two different weakly zero-3-sum-free sets of size $2^{r-1} + 1$:

- the zero element together with all elements with last component 1 and
- the zero element together with all elements with an odd number of 1 components.

Therefore, $\tau\hat{}(\mathbb{Z}_2^r, 3)$ is at least $2^{r-1}+1$. On the other hand, by Proposition F.69 and Theorem E.60, $\tau\hat{}(\mathbb{Z}_2^r, 3)$ is at most $2^{r-1}+1$. Therefore:

Theorem F.75 *For all positive integers r we have $\tau\hat{}(\mathbb{Z}_2^r, 3) = 2^{r-1}+1$.*

(Edwards proved Theorem F.75 independently in [71].)

Comparing the results thus far in this section to the corresponding results in Section E.3.1, we may get the impression that $\tau\hat{}(G,h)$ and $\chi\hat{}(G,h)$ are always very close to each other, and in fact, tend to only differ by one (cf. Proposition F.69). Our next example shows that, in fact, $\tau\hat{}(G,h)$ and $\chi\hat{}(G,h)$ can be arbitrarily far from one another.

Suppose that A is a weak zero-4-sum-free set in $\mathbb{Z}_2^r$; we will show that it is then a weak Sidon set as well. Indeed, if

$$a_1 + a_2 = a_3 + a_4$$

for some elements $a_1, a_2, a_3, a_4 \in \mathbb{Z}_2^r$ with $a_1 \neq a_2$ and $a_3 \neq a_4$, then we also have

$$a_1 + a_2 + a_3 + a_4 = 0,$$

which implies that at least two of the elements are equal, but that leads to $\{a_1, a_2\} = \{a_3, a_4\}$. We can show the same way that the converse holds as well, and therefore:

Proposition F.76 *A subset of the elementary abelian 2-group $\mathbb{Z}_2^r$ is a weak zero-4-sum-free set if, and only if, it is a weak Sidon set. Consequently,*

$$\tau\hat{}(\mathbb{Z}_2^r, 4) = \sigma\hat{}(\mathbb{Z}_2^r, 2).$$

Now recall that, by Proposition C.56, if an m-subset of $\mathbb{Z}_2^r$ is a weak Sidon set, then

$$2^r \geq \binom{m}{2};$$

consequently,

$$\tau\hat{}(\mathbb{Z}_2^r, 4) \leq 2^{(r+1)/2} + 1.$$

On the other hand, we have

$$\chi\hat{}(\mathbb{Z}_2^r, 4) \geq \chi(\mathbb{Z}_2^r, 4) = v_1(2^r, 4) + 1 = 2^{r-1} + 1.$$

Therefore:

Proposition F.77 *We have*

$$\lim_{r \to \infty} (\chi\hat{}(\mathbb{Z}_2^r, 4) - \tau\hat{}(\mathbb{Z}_2^r, 4)) = \infty.$$

We have the following open questions:

Problem F.78 *For positive integer r and each h with $3 \leq h \leq 2^{r-1} - 2$, find $\tau\hat{}(\mathbb{Z}_2^r, h)$.*

Problem F.79 *For each abelian group G of order n and each h with*

$$3 \leq h \leq \frac{n + |\mathrm{Ord}(G,2)| - 3}{2},$$

find $\tau\hat{}(G,h)$.

We can develop some useful lower bounds for $\tau\hat{}(G,h)$ by constructing explicit weak zero-h-sum-free sets as follows. We start by considering cyclic groups. The most general result we have thus far is as follows:

Proposition F.80 *For positive integers n and $h \leq n$, let r be the nonnegative remainder of $(h^2-h-2)/2$ when divided by* $\gcd(n,h)$. *We have*

$$\tau\hat{}(\mathbb{Z}_n, h) \geq \left\lfloor \frac{n+h^2-r-2}{h} \right\rfloor.$$

We should note that $(h^2-h-2)/2$ is an integer for all h. For the proof, see page 360.

Since r is at most $\gcd(n,h)-1$, we have the following immediate consequence:

Corollary F.81 *For all positive integers n and $h \leq n$,*

$$\tau\hat{}(\mathbb{Z}_n, h) \geq \left\lfloor \frac{n+h^2-\gcd(n,h)-1}{h} \right\rfloor.$$

Furthermore, when n is divisible by h, then $\gcd(n,h)=h$; if h is even, then

$$\frac{h^2-h-2}{2} = \left(\frac{h}{2}-1\right)\cdot h + \frac{h-2}{2},$$

so $r=(h-2)/2$, and we get the following:

Corollary F.82 *If h is even and n is divisible by h, then*

$$\tau\hat{}(\mathbb{Z}_n, h) \geq \frac{n}{h} + h - 1.$$

Note that when n is divisible by h, then

$$\left\lfloor \frac{n+h^2-\gcd(n,h)-1}{h} \right\rfloor = \frac{n}{h} + h - 2,$$

so Corollary F.82 is stronger than Corollary F.81 in this case.

We can do very slightly better when n is divisible by h^2 and $h > 2$. This was first observed and proved by Yager-Elorriaga in [200] for $h=3$ and $h=4$; we provide a general proof.

Proposition F.83 *For positive integers n and $h \geq 3$ for which n is divisible by h^2 we have*

$$\tau\hat{}(\mathbb{Z}_n, h) \geq \frac{n}{h} + h.$$

For the proof, see page 361.

There are some additional instances where we can do better—we present these (rather easy) results below.

First, we observe that the set

$$A = \{\pm 1, \pm 2, \dots, \pm (h+1)/2\}$$

is weakly zero-h-sum-free when h is odd and $n \geq h+2$; indeed, we have $h\hat{}A = A$. The similar, but slightly less obvious set

$$A = \{0, 1, \pm 2, \pm 3, \dots, \pm h/2, h/2+1\}$$

works when h is even and $n \geq h+3$, since in this case we have

$$h\hat{}A = \{1, 2, \dots, h/2+2, h/2+4, h/2+5, h+2\}.$$

This gives the following.

Proposition F.84 *Suppose that $n \geq h+2$ when h is odd and $n \geq h+3$ when h is even. Then $\tau\hat{~}(\mathbb{Z}_n, h) \geq h+1$.*

Another simple construction applies when n is even and h is odd: clearly, the set

$$\{1, 3, 5, \ldots, n-1\}$$

is then weakly zero-h-sum-free. As a generalization, assume that n is divisible by some positive integer d, and that, for some nonnegative integer k, none of $h, h+1, \ldots, h+k$ is divisible by d. Consider then the set $A = A_1 \cup A_2$ where

$$A_1 = \{1, d+1, 2d+1, \ldots, (n/d-1)\, d+1\}$$

and

$$A_2 = \{2, d+2, 2d+2, \ldots, (k-1)d+2\}.$$

To prove that A is weakly zero-h-sum-free, consider an element x of $h\hat{~}A$ where h_1 terms come from A_1 and h_2 terms come from A_2. This sum is then of the form

$$x = (q_1 d + h_1) + (q_2 d + 2h_2) = (q_1 + q_2)d + h + h_2$$

for some integers q_1 and q_2. Since $h \leq h + h_2 \leq h + k$, by assumption, $h + h_2$ is not divisible by d, and therefore x is not divisible by d. But then it cannot be divisible by n either, proving that $0 \not\in h\hat{~}A$. We just proved the following:

Proposition F.85 *Let n and $h \leq n$ be positive integers, and let d be a positive divisor of n for which none of $h, h+1, \ldots, h+k$ is divisible by d. We then have*

$$\tau\hat{~}(\mathbb{Z}_n, h) \geq \frac{n}{d} + k.$$

The following result is not difficult to establish either. For fixed positive integers n_1, n_2, and h, consider the set

$$A = \{0, 1, 2, \ldots, \lfloor n_1/h \rfloor\}$$

in $\mathbb{Z}_{n_1}$ and any weakly zero-h-sum-free set B in $\mathbb{Z}_{n_2}$. We show that

$$A \times B = \{(a, b) \mid a \in A, b \in B\}$$

is weakly zero-h-sum-free in $\mathbb{Z}_{n_1} \times \mathbb{Z}_{n_2}$.

Indeed, if the sum of h elements $(a_1, b_1), \ldots, (a_h, b_h)$ of $A \times B$ were to equal $(0, 0)$, then, since

$$0 \leq a_1 + \cdots + a_h \leq h \cdot \lfloor n_1/h \rfloor \leq n_1,$$

either $a_1 = \cdots = a_h = 0$ or (n_1 is divisible by h and) $a_1 = \cdots = a_h = n_1/h$. But if all first coordinates are equal, then all second coordinates must be distinct, but that would mean that we have h distinct elements in B that sum to 0 in $\mathbb{Z}_{n_2}$, which is a contradiction. Thus, we proved the following result.

Proposition F.86 *For positive integers n_1, n_2, and h we have*

$$\tau\hat{~}(\mathbb{Z}_{n_1} \times \mathbb{Z}_{n_2}, h) \geq (\lfloor n_1/h \rfloor + 1) \cdot \tau\hat{~}(\mathbb{Z}_{n_2}, h).$$

Note that Proposition F.86 applies even to a cyclic group, as long as its order has more than one prime divisor.

Relying on a computer program, Yager-Elorriaga in [200] exhibited the values of $\tau\hat{}(\mathbb{Z}_n, h)$ for $h \in \{3,4\}$ and all $n \leq 30$; Blyler in [40] extended this to $h \leq 10$—we provide these data in the table below. Most entries are in agreement with the highest of the values that Propositions F.80, F.83, F.84, F.85, or F.86 yield. (None of these propositions is superfluous.) For some entries this is not the case; we marked these entries in the table by a * and listed them here with an exemplary set.

$\tau\hat{}(\mathbb{Z}_{10}, 3) = 6$	$\{1, 2, 4, 6, 8, 9\}$
$\tau\hat{}(\mathbb{Z}_{15}, 5) = 8$	$\{1, 2, 4, 5, 7, 10, 11, 14\}$
$\tau\hat{}(\mathbb{Z}_{15}, 6) = 8$	$\{0, 1, 2, 3, 4, 6, 12, 13\}$
$\tau\hat{}(\mathbb{Z}_{27}, 6) = 10$	$\{0, 2, 4, 7, 9, 11, 13, 18, 20, 22\}$
$\tau\hat{}(\mathbb{Z}_{21}, 7) = 9$	$\{0, 1, 2, 3, 4, 5, 7, 8, 9\}$
$\tau\hat{}(\mathbb{Z}_{20}, 8) = 10$	$\{0, 1, 2, 3, 4, 5, 6, 8, 9, 18\}$
$\tau\hat{}(\mathbb{Z}_{28}, 8) = 11$	$\{0, 1, 2, 3, 4, 5, 6, 8, 24, 25, 26\}$
$\tau\hat{}(\mathbb{Z}_{21}, 9) = 11$	$\{0, 1, 2, 3, 4, 5, 6, 7, 10, 12, 13\}$
$\tau\hat{}(\mathbb{Z}_{27}, 9) = 12$	$\{0, 1, 2, 3, 9, 10, 11, 12, 18, 19, 20, 21\}$
$\tau\hat{}(\mathbb{Z}_{25}, 10) = 12$	$\{0, 1, 2, 3, 4, 5, 6, 7, 8, 10, 13, 15\}$

n	$\tau\hat{\ }(\mathbb{Z}_n,3)$	$\tau\hat{\ }(\mathbb{Z}_n,4)$	$\tau\hat{\ }(\mathbb{Z}_n,5)$	$\tau\hat{\ }(\mathbb{Z}_n,6)$	$\tau\hat{\ }(\mathbb{Z}_n,7)$	$\tau\hat{\ }(\mathbb{Z}_n,8)$	$\tau\hat{\ }(\mathbb{Z}_n,9)$	$\tau\hat{\ }(\mathbb{Z}_n,10)$
1	1	1	1	1	1	1	1	1
2	2	2	2	2	2	2	2	2
3	2	3	3	3	3	3	3	3
4	3	4	4	4	4	4	4	4
5	4	4	4	5	5	5	5	5
6	4	4	5	6	6	6	6	6
7	4	5	6	6	6	7	7	7
8	5	5	6	7	7	8	8	8
9	6	5	6	7	8	8	8	9
10	6*	5	6	7	8	8	9	10
11	6	6	6	7	8	9	10	10
12	6	6	7	7	8	9	10	10
13	6	6	7	7	8	9	10	11
14	7	6	7	8	8	9	10	11
15	8	7	8*	8*	8	9	10	11
16	8	8	8	8	9	9	10	11
17	8	7	8	8	9	9	10	11
18	9	7	9	8	9	9	10	11
19	8	8	8	8	9	10	10	11
20	10	8	10	9	10	10*	11	11
21	8	8	8	8	9*	10	11*	11
22	11	8	11	9	11	10	11	12
23	10	9	9	9	10	10	11	12
24	12	9	12	9	12	10	12	12
25	10	9	10	9	10	10	11	12*
26	13	9	13	10	13	10	13	12
27	12	10	10	10*	10	11	12*	12
28	14	10	14	10	14	11*	14	12
29	12	10	10	10	10	11	12	12
30	15	11	15	10	15	11	15	12

After these lower bounds, we mention some exact results. First, the value of $\tau\hat{\ }(\mathbb{Z}_p, h)$ for p prime immediately follows from combining Propositions F.69 and F.80 and Theorem E.65:

Theorem F.87 *Suppose that p is a positive prime and $1 \leq h \leq p-1$. We then have*

$$\tau\hat{\ }(\mathbb{Z}_p, h) = \left\lfloor \frac{p-2}{h} \right\rfloor + h.$$

Next, we have:

Theorem F.88 *If $n \geq 12$ is even and $3 \leq h \leq n-1$ is odd, then*

$$\tau\hat{}(\mathbb{Z}_n, h) = \chi\hat{}(\mathbb{Z}_n, h) - 1 = \begin{cases} n-1 & \text{if } h = 1; \\ n/2 & \text{if } 3 \leq h \leq n/2 - 2; \\ n/2+1 & \text{if } h = n/2 - 1; \\ h+1 & \text{if } n/2 \leq h \leq n-2; \\ n-1 & \text{if } h = n-1. \end{cases}$$

The proof can be found on page 362.

We pose the following problems.

Problem F.89 *Find the exact formulas for $\tau\hat{}(\mathbb{Z}_n, 3)$ and $\tau\hat{}(\mathbb{Z}_n, 5)$ for all odd n.*

Problem F.90 *Find $\tau\hat{}(\mathbb{Z}_n, 4)$.*

Regarding noncyclic groups, we can use the same argument as the one provided for Theorem F.12 to prove the following:

Theorem F.91 *If h is relatively prime to n, then*

$$\tau\hat{}(G, h) = \chi\hat{}(G, h) - 1.$$

Let us now turn to elementary abelian groups. We have already considered the elementary abelian 2-group $\mathbb{Z}_2^r$, so we move on to the elementary abelian 3-groups. Pursuing the determination of $\tau\hat{}(\mathbb{Z}_3^r, 3)$ is attracting much attention as it is related to finite geometry. For example:

- $\tau\hat{}(\mathbb{Z}_3^r, 3)$ is the maximum size of a cap (a collection of points without any three being collinear) in the affine space $\mathrm{AG}(r, 3)$, and
- $2 \cdot \tau\hat{}(\mathbb{Z}_3^r, 3)$ is the maximum number of points in the integer lattice $\mathbb{Z}^r$ so that the centroid of no three of them is a lattice point.

We elaborated more on these questions in an "appetizer" section, see page 56.

The following table summarizes all values of $\tau\hat{}(\mathbb{Z}_3^r, 3)$ that are known (see [97] by Gao and Thangadurai and its references for the first five entries and [175] by Potechin for the last):

r	1	2	3	4	5	6
$\tau\hat{}(\mathbb{Z}_3^r, 3)$	2	4	9	20	45	112

We also know that the sets of maximum size for $r \leq 6$ are essentially unique.

Even good bounds are difficult to achieve for $\tau\hat{}(\mathbb{Z}_3^r, 3)$—we summarize what is currently known. Starting with lower bounds, observe that the set $\{0,1\}^r$ is clearly weakly zero-3-sum-free in $\mathbb{Z}_3^r$, hence

$$\tau\hat{}(\mathbb{Z}_3^r, 3) \geq 2^r.$$

A better and very nice lower bound can be developed as follows. For $i = 0, 1, \ldots, r-1$, consider the collection of elements A_i of $\mathbb{Z}_3^r$ that contain exactly i 0-components, and whose remaining $r - i$ components are all 1, and let $-A_i$ denote the negatives of the elements of

A_i (which contain exactly i 0-components, and whose remaining $r-i$ components are all 2). (We will not use A_r.) Note that

$$|A_i \cup -A_i| = 2 \cdot \binom{r}{i}$$

for each $i = 0, 1, \ldots, r-1$. Bui (see [45]) proved the following interesting result: If I is a subset of $\{0, 1, \ldots, r-1\}$ for which the equation

$$i_1 + i_2 - i_3 = r$$

has no solution with $i_1, i_2, i_3 \in I$, then $A = \cup_{i \in I}(A_i \cup -A_i)$ is weakly zero-3-sum-free. (Note that we had to exclude r from I as $i_1 = i_2 = i_3 = r$ does yield a solution to the equation.) For example, for $r = 3$, the index set $I = \{0, 1\}$ yields no solution for

$$i_1 + i_2 - i_3 = 3,$$

and, correspondingly, the set $A = \pm A_0 \cup \pm A_1$, consisting of the eight elements

$$(1,1,1), (2,2,2), (0,1,1), (0,2,2), (1,0,1), (2,0,2), (1,1,0), (2,2,0),$$

is, as can be easily verified, weakly zero-3-sum-free in $\mathbb{Z}_3^3$: no three of them add to $(0,0,0)$. (The index set $I = \{0, 2\}$ would work as well.)

We can select the index set I to maximize $|A|$ by positioning it in the "middle" of $\{0, 1, \ldots, r-1\}$. More precisely, if $r = 3q + s$ with a remainder $s = 0, 1, 2$, then letting

$$I = \{q+s-1, q+s, q+s+1, \ldots, 2q+s-1\}$$

works, since for any $i_{,1}, i_2, i_3 \in I$, we have $i_1 + i_2 - i_3$ at most equal to

$$(2q+s-1) + (2q+s-1) - (q+s-1) = 3q+s-1 = r-1 < r,$$

and it yields an optimal $|A|$. This results in the following lower bound:

Theorem F.92 (Bui; cf. [45]) *Let r be any positive integer, and write $r = 3q + s$ with $s = 0, 1,$ or 2. Then we have*

$$\tau\hat{}(\mathbb{Z}_3^r, 3) \geq 2 \cdot \sum_{i=q+s-1}^{2q+s-1} \binom{r}{i}.$$

Theorem F.92 gives a remarkably good lower bound for the values of $\tau\hat{}(\mathbb{Z}_3^r, 3)$; for example, for $r = 4$, 5, and 6, we get that $\tau\hat{}(\mathbb{Z}_3^r, 3)$ is at least 20 (the actual value, see above), 40, and 82, respectively.

It would be very interesting to see if Bui's idea generalizes to other settings; we pose the following problems.

Problem F.93 *Find a lower bound for $\tau\hat{}(\mathbb{Z}_k^r, 3)$ for $k \geq 4$.*

Problem F.94 *Find a lower bound for $\tau\hat{}(\mathbb{Z}_3^r, h)$ for $h \geq 4$.*

Beyond the lower bound of Theorem F.92, Edel in [70], improving on a result of Frankl, Graham, and Rödl in [84], has shown that there are infinitely many values of r for which

$$\tau\hat{}(\mathbb{Z}_3^r, 3) > 2.2^r.$$

(Note that the lower bound of Theorem F.92 is less than 2^{r+1}.)

An easy upper bound can be derived from the fact that if A is weakly zero-3-sum-free in $\mathbb{Z}_3^r$, then it is disjoint from

$$B = -a_1 - (A \setminus \{a_1\})$$

for any fixed $a_1 \in A$. Indeed, if $-a_1 - a_i = a_j$ for some $a_i \in A \setminus \{a_1\}$ and $a_j \in A$, then $a_1 + a_i + a_j = 0$, so the fact that A is weakly zero-3-sum-free implies that these three elements of A cannot be all distinct, but in $\mathbb{Z}_3^r$ this actually implies that all three are equal, contradicting $a_i \in A \setminus \{a_1\}$. Since $|B| = |A| - 1$, we get

$$\tau\hat{\ }(\mathbb{Z}_3^r, 3) \leq (3^r + 1)/2.$$

This bound was improved to

$$\tau\hat{\ }(\mathbb{Z}_3^r, 3) \leq (r+1) \cdot 3^r/r^2$$

by Bierbrauer and Edel in [39], where the upper bound is of approximate size $3^r/r$ for large r. More recently, Bateman and Katz proved in their lengthy paper [32] that we actually have

$$\lim_{r \to \infty} \tau\hat{\ }(\mathbb{Z}_3^r, 3)/(3^r/r) = 0.$$

We offer the following very difficult problem:

Problem F.95 *Find better lower and upper bounds for* $\tau\hat{\ }(\mathbb{Z}_3^r, 3)$.

The case when h equals the exponent κ of the group, as with the elementary abelian 3-group and $h = 3$ above, has attracted much attention: the value of $\tau\hat{\ }(G, \kappa) + 1$ (the smallest integer such that each subset of G of that cardinality has a subset size κ whose elements sum to 0) is called the *Harborth constant* of G. Below we summarize what is known about $\tau\hat{\ }(G, \kappa)$.

It is obvious that for every G of order n and exponent κ we have

$$\kappa - 1 \leq \tau\hat{\ }(G, \kappa) \leq n;$$

we can easily classify cases when either inequality becomes an equality as follows. For the elementary abelian 2-group, we have $\tau\hat{\ }(\mathbb{Z}_2^r, 2) = 2^r$ by Proposition F.70. Furthermore, for the cyclic group, by Proposition F.71, we have $\tau\hat{\ }(\mathbb{Z}_n, n) = n$ if n is even and $\tau\hat{\ }(\mathbb{Z}_n, n) = n-1$ if n is odd. Next, we show that in all other cases,

$$\kappa \leq \tau\hat{\ }(G, \kappa) \leq n - 1.$$

Suppose that $G \cong G_1 \times \mathbb{Z}_\kappa$, where $\kappa \geq 3$ (so G is not isomorphic to an elementary abelian 2-group) and $|G_1| \geq 2$ (so G is not cyclic). To prove our claim, we need to find a subset of size κ whose elements don't sum to zero, and another subset of size κ whose elements do sum to zero. Note that the elements in $\{0\} \times \mathbb{Z}_\kappa$ sum to zero if κ is odd, and don't sum to zero if κ is even. Furthermore, for any $g \in G_1 \setminus \{0\}$, if κ is odd, then the elements in $\{0\} \times (\mathbb{Z}_\kappa \setminus \{0\}) \cup \{(g, 0)\}$ don't sum to zero, and if κ is even, then the elements in $\{0\} \times (\mathbb{Z}_\kappa \setminus \{0, \kappa/2\}) \cup \{(g, 1), (-g, \kappa - 1)\}$ sum to zero. We thus proved:

Proposition F.96 *For all groups G of order n and exponent κ we have*

$$\kappa - 1 \leq \tau\hat{\ }(G, \kappa) \leq n.$$

Furthermore:

- *the lower bound holds if, and only if, G is cyclic of odd order, and*
- *the upper bound holds if, and only if, G is cyclic of even order or G is an elementary abelian 2-group.*

We should note that the claim regarding the upper bound appeared by Gao and Geroldinger in [91] (see Lemma 10.1).

Regarding exact values, we first consider $\tau\hat{}(\mathbb{Z}_k^r, k)$. As Kemnitz observed in [128], the set

$$A = \{0,1\}^{r-1} \times \{0,1,\ldots,k-2\},$$

that is, the collection of elements whose first $r-1$ components are 0 or 1 and whose last component is not $k-1$, is zero-k-sum-free: indeed, the sum of k distinct elements in A can only be the zero element of $\mathbb{Z}_k^r$ if they agree in each of their first $r-1$ components, but then their last components must be distinct, which is impossible since we only have $k-1$ choices. We can do slightly better when k is even: k distinct elements of

$$A = \{0,1\}^{r-1} \times \mathbb{Z}_k$$

again must share their first $r-1$ components, but then their last components must be distinct and thus add to

$$0+1+\cdots+k-1 = k/2 \neq 0$$

in $\mathbb{Z}_k$. Thus we get:

Proposition F.97 (Kemnitz; cf. [128]) *For each $k \geq 2$ and $r \geq 1$ we have*

$$\tau\hat{}(\mathbb{Z}_k^r, k) \geq \begin{cases} (k-1)\cdot 2^{r-1} & \text{if } k \text{ odd;} \\ k \cdot 2^{r-1} & \text{if } k \text{ even.} \end{cases}$$

Note that, by Proposition F.71, equality holds for $r=1$, and it has been conjectured that equality holds for $r=2$ as well:

Conjecture F.98 (Gao and Thangadurai; cf. [97]) *For each $k \geq 2$ we have*

$$\tau\hat{}(\mathbb{Z}_k^2, k) = \begin{cases} 2k-2 & \text{if } k \text{ odd;} \\ 2k & \text{if } k \text{ even.} \end{cases}$$

Kemnitz in [128] proved that Conjecture F.98 holds for $k \in \{2,3,5,7\}$, and in [97] Gao and Thangadurai established Conjecture F.98 for prime values of k with $k \geq 67$; this has been improved somewhat by Gao, Geroldinger, and Schmid:

Theorem F.99 (Gao, Geroldinger, and Schmid; cf. [92]) *If $p \geq 47$ is a prime, then*

$$\tau\hat{}(\mathbb{Z}_p^2, p) = 2p-2.$$

Furthermore, in [92] the authors also determine all weakly zero-p-sum-free subsets of $\mathbb{Z}_p$ of size $2p-2$ (for $p \geq 47$).

Additionally, Gao and Thangadurai showed in [97] that $\tau\hat{}(\mathbb{Z}_4^2, 4) = 8$, and Schmid and co-authors verified that $\tau\hat{}(\mathbb{Z}_6^2, 6) = 12$; cf. [184].

We pose the following problems:

Problem F.100 *Prove that Conjecture F.98 holds for prime numbers $11 \leq p \leq 43$.*

Problem F.101 *Prove that Conjecture F.98 holds for small composite numbers $k \geq 8$.*

However, we know of at least one case when the inequality is strict in Proposition F.97: We have $\tau\hat{}(\mathbb{Z}_3^3, 3) \geq 9$, since

$$\{(0,0,0),(1,0,0),(0,1,0),(0,0,1),(1,1,0),(1,1,1),(1,2,1),(2,0,1),(0,1,2)\}$$

is weakly zero-3-sum-free in $\mathbb{Z}_3^3$.

Let us now examine general noncyclic groups, in particular groups of rank 2. Let n_1 and n_2 be positive integers with $n_1 \geq 2$ and $n_1 | n_2$, and consider $\mathbb{Z}_{n_1} \times \mathbb{Z}_{n_2}$. It is easy to see that the set

$$(\{0\} \times A_2) \cup (\{1\} \times A_2')$$

is weakly zero-n_2-sum-free for all A_2 and A_2' of $\mathbb{Z}_{n_2}$ when $|A_2| = n_2 - 1$ and $|A_2'| = n_1 - 1$. And, like we mentioned above, when n_2 is even, we may replace A_2 by all of $\mathbb{Z}_{n_2}$. This yields the following:

Proposition F.102 *Let n_1 and n_2 be integers greater than 1 so that n_2 is divisible by n_1. We then have*

$$\tau\hat{}(\mathbb{Z}_{n_1} \times \mathbb{Z}_{n_2}, n_2) \geq n_1 + n_2 - 2;$$

if n_2 is even, we have

$$\tau\hat{}(\mathbb{Z}_{n_1} \times \mathbb{Z}_{n_2}, n_2) \geq n_1 + n_2 - 1.$$

The question then arises whether we can do better than Proposition F.102; we exhibit two cases when we can: one for $n_1 = 2$, the other for $n_1 = 3$.

For the case when $n_1 = 2$, we follow the construction of Marchan, Ordaz, Ramos, and Schmid; cf. [150]. Suppose that $n_2 = 2k$ with k odd, in which case $\mathbb{Z}_2 \times \mathbb{Z}_{n_2}$ is isomorphic to $\mathbb{Z}_2 \times \mathbb{Z}_2 \times \mathbb{Z}_k$. We define the subsets B_1 and B_2 of $\mathbb{Z}_k$ as

$$B_1 = \{0, 1, 2, \ldots, (k-1)/2\}$$

and

$$B_2 = \{(k+1)/2, (k+1)/2 + 1, \ldots, k-1, 0\},$$

and set

$$A = (\{0\} \times \{0\} \times B_1) \cup (\{0\} \times \{1\} \times B_2) \cup (\{1\} \times \{0\} \times B_2) \cup (\{1\} \times \{1\} \times B_1).$$

Then A is a subset of $\mathbb{Z}_2 \times \mathbb{Z}_2 \times \mathbb{Z}_k$ of size $2k+2$, and it is easy to verify that the sum of these $2k+2$ elements equals $(0,0,0)$. Therefore, having k distinct elements of A adding to $(0,0,0)$ is equivalent to having two of them add to $(0,0,0)$, but this is clearly not the case. Therefore, when $n_1 = 2$ and $n_2 \equiv 2 \mod 4$, we have

$$\tau\hat{}(\mathbb{Z}_{n_1} \times \mathbb{Z}_{n_2}, n_2) \geq n_1 + n_2.$$

It turns out that we cannot do better:

Theorem F.103 (Marchan, Ordaz, Ramos, and Schmid; cf. [150]) *For each positive integer k,*

$$\tau\hat{}(\mathbb{Z}_2 \times \mathbb{Z}_{2k}, 2k) = \begin{cases} 2k+2 & \text{if } k \text{ odd;} \\ 2k+1 & \text{if } k \text{ even.} \end{cases}$$

Let us turn now to $n_1 = 3$. We already know from Proposition F.102 that

$$\tau\hat{}(\mathbb{Z}_3 \times \mathbb{Z}_{3k}, 3k) \geq 3k + 2$$

when k is even; we now show that the same bound holds when k is odd and $k \geq 2$ (for $k = 1$ we have $\tau\hat{}(\mathbb{Z}_3^2, 3) = 4$; cf. Conjecture F.98 above). Kiefer in [130] presents the following subset of $\mathbb{Z}_3 \times \mathbb{Z}_{3k}$:

$$A = (\{0\} \times (\mathbb{Z}_{3k} \setminus \{3k-2, 3k-1\})) \cup \{(1,0), (1,1), (1, 3k-6), (2,0)\}.$$

Then A has size $3k + 2$ (note that $k \geq 3$), and these elements add to $(2, 3k - 2)$. As with the previous example, for k distinct elements to add to $(0,0)$, we must have two distinct elements that add to $(2, 3k-2)$, but one can quickly see that that is impossible. This yields:

Proposition F.104 (Kiefer; cf. [130]) *For all values of $k > 1$ we have*

$$\tau\hat{}(\mathbb{Z}_3 \times \mathbb{Z}_{3k}, 3k) \geq 3k + 2.$$

It turns out that one can do a bit better for $k = 3$. Namely, consider the set

$$A = \{(0,0), (0,1), (0,3), (0,4), (0,6), (0,7), (1,0), (1,1), (1,3), (1,4), (1,6), (1,7)\}$$

in $\mathbb{Z}_3 \times \mathbb{Z}_9$; note that A can be rewritten as

$$A = \{0,1\} \times (\{0,3,6\} + \{0,1\}).$$

Since the twelve elements add to $(0,6)$, A is weakly zero-9-sum-free if, and only if, $(0,6) \not\in 3\hat{}A$. Indeed, for three distinct elements of A to add to $(0,6)$, the three first components would need to agree (all 0s or all 1s), but then the second components must be distinct. There are only two possibilities for the second components to add to a number divisible by 3: $0+3+6$ or $1+4+7$, but neither sum equals 6 in $\mathbb{Z}_9$. Therefore,

$$\tau\hat{}(\mathbb{Z}_3 \times \mathbb{Z}_9, 9) \geq 12.$$

Schmid writes in [184] that he and co-authors have proved that the lower bound in Proposition F.104 holds for large enough primes, but the general question remains open:

Problem F.105 *Evaluate $\tau\hat{}(\mathbb{Z}_3 \times \mathbb{Z}_{3k}, 3k)$ for each positive integer k.*

And, more generally:

Problem F.106 *Evaluate $\tau\hat{}(\mathbb{Z}_{n_1} \times \mathbb{Z}_{n_2}, n_2)$ for positive integers $n_1 \geq 4$, $n_1 | n_2$, and $n_2 > n_1$.*

More generally still:

Problem F.107 *Evaluate $\tau\hat{}(G, \kappa)$ for each noncyclic group G of exponent κ.*

F.3.2 Limited number of terms

In this section we investigate, for a given group G and positive integer t, the quantity

$$\tau\hat{}(G, [1,t]) = \max\{|A| \mid A \subseteq G, 0 \not\in [1,t]\hat{}A\},$$

that is, the maximum size of a weak zero-$[1,t]$-sum-free subset of G.

As before, we can easily verify the following:

Proposition F.108 *In a group of order n, we have*

$$\tau\hat{}(G,[1,1]) = n-1$$

and

$$\tau\hat{}(G,[1,2]) = \frac{n+|\mathrm{Ord}(G,2)|-1}{2};$$

as a special case, for the cyclic group of order n we have

$$\tau\hat{}(\mathbb{Z}_n,[1,2]) = \left\lfloor \frac{n}{2} \right\rfloor.$$

We have the following general upper bound for $\tau\hat{}(G,[1,t])$:

Proposition F.109 *For every G and $t \geq 2$, we have*

$$\tau\hat{}(G,[1,t]) \leq \tau\hat{}(G,t) - 1.$$

Indeed, if A is a weak zero-$[1,t]$-sum-free in G of size $\tau\hat{}(G,[1,t])$, then $0 \not\in A$, so $A \cup \{0\}$ has size $\tau\hat{}(G,[1,t]) + 1$. Furthermore, $0 \not\in (t-1)\hat{}A$ and $0 \not\in t\hat{}A$, which implies that $0 \not\in t\hat{}(A \cup \{0\})$. This proves that $\tau\hat{}(G,t)$ is at least one more than $\tau\hat{}(G,[1,t])$, as claimed. As Propositions F.70 and F.108 show, we have equality in Proposition F.109 for $t = 2$.

Let us now consider $t \geq 3$. Clearly, if m is a positive integer for which

$$m + (m-1) + \cdots + (m-t+1) \leq n-1,$$

then the set

$$A = \{1, 2, \ldots, m\}$$

is a weak zero-$[1,t]$-sum-free subset of the cyclic group $\mathbb{Z}_n$; this gives:

Proposition F.110 *For all positive integers n and t we have*

$$\tau\hat{}(\mathbb{Z}_n,[1,t]) \geq \left\lfloor \frac{n-1}{t} + \frac{t-1}{2} \right\rfloor.$$

(Observe that, by Proposition F.108, equality holds for $t = 1$ and $t = 2$.)

We are able to do better in certain cases. Consider, for example, the case when $n-2$ is divisible by 6, and define the subset A of $\mathbb{Z}_n$ as

$$A = \{1, 2, \ldots, (n-2)/6\} \cup \{n/2, n/2+1, \ldots, (2n+2)/3\}.$$

(Note that when $n-2$ is divisible by 6, then n is even and $2n+2$ is divisible by 6.) An easy computation shows that $0 \not\in [1,3]\hat{}A$; furthermore, A contains $(n+4)/3$ elements. Therefore, we get:

Proposition F.111 *If $n-2$ is divisible by 6, then*

$$\tau\hat{}(\mathbb{Z}_n,[1,3]) \geq (n+4)/3.$$

In a similar manner, Yin proved that the set

$$A = \{1, 2, \ldots, n/6\} \cup \{n/2, n/2+1, \ldots, 2n/3\}$$

is a weak zero-$[1,3]$-sum-free set in $\mathbb{Z}_n$ when n is divisible by 6, and that the set

$$\{1, 2, \ldots, \lfloor (n+2)/8 \rfloor\} \cup \{n/2, n/2+1, \ldots, n/2 + \lfloor (n+6)/8 \rfloor\}$$

is a weak zero-$[1,4]$-sum-free set in $\mathbb{Z}_n$ when $n-2$ is divisible by 4. These results yield:

Proposition F.112 (Yin; cf. [202]) *1. If n is divisible by 6, then*

$$\tau\hat{}(\mathbb{Z}_n, [1,3]) \geq n/3 + 1.$$

2. If $n-2$ is divisible by 4, then

$$\tau\hat{}(\mathbb{Z}_n, [1,4]) \geq (n+6)/4.$$

We should note that Proposition F.111 and both parts of Proposition F.112 yield values one larger than Proposition F.110 does.

For an upper bound, we have the following conjecture of Hamidoune (see Conjecture 1.1 in [112]):

Conjecture F.113 (Hamidoune; cf. [112]) *For all integers $n \geq t$ we have*

$$\tau\hat{}(\mathbb{Z}_n, [1,t]) \leq (n-2)/t + t - 1.$$

In particular, for $t = 3$ we believe that

$$\tau\hat{}(\mathbb{Z}_n, [1,3]) \leq (n+4)/3,$$

and Proposition F.111 above shows that this inequality, if true, is sharp.

It is a very interesting question to see if Conjecture F.113 is sharp in general:

Problem F.114 *Find all pairs of integers t and n for which a weak zero-$[1,t]$-sum-free subset of size $\lfloor (n-2)/t \rfloor + t - 1$ exists in the cyclic group $\mathbb{Z}_n$.*

A less exact upper bound was established by Alon in [1]: if n is large enough (depending on t and any fixed positive real number ϵ), then

$$\tau\hat{}(\mathbb{Z}_n, [1,t]) \leq (1/t + \epsilon) \cdot n.$$

Lev generalized this to arbitrary finite abelian groups as follows:

Theorem F.115 (Lev; cf. [143]) *For any t and positive real number ϵ, there is a constant $n_0(t,\epsilon)$ so that*

$$\tau\hat{}(G, [1,t]) \leq (1/t + \epsilon) \cdot n$$

holds for every $n \geq n_0(t,\epsilon) \cdot |\mathrm{Ord}(G,2)|$.

We have very few exact values known for $\tau\hat{}(G,[1,t])$, particularly when G is not cyclic. We mention the following result:

Proposition F.116 (Yin; cf. [202]) *For all positive integers r, we have $\tau\hat{}(\mathbb{Z}_2^r, [1,3]) = 2^{r-1}$.*

We present the short proof on page 363.

For the group $\mathbb{Z}_3^3$, Bhowmik and Schlage-Puchta computed in [34] three values:

$$\tau\hat{}(\mathbb{Z}_3^3, [1,3]) = 8, \ \tau\hat{}(\mathbb{Z}_3^3, [1,4]) = 7, \ \tau\hat{}(\mathbb{Z}_3^3, [1,5]) = 7.$$

In [35], the same authors prove that any weak zero-$[1,3]$-free subset of size 8 in $\mathbb{Z}_3^3$ is of the form

$$\{a_1, a_2, a_3, a_1+a_2, a_1+a_2+a_3, a_1+2a_2+a_3, 2a_1+a_3, a_2+2a_3\}.$$

The general problem of finding the exact value of $\tau\hat{}(G,[1,t])$ is still wide open:

Problem F.117 *Find $\tau\hat{}(G,[1,t])$ for all groups G and integers $t \geq 3$.*

F.3.3 Arbitrary number of terms

Here we investigate the maximum value of m for which there exists an m-subset A of G that is *zero-sum-free* in G; that is, for which its sumset

$$\Sigma^* A = \cup_{h=1}^{\infty} h\hat{\ } A$$

does not contain zero:

$$\tau\hat{\ }(G, \mathbb{N}) = \max\{|A| \mid A \subseteq G, 0 \notin \Sigma^* A\}.$$

The quantity $\tau\hat{\ }(G, \mathbb{N}) + 1$, that is, the smallest integer for which for any subset A of G of that size (or more) one has $0 \in \Sigma^* A$, is called the *Olson's constant* of G, and has been the subject of much attention since Erdős and Heilbronn first discussed it in [80] in 1964. (The term "Olson's constant" was coined at a 1994 meeting in Venezuela to honor Olson who contributed tremendously to this and other questions in additive combinatorics.)

Let us consider the cyclic group first. It is easy to see that if $1 + 2 + \cdots + m < n$, then the set $\{1, 2, \ldots, m\}$ is zero-sum-free in $\mathbb{Z}_n$, and therefore we get:

Proposition F.118 *For all positive integers n we have*

$$\tau\hat{\ }(\mathbb{Z}_n, \mathbb{N}) \geq \left\lfloor (\sqrt{8n - 7} - 1)/2 \right\rfloor.$$

Selfridge (see Problem C.15 in [105]; cf. also page 95 in [79]) offered two other constructions that are sometimes better than the one above. First, observe that there is no harm replacing the element 2 with -2, so when $1 + 3 + 4 + \cdots + m < n$, then the set $\{1, -2, 3, \ldots, m\}$ is zero-sum-free in $\mathbb{Z}_n$. (For $-2 \notin \{1, 3, 4, \ldots, m\}$ we must have $n \geq 6$.) We thus get:

Proposition F.119 (Selfridge; cf. Problem C.15 in [105]) *For all positive integers $n \geq 6$ we have*

$$\tau\hat{\ }(\mathbb{Z}_n, \mathbb{N}) \geq \left\lfloor (\sqrt{8n + 9} - 1)/2 \right\rfloor.$$

The other construction of Selfridge assumes that n is even, and considers the set

$$A = \begin{cases} \{1, 2, \ldots, (m-1)/2\} \cup \{n/2, n/2 + 1, \ldots, n/2 + (m-1)/2\} & \text{if } m \text{ is odd;} \\ \{1, 2, \ldots, m/2 - 1\} \cup \{n/2, n/2 + 1, \ldots, n/2 + m/2\} & \text{if } m \text{ is even.} \end{cases}$$

(When $m = 1$, we simply take $A = \{n/2\}$.)

Clearly, A is zero-sum-free in $\mathbb{Z}_n$ when m is odd and

$$2 \cdot (1 + 2 + \cdots + (m-1)/2) < n/2,$$

or when m is even and

$$2 \cdot (1 + 2 + \cdots + (m/2 - 1)) + m/2 < n/2.$$

Therefore, we can say that

$$\tau\hat{\ }(\mathbb{Z}_n, \mathbb{N}) \geq \begin{cases} \left\lfloor \sqrt{2n - 3} \right\rfloor & \text{if this value is odd;} \\ \left\lfloor \sqrt{2n - 4} \right\rfloor & \text{if this value is even.} \end{cases}$$

(Note that these two conditions are not mutually exclusive, but, as the next few lines demonstrate, they do cover all cases.) Observe that $\left\lfloor \sqrt{2n-3} \right\rfloor$ and $\left\lfloor \sqrt{2n-4} \right\rfloor$ are only unequal when $2n - 3$ happens to be a square number, but when that is the case, then it must be odd (and thus $\left\lfloor \sqrt{2n-3} \right\rfloor$ is odd as well). Therefore, we can conclude:

Proposition F.120 (Selfridge; cf. Problem C.15 in [105]) *For all positive even integers n we have*

$$\tau\hat{}(\mathbb{Z}_n, \mathbb{N}) \geq \left\lfloor \sqrt{2n-3} \right\rfloor.$$

We can also verify that Proposition F.119 never trumps Proposition F.120; indeed, for all $n \geq 14$, we have

$$\sqrt{2n-3} \geq (\sqrt{8n+9}-1)/2.$$

(For $n \in \{6, 8, 10, 12\}$ the two floors are equal.) We are not aware of any even n for which Proposition F.120 is not exact, and we have:

Conjecture F.121 (Selfridge; cf. Problem C.15 in [105]) *For all positive even integers n we have*

$$\tau\hat{}(\mathbb{Z}_n, \mathbb{N}) = \left\lfloor \sqrt{2n-3} \right\rfloor.$$

Subocz in [193] verified Conjecture F.121 for all even $n \leq 64$.

Problem F.122 *Prove (or disprove) Conjecture F.121.*

The case when n is odd seems more complicated. From Subocz's paper [193] we see that Proposition F.119 gives the correct value of $\tau\hat{}(\mathbb{Z}_n, \mathbb{N})$ for all odd n with $7 \leq n \leq 63$, except for $n = 25$; that exception is demonstrated by the fact that

$$\{1, 6, 11, 16, 21\} \cup \{5, 10\}$$

is zero-sum-free in $\mathbb{Z}_{25}$.

After a variety of partial results, the case when n is prime is now settled: As Balandraud pointed out in [25] (cf. [26]) and as we now explain, for a prime p, the value of $\tau\hat{}(\mathbb{Z}_p, \mathbb{N})$ easily follows from his Theorem D.120. The cases of $p \in \{2, 3, 5\}$ can be easily evaluated, so we assume that $p \geq 7$. Given Proposition F.119, we only need to prove that if $A \subseteq \mathbb{Z}_p$ has size

$$|A| \geq \left\lfloor (\sqrt{8p+9}-1)/2 \right\rfloor + 1 > (\sqrt{8p+9}-1)/2,$$

then A is not zero-sum-free in $\mathbb{Z}_p$.

Clearly, if A is not asymmetric, that is, if A and $-A$ are not disjoint, then $0 \in 2\hat{}A$, and A is not zero-sum-free. On the other hand, if A is asymmetric, then $|A| \leq (p-1)/2$, and by Theorem D.120, $\Sigma^* A$ has size at least

$$\min\{p, |A| \cdot (|A|+1)/2\} \geq \min\left\{p, (\sqrt{8p+9}-1) \cdot (\sqrt{8p+9}+1)/8\right\} = \min\{p, p+1\} = p.$$

But this means that $\Sigma^* A = \mathbb{Z}_p$, thus A is not zero-sum-free.

Theorem F.123 (Balandraud; cf. [25], [26]) *We have $\tau\hat{}(\mathbb{Z}_2, \mathbb{N}) = 1$, $\tau\hat{}(\mathbb{Z}_3, \mathbb{N}) = 1$, and $\tau\hat{}(\mathbb{Z}_5, \mathbb{N}) = 2$; furthermore, for every prime $p \geq 7$, we have*

$$\tau\hat{}(\mathbb{Z}_p, \mathbb{N}) = \left\lfloor (\sqrt{8p+9}-1)/2 \right\rfloor.$$

Perhaps it is worth pointing out that the $+9$ in the expression above can be omitted, since for every prime $p \geq 7$,

$$\left\lfloor (\sqrt{8p+9}-1)/2 \right\rfloor = \left\lfloor \sqrt{2p} - 1/2 \right\rfloor.$$

To see this, note that for $(\sqrt{8p+c}-1)/2$ to equal an integer k, one must have $c \equiv 1 \mod 8$, and if $(\sqrt{8p+9}-1)/2$ or $(\sqrt{8p+1}-1)/2$ equals k, then p equals $(k+2)(k-1)/2$ or $k(k+1)/2$, but neither of these quantities can equal a prime $p \geq 7$. Dropping the $+9$ from our formula also has the advantage that the cases of $p \in \{2, 3, 5\}$ do not need to be separated, thus we get:

Corollary F.124 (Balandraud; cf. [25], [26]) *For every prime p, we have*

$$\tau\hat{}(\mathbb{Z}_p, \mathbb{N}) = \left\lfloor \sqrt{2p} - 1/2 \right\rfloor.$$

The question of finding the exact value of $\tau\hat{}(\mathbb{Z}_n, \mathbb{N})$ for composite n remains open:

Problem F.125 *Evaluate $\tau\hat{}(\mathbb{Z}_n, \mathbb{N})$ for odd composite n. In particular, find all instances when*

$$\tau\hat{}(\mathbb{Z}_n, \mathbb{N}) > \left\lfloor (\sqrt{8n+9} - 1)/2 \right\rfloor.$$

Let us now move on to noncyclic groups. First, following Gao, Ruzsa, and Thangadurai (see [96]), we establish a recursive lower bound for $\tau\hat{}(G, \mathbb{N})$. Suppose that G has exponent κ and $G \cong G_1 \times \mathbb{Z}_\kappa$. Let A_1 be a zero-sum-free set in G_1 and, if possible, let A_2 be any subset of G_1 of size $\kappa - 1$; if $|G_1| < \kappa - 1$, then let $A_2 = G_1$. It is then easy to see that

$$A = (A_1 \times \{0\}) \cup (A_2 \times \{1\})$$

is zero-sum-free in $G_1 \times \mathbb{Z}_\kappa$. Therefore:

Proposition F.126 *For every finite abelian group G_1 and positive integer κ we have*

$$\tau\hat{}(G_1 \times \mathbb{Z}_\kappa, \mathbb{N}) \geq \tau\hat{}(G_1, \mathbb{N}) + \min\{|G_1|, \kappa - 1\}.$$

Consequently, we have the following lower bound for groups of the form $\mathbb{Z}_k^r$:

Corollary F.127 (Gao, Ruzsa, and Thangadurai; cf. [96]) *For all $k \geq 2$ and $r \geq 2$, we have*

$$\tau\hat{}(\mathbb{Z}_k^r, \mathbb{N}) \geq \tau\hat{}(\mathbb{Z}_k^{r-1}, \mathbb{N}) + k - 1.$$

Furthermore, Gao, Ruzsa, and Thangadurai in [96] proved that when k is a very large prime and $r = 2$, then equality holds in Corollary F.127; the requirement on the prime being large was then greatly reduced by Bhowmik and Schlage-Puchta in [36], though it is still formidable:

Theorem F.128 (Bhowmik and Schlage-Puchta; cf. [36]) *For every prime $p >$ 6000, we have*

$$\tau\hat{}(\mathbb{Z}_p^2, \mathbb{N}) = \tau\hat{}(\mathbb{Z}_p, \mathbb{N}) + p - 1.$$

Combining this with Corollary F.124 yields:

Corollary F.129 (Bhowmik and Schlage-Puchta; cf. [36]) *For every prime $p >$ 6000, we have*

$$\tau\hat{}(\mathbb{Z}_p^2, \mathbb{N}) = p + \left\lfloor \sqrt{2p} - 1/2 \right\rfloor - 1.$$

The obvious question here is whether the bound on p can be reduced:

Problem F.130 *Prove that the conclusion of Corollary F.129 (or, equivalently, of Theorem F.128) holds for primes $p < 6000$.*

In fact, we believe that the same equality holds for composite values of k as well:

Conjecture F.131 (Gao, Ruzsa, and Thangadurai; cf. [96]) *For all values of $k \geq 2$, we have*

$$\tau\hat{}(\mathbb{Z}_k^2, \mathbb{N}) = \tau\hat{}(\mathbb{Z}_k, \mathbb{N}) + k - 1.$$

Problem F.132 *Prove (or disprove) Conjecture F.131.*

The paper [96] actually had a stronger version of Conjecture F.131: the authors conjectured that equality always holds in Corollary F.127. However, as it was first pointed out by Gao and Geroldinger in [91], for every prime power $k \geq 3$ there exists an $r \geq 2$ for which the claim fails. Furthermore, as Ordaz et al. proved in [168] and as we review below, even for $r = 3$, equality in Corollary F.127 can only hold for at most finitely many prime values of k (possibly only for $k = 2$).

Ordaz at al. proved lower bounds for $\tau\hat{}(\mathbb{Z}_k^r, \mathbb{N})$ for all $r \geq 3$:

Theorem F.133 (Ordaz, Phillipp, Santos, and Schmid; cf. [168]) *For all integers $k \geq 2$ we have*

$$\tau\hat{}(\mathbb{Z}_k^3, \mathbb{N}) \geq \begin{cases} 3(k-1) & \text{if } k \leq 3; \\ 2k + \left\lfloor \left(\sqrt{8k-31} - 1\right)/2 \right\rfloor & \text{if } k \geq 4. \end{cases}$$

Theorem F.134 (Ordaz, Phillipp, Santos, and Schmid; cf. [168]) *For all integers $k \geq 2$ and $r \geq 4$ we have*

$$\tau\hat{}(\mathbb{Z}_k^r, \mathbb{N}) \geq \begin{cases} r(k-1) & \text{if } k \leq r+1; \\ (r-1)k + \left\lfloor \left(\sqrt{8(k-r)+1} - 1\right)/2 \right\rfloor & \text{if } k \geq r+2. \end{cases}$$

(Observe that there are slight discrepancies between the formulae as well as the conditions of Theorems F.133 and F.134.)

Combining Theorem F.133 and Corollary F.129, for a prime $p > 6000$ we thus get:

$$\tau\hat{}(\mathbb{Z}_p^3, \mathbb{N}) \geq 2p + \left\lfloor \left(\sqrt{8p-31} - 1\right)/2 \right\rfloor > 2p + \left\lfloor \sqrt{2p} - 1/2 \right\rfloor - 2 = \tau\hat{}(\mathbb{Z}_p^2, \mathbb{N}) + p - 1,$$

which shows that equality in Corollary F.127 does not hold for primes $p > 6000$. This disproves the conjecture in [96] that we mentioned above.

The authors of [168] also believe (though shy away from a conjecture) that equality holds everywhere in Theorems F.133 and F.134; we thus have the following interesting question:

Problem F.135 *Decide whether equality always holds in Theorems F.133 and F.134.*

We do know that equality holds in Theorems F.133 and F.134 for $k \in \{2, 3, 4, 5\}$ and in Theorem F.133 when $k \in \{6, 7\}$:

Theorem F.136 (Subocz; cf. [193]) *For every $r \geq 3$ we have $\tau\hat{}(\mathbb{Z}_2^r, \mathbb{N}) = r$ and $\tau\hat{}(\mathbb{Z}_3^r, \mathbb{N}) = 2r$.*

Theorem F.137 (Ordaz, Phillipp, Santos, and Schmid; cf. [168]) *For every $r \geq 4$ we have $\tau\hat{}(\mathbb{Z}_4^r, \mathbb{N}) = 3r$ and $\tau\hat{}(\mathbb{Z}_5^r, \mathbb{N}) = 4r$. Furthermore, $\tau\hat{}(\mathbb{Z}_4^3, \mathbb{N}) = 8$, $\tau\hat{}(\mathbb{Z}_5^3, \mathbb{N}) = 11$, $\tau\hat{}(\mathbb{Z}_6^3, \mathbb{N}) = 13$, and $\tau\hat{}(\mathbb{Z}_7^3, \mathbb{N}) = 16$.*

We also have the following upper bound:

Theorem F.138 (Ordaz, Phillipp, Santos, and Schmid; cf. [168]) *For every $r \geq 1$ and prime power q,*

$$\tau\hat{}(\mathbb{Z}_q^r, \mathbb{N}) \leq r(q-1).$$

Combining Theorems F.138 and F.134 yields:

Theorem F.139 (Ordaz, Phillipp, Santos, and Schmid; cf. [168]) *For every $r \geq 4$ and prime power q for which $q \leq r+1$,*

$$\tau\hat{}(\mathbb{Z}_q^r, \mathbb{N}) = r(q-1).$$

We mention some other exact results:

Theorem F.140 (Ordaz, Phillipp, Santos, and Schmid; cf. [168]) *For every $r_1, r_2 \geq 1$, we have*

$$\tau\hat{}(\mathbb{Z}_2^{r_1} \times \mathbb{Z}_4^{r_2}, \mathbb{N}) = r_1 + 3r_2.$$

Theorem F.141 (Ordaz, Phillipp, Santos, and Schmid; cf. [168]) *Suppose that G has invariant factorization*

$$\mathbb{Z}_p^{r_1} \times \mathbb{Z}_{p^2}^{r_2} \times \cdots \times \mathbb{Z}_{p^k}^{r_k}$$

where p is prime, $r_1, \ldots, r_k \geq 0$, $r_k > 0$, and $r_1 + \cdots + r_k \geq p^k$. Then we have

$$\tau\hat{}(G, \mathbb{N}) = r_1(p-1) + r_2(p^2-1) + \cdots + r_k(p^k-1).$$

(This last result on $\tau\hat{}(G, \mathbb{N})$ for p-groups, under stronger assumptions on the rank of G, was given by Gao and Geroldinger as Corollary 7.4 in [90].)

Note that by Theorem F.141, it suffices to determine $\tau\hat{}(\mathbb{Z}_3^{r_1} \times \mathbb{Z}_9^{r_2}, \mathbb{N})$ for $r_1 + r_2 \leq 8$:

Problem F.142 *Evaluate $\tau\hat{}(\mathbb{Z}_3^{r_1} \times \mathbb{Z}_9^{r_2}, \mathbb{N})$ for all $r_1, r_2 \geq 1$ with $r_1 + r_2 \leq 8$.*

Problem F.143 *Evaluate $\tau\hat{}(\mathbb{Z}_6^r, \mathbb{N})$ for all $r \geq 4$.*

Turning now to general finite abelian groups, we first pose a 1973 conjecture of Erdős:

Conjecture F.144 (Erdős; cf. [78]) *For every finite abelian group of order n, we have*

$$\tau\hat{}(G, \mathbb{N}) < \sqrt{2n}.$$

According to Corollary F.129, Conjecture F.144 holds for all prime values of $n > 6000$; furthermore, we see from Subocz's tables in [193] that the conjecture holds for all groups of order $n \leq 50$ and all cyclic groups of order $n \leq 64$.

In 1975, Olson proved the following:

Theorem F.145 (Olson; cf. [167]) *For every G we have*

$$\tau\hat{}(G, \mathbb{N}) < 3\sqrt{n}.$$

The best current general result is the following:

Theorem F.146 (Hamidoune and Zémor; cf. [115]) *There exists a positive real number C for which*

$$\tau\hat{}(G, \mathbb{N}) < \sqrt{2n} + C \cdot \sqrt[3]{n} \ln n$$

holds for every finite abelian group of order n.

We also mention the following interesting conjecture:

Conjecture F.147 (Subocz; cf. [193]) *For every finite abelian group of order n, we have*

$$\tau\hat{}(G, \mathbb{N}) \leq \tau\hat{}(\mathbb{Z}_n, \mathbb{N}).$$

We should note that the difference between $\tau\hat{}(G,\mathbb{N})$ (for $|G| = n$) and $\tau\hat{}(\mathbb{Z}_n,\mathbb{N})$ can be arbitrarily large: For example, we have $\tau\hat{}(\mathbb{Z}_2^r,\mathbb{N}) = r$ (see Theorem F.136) but $\tau\hat{}(\mathbb{Z}_{2^r},\mathbb{N}) \sim 2^{(r+1)/2}$ (see Proposition F.120).

In closing this section, we mention that results regarding the structure of zero-sum-free subsets A of $\mathbb{Z}_p$ (with p prime) for which $|A|$ is close to $\tau\hat{}(\mathbb{Z}_p,\mathbb{N})$ are discussed by Deshouillers and Prakash in [64]; by Nguyen, Szemerédi, and Vu in [163]; and by Nguyen and Vu in [164]. For example, the analogue of Theorem E.106 is:

Theorem F.148 (Nguyen and Vu; cf. [164]) *There is a positive constant c, so that whenever A is a zero-sum-free subset of $\mathbb{Z}_p$ for an odd prime p, then there is a subset A' of A of size at most $cp^{6/13}\log p$ and an element $b \in \{1,\dots,p-1\}$, for which*

$$||b \cdot (A \setminus A')|| < p.$$

Zero-sum-free subsets of $\mathbb{Z}_p^2$ (with $p > 6000$ prime) that have size exactly $\tau\hat{}(\mathbb{Z}_p^2,\mathbb{N})$ are classified by Bhowmik and Schlage-Puchta in [36] and [37].

We offer the following difficult problems:

Problem F.149 *Prove Conjecture F.144, or at least improve on the bound in Theorem F.146.*

Problem F.150 *for each group G, classify all zero-sum-free sets in G of size exactly $\tau\hat{}(G,\mathbb{N})$.*

In particular, one may be able to improve on Theorem F.148:

Problem F.151 *For any odd prime p, classify all zero-sum-free sets in $\mathbb{Z}_p$ of size exactly $\tau\hat{}(\mathbb{Z}_p,\mathbb{N}) = \lfloor\sqrt{2p} - 1/2\rfloor$.*

F.4 Restricted signed sumsets

In this section we investigate the quantity

$$\tau\hat{\pm}(G,H) = \max\{|A| \mid A \subseteq G, 0 \notin H\hat{\pm}A\}$$

(if there is no subset A for which $0 \notin H\hat{\pm}A$, we let $\tau\hat{\pm}(G,H) = 0$). Clearly, we always have $\tau\hat{\pm}(G,H) = 0$ when $0 \in H$; however, when $0 \notin H$ and $n \geq 2$, then $\tau\hat{\pm}(G,H) \geq 1$: for any $a \in G \setminus \{0\}$, for the one-element set $A = \{a\}$ we obviously have $0 \notin H\hat{\pm}A$.

We consider three special cases: when H consists of a single positive integer h, when H consists of all positive integers up to some value t, and when $H = \mathbb{N}$.

F.4.1 Fixed number of terms

Here we investigate, for a given group G and positive integer h, the quantity

$$\tau\hat{\pm}(G,h) = \max\{|A| \mid A \subseteq G, 0 \notin h\hat{\pm}A\}.$$

As before, we can easily verify the following:

Proposition F.152 *In a group of order n, we have*

$$\tau\hat{\pm}(G,1) = n - 1;$$

$$\tau_{\hat{\pm}}(G,2) = \frac{n + |\mathrm{Ord}(G,2)| + 1}{2};$$

and

$$\tau_{\hat{\pm}}(G,h) = n$$

for every $h \geq n+1$.

Next, we evaluate $\tau_{\hat{\pm}}(G,n)$ as follows. Note that, trivially, $\tau_{\hat{\pm}}(G,n) \geq n-1$. Recall also that on page 247 we found that when G has exponent κ and $G \cong G_1 \times \mathbb{Z}_\kappa$, then the elements of $G_1 \times \mathbb{Z}_\kappa$ sum to $(0,0)$ if $|G_1| = n/\kappa$ is even or if κ is odd, and to $(0, \kappa/2)$ when n/κ is odd and κ is even. We can thus also observe that when κ is divisible by 4 and n/κ is odd, then adding all elements of $G_1 \times \mathbb{Z}_\kappa$, except for the element $(0, \kappa/4)$ that we instead subtract, we get $(0,0)$. That leaves us with the case when $\kappa \equiv 2 \bmod 4$ and n/κ is odd. But in this case, no signed sum of the n elements results in $(0,0)$, since the second component of the signed sum will differ from the second component of the sum (which is odd) in an even number (if we subtract an element instead of adding it, the sum gets reduced by an even value). Therefore, we proved:

Proposition F.153 *Suppose that G has order n and exponent κ. Then*

$$\tau_{\hat{\pm}}(G,n) = \begin{cases} n & \textit{if } \kappa \equiv 2 \bmod 4 \textit{ and } n/\kappa \textit{ is odd;} \\ n-1 & \textit{otherwise.} \end{cases}$$

For $3 \leq h \leq n-1$, the value of $\tau_{\hat{\pm}}(G,h)$ is not known in general.

The values of $\tau_{\hat{\pm}}(G,h)$ behave quite differently for even and odd values of h. For example, let us consider $h = 4$ first. Note that, if for a set $A \subseteq G$, we have $0 \notin 4_{\hat{\pm}}A$, then A must be a weak Sidon set in G: Indeed, if we had

$$a_1 + a_2 = a_3 + a_4$$

for some $a_1, a_2, a_3, a_4 \in A$ with $a_1 \neq a_2$ and $a_3 \neq a_4$, then the four elements cannot be all distinct as then we would have $0 \in 4_{\hat{\pm}}A$. Therefore, we must have, for example, $a_1 \in \{a_3, a_4\}$, which proves that A is a weak Sidon set in G. Therefore, by Corollary C.69, we get the following upper bound:

Proposition F.154 *For every G of order n, we have*

$$\tau_{\hat{\pm}}(G,4) \leq \lfloor \sqrt{2n} \rfloor - 1.$$

For odd values of h, however, we may have lower bounds for $\tau_{\hat{\pm}}(G,h)$ that are linear in n. Consider first the cyclic group $\mathbb{Z}_n$ with n even. Note that, when h is odd, for the set

$$A = \{1, 3, 5, \ldots, n-1\}$$

we have $h_{\hat{\pm}}A \subseteq A$; in particular, $0 \notin h_{\hat{\pm}}A$. This shows that

$$\tau_{\hat{\pm}}(\mathbb{Z}_n, h) \geq n/2.$$

On the other hand,

$$\tau_{\hat{\pm}}(G,h) \leq \tau^{\hat{}}(G,h),$$

so by Theorem F.88, when $n \geq 12$ and $3 \leq h \leq n/2 - 2$, then

$$\tau_{\hat{\pm}}(\mathbb{Z}_n, h) \leq n/2.$$

Therefore:

Theorem F.155 *When n is even, h is odd, and $3 \leq h \leq n/2-2$, then*

$$\tau\hat{\pm}(\mathbb{Z}_n, h) = n/2.$$

For the case when n and h are both odd, Collins has the following lower bound:

Proposition F.156 (Collins; cf. [58]) *When n and $h \leq n$ are both odd, we have*

$$\tau\hat{\pm}(\mathbb{Z}_n, h) \geq \left\lfloor \frac{n}{h} + \frac{h^2-3}{2h} \right\rfloor.$$

We present a proof for Proposition F.156 on page 363.

We believe that Proposition F.156 is quite accurate. For example, when p is prime, we know from Theorem F.87 that

$$\tau\hat{\pm}(\mathbb{Z}_p, h) \leq \tau\hat{}(\mathbb{Z}_p, h) = \left\lfloor \frac{p-2}{h} \right\rfloor + h.$$

In particular, for $h = 3$ we get the following:

Corollary F.157 *For every prime p, we have*

$$(p+1)/3 \leq \tau\hat{\pm}(\mathbb{Z}_p, 3) \leq (p+7)/3.$$

We offer the following intriguing problems:

Problem F.158 *Find $\tau\hat{\pm}(\mathbb{Z}_p, 3)$ for each prime p.*

Problem F.159 *Find $\tau\hat{\pm}(\mathbb{Z}_p, h)$ for each prime p and odd $h \leq p$.*

Problem F.160 *For each odd n and odd h with $3 \leq h \leq n-1$, evaluate $\tau\hat{\pm}(\mathbb{Z}_n, h)$.*

Problem F.161 *For each even n and odd h with $n/2-1 \leq h \leq n-1$, evaluate $\tau\hat{\pm}(\mathbb{Z}_n, h)$.*

Problem F.162 *For given n and even h, evaluate, or at least find a good lower bound for $\tau\hat{\pm}(\mathbb{Z}_n, h)$, in particular for $\tau\hat{\pm}(\mathbb{Z}_n, 4)$.*

Let us now turn to noncyclic groups. The general question, of course, is as follows:

Problem F.163 *For each G and h, evaluate, or at least find a good lower bound for $\tau\hat{\pm}(G, h)$, in particular for $\tau\hat{\pm}(G, 3)$.*

The case of $h = \kappa$ attracts special interest. Recall that the value $\tau\hat{}(G, \kappa)+1$ (the smallest integer such that each subset of G of that cardinality has a subset size κ whose elements sum to 0) is called the Harborth constant of G; analogously, here we define the *signed Harborth constant of* G as the value of $\tau\hat{\pm}(G, \kappa) + 1$. This value is determined in Proposition F.153 above for cyclic groups; we have the following additional result:

Theorem F.164 (Marchan, Ordaz, Ramos, and Schmid; cf. [150]) *For a positive integer k, we have*

$$\tau\hat{\pm}(\mathbb{Z}_2 \times \mathbb{Z}_{2k}, 2k) = \begin{cases} 4 & \text{if } k \in \{1,2\}; \\ 2k+1 & \text{if } k \geq 3. \end{cases}$$

Furthermore, in [151] the same authors determined all subsets A of size $2k+1$ in $\mathbb{Z}_2 \times \mathbb{Z}_{2k}$ for which $0 \not\in (2k)\hat{\pm}A$.

We pose the following problem:

Problem F.165 *Find $\tau\hat{\pm}(G, \kappa)$ for each noncyclic group G of exponent $\kappa \geq 3$.*

If $\kappa = 2$, then the answer is given by Proposition F.152:

$$\tau\hat{\pm}(\mathbb{Z}_2^r, 2) = 2^r$$

for every $r \in \mathbb{N}$.

F.4.2 Limited number of terms

The analogue of a t-independent set for restricted addition is called a *weak t-independent set*. In particular, in this section we investigate, for a given group G and positive integer t, the quantity

$$\tau\hat{\pm}(G,[1,t]) = \max\{|A| \mid A \subseteq G, 0 \not\in [1,t]\hat{\pm}A\},$$

that is, the maximum size of a weak t-independent subset of G.

For $t=1$, we see that A is weakly 1-independent if, and only if, $0 \not\in A$. We can easily determine the value of $\tau\hat{\pm}(G,[1,2])$ as well. Obviously, a weak 2-independent set A cannot contain 0; furthermore, for each element g of G, A cannot contain both g and $-g$, unless $g=-g$. So, to get a maximum weak 2-independent set in G, take exactly one of each element or its negative in $G \setminus \mathrm{Ord}(G,2) \setminus \{0\}$, and take all elements of $\mathrm{Ord}(G,2)$. We can summarize these findings as follows:

Proposition F.166 *For all finite abelian groups G of order n we have*

$$\tau\hat{\pm}(G,[1,1]) = n-1$$

and

$$\tau\hat{\pm}(G,[1,2]) = \frac{n+|\mathrm{Ord}(G,2)|-1}{2}.$$

As a special case, for the cyclic group of order n we have

$$\tau\hat{\pm}(\mathbb{Z}_n,[1,2]) = \lfloor n/2 \rfloor.$$

For $t \geq 3$, the value of $\tau\hat{\pm}(G,[1,t])$ is not known. In [24], Bajnok and Ruzsa proved the following general bounds:

Theorem F.167 (Bajnok and Ruzsa; cf. [24]) *For every G and $t \geq 2$ we have*

$$\left(t!/2^t \cdot n\right)^{1/t} - t/2 < \tau\hat{\pm}(G,[1,t]) < \left(\lfloor t/2 \rfloor! \cdot n\right)^{1/\lfloor t/2 \rfloor} + t/2.$$

In particular, for $t \geq 4$, $\tau\hat{\pm}(G,[1,t])$ is not a linear function of n.

For $t=3$, the upper bound in Theorem F.167 is trivial; we can do better as follows. Observe that if A is a weak 3-independent set in G, then the sets $\{0\}$, A, and $2\hat{}A$ must be pairwise disjoint: indeed, if we were to have, say, $a_1+a_2=a_3$ for some $a_1,a_2,a_3 \in A$ with $a_1 \neq a_2$, then we would have to have $a_1=a_3$ or $a_2=a_3$, but that would imply that one of the elements is 0, a contradiction. Note also that for each $a_0 \in A$,

$$\{a_0+a \mid a \in A, a \neq a_0\} \subseteq 2\hat{}A,$$

so $2\hat{}A$ has size at least $|A|-1$. Therefore,

$$n \geq |\{0\} \cup A \cup 2\hat{}A| \geq 1+|A|+|A|-1 = 2|A|,$$

which yields the following upper bound:

Proposition F.168 *For all G of order n we have*

$$\tau\hat{\pm}(G,[1,3]) \leq n/2.$$

A better upper bound for $\tau\hat{\pm}(G,[1,3])$ can be derived in the case when n is odd. Note that, in this case, if A is weakly 3-independent, then A and $-A$ must be disjoint. (This may be false if n is even: the elements of order 2 may belong to both A and $-A$.) Furthermore, the set $A \cup (-A)$ is weakly zero-3-sum-free: the sum of three distinct elements of $A \cup (-A)$ cannot be zero as it is either an element of $A \cup (-A)$ or the signed sum of three distinct elements of A. Therefore, $A \cup (-A)$ has size at most equal to $\tau\hat{}(G,3)$, which yields:

Proposition F.169 *For all G of odd order n we have*

$$\tau\hat{\pm}(G,[1,3]) \leq \tau\hat{}(G,3)/2.$$

For $t = 3$ we can also derive a much better lower bound for $\tau\hat{\pm}(G,[1,3])$ than what Theorem F.167 gives based on the idea that if A is a (regular) 3-independent set in a group G_2, then $G_1 \times A$ is a weak 3-independent set in $G_1 \times G_2$ for any group G_1:

Proposition F.170 *For all groups G_1 and G_2 we have*

$$\tau\hat{\pm}(G_1 \times G_2,[1,3]) \geq |G_1| \cdot \tau_{\pm}(G_2,[1,3]).$$

Since any group G of order n and exponent κ is isomorphic to $G_1 \times \mathbb{Z}_\kappa$ for some group G_1 of order n/κ, Proposition F.170 implies:

Corollary F.171 *For every group G of order n and exponent κ we have*

$$\tau\hat{\pm}(G,[1,3]) \geq n/\kappa \cdot \tau_{\pm}(\mathbb{Z}_\kappa,[1,3]).$$

Recall that the value of $\tau_{\pm}(\mathbb{Z}_\kappa,[1,3])$ was explicitly evaluated in Theorem F.41.

While we have the various bounds just discussed, exact values for $\tau\hat{\pm}(G,[1,3])$ are not known even for cyclic groups:

Problem F.172 *Find the exact values of $\tau\hat{\pm}(\mathbb{Z}_n,[1,3])$.*

Let us now consider noncyclic groups, in particular, groups of the form $\mathbb{Z}_k^r$.

First, we consider $\mathbb{Z}_2^r$. It is not hard to see that the set $\mathbb{Z}_2^{r-1} \times \{1\}$ is weakly 3-independent in $\mathbb{Z}_r^2$ (note that no two distinct elements add to zero), and therefore

$$\tau\hat{\pm}(\mathbb{Z}_2^r,[1,3]) \geq 2^{r-1}.$$

Together with Proposition F.168, this implies:

Proposition F.173 *For every $r \geq 1$ we have*

$$\tau\hat{\pm}(\mathbb{Z}_2^r,[1,3]) = 2^{r-1}.$$

We can obtain a large weakly 3-independent set in $\mathbb{Z}_3^r$ by using Bui's idea described on page 253: (using notations described there) the set $\cup_{i \in I} A_i$ is weakly 3-independent in $\mathbb{Z}_3^r$, yielding the following lower bound.

Theorem F.174 *Let r be any positive integer, and write $r = 3q + s$ with $s = 0, 1,$ or 2. Then we have*

$$\tau\hat{\pm}(\mathbb{Z}_3^r,[1,3]) \geq \sum_{i=q+s-1}^{2q+s-1} \binom{r}{i}.$$

The lower bounds provided by Theorem F.174 are quite good. Indeed, by Proposition F.169 above and considering the values for $\tau\hat{}(\mathbb{Z}_3^r,3)$ on page 253, we see that they give the exact values for $r \leq 4$:

$$\tau\hat{\pm}(\mathbb{Z}_3,[1,3]) = 1,\ \tau\hat{\pm}(\mathbb{Z}_3^2,[1,3]) = 2,\ \tau\hat{\pm}(\mathbb{Z}_3^3,[1,3]) = 4,\ \tau\hat{\pm}(\mathbb{Z}_3^4,[1,3]) = 10.$$

For $k \geq 4$, we know no better lower bounds for $\tau\hat{\pm}(\mathbb{Z}_k^r,[1,3])$ than what Corollary F.171 above provides.

We pose the following problems:

Problem F.175 *Find the exact values of* $\tau\hat{_\pm}(\mathbb{Z}_3^r, [1,3])$ *for* $r \geq 5$.

Problem F.176 *Find the exact values of* $\tau\hat{_\pm}(\mathbb{Z}_4^r, [1,3])$, $\tau\hat{_\pm}(\mathbb{Z}_5^r, [1,3])$, *etc.*

Problem F.177 *Find* $\tau\hat{_\pm}(G, [1,3])$ *for other noncyclic groups* G.

Of course, the ultimate goal is to answer this question:

Problem F.178 *Find* $\tau\hat{_\pm}(G, [1,t])$ *for all groups* G *and integers* $t \geq 3$.

As a modest step toward Problem F.178, we establish:

Proposition F.179 *If* n *and* t *are positive integers so that* $2^{t-1} \leq n < 2^t$, *then*

$$\tau\hat{_\pm}(\mathbb{Z}_n, [1,t]) = t - 1.$$

The short proof of Proposition F.179 is on page 365.

F.4.3 Arbitrary number of terms

In this subsection we are trying to evaluate

$$\tau\hat{_\pm}(G, \mathbb{N}) = \max\{|A| \mid A \subseteq G, 0 \not\in \cup_{h=1}^{\infty} h\hat{_\pm}A\}.$$

A subset A of G for which

$$0 \not\in \cup_{h=1}^{\infty} h\hat{_\pm}A$$

holds is called a *dissociated* subset of G. Recall that, by Proposition C.77, our condition is equivalent to

$$|\Sigma A| = 2^{|A|}.$$

Indeed, we have

$$\tau\hat{_\pm}(G, \mathbb{N}) = \sigma\hat{_\pm}(G, \mathbb{N}_0).$$

The case when G is cyclic is easy. Suppose that $G = \mathbb{Z}_n$, and let

$$m = \lfloor \log_2 n \rfloor;$$

note that we then have

$$2^m \leq n < 2^{m+1}.$$

According to Proposition F.179, we have

$$\tau\hat{_\pm}(\mathbb{Z}_n, [1, m+1]) = m.$$

In particular, there is an m-subset A of $\mathbb{Z}_n$ (namely, as the proof of Proposition F.179 shows,

$$A = \{1, 2, \ldots, 2^{m-1}\}$$

works) for which $0 \not\in [1, m+1]\hat{_\pm}A$. But, since $|A| = m$,

$$[1, m+1]\hat{_\pm}A = \cup_{h=1}^{\infty} h\hat{_\pm}A,$$

which proves that $\tau\hat{_\pm}(\mathbb{Z}_n, \mathbb{N}) \geq m$.

On the other hand, clearly,

$$\tau\hat{_\pm}(\mathbb{Z}_n, \mathbb{N}) \leq \tau\hat{_\pm}(\mathbb{Z}_n, [1, m+1]) = m.$$

Thus we have:

Proposition F.180 *For every positive integer n,*

$$\tau\hat{\pm}(\mathbb{Z}_n, \mathbb{N}) = \lfloor \log_2 n \rfloor.$$

We do not quite have the value of $\tau\hat{\pm}(G, \mathbb{N})$ for noncyclic groups G, but we can find lower and upper bounds as follows. For an upper bound, note that if A is dissociated in G, then $|\Sigma A| = 2^m$, so $\tau\hat{\pm}(G, \mathbb{N}) \leq \lfloor \log_2 n \rfloor$.

For a lower bound, let us suppose that G is of type $(n_1, \ldots, n_r)$; that is, we have integers $n_1, \ldots, n_r$ so that $2 \leq n_1$ and n_i is a divisor of n_{i+1} for each $i = 1, 2, \ldots, r-1$, and for which

$$G \cong \mathbb{Z}_{n_1} \times \cdots \mathbb{Z}_{n_r}.$$

Let $m_i = \lfloor \log_2 n_i \rfloor$. It is easy to see that the set

$$\left(\{1, 2, \ldots, 2^{m_1 - 1}\} \times \{0\}^{r-1}\right) \cup \cdots \cup \left(\{0\}^{r-1} \times \{1, 2, \ldots, 2^{m_r - 1}\}\right)$$

is dissociated in G, and thus

$$\tau\hat{\pm}(G, \mathbb{N}) \geq m_1 + \cdots + m_r.$$

Therefore, we get:

Proposition F.181 *Suppose that G is an abelian group of type $(n_1, \ldots, n_r)$ and order n. We then have*

$$\lfloor \log_2 n_1 \rfloor + \cdots + \lfloor \log_2 n_r \rfloor \leq \tau\hat{\pm}(G, \mathbb{N}) \leq \lfloor \log_2 n \rfloor.$$

Considering that

$$\log_2 n_1 + \cdots + \log_2 n_r = \log_2(n_1 \cdots n_r) = \log_2 n,$$

we can say that the lower and upper bounds in Proposition F.181 are close. However, they are not equal: For example, for the groups $\mathbb{Z}_6^2$ or $\mathbb{Z}_7^2$, the lower bound equals 4 and the upper bound equals 5, and, indeed, the set

$$\{(0,1), (1,0), (1,2), (1,4), (3,2)\}$$

(for example) is dissociated in both groups.

Hence the general question remains open:

Problem F.182 *Find the value of $\tau\hat{\pm}(G, \mathbb{N})$ for noncyclic groups G.*

Let us now turn to the inverse problem of classifying all dissociated subsets of G of maximum size. We consider cyclic groups first; in particular, the cyclic group of order $n = 2^k$ where $k \in \mathbb{N}$—note that, by Proposition F.180, we have $\tau\hat{\pm}(\mathbb{Z}_{2^k}, \mathbb{N}) = k$.

We provide the following recursive construction for a collection $\mathcal{A}_k$ of k-subsets of $\mathbb{Z}_{2^k}$:

- We let $\mathcal{A}_1$ consist of the single subset $\{1\}$ of $\mathbb{Z}_2$.
- Suppose that $\mathcal{A}_j$ is already constructed for some positive integer j. For a given member $A_j = \{a_1, \ldots, a_j\}$ of $\mathcal{A}_j$ and for a given element $\epsilon = (\epsilon_1, \ldots, \epsilon_j)$ of $\mathbb{Z}_2^j$, we define the set

  $$A_{j+1}(A_j, \epsilon) = \{2^j\} \cup \{a_1 + \epsilon_1 \cdot 2^j, \ldots, a_j + \epsilon_j \cdot 2^j\};$$

  and then set

  $$\mathcal{A}_{j+1} = \{A_{j+1}(A_j, \epsilon) \mid A_j \in \mathcal{A}_j \text{ and } \epsilon \in \mathbb{Z}_2^j\}.$$

So, for $k = 2$, we get

$$\mathcal{A}_2 = \{\{2\} \cup \{1 + 0 \cdot 2\}, \{2\} \cup \{1 + 1 \cdot 2\} = \{\{1, 2\}, \{2, 3\}\};$$

and for $k = 3$, we have

$$\begin{aligned} \mathcal{A}_3 &= \{\{4\} \cup \{1 + 0 \cdot 4, 2 + 0 \cdot 4\}, \{4\} \cup \{1 + 0 \cdot 4, 2 + 1 \cdot 4\}, \{4\} \cup \{1 + 1 \cdot 4, 2 + 0 \cdot 4\}, \\ &\quad \{4\} \cup \{1 + 1 \cdot 4, 2 + 1 \cdot 4\}, \{4\} \cup \{2 + 0 \cdot 4, 3 + 0 \cdot 4\}, \{4\} \cup \{2 + 0 \cdot 4, 3 + 1 \cdot 4\}, \\ &\quad \{4\} \cup \{2 + 1 \cdot 4, 3 + 0 \cdot 4\}, \{4\} \cup \{2 + 1 \cdot 4, 3 + 1 \cdot 4\}\} \\ &= \{\{1, 2, 4\}, \{1, 4, 6\}, \{2, 4, 5\}, \{4, 5, 6\}, \{2, 3, 4\}, \{2, 4, 7\}, \{3, 4, 6\}, \{4, 6, 7\}\}. \end{aligned}$$

We have the following conjecture for dissociated subsets of $\mathbb{Z}_{2^k}$ of maximum size

$$\tau \hat{_\pm}(\mathbb{Z}_{2^k}, \mathbb{N}) = k.$$

Conjecture F.183 *A k-subset A of $\mathbb{Z}_{2^k}$ is dissociated if, and only if, $A \in \mathcal{A}_k$.*

Using the computer program [120], we have verified Conjecture F.183 for $k \in \{1, 2, 3, 4\}$. Note that Conjecture F.183 implies that there are exactly $2^{(k^2-k)/2}$ dissociated subsets of $\mathbb{Z}_{2^k}$ of maximum size.

Problem F.184 *Prove (or disprove) Conjecture F.183.*

Observe that, if Conjecture F.183 is true, then every dissociated subset of $\mathbb{Z}_{2^k}$ of maximum size contains the element 2^{k-1}. As a modest step toward Conjecture F.183, we state:

Conjecture F.185 *Every dissociated subset of $\mathbb{Z}_{2^k}$ of size k contains the element 2^{k-1}.*

Problem F.186 *Prove (or disprove) Conjecture F.185.*

More generally:

Problem F.187 *For each positive integer n, classify all dissociated subsets of $\mathbb{Z}_n$ of maximum size $\lfloor \log_2 n \rfloor$.*

There is an interest in finding the largest dissociated subsets in any subset of G, not just G itself. For a subset A of G, we let

$$\dim A = \max\{|B| \mid B \subseteq A, B \text{ is dissociated}\}$$

denote the *dissociativity dimension* of A in G. (We mention in passing that there are other notions for the dimension of a set: cf. [50] and [185].) Of course,

$$\dim G = \tau \hat{_\pm}(G, \mathbb{N}).$$

As a (less tight) generalization of Proposition F.181, Lev and Yuster proved the following:

Proposition F.188 (Lev and Yuster; cf. [145]) *For any subset A of G, we have*

$$r_A \leq \dim A \leq \lfloor r_A \cdot \log_2 \kappa \rfloor,$$

where κ is the exponent of G and r_A is the rank of the subgroup $\langle A \rangle$ generated by A.

For any group G of order n and for any positive integer $m \leq n$, we introduce the following quantity:

$$\dim(G,m) = \min\{\dim A \mid A \subseteq G, |A| = m\}.$$

Our question is as follows:

Problem F.189 *For every group G of order n and for any positive integer $m \leq n$, find $\dim(G,m)$.*

Clearly, $\dim(G,1) = 0$ in every group G, since the only dissociated subset of $\{0\}$ is the empty set. It is also easy to see that $\dim(G,2) = 1$ (consider a set $\{0,g\}$ for some non-zero element g). For $\dim(G,3)$, if G has exponent 3 or more, we can take the set $\{0,g,-g\}$ for any $g \in G$ of order 3 or more, so $\dim(G,3) = 1$; if G is the elementary abelian 2-group, then $\dim(G,3) = 2$. We have the following conjecture:

Conjecture F.190 *For all positive integers n and $m \leq n$, we have*

$$\dim(\mathbb{Z}_n, m) = \lfloor \log_2 m \rfloor.$$

For example, we have $\dim(\mathbb{Z}_{10}, 7) = 2$, since each 4-subset (and thus each 7-subset) of $\mathbb{Z}_n$ (when $n \geq 4$) has dimension at least 2 (only subsets of a set of the form $\{0,g,-g\}$ have dimension less than 2), and the set

$$\{0,1,2,3,7,8,9\}$$

has dimension 2.

For $m = n$, Conjecture F.190 becomes Proposition F.180. Furthermore, when $n = 2^k$ for some $k \in \mathbb{N}$, then for $m = n-1$, Conjecture F.190 says that

$$\dim(\mathbb{Z}_{2^k}, 2^k - 1) = k - 1;$$

in other words, one can find a subset of size $2^k - 1$ in $\mathbb{Z}_{2^k}$, which does not have a dissociated subset of size k. Observe that, if Conjecture F.185 is true, then the set $\mathbb{Z}_{2^k} \setminus \{2^{k-1}\}$ is one such set.

Problem F.191 *Prove (or disprove) Conjecture F.190.*

Chapter G

Sum-free sets

Recall that for a given finite abelian group G, m-subset $A = \{a_1, \ldots, a_m\}$ of G, $\Lambda \subseteq \mathbb{Z}$, and $H \subseteq \mathbb{N}_0$, we defined the sumset of A corresponding to Λ and H as

$$H_\Lambda A = \{\lambda_1 a_1 + \cdots + \lambda_m a_m \mid (\lambda_1, \ldots, \lambda_m) \in \Lambda^m(H)\}$$

where the index set $\Lambda^m(H)$ is defined as

$$\Lambda^m(H) = \{(\lambda_1, \ldots, \lambda_m) \in \Lambda^m \mid |\lambda_1| + \cdots + |\lambda_m| \in H\}.$$

In this chapter we consider, for G, H, and Λ, *H-sum-free subsets* of G over Λ; that is, subsets A of G for which

$$(h_1)_\Lambda A \cap (h_2)_\Lambda A = \emptyset$$

for any two distinct elements h_1 and h_2 of H.

Let us make some preliminary observations. First of all, if H contains fewer than two elements, then every subset A of G is H-sum-free over Λ. Second, if $H = \{0, h\}$ for some positive integer h, then for a subset to be H-sum-free is the same as it being zero-h-sum-free, a property we studied in Chapter F. More generally, if $0 \in H$, then a set A being H-sum-free implies that A is zero-H'-sum-free for $H' = H \setminus \{0\}$.

Furthermore, the sum-free property is clearly a weakening of the Sidon property studied in Chapter C: if linear combinations corresponding to different elements of the entire index set $\Lambda^m(H)$ are distinct, then they are certainly distinct when corresponding to distinct h_1 and h_2 of H.

While the sum-free property is thus closely related to properties discussed elsewhere in the book, it offers unique opportunities for the study of interesting and well-known questions in additive combinatorics.

In this chapter we attempt to find $\mu_\Lambda(G, H)$, the maximum possible size of an H-sum-free set over Λ in a given finite abelian group G. If no H-sum-free set exists, we put $\mu_\Lambda(G, H) = 0$. With this notation, our observations above can be stated as follows:

Proposition G.1 *Let G and $\Lambda \subseteq \mathbb{Z}$ be arbitrary.*
If $|H| \leq 1$, then $\mu_\Lambda(G, H) = n$.
If $0 \in H$, then

$$\mu_\Lambda(G, H) \leq \tau_\Lambda(G, H \setminus \{0\});$$

in particular, if $H = \{0, h\}$, then

$$\mu_\Lambda(G, H) = \tau_\Lambda(G, h).$$

Proposition G.2 *For all G, $\Lambda \subseteq \mathbb{Z}$, and $H \subseteq \mathbb{N}_0$ we have*

$$\mu_\Lambda(G, H) \geq \sigma_\Lambda(G, H).$$

In the following sections we attempt to find $\mu_\Lambda(G, H)$ for special coefficient sets Λ.

G.1 Unrestricted sumsets

Our goal in this section is to investigate the maximum possible size of an H-sum-free set over the set of nonnegative integers, that is, the quantity

$$\mu(G, H) = \max\{|A| \mid A \subseteq G; h_1, h_2 \in H; h_1 \neq h_2 \Rightarrow h_1 A \cap h_2 A = \emptyset\}.$$

Clearly, we always have $\mu(G, H) = 0$ whenever H contains two elements h_1 and h_2 whose difference is a multiple of the exponent κ of G, since for any element $a \in G$, we then have $h_1 a = h_2 a$. However, when distinct elements of H leave different remainders when divided by κ, then $\mu(G, H) \geq 1$: for any $a \in G$ with order κ, at least the one-element set $A = \{a\}$ will be H-sum-free.

It is often useful to consider G of the form $G_1 \times G_2$. (We may do so even when G is cyclic if its order has at least two different prime divisors.) It is not hard to see that, if $A_1 \subseteq G_1$ is H-sum-free in G_1, then

$$A = \{(a, g) \mid a \in A_1, g \in G_2\}$$

is H-sum-free in G. Indeed, if we were to have

$$(a_1, g_1) + \cdots + (a_{h_1}, g_{h_1}) = (a'_1, g'_1) + \cdots + (a'_{h_2}, g'_{h_2})$$

for some $h_1, h_2 \in H$, $h_1 \neq h_2$, and $(a_i, g_i), (a'_i, g'_i) \in A$, then the equation for the sum of the first coordinates contradicts the fact that A_1 is H-sum-free in G_1. Thus, we have the following.

Proposition G.3 *For all finite abelian groups G_1 and G_2 and for all $H \subseteq \mathbb{N}_0$ we have*

$$\mu(G_1 \times G_2, H) \geq \mu(G_1, H) \cdot |G_2|.$$

Below we consider two special cases: when H consists of two distinct positive integers (by Proposition G.1, the case when one of the integers equals 0 is identical to Section F.1.1), and when H consists of all positive integers up to some value s. The cases when $H = \mathbb{N}_0$ or $H = \mathbb{N}$, as we just mentioned, yield no H-sum-free sets.

G.1.1 Fixed number of terms

Suppose that k and l are distinct positive integers; without loss of generality, we assume that $k > l$. Sets satisfying the condition $(kA) \cap (lA) = \emptyset$ are called *(k, l)-sum-free sets.* Here we intend to determine the maximum value of m for which G contains a (k, l)-sum-free subset of size m—this value is denoted here by $\mu(G, \{k, l\})$.

There are several similarities between (k, l)-sum-free sets and zero-h-sum-free sets; in fact, in some respects, one might consider a zero-h-sum-free set to be $(h, 0)$-sum-free. Therefore, our discussion in this section will be similar to that in Section F.1.1. This similarity has its limits, however—see, for example, our comments below before Problem G.12 and after Conjecture G.15.

Since k and l are positive integers with $k > l$, the first case to discuss is when $k = 2$ and $l = 1$; a $(2, 1)$-sum-free set in G is simply referred to as a *sum-free set*.

For the cyclic group $\mathbb{Z}_n$, we can find explicit sum-free sets as follows. For every n, the integers that are between $n/3$ and $2n/3$ form a sum-free set; more precisely, the set

$$\left\{\left\lfloor \tfrac{n}{3} \right\rfloor, \left\lfloor \tfrac{n}{3} \right\rfloor + 1, \left\lfloor \tfrac{n}{3} \right\rfloor + 2, \dots, 2\left\lfloor \tfrac{n}{3} \right\rfloor - 1\right\}$$

is sum-free in $\mathbb{Z}_n$. (The integers between $n/6$ and $n/3$, together with those between $2n/3$ and $5n/6$, with the endpoints carefully chosen, provide another sum-free set in $\mathbb{Z}_n$.)

Like in the case of zero-3-sum-free sets, we can do better when n has a prime divisor p which is congruent to 2 mod 3. We see that the set

$$\{(p+1)/3 + pi_1 + i_2 \mid i_1 = 0, 1, \dots, n/p - 1, \ i_2 = 0, 1, \dots, (p-2)/3\}$$

is sum-free.

These examples show that we have

$$\mu(\mathbb{Z}_n, \{2,1\}) \geq \begin{cases} \left(1 + \frac{1}{p}\right) \frac{n}{3} & \text{if } n \text{ has prime divisors congruent to 2 mod 3,} \\ & \text{and } p \text{ is the smallest such divisor,} \\ \left\lfloor \frac{n}{3} \right\rfloor & \text{otherwise;} \end{cases}$$

and Diananda and Yap proved in 1969 (see [65]) that equality holds. Thus, using our function introduced on page 76, we have:

Theorem G.4 (Diananda and Yap; cf. [65]) *For all positive integers n, we have*

$$\mu(\mathbb{Z}_n, \{2,1\}) = v_1(n, 3).$$

More generally, we have the following result.

Theorem G.5 (Bajnok; cf. [13]) *Suppose that k and l are positive integers and $k > l$. Then we have*

$$v_{k-l}(n, k+l) \leq \mu(\mathbb{Z}_n, \{k,l\}) \leq v_1(n, k+l).$$

Of course, when the lower and upper bounds coincide, we get equality:

Corollary G.6 *If $k - l$ and n are relatively prime, then*

$$\mu(\mathbb{Z}_n, \{k,l\}) = v_1(n, k+l).$$

In particular, for all positive integers n and l, we have

$$\mu(\mathbb{Z}_n, \{l+1,l\}) = v_1(n, 2l+1).$$

We note that Corollary G.6 was established in [114] by Hamidoune and Plagne.

As in Theorem F.8, we can determine the size of the largest (k,l)-sum-free set in cyclic groups of prime order:

Theorem G.7 *The size of the largest (k,l)-sum-free set in the cyclic group of prime order p is*

$$\mu(\mathbb{Z}_p, \{k,l\}) = v_{k-l}(p, k+l) = \begin{cases} 0 & \text{if } p | (k-l), \\ \left\lfloor \frac{p-2}{k+l} \right\rfloor + 1 & \text{otherwise.} \end{cases}$$

The value of $\mu(G, \{k,l\})$ for groups of composite order is not known in general. By the following result, it suffices to consider arithmetic progressions in order to determine $\mu(\mathbb{Z}_n, \{k,l\})$:

Theorem G.8 (Bajnok; cf. [13]) *For a given divisor d of n, let $\alpha(\mathbb{Z}_d, \{k,l\})$ be the maximum size of a (k,l)-sum-free arithmetic progression in $\mathbb{Z}_d$. Then*

$$\mu(\mathbb{Z}_n, \{k,l\}) = \max\left\{\alpha(\mathbb{Z}_d, \{k,l\}) \cdot \frac{n}{d} \mid d \in D(n)\right\}.$$

Using Theorem G.8, we can compute $\mu(\mathbb{Z}_n, \{3,1\})$ and $\mu(\mathbb{Z}_n, \{4,1\})$:

Theorem G.9 (Bajnok; cf. [13]) *For all positive integers n, we have*

$$\mu(\mathbb{Z}_n, \{3,1\}) = v_2(n,4).$$

Theorem G.10 (Butterworth; cf. [47]) *For all positive integers n, we have*

$$\mu(\mathbb{Z}_n, \{4,1\}) = v_3(n,5).$$

Recall from page 79 that

$$v_2(n,4) = \begin{cases} \left(1+\frac{1}{p}\right)\frac{n}{4} & \text{if } n \text{ has prime divisors congruent to 3 mod 4,} \\ & \text{and } p \text{ is the smallest such divisor,} \\ \left\lfloor \frac{n}{4} \right\rfloor & \text{otherwise;} \end{cases}$$

the value of $v_3(n,5)$ is more complicated, see page 79.

Problem G.11 *Use Theorem G.8 to compute $\mu(\mathbb{Z}_n, \{k,l\})$ for other choices of k and l.*

It may seem from our results thus far that $\mu(\mathbb{Z}_n, \{k,l\})$ is given by $v_{k-l}(n, k+l)$. While this seems to be often the case, there are instances when

$$\mu(\mathbb{Z}_n, \{k,l\}) > v_{k-l}(n, k+l);$$

for example, $\mu(\mathbb{Z}_9, \{5,2\}) = 2$ (e.g., the set $\{1,2\}$ is $(5,2)$-sum-free in $\mathbb{Z}_9$) while $v_3(9,7) = 1$, and $\mu(\mathbb{Z}_{16}, \{5,1\}) = 3$ (e.g., the set $\{1,2,3\}$ is $(5,1)$-sum-free in $\mathbb{Z}_{16}$) while $v_4(16,6) = 2$. (These examples point to a difference between sum-free sets and zero-sum-free sets; cf. Conjecture F.5.) It is a very interesting question to find others:

Problem G.12 *Find other cases when $\mu(\mathbb{Z}_n, \{k,l\}) > v_{k-l}(n, k+l)$.*

The general question regarding cyclic group is, of course:

Problem G.13 *Evaluate $\mu(\mathbb{Z}_n, \{k,l\})$ for all positive integers n, k, and l.*

Turning to the case of noncyclic groups, we first state the following consequence of Proposition G.3:

Corollary G.14 *For every group G of order n and exponent κ we have*

$$\mu(G, \{k,l\}) \geq \mu(\mathbb{Z}_\kappa, \{k,l\}) \cdot \frac{n}{\kappa}.$$

We (somewhat hesitantly) believe that equality holds:

Conjecture G.15 *For every group G of order n and exponent κ we have*

$$\mu(G, \{k,l\}) = \mu(\mathbb{Z}_\kappa, \{k,l\}) \cdot \frac{n}{\kappa}.$$

We should point out that Conjecture G.15, if true, points to a difference in behavior between sum-free sets and zero-sum-free sets, since we definitely may have

$$\tau(G,h) \neq \tau(\mathbb{Z}_\kappa, h) \cdot \frac{n}{\kappa};$$

see, for example, Proposition F.13 and Theorem F.14. If Conjecture G.15 is true, it is likely to be extremely difficult to prove; we offer the challenging problem:

Problem G.16 *Prove (or disprove) Conjecture G.15.*

We have the following partial result:

Theorem G.17 (Bajnok; cf. [13]) *Conjecture G.15 holds whenever the exponent κ of G possesses at least one divisor d that is not congruent to any integer between 1 and* $\gcd(d, k-l)$ *(inclusive) mod $k+l$.*

Thus, for example, if the exponent—or, equivalently, the order—of G is divisible by 3 or has at least one (prime) divisor congruent to 2 mod 3, then $\mu(G, \{2,1\})$ is determined. The case when all (prime) divisors of G are congruent to 1 mod 3 was unsolved for four decades, until in 2005 in the breakthrough paper [101] Green and Ruzsa proved that Conjecture G.15 holds for sum-free sets:

Theorem G.18 (Green and Ruzsa; cf. [101]) *Let κ be the exponent of G. Then*

$$\mu(G, \{2,1\}) = \mu(\mathbb{Z}_\kappa, \{2,1\}) \cdot \frac{n}{\kappa} = v_1(\kappa, 3) \cdot \frac{n}{\kappa}.$$

We should mention that the proof in [101] relies, in part, on a computer program.

For other k and l, the value of $\mu(G, \{k,l\})$ is not known in general, though we have the following bounds:

Theorem G.19 (Bajnok; cf. [13]) *Suppose that G is an abelian group of order n and exponent κ. Then, for all positive integers k and l with $k > l$ we have*

$$v_{k-l}(\kappa, k+l) \cdot \frac{n}{\kappa} \leq \mu(G, \{k,l\}) \leq v_1(n, k+l).$$

Problem G.20 *Determine $\mu(G, \{k,l\})$ for arbitrary finite abelian groups. In particular, find $\mu(G, \{3,1\})$ for all noncyclic groups G.*

We have the value of $\mu(G, \{k,l\})$ when G is (isomorphic to) an elementary abelian 2-group. Note that, when k and l have the same parity, then $ka = la$ for all $a \in \mathbb{Z}_2^r$, so $\mu(\mathbb{Z}_2^r, \{k,l\}) = 0$ in this case. When k and l have opposite parity, then the set $\{1\} \times \mathbb{Z}_2^{r-1}$ is (k,l)-sum-free in $\mathbb{Z}_2^r$, so $\mu(\mathbb{Z}_2^r, \{k,l\}) \geq 2^{r-1}$. On the other hand, for every subset A of $\mathbb{Z}_2^r$, both kA and lA have size at least $|A|$, so they cannot be disjoint when A is (k,l)-sum-free and $|A| > 2^{r-1}$. Therefore:

Proposition G.21 *For all positive integers r, k, and l with $k > l$, we have*

$$\mu(\mathbb{Z}_2^r, \{k,l\}) = \begin{cases} 0 & \text{if } k \equiv l \bmod 2; \\ 2^{r-1} & \text{if } k \not\equiv l \bmod 2. \end{cases}$$

We can observe that, by Propositions G.21 and F.14, we have

$$\mu(\mathbb{Z}_2^r, \{k,l\}) = \tau(\mathbb{Z}_2^r, k+l)$$

for all parameters.

We now turn to the inverse problem of classifying all (k,l)-sum-free subsets A of G of maximum size $|A| = \mu(G, \{k,l\})$.

Recall that arithmetic progressions play a fundamental role in providing (k,l)-sum-free sets (see Theorem G.8) and, in particular, in providing sum-free sets. With this in mind, we start by determining all sum-free arithmetic progressions in cyclic groups that have maximum size.

Proposition G.22 *Suppose that A is a sum-free arithmetic progression in $\mathbb{Z}_n$ of size*

$$|A| = \mu(\mathbb{Z}_n, \{2,1\}) = v_1(n,3).$$

Then one of the following possibilities must hold:

1. *n is even and*
$$A = \{1, 3, \ldots, n-1\}.$$

2. *n is divisible by 3, has no prime divisors congruent to 2 mod 3, and there exists an integer b relatively prime to n for which*

   (a)
$$b \cdot A = \{1, 4, 7, \ldots, n-2\}$$
   or

   (b)
$$b \cdot A = \{n/3, n/3+1, \ldots, 2n/3-1\}.$$

3. *n is equal to a prime p that is congruent to 2 mod 3, and there exists an integer b relatively prime to p for which*
$$b \cdot A = \{(p+1)/3, (p+1)/3+1, \ldots, 2(p+1)/3-1\}.$$

4. *n is congruent to 1 mod 3, has no prime divisors congruent to 2 mod 3, and there exists an integer b relatively prime to n for which*

   (a)
$$b \cdot A = \{(n-1)/3, (n-1)/3+1, \ldots, 2(n-1)/3-1\}$$
   or

   (b)
$$b \cdot A = \{(n-1)/3+1, (n-1)/3+2, \ldots, 2(n-1)/3\}.$$

(We should add that we may assume that the integer b of Case 2 (a) equals 1 or 2, or, equivalently, 1 or -1.) The proof of Proposition G.22 starts on page 366.

To illuminate Proposition G.22, we discuss an example in detail. Recall that for each n, we mentioned that the elements between $n/6$ and $n/3$, together with the elements between $2n/3$ and $5n/6$, with the endpoints carefully chosen, form a sum-free set in $\mathbb{Z}_n$. Is this an arithmetic progression, and, if it has size $v_1(n,3)$, is it included in Proposition G.22? The answers to both questions are affirmative. For example, if $n = 27$, then $v_1(n,3) = n/3 = 9$, and the sum-free set in question is

$$A = \{5,6,7,8,9\} \cup \{19,20,21,22\}.$$

It is easy to verify that A is sum-free in $\mathbb{Z}_{27}$, and we see that it is an arithmetic progression, since

$$A = \{5, 5+14, 5+2\cdot 14, \ldots, 5+8\cdot 14\}.$$

We can verify that it is included in case 2 (b) of Proposition G.22, by noting that 25 and 27 are relatively prime, and

$$25 \cdot A = \{9,10,11,12,13,14,15,16,17\}.$$

The classification of sum-free sets of maximum size has been known for several decades for any G whose order has a prime divisor congruent to 2 mod 3 or whose order is divisible by 3. It turns out that all such sets are arithmetic progressions of cosets of a subgroup of G. Namely, we have the following two results:

Theorem G.23 (Diananda and Yap; cf. [65]; see also Theorem 7.8 in [199]) *Suppose that the order n of G has prime divisors congruent to 2 mod 3, and p is the smallest of them. If A is a sum-free set of size*

$$|A| = \mu(G,\{2,1\}) = \left(1+\frac{1}{p}\right)\cdot\frac{n}{3}$$

in G, then there is a subgroup H in G, so that G/H is cyclic of order p, and A is the union of $(p+1)/3$ cosets of H that form an arithmetic progression.

It is worth noting that, as a special case of Theorem G.23, we get cases 1 and 3 of Proposition G.22 when G is cyclic of even order or cyclic of prime order p (with $p \equiv 2$ mod 3), respectively.

Theorem G.24 (Street; cf. [191, 192]; see also Theorem 7.9 in [199]) *Suppose that the order n of G is divisible by 3 and has no prime divisors congruent to 2 mod 3. If A is a sum-free set of size*

$$|A| = \mu(G,\{2,1\}) = n/3$$

in G, then there is a divisor k of $n/3$ and a subgroup H in G, so that G/H is cyclic of order $3k$, and A is the union of k cosets of H that form an arithmetic progression.

This time, observe that when G is cyclic, then for $k = 1$ and $k = n/3$ we get cases 2 (a) and 2 (b) of Proposition G.22, respectively.

This leaves us with the task of characterizing sum-free sets of maximum size in G when the order of G has only prime divisors congruent to 1 mod 3. As it turns out, in this case not all such sets are arithmetic progressions of cosets of a subgroup, though the one other type of set is not far from it. We first present the result for cyclic groups:

Theorem G.25 (Yap; cf. [201]) *Suppose that all prime divisors of n are congruent to 1 mod 3. If A is a sum-free set of size*

$$|A| = \mu(\mathbb{Z}_n,\{2,1\}) = (n-1)/3$$

in $\mathbb{Z}_n$, then there exists an integer b relatively prime to n for which

$$b\cdot A = \{(n-1)/3, (n-1)/3+1, \ldots, 2(n-1)/3-1\},$$

$$b \cdot A = \{(n-1)/3+1, (n-1)/3+2, \ldots, 2(n-1)/3\},$$

or

$$b \cdot A = \{(n-1)/3, (n-1)/3+2, (n-1)/3+3, \ldots, 2(n-1)/3-1, 2(n-1)/3+1\}.$$

Observe that the third set listed is two elements short of being an arithmetic progression; note also that this possibility can only occur if $n > 7$.

The classification for noncyclic groups is presented in the article [28] by Balasubramanian, Prakash, and Ramana. Rather than stating the exact result, which is complicated, we only discuss the situation in groups of rank 2.

How can one find sum-free sets of maximum size in $\mathbb{Z}_{n_1} \times \mathbb{Z}_{n_2}$? The most obvious idea is to follow Proposition G.3; for example, if A is a maximum-size sum-free set in $\mathbb{Z}_{n_2}$, then, by Theorem G.18,

$$\mathbb{Z}_{n_1} \times A$$

is a maximum-size sum-free set in $\mathbb{Z}_{n_1} \times \mathbb{Z}_{n_2}$. The following construction is less obvious: we can verify that, with an arbitrary subgroup H of $\mathbb{Z}_{n_1}$, the set $A = A_1 \cup A_2 \cup A_3$, where

$$\begin{aligned} A_1 &= H \times \{(n_2-1)/3\} \\ A_2 &= (\mathbb{Z}_{n_1} \setminus H) \times \{2(n_2-1)/3\} \\ A_3 &= \mathbb{Z}_{n_1} \times \{(n_2-1)/3+1, (n_2-1)/3+2, \ldots, 2(n_2-1)/3-1\} \end{aligned}$$

is a maximum-size sum-free set in $\mathbb{Z}_{n_1} \times \mathbb{Z}_{n_2}$. To see that A is sum-free, observe that, considering the components in $\mathbb{Z}_{n_2}$ alone, for elements $g_1, g_2, g_3 \in A$ to satisfy $g_1 + g_2 = g_3$, we must have $g_1, g_2 \in A_1$ and $g_3 \in A_2$, but then the first component of $g_1 + g_2$ is in H while that is not the case for g_3.

The following construction is similar: let H again be an arbitrary subgroup of $\mathbb{Z}_{n_1}$, and let $A = A_1 \cup A_2 \cup A_3$, where

$$\begin{aligned} A_1 &= H \times \{(n_2-1)/3, 2(n_2-1)/3+1\} \\ A_2 &= (\mathbb{Z}_{n_1} \setminus H) \times \{(n_2-1)/3+1, 2(n_2-1)/3\} \\ A_3 &= \mathbb{Z}_{n_1} \times \{(n_2-1)/3+2, (n_2-1)/3+3, \ldots, 2(n_2-1)/3-1\}. \end{aligned}$$

We can again verify that A is a maximum-size sum-free set in $\mathbb{Z}_{n_1} \times \mathbb{Z}_{n_2}$.

The result of Balasubramanian, Prakash, and Ramana in [28] says that the three types of sets just described provide "essentially" the only types of sum-free sets in $\mathbb{Z}_{n_1} \times \mathbb{Z}_{n_2}$ of maximum size—see [28] for the precise statement for groups of arbitrary rank. (We should add that the result for elementary abelian groups was proved by Street in 1971; see [178] or Theorem 7.21 in [199].)

Let us turn now to the classification of maximum-size (k,l)-sum-free sets for $(k,l) \neq (2,1)$. While we may suspect that the sets come in even more variety than they did for sum-free sets, we find that this is not the case when the group is of prime order: all maximum-sized sets are arithmetic progressions:

Theorem G.26 (Plagne; cf. [171]) *Let p be a positive prime, and k and l be positive integers with $k > l$ and $k \geq 3$; assume also that p does not divide $k-l$. If A is a (k,l)-sum-free set of size*

$$|A| = \mu(\mathbb{Z}_p, \{k,l\}) = \left\lfloor \frac{p-2}{k+l} \right\rfloor + 1$$

in $\mathbb{Z}_p$, then A is an arithmetic progression.

(We note that Theorem G.26 was conjectured and proved in part by Bier and Chin in [38].)

We can go a step beyond the statement of Theorem G.26 and determine the maximum-sized (k,l)-sum-free sets in $\mathbb{Z}_p$ more explicitly:

Theorem G.27 *Let p be a positive prime, and k and l be positive integers with $k>l$ and $k \geq 3$; assume also that p does not divide $k-l$. We let M and r denote, respectively, the quotient and the remainder of $p-2$ when divided by $k+l$. For $i=1,2,\ldots,\lfloor r/2 \rfloor+1$, let a_i be the (unique) solution of*

$$(k-l)a_i = lM + i$$

in $\mathbb{Z}_p$. If A is a (k,l)-sum-free set of size

$$|A| = \mu(\mathbb{Z}_p, \{k,l\}) = M+1$$

in $\mathbb{Z}_p$, then there is an $i \in \{1,2,\ldots,\lfloor r/2 \rfloor +1\}$ and an integer b relatively prime to p for which

$$b \cdot A = \{a_i, a_i+1, \ldots, a_i+M\}.$$

As an example, let us consider the case of maximum-sized $(4,1)$-sum-free sets in $\mathbb{Z}_{19}$. Note that $M = \lfloor (19-2)/(4+1) \rfloor = 3$, so $\mu(\mathbb{Z}_{19}, \{4,1\}) = 4$, and that $r=2$ and thus $\lfloor r/2 \rfloor + 1 = 2$. Furthermore, solving the equations

$$(4-1)\cdot a_i = 1 \cdot 3 + i$$

for $i=1,2$ in $\mathbb{Z}_{19}$, we find that $a_1 = 14$ and $a_2 = 8$. Therefore, Theorem G.27 claims that any $(4,1)$-sum-free set of maximum size 4 in $\mathbb{Z}_{19}$ is a dilate of either $\{14,15,16,17\}$ or $\{8,9,10,11\}$. The easy proof—relying, of course, on Theorem G.26—is on page 368.

We do not have a classification of maximum-sized (k,l)-sum-free sets with $k \geq 3$ in groups of composite order:

Problem G.28 *Let n, k, and l be positive integers with $k>l$ and $k \geq 3$. Classify all (k,l)-sum-free sets of size $\mu(\mathbb{Z}_n, \{k,l\})$ in the cyclic group $\mathbb{Z}_n$.*

A particularly special (k,l)-sum-free set is one where no elements are "wasted": we say that a (k,l)-sum-free set A in G is *complete* if kA and lA partition G; in other words, not only do we have $kA \cap lA = \emptyset$ (i.e., A is (k,l)-sum-free), but also $kA \cup lA = G$.

Before going any further, a word of caution: While the definition of $A \subseteq G$ being (k,l)-sum-free can be given either as $kA \cap lA = \emptyset$ or as $kA - lA \subseteq G \setminus \{0\}$, the notions that for a (k,l)-sum-free set $kA \cup lA = G$ holds or that $kA - lA = G \setminus \{0\}$ holds are not equivalent (though both notions express some sort of "un-wastefulness")! We have counter-examples in both directions:

- If $A = \{1,3,5,7,9\} \subseteq \mathbb{Z}_{10}$, then $2A = \{0,2,4,6,8\}$, so A is sum-free set in $\mathbb{Z}_{10}$; $A \cup 2A = G$, thus A is complete, yet $2A - A = A \neq \mathbb{Z}_{10} \setminus \{0\}$.

- If $A = \{1,3,9\} \subseteq \mathbb{Z}_{13}$, then $2A = \{2,4,5,6,10,12\}$, so A is sum-free in $\mathbb{Z}_{13}$; $2A - A = \mathbb{Z}_{13} \setminus \{0\}$, yet $A \cup 2A \neq G$ thus A is not complete.

While (k,l)-sum-free sets A in G for which $kA - lA = G \setminus \{0\}$ may also be of interest, we here define them as complete when $kA \cup lA = G$.

We may use our classification of maximum-size sum-free sets (Theorems G.23, G.24, and G.25) to determine which of these sets are complete sum-free sets. Consider first the case when the order n of G has prime divisors congruent to 2 mod 3; let p be the smallest such divisor. From Theorem G.23 we see that a maximum-size sum-free set in G has size $(p+1)/3 \cdot n/3$ and is of the form

$$A = (a+H) \cup (a+d+H) \cup \cdots \cup (a+(p-2)/3d+H)$$

for a subgroup $H \leq G$ of order n/p and elements $a, d \in G$. Then

$$2A = (2a+H) \cup (2a+d+H) \cup \cdots \cup (2a+2(p-2)/3d+H),$$

so $2A$ has size $(2p-1)/3 \cdot n/p$. Therefore, $|A|+|2A| = n$, which (since A and $2A$ are disjoint) can only be if $A \cup 2A = G$, hence A is complete.

Next, assume that n is divisible by 3, but has no divisors congruent to 2 mod 3. By Theorem G.24, a maximum-size sum-free set in G has size $n/3$ and is of the form

$$A = (a+H) \cup (a+d+H) \cup \cdots \cup (a+(k-1)d+H)$$

for a divisor k of $n/3$, a subgroup $H \leq G$ of order $n/(3k)$, and elements $a, d \in G$. Then

$$2A = (2a+H) \cup (2a+d+H) \cup \cdots \cup (2a+(2k-2)d+H),$$

so $2A$ has size $(2k-1) \cdot n/(3k)$. Therefore, $|A| + |2A| = (3k-1)/(3k) \cdot n < n$, so A is not complete.

Suppose now that n has only prime divisors congruent to 1 mod 3. By Theorem G.25, if A is a sum-free set of maximum size $(n-1)/3$ in $\mathbb{Z}_n$, then there exists an integer b relatively prime to n for which $b \cdot A$ is equal to

$$A_1 = \{(n-1)/3, (n-1)/3+1, \ldots, 2(n-1)/3-1\},$$

$$A_2 = \{(n-1)/3+1, (n-1)/3+2, \ldots, 2(n-1)/3\},$$

or

$$A_3 = \{(n-1)/3, (n-1)/3+2, (n-1)/3+3, \ldots, 2(n-1)/3-1, 2(n-1)/3+1\}.$$

Since we have $|b \cdot A| = |A|$ and $|2(b \cdot A)| = |2A|$, it suffices to examine sets A_1, A_2, and A_3. We get

$$2A_1 = \{2(n-1)/3, 2(n-1)/3+1, \ldots, 4(n-1)/3-2\},$$

$$2A_2 = \{2(n-1)/3+2, 2(n-1)/3+3, \ldots, 4(n-1)/3\},$$

and

$$2A_3 = \{2(n-1)/3\} \cup \{2(n-1)/3+2, 2(n-1)/3+3, \ldots, 4(n-1)/3\} \cup \{4(n-1)/3+2\}.$$

(All elements listed are considered mod n, of course.) Therefore, $2A_1$ and $2A_2$ both have size $2(n-1)/3-1$, so A_1 and A_2 are not complete. However, $|2A_3| = 2(n-1)/3+1$, so $|A_3 + |2A_3| = n$ and A_3 is complete.

We can summarize our findings as follows:

Theorem G.29 *Let A be a maximum-sized sum-free set in G; as usual, let $|G| = n$.*

1. *If n has a prime divisor congruent to 2 mod 3, then A is a complete sum-free set in G.*

2. *If n is divisible by 3 but has no prime divisors congruent to 2 mod 3, then A is not a complete sum-free set in G.*

3. If n has only prime divisors congruent to 1 mod 3 and G is cyclic, then A is a complete sum-free set in G if, and only if, there is an integer b relatively prime to n for which

$$b \cdot A = \{(n-1)/3, (n-1)/3+2, (n-1)/3+3, \ldots, 2(n-1)/3-1, 2(n-1)/3+1\}.$$

This leaves us with the following problem:

Problem G.30 *For each noncyclic group G whose order has only prime divisors congruent to 1 mod 3, determine which sum-free sets A in G of maximum size $|A| = (n-1)/3$ are complete.*

Problem G.30 is probably not difficult given the classification of the sum-sets in question that we presented above (see page 282).

Turning to the case when $k \geq 3$, we use Theorem G.27 to analyze maximum (k,l)-sum-free sets in cyclic groups of prime order p. According to Theorem G.27, it suffices to analyze

$$A = \{a_i, a_i+1, \ldots, a_i+M\}$$

where $M = \lfloor (p-2)/(k+l) \rfloor$, $r = p-2-(k+l)M$, $i = 1, 2, \ldots, \lfloor r/2 \rfloor + 1$, and a_i is the unique solution to the equation $(k-l)a_i = lM + i$ in $\mathbb{Z}_p$.

We find that

$$\begin{aligned} kA &= \{ka_i, ka_i+1, \ldots, ka_i+kM\} \\ lA &= \{la_i, la_i+1, \ldots, la_i+lM\}. \end{aligned}$$

Now by definition, A is complete when

$$|kA| + |lA| = (kM+1) + (lM+1) = p,$$

and this holds if, and only if, $M = (p-2)/(k+l)$, that is, $p-2$ is divisible by $k+l$. In this case, $r = 0$, and thus i can only equal 1. We get:

Theorem G.31 *Let p be a positive prime, and k and l be positive integers with $k > l$ and $k \geq 3$. Let A be a maximum-size (k,l)-sum-free set in $\mathbb{Z}_p$. Then A is complete if, and only if, $p-2$ is divisible by $k+l$, and A is a dilate of the set*

$$A = \{a, a+1, \ldots, a+M\},$$

where $M = (p-2)/(k+l)$ and a is the unique solution to the equation $(k-l)a = lM+1$ in $\mathbb{Z}_p$.

As an example, we can compute explicitly that, when $p-2$ is divisible by 5, then any complete $(4,1)$-sum-free set in $\mathbb{Z}_p$ is a dilate of the set

$$\{(2p+1)/5, (2p+1)/5+1, \ldots, (3p-1)/5\}.$$

The similar question is not yet solved in cyclic groups of composite order:

Problem G.32 *Let n, k, and l be positive integers with $k > l$ and $k \geq 3$. Classify all complete (k,l)-sum-free sets of maximum size $\mu(\mathbb{Z}_n, \{k,l\})$ in the cyclic group $\mathbb{Z}_n$.*

Of course, a complete (k,l)-sum-free set need not have maximum size $\mu(\mathbb{Z}_n, \{k,l\})$. For example, when p is any (not necessarily the smallest) prime divisor of n with $p \equiv 2$ mod 3, then the set

$$A = \{(p+1)/3 + i + pj \mid i = 0, 1, \ldots, (p-2)/3; j = 0, 1, \ldots, n/p-1\}$$

is a complete sum-free set in $\mathbb{Z}_n$, since

$$2A = \{2(p+1)/3 + i + pj \mid i = 0, 1, \ldots, (2p-4)/3; j = 0, 1, \ldots, n/p - 1\}$$

and thus $A \cap 2A = \emptyset$ but $A \cup 2A = \mathbb{Z}_n$.

The work [117] of Haviv and Levy generated many other examples of complete sum-free sets that also have the additional property that they are symmetric (that is, subsets A of G for which $A = -A$). Their results are somewhat complicated, so we just mention the following special case:

Theorem G.33 (Haviv and Levy; cf. [117]) *Let A be a complete sum-free subset of $\mathbb{Z}_p$ for some prime p, and suppose that A has size $\mu(\mathbb{Z}_p, \{2, 1\}) - 2$. If p is sufficiently large, then there is an element $b \in \mathbb{Z}_p \setminus \{0\}$ for which:*

- *If $p \equiv 2$ mod 3, then*

$$b \cdot A = \left\{\frac{p-5}{3} + i \mid i \in T\right\} \cup \left[\frac{p+13}{3}, \frac{2p-13}{3}\right] \cup \left\{\frac{2p+5}{3} - i \mid i \in T\right\}$$

where $T = \{0, 2, 4\}$ or $T = \{0, 3, 4\}$.

- *If $p \equiv 1$ mod 3, then*

$$b \cdot A = \left\{\frac{p-7}{3} + i \mid i \in T\right\} \cup \left[\frac{p+17}{3}, \frac{2p-17}{3}\right] \cup \left\{\frac{2p+7}{3} - i \mid i \in T\right\}$$

where $T = \{0, 2, 4, 6\}$, $T = \{0, 3, 5, 6\}$, $T=\{0,4,5,6\}$, or $T = \{1, 2, 6, 7\}$.

The classification of all complete (k, l)-sum-free sets remains open:

Problem G.34 *Find all complete sum-free—or, more generally, complete (k, l)-sum-free—sets in finite abelian groups that do not have maximum size $\mu(\mathbb{Z}_n, \{k, l\})$.*

Actually, Problem G.34 stays intriguing even when dropping the requirement that the set be complete: we may just ask for (k, l)-sum-free sets that do not have maximum size $\mu(\mathbb{Z}_n, \{k, l\})$ but are *maximal*—that is, they cannot be enlarged without losing the (k, l)-sum-free property. One then may ask for the possible cardinalities of maximal sum-free sets:

Problem G.35 *For each group G and for all positive integers k, l with $k > l$, find all possible values of m for which a maximal (k, l)-sum-free set of size m exist.*

In particular, for positive integers i, we let $M_i(G, \{k, l\})$ denote the i-th largest cardinality of a maximal (k, l)-sum-free set in G (for values of i for which this exists), and $M(G, \{k, l\})$ denote the size of the smallest maximal (k, l)-sum-free set in G. Note that

$$M_1(G, \{k, l\}) = \mu(G, \{k, l\}).$$

We then may ask:

Problem G.36 *Find the i-th largest size $M_i(G, \{k, l\})$ of maximal (k, l)-sum-free set in G. In particular, find $M_2(G, \{k, l\})$.*

Problem G.37 *Find the size $M(G, \{k, l\})$ of the smallest maximal (k, l)-sum-free set in G.*

As an example, consider $G = \mathbb{Z}_{11}$, where $\mu(\mathbb{Z}_{11}, \{2,1\}) = 4$. The set $A = \{1,3,5\}$ is sum-free in G, but $A \cup \{a\}$ is not sum-free for any $a \in G \setminus A$. We can also verify that each set of size two can be enlarged so that it stays sum-free, hence

$$M_2(\mathbb{Z}_{11}, \{2,1\}) = M(\mathbb{Z}_{11}, \{2,1\}) = 3.$$

Clark and Pedersen have the following result in elementary abelian 2-groups:

Theorem G.38 (Clark and Pedersen; cf. [57]) *For each positive integer $r \geq 4$, we have*

$$M_2(\mathbb{Z}_2^r, \{2,1\}) = 5 \cdot 2^{r-4}.$$

Clark and Pedersen also computed all values of $M_i(\mathbb{Z}_2^r, \{2,1\})$ (that exist) and $M(\mathbb{Z}_2^r, \{2,1\})$ for $r \leq 6$ and possibly all for $7 \leq r \leq 10$. From their data, we may conjecture the following:

Conjecture G.39 *For each positive integer $r \geq 4$ and $2 \leq i \leq r-2$, we have*

$$M_i(\mathbb{Z}_2^r, \{2,1\}) = 2^{r-2} + 2^{r-2-i},$$

and

$$M(\mathbb{Z}_2^r, \{2,1\}) = 2^{\lceil r/2 \rceil} + 2^{\lfloor r/2 \rfloor} - 3.$$

We note that our formula does not hold for $i = 1$, but by Theorem G.38, it holds for all $r \geq 4$ and $i = 2$. Furthermore, from Corollary 5 in [57] (with $s = 2$ and $t = r - 2 - i$) we also know that maximal sum-free sets of size $2^{r-2} + 2^{r-2-i}$ exist in $\mathbb{Z}_2^r$ for each $r \geq 4$ and $2 \leq i \leq r-2$, and (with $s = \lfloor r/2 \rfloor$ and $t = 0$) that maximal sum-free sets of size $2^{\lceil r/2 \rceil} + 2^{\lfloor r/2 \rfloor} - 3$ exist in $\mathbb{Z}_2^r$ for each $r \geq 4$. However, it has not been established yet that these values equal $M_i(\mathbb{Z}_2^r, \{2,1\})$ and $M(\mathbb{Z}_2^r, \{2,1\})$, respectively.

Problem G.40 *Prove Conjecture G.39.*

We now turn to a related problem that was first investigated by Erdős in [77]. Namely, we wish to examine the size of the largest (k,l)-sum-free subset of a given $A \subseteq G$: we denote this quantity by $\mu(G, \{k,l\}, A)$. We can then ask for the "worst-case scenario": For each positive integer $m \leq n$, find the m-subsets A_0 of G so that

$$\mu(G, \{k,l\}, A_0) \leq \mu(G, \{k,l\}, A)$$

for all m-subsets A of G; we then let $\mu(G, \{k,l\}, m) = \mu(G, \{k,l\}, A_0)$.

Problem G.41 *For each group G and all positive integers k, l, and m (with $k > l$ and $m \leq n$), find $\mu(G, \{k,l\}, m)$.*

Of course, we have

$$\mu(G, \{k,l\}, n) = \mu(G, \{k,l\}),$$

but values of $\mu(G, \{k,l\}, m)$ for $m < n$ are not known (cf. the preprint [196] by Tao and Vu for the history of this problem and for more information). We have the following general lower bound for the case of sum-free sets:

Theorem G.42 (Alon and Kleitman; cf. [2]) *For every group G and positive integer m (with $m \leq n$) we have*

$$\mu(G, \{2,1\}, m) \geq \lfloor 2m/7 \rfloor.$$

We note that the bound here is tight in the sense that, by Theorem G.18, it is achieved by the group $\mathbb{Z}_7^r$ when $m = n = 7^r$.

As a step toward solving Problem G.41, we may ask for the smallest value of m for which $\mu(G, \{k, l\}, m) = \mu(G, \{k, l\})$:

Problem G.43 *For each group G and all positive integers k and l with $k > l$, find the smallest value $T(G, \{k, l\})$ of m for which $\mu(G, \{k, l\}, m) = \mu(G, \{k, l\})$.*

When p is an odd prime and $p \equiv 2 \mod 3$, then by Theorem G.23 and Proposition G.22 (part 3), we know that $\mathbb{Z}_p$ has exactly $(p-1)/2$ sum-free subsets of size $\mu(\mathbb{Z}_p, \{2, 1\}) = (p+1)/3$. Furthermore, by the result of Chin in [55], these sets form a *block design*; that is, every nonzero element of $\mathbb{Z}_p$ is contained in the same number (in our case, $(p+1)/6$) of these $(p-1)/2$ sets. Since for $p > 5$, the value of $2 \cdot (p+1)/6$ is less than $(p-1)/2$, deleting two nonzero elements of $\mathbb{Z}_p$ results in a set of size $p-2$ that still contains one of the $(p-1)/2$ sum-free sets of maximum size. Therefore, we have

$$\mu(\mathbb{Z}_p, \{2, 1\}, p-2) = \mu(\mathbb{Z}_p, \{2, 1\}),$$

which yields:

Theorem G.44 *For an odd prime $p > 5$ with $p \equiv 2$ mod 3, we have*

$$T(\mathbb{Z}_p, \{2, 1\}) \leq p - 2.$$

As an example, let us consider the group $\mathbb{Z}_{11}$; by Theorem G.44, $T(\mathbb{Z}_{11}, \{2, 1\})$ is at most 9. We know that $\mu(\mathbb{Z}_p, \{2, 1\}) = 4$ and that there are five sum-free subsets of size four:

$$\{4, 5, 6, 7\}, \ \{8, 10, 1, 3\}, \ \{1, 4, 7, 10\}, \ \{5, 9, 2, 6\}, \ \{9, 3, 8, 2\}.$$

We see that none of these five sets are contained in

$$A = \mathbb{Z}_{11} \setminus \{2, 4, 8\},$$

and thus we get $T(\mathbb{Z}_{11}, \{2, 1\}) = 9$. We do not know the value of $T(\mathbb{Z}_p, \{2, 1\})$ in general, so, as a special case of Problem G.43, we ask:

Problem G.45 *Find the value of $T(\mathbb{Z}_p, \{2, 1\})$ for every odd prime p with $p \equiv 2$ mod 3.*

Our next pursuit is the following related problem. For a given m-subset A of a given group G, we may ask for the number of ordered pairs (a_1, a_2) with $a_1, a_2 \in A$ for which $a_1 + a_2 \in A$; we denote this quantity by $P(G, A)$. (Note that, when $a_1 + a_2 \in A$, then $a_2 + a_1 \in A$, so when a_1 and a_2 are distinct, both (a_1, a_2) and (a_2, a_1) contribute toward the count.) We then may ask for the minimum value of $P(G, A)$ among all m-subsets A of G; we denote this quantity by $P(G, m)$. Observe that, by definition, we have $P(G, m) = 0$ for each $m \leq \mu(G, \{2, 1\})$, but $P(G, m) \geq 1$ for each $m \geq \mu(G, \{2, 1\}) + 1$.

Problem G.46 *For each group G and all positive integers m with*

$$\mu(G, \{2, 1\}) + 1 \leq m \leq n,$$

find $P(G, m)$.

Consider, as an example, the cyclic group $\mathbb{Z}_p$ of prime order $p \geq 3$, where

$$\mu(\mathbb{Z}_p, \{2,1\}) = \lfloor (p+1)/3 \rfloor .$$

As we have already observed, the "middle" third of the elements, namely, the set

$$\{\lfloor (p+1)/3 \rfloor, \lfloor (p+1)/3 \rfloor + 1, \dots, 2\lfloor (p+1)/3 \rfloor - 1\}$$

forms a sum-free set of $\mathbb{Z}_p$ of maximum size. Suppose that m is a positive integer so that

$$m > \lfloor (p+1)/3 \rfloor .$$

We enlarge our set to

$$A = \{\lfloor (p+1-m)/2 \rfloor, \lfloor (p+1-m)/2 \rfloor + 1, \dots, \lfloor (p+1-m)/2 \rfloor + m - 1\} .$$

Then A has size m, and it is still in the "middle" (when m is even, it is symmetrical in that its first and last elements add to p, and if m is odd, then it is as close to being symmetrical as possible). A short calculation finds that

$$P(\mathbb{Z}_p, A) = \lfloor (3m-p)^2/4 \rfloor .$$

A recent result by Samotij and Sudakov says that one cannot do better:

Theorem G.47 (Samotij and Sudakov; cf. [182], [183]) *Suppose that p is an odd prime, and A is a subset of $\mathbb{Z}_p$ of size $m > \lfloor (p+1)/3 \rfloor$. Then*

$$P(\mathbb{Z}_p, m) = \lfloor (3m-p)^2/4 \rfloor .$$

Furthermore, $P(\mathbb{Z}_p, A) = P(\mathbb{Z}_p, m)$ holds if, and only if, there is an integer b, relatively prime to p, for which

$$b \cdot A = \{\lfloor (p+1-m)/2 \rfloor, \lfloor (p+1-m)/2 \rfloor + 1, \dots, \lfloor (p+1-m)/2 \rfloor + m - 1\} .$$

(We note that the statement of this result is stated erroneously in [182]; see [183] for the corrected version.)

In [182] the authors evaluate $P(G, m)$ for the elementary abelian group G, and state some partial results for other groups, but warn that finding $P(G, m)$ for all G and m "would be rather difficult."

As a generalization, for a positive integer k we define

$$P(G, k, A) = |\{(a_1, \dots, a_k) \in A^k \mid a_1 + \cdots a_k \in A\}|$$

and

$$P(G, k, m) = \min\{P(G, k, A) \mid A \in G, |A| = m\}.$$

Note that we have $P(G, 2, A) = P(G, A)$ and $P(G, 2, m) = P(G, m)$.

Problem G.48 *Evaluate $P(G, k, m)$ for all groups G and positive integers k and m.*

Of course, we are also interested in partial progress toward Problem G.48, such as the following:

Problem G.49 *Evaluate $P(\mathbb{Z}_p, k, m)$ for prime values of p and $k > l$ with $k \geq 3$.*

G.1.2 Limited number of terms

Given a group G and a nonnegative integer s, here we are interested in finding the maximum possible size of an $[0, s]$-sum-free set in G, that is, the quantity

$$\mu(G, [0, s]) = \max\{|A| \mid A \subseteq G; 0 \leq l < k \leq s \Rightarrow kA \cap lA = \emptyset\}.$$

Note that in a group G of exponent κ, we have $0g = \kappa g$ for any $g \in G$, so no subset of G is $[0, s]$-sum-free if $s \geq \kappa$. Furthermore, the cases of $s = 0$ and $s = 1$ are easy: any subset A of G is $[0, 0]$-sum-free, and $A \subseteq G$ is $[0, 1]$-sum-free if, and only if, $0 \not\in A$. We thus have:

Proposition G.50 *If G is of exponent κ and $s \geq \kappa$, then $\mu(G, [0, s]) = 0$. Furthermore, $\mu(G, [0, 0]) = n$ and $\mu(G, [0, 1]) = n - 1$.*

According to Proposition G.50, we can assume that $2 \leq s \leq \kappa - 1$.

We have two explicit constructions for $[0, s]$-sum-free sets in $\mathbb{Z}_n$. The first comes from Zajaczkowski (see [204]): Let

$$m = \left\lfloor \frac{n - s - 1}{s^2} \right\rfloor + 1,$$

and consider

$$A = \{1 + is \mid i = 0, 1, \dots, m - 1\}$$

in $\mathbb{Z}_n$. Note that the elements of hA are congruent to h mod s; therefore, kA and lA are pairwise distinct for all $1 \leq l < k \leq s$. Furthermore, $0 \not\in sA$, since

$$s \cdot (1 + (m - 1)s) = s + s^2 \cdot \left\lfloor \frac{n - s - 1}{s^2} \right\rfloor \leq s + (n - s - 1) < n.$$

Therefore, A is indeed $[0, s]$-sum-free set in $\mathbb{Z}_n$.

For our second construction, let m be as above, and set

$$a = (s - 1)(m - 1) + 1.$$

Then for

$$A = \{a, a + 1, \dots, a + m - 1\}$$

we see that

$$hA = \{ha, ha + 1, \dots, ha + h(m - 1)\}.$$

Note that for each $h = 0, 1, 2, \dots, s - 1$ we have

$$ha + h(m - 1) < (h + 1)a,$$

since this inequality is equivalent to $h(m - 1) < a$, which holds as

$$h(m - 1) \leq (s - 1)(m - 1) = a - 1.$$

Furthermore,

$$sa + s(m - 1) = s((s - 1)(m - 1) + 1) + s(m - 1) = s^2(m - 1) + s = s^2 \cdot \left\lfloor \frac{n - s - 1}{s^2} \right\rfloor + s < n.$$

Therefore, A is $[0, s]$-sum-free set in $\mathbb{Z}_n$.

From both constructions we get:

Proposition G.51 (Zajaczkowski; cf. [204]) *For all positive integers n and s, we have*

$$\mu(\mathbb{Z}_n, [0,s]) \geq \left\lfloor \frac{n-s-1}{s^2} \right\rfloor + 1.$$

Recall that when the exponent (in this case, the order) of the group G is s or less, then no set is $[0,s]$-sum-free in G. Observe that

$$\left\lfloor \frac{n-s-1}{s^2} \right\rfloor + 1$$

equals 0 when $n \leq s$ but is positive when $n \geq s+1$.

In many cases, we can do better than Proposition G.51. Let d be any positive divisor of n, and let H be the subgroup of order n/d in $\mathbb{Z}_n$. Note that if A is a $[0,s]$-sum-free set in $\mathbb{Z}_d$, then $A+H$ is a $[0,s]$-sum-free set in $\mathbb{Z}_n$. Therefore:

Proposition G.52 *For all positive integers n and s, we have*

$$\mu(\mathbb{Z}_n, [0,s]) \geq \mu(\mathbb{Z}_d, [0,s]) \cdot \frac{n}{d}.$$

Combining Propositions G.51 and G.52 yields:

Corollary G.53 *For all positive integers n and s, we have*

$$\mu(\mathbb{Z}_n, [0,s]) \geq \max\left\{ \left(\left\lfloor \frac{d-s-1}{s^2} \right\rfloor + 1 \right) \cdot \frac{n}{d} \mid d \in D(n) \right\}.$$

(Zajaczkowski proved a special case of this for $s=2$ in [204].)

We can analyze the bound of Corollary G.53 for small values of s as follows. First, note that for $s=1$, Corollary G.53 becomes

$$\mu(\mathbb{Z}_n, [0,1]) \geq \max\left\{ (d-1) \cdot \frac{n}{d} \mid d \in D(n) \right\} = n-1,$$

and from Proposition G.50 we know that equality holds.

For $s=2$, we get

$$\mu(\mathbb{Z}_n, [0,2]) \geq \max\left\{ \left\lfloor \frac{d+1}{4} \right\rfloor \cdot \frac{n}{d} \mid d \in D(n) \right\}.$$

We separate two cases. Assume first that n has no divisors that are congruent to 3 mod 4. In this case, for all $d \in D(n)$ we have

$$\left\lfloor \frac{d+1}{4} \right\rfloor \cdot \frac{n}{d} \leq \frac{d}{4} \cdot \frac{n}{d} = \frac{n}{4};$$

since the left-hand side is an integer, it can be at most $\lfloor n/4 \rfloor$. On the other hand, using $d=n$ we see that

$$\max\left\{ \left\lfloor \frac{d+1}{4} \right\rfloor \cdot \frac{n}{d} \mid d \in D(n) \right\} \geq \left\lfloor \frac{n+1}{4} \right\rfloor = \left\lfloor \frac{n}{4} \right\rfloor,$$

so Corollary G.53 gives

$$\mu(\mathbb{Z}_n, [0,2]) \geq \left\lfloor \frac{n}{4} \right\rfloor.$$

Suppose now that n has divisors that are congruent to 3 mod 4. Then for each such divisor d, we get

$$\left\lfloor \frac{d+1}{4} \right\rfloor \cdot \frac{n}{d} = \frac{d+1}{4} \cdot \frac{n}{d} = \left(1+\frac{1}{d}\right)\frac{n}{4};$$

this quantity is greatest when d is the smallest such divisor, namely, it is the smallest prime divisor of n that is congruent to 3 mod 4. In summary, for $s=2$ Corollary G.53 yields (see page 79):

Corollary G.54 *For all positive integers n, we have*

$$\mu(\mathbb{Z}_n,[0,2]) \geq v_2(n,4) = \begin{cases} \left(1+\frac{1}{p}\right)\frac{n}{4} & \text{if } n \text{ has prime divisors congruent to 3 mod 4,} \\ & \text{and } p \text{ is the smallest such divisor;} \\ \left\lfloor \frac{n}{4} \right\rfloor & \text{otherwise.} \end{cases}$$

We believe that equality holds in Corollary G.54:

Conjecture G.55 *For all positive integers n, we have*

$$\mu(\mathbb{Z}_n,[0,2]) = v_2(n,4).$$

Zajaczkowski verified Conjecture G.55 for all $n \leq 20$ (cf. [204]).

Problem G.56 *Prove Conjecture G.55.*

Moving on to $s=3$, we see that Corollary G.53 becomes:

Corollary G.57 *For all positive integers n, we have*

$$\mu(\mathbb{Z}_n,[0,3]) \geq \max\left\{\left\lfloor \frac{d+5}{9} \right\rfloor \cdot \frac{n}{d} \mid d \in D(n)\right\}.$$

(Note that the bound here is not always equal to $v_3(n,9)$.) In particular, when n only has divisors that are congruent to 0, 1, 2, or 3 mod 9, then we only get

$$\mu(\mathbb{Z}_n,[0,3]) \geq \lfloor n/9 \rfloor.$$

It is not clear if we can do better:

Problem G.58 *Improve, if possible, on the lower bound of Corollary G.57.*

As an example, we prove that for $n=11$, we cannot improve on the bound of Corollary G.57, which yields only

$$\mu(\mathbb{Z}_{11},[0,3]) \geq 1.$$

We argue as follows. Suppose, indirectly, that $\mathbb{Z}_{11}$ contains a 2-subset $A=\{a,b\}$ that is $[0,3]$-sum-free. This means that the sets

$$\begin{aligned} 0A &= \{0\} \\ 1A &= \{a,b\} \\ 2A &= \{2a,a+b,2b\} \\ 3A &= \{3a,2a+b,a+2b,3b\} \end{aligned}$$

are pairwise disjoint. Note also that the elements listed in each set are all distinct; for example, $2a+b=3b$ would yield $2a=2b$, which in $\mathbb{Z}_{11}$ can only happen if $a=b$. Therefore, the ten elements listed are ten distinct elements of $\mathbb{Z}_{11}$; in other words, there is a unique element $c \in \mathbb{Z}_{11}$ so that

$$[0,3]A = \mathbb{Z}_{11} \setminus \{c\}.$$

Summing the elements on the two sides, we get

$$10a + 10b = (0 + 1 + \cdots + 10) - c,$$

or

$$10a + 10b = -c,$$

from which $c = a + b$. But that is a contradiction, since $a + b \in A$ and thus $c \in [0,3]A$. Therefore,

$$\mu(\mathbb{Z}_{11}, [0,3]) = 1.$$

We know even less about the cases of $s \geq 4$:

Problem G.59 *Improve, if possible, on the lower bound of Corollary G.53 for $s \geq 4$.*

We state the following challenging problems:

Problem G.60 *Find the exact value of $\mu(\mathbb{Z}_n, [0,s])$ for all n and a given $s \geq 4$.*

Problem G.61 *Find the exact value of $\mu(G, [0,2])$ for noncyclic groups G.*

Problem G.62 *Find the exact value of $\mu(G, [0,s])$ for noncyclic groups G and $s \geq 3$.*

G.1.3 Arbitrary number of terms

Here we ought to consider

$$\mu(G,H) = \max\{|A| \mid A \subseteq G; h_1, h_2 \in H; h_1 \neq h_2 \Rightarrow h_1 A \cap h_2 A = \emptyset\}$$

for the case when H is the set of all nonnegative or all positive integers. However, as we have already mentioned, we have $\mu(G,H) = 0$ whenever H contains two elements whose difference is a multiple of the exponent of the group.

G.2 Unrestricted signed sumsets

G.2.1 Fixed number of terms

G.2.2 Limited number of terms

G.2.3 Arbitrary number of terms

Here we ought to consider

$$\mu_{\pm}(G,H) = \max\{|A| \mid A \subseteq G; h_1, h_2 \in H; h_1 \neq h_2 \Rightarrow (h_1)_{\pm} A \cap (h_2)_{\pm} A = \emptyset\}$$

for the case when H is the set of all nonnegative or all positive integers. However, as we have already mentioned, we have $\mu_{\pm}(G,H) = 0$ whenever H contains two elements whose difference is a multiple of the exponent of the group.

G.3 Restricted sumsets

Our goal in this section is to investigate the maximum possible size of a weak H-sum-free set, that is, the quantity

$$\mu\hat{}(G,H) = \max\{|A| \mid A \subseteq G; h_1, h_2 \in H; h_1 \neq h_2 \Rightarrow h_1\hat{}A \cap h_2\hat{}A = \emptyset\}.$$

Clearly, for all G of order at least 2, we have $\mu\hat{}(G,H) \geq 1$: for any $a \in G \setminus \{0\}$, at least the one-element set $A = \{a\}$ will be weakly H-sum-free.

It is important to note that Proposition G.3 does not carry through for $\mu\hat{}(G,H)$: for example, $\{1,2,4\}$ is weakly $\{1,2\}$-sum-free in $\mathbb{Z}_{10}$ (the sum of any two distinct elements of A is in $G \setminus A$), but $\{1,2,4\} \times \mathbb{Z}_{10}$ is not weakly $\{2,1\}$-sum-free in $\mathbb{Z}_{10}^2$ since, for example,

$$(1,3) + (1,4) = (2,7).$$

Below we consider three special cases: when H consists of two distinct positive integers (by Proposition G.1, the case when one of the integers equals 0 is identical to Section F.3.1), when H consists of all nonnegative integers up to some value s, and when $H = \mathbb{N}_0$.

G.3.1 Fixed number of terms

The analogue of a (k,l)-sum-free set for restricted addition is called a weak (k,l)-sum-free set; namely, subsets A of G satisfying the condition $(k\hat{}A) \cap (l\hat{}A) = \emptyset$ are called *weak (k,l)-sum-free sets*. Here we investigate, for a given group G and positive integers k and l with $k > l$, the quantity

$$\mu\hat{}(G,\{k,l\}) = \max\{|A| \mid A \subseteq G, (k\hat{}A) \cap (l\hat{}A) = \emptyset\},$$

that is, the maximum size of a weak (k,l)-sum-free set in G.

Since for $k \geq n+1$ we trivially have $\mu\hat{}(G,\{k,l\}) = n$, we assume that $l < k \leq n$.

Note that, if A is (k,l)-sum-free, then it is also weakly (k,l)-sum-free, so by Theorem G.19, we have the following lower bound:

Proposition G.63 *Suppose that G is an abelian group of order n and exponent κ. Then, for all positive integers k and l with $k > l$ we have*

$$\mu\hat{}(G,\{k,l\}) \geq \mu(G,\{k,l\}) \geq v_{k-l}(\kappa, k+l) \cdot \frac{n}{\kappa}.$$

There are a variety of cases when $\mu\hat{}(G,\{k,l\})$ is strictly more than $\mu(G,\{k,l\})$; we discuss some of these below.

First, we investigate intervals in the cyclic group, that is, arithmetic progressions in $\mathbb{Z}_n$ whose common difference is 1. For a fixed element $a \in \mathbb{Z}_n$ and positive integer $m \leq n$, the set

$$A = \{a, a+1, \ldots, a+m-1\}$$

is called an *interval* of length $m-1$ (we say that A has size m and length $m-1$). For example, $\{3,4,5,6\}$ and $\{5,6,0,1\}$ are both intervals of length 3 (size 4) in $\mathbb{Z}_7$.

The following result exhibits a formula for the maximum size $\gamma\hat{}(\mathbb{Z}_n,\{k,l\})$ of a weak (k,l)-sum-free interval in $\mathbb{Z}_n$. We introduce some notations:

$$\delta = \gcd(n, k-l),$$

$$J = k^2 + l^2 - (k+l),$$

$$M = \lfloor (n+J-2)/(k+l) \rfloor,$$
$$K = kM - J/2 + 1,$$
$$L = lM - J/2 + 1.$$

Observe that

$$K + L = (k+l)\lfloor (n+J-2)/(k+l) \rfloor - J + 2 \leq n.$$

Proposition G.64 *Let n, k, and l be positive integers with $l < k \leq n$, and let $\gamma\hat{}(\mathbb{Z}_n, \{k,l\})$ be the maximum size of a weak (k,l)-sum-free interval in $\mathbb{Z}_n$. With the notations just introduced,*

$$\gamma\hat{}(\mathbb{Z}_n, \{k,l\}) = \begin{cases} M+1 & \text{if } L/\delta \leq \lfloor (n-K)/\delta \rfloor; \\ \\ M & \text{otherwise.} \end{cases}$$

The proof of Proposition G.64 is on page 369.

We have the following more explicit consequence of Proposition G.64 (see the corresponding Corollary F.81 regarding weak zero-h-sum-free sets):

Corollary G.65 *For positive integers n, k, and l with $l < k \leq n$ we have*

$$\mu\hat{}(\mathbb{Z}_n, \{k,l\}) \geq \left\lfloor \frac{n + k^2 + l^2 - \gcd(n, k-l) - 1}{k+l} \right\rfloor.$$

We provide two proofs: one using Proposition G.64 and another that is more direct; see page 370.

We can use Corollary G.65 (as well as Theorem D.51) to find the value of $\mu\hat{}(\mathbb{Z}_p, \{k,l\})$ for all k, l, and prime values of p. By Corollary G.65, we get

$$\mu\hat{}(\mathbb{Z}_p, \{k,l\}) \geq \left\lfloor \frac{p + k^2 + l^2 - 2}{k+l} \right\rfloor.$$

To see that equality holds, we employ Theorem D.51 to conclude that for a weak (k,l)-sum-free set $A \subseteq \mathbb{Z}_p$ we have

$$p \geq |k\hat{}A| + |l\hat{}A| \geq (k|A| - k^2 + 1) + (l|A| - l^2 + 1),$$

from which

$$|A| \leq \frac{p + k^2 + l^2 - 2}{k+l}$$

follows. Therefore:

Theorem G.66 *For all primes p and positive integers k and l with $l < k \leq p$ we have*

$$\mu\hat{}(\mathbb{Z}_p, \{k,l\}) = \left\lfloor \frac{p + k^2 + l^2 - 2}{k+l} \right\rfloor.$$

The value of $\mu\hat{}(\mathbb{Z}_n, \{k,l\})$ for composite n is harder to find, and is known for all n only for $(k,l) = (2,1)$ (the case of *weak sum-free sets*)—we present this result next.

By combining Proposition G.63 and Corollary G.65, we get:

$$\mu\hat{}(\mathbb{Z}_n, \{k,l\}) \geq \max\left\{ \mu(\mathbb{Z}_n, \{k,l\}), \left\lfloor \frac{n + k^2 + l^2 - \gcd(n, k-l) - 1}{k+l} \right\rfloor \right\}.$$

For $k=2$ and $l=1$, this translates to

$$\mu\hat{}(\mathbb{Z}_n,\{2,1\}) \geq \max\left\{\mu(\mathbb{Z}_n,\{2,1\}), \left\lfloor \tfrac{n}{3} \right\rfloor + 1\right\}.$$

The value of $\mu(\mathbb{Z}_n,\{2,1\})$ is given by Corollary G.6 as $v_1(n,3)$; on page 277 we see that

$$v_1(n,3) = \begin{cases} \left(1+\frac{1}{p}\right)\frac{n}{3} & \text{if } n \text{ has prime divisors congruent to 2 mod 3,} \\ & \text{and } p \text{ is the smallest such divisor,} \\ \left\lfloor \frac{n}{3} \right\rfloor & \text{otherwise.} \end{cases}$$

We should also note that, for any prime divisor p of n which is congruent to 2 mod 3, we have

$$\left(1+\frac{1}{p}\right)\frac{n}{3} \geq \left\lfloor \frac{n}{3} \right\rfloor + 1.$$

Indeed, if $n \geq 3p$, then

$$\left(1+\frac{1}{p}\right)\frac{n}{3} \geq \left(1+\frac{1}{n/3}\right)\frac{n}{3} = \frac{n}{3} + 1 \geq \left\lfloor \frac{n}{3} \right\rfloor + 1;$$

if $n = 2p$, then

$$\left(1+\frac{1}{p}\right)\frac{n}{3} = \left(1+\frac{1}{n/2}\right)\frac{n}{3} = \frac{n-1}{3} + 1 = \left\lfloor \frac{n}{3} \right\rfloor + 1;$$

and if $n = p$, then

$$\left(1+\frac{1}{p}\right)\frac{n}{3} = \left(1+\frac{1}{n}\right)\frac{n}{3} = \frac{n-2}{3} + 1 = \left\lfloor \frac{n}{3} \right\rfloor + 1.$$

This proves that

$$\mu\hat{}(\mathbb{Z}_n,\{2,1\}) \geq \begin{cases} \left(1+\frac{1}{p}\right)\frac{n}{3} & \text{if } n \text{ has prime divisors congruent to 2 mod 3,} \\ & \text{and } p \text{ is the smallest such divisor,} \\ \left\lfloor \frac{n}{3} \right\rfloor + 1 & \text{otherwise.} \end{cases}$$

It turns out that equality holds:

Theorem G.67 (Zannier; cf. [205]) *For all positive integers we have*

$$\mu\hat{}(\mathbb{Z}_n,\{2,1\}) = \begin{cases} \left(1+\frac{1}{p}\right)\frac{n}{3} & \textit{if } n \textit{ has prime divisors congruent to 2 mod 3,} \\ & \textit{and } p \textit{ is the smallest such divisor,} \\ \left\lfloor \frac{n}{3} \right\rfloor + 1 & \textit{otherwise.} \end{cases}$$

We present Zannier's short and elegant proof on page 372.

Let us turn to the case of $k=3$ and $l=1$. Our considerations above now yield

$$\mu\hat{}(\mathbb{Z}_n,\{3,1\}) \geq \max\left\{\mu(\mathbb{Z}_n,\{3,1\}), \left\lfloor \frac{n+9-\gcd(n,2)}{4} \right\rfloor\right\}.$$

Recall that by Theorem G.9 and page 79,

$$\mu(\mathbb{Z}_n, \{3,1\}) = v_2(n,4) = \begin{cases} \left(1+\frac{1}{p}\right)\frac{n}{4} & \text{if } n \text{ has prime divisors congruent to 3 mod 4,} \\ & \text{and } p \text{ is the smallest such divisor,} \\ \left\lfloor \frac{n}{4} \right\rfloor & \text{otherwise.} \end{cases}$$

Note that, unless n is divisible by 4, we have

$$\left\lfloor \frac{n+9-\gcd(n,2)}{4} \right\rfloor = \left\lfloor \frac{n}{4} \right\rfloor + 2,$$

and thus for these values of n we have

$$\mu\hat{}(\mathbb{Z}_n, \{3,1\}) \geq \lfloor n/4 \rfloor + 2.$$

We are able to make the same claim even when n is divisible by 4 as long as it is not divisible by 8, since we may apply the stronger Proposition G.64 to conclude exactly this when

$$\frac{n/4-1}{2} \leq \left\lfloor \frac{n/4-1}{2} \right\rfloor$$

holds, or, equivalently, when n leaves a remainder of 4 mod 8.

This leaves us with the cases when n is divisible by 8. Ziegler in [206] noticed that in some subcases one still has a weak $(3,1)$-sum-free set in $\mathbb{Z}_n$ of size $n/4+2$ (even when n has no prime divisors congruent to 3 mod 4). For example, there are several such sets of size 4 in $\mathbb{Z}_8$, including $\{0,1,2,4\}$ and $\{0,3,4,6\}$. Furthermore, he observed that when n is divisible by 16, then

$$\left\{\frac{n}{16}, \frac{n}{16}+1, \ldots, \frac{3n}{16}\right\} \cup \left\{\frac{9n}{16}, \frac{9n}{16}+1, \ldots, \frac{11n}{16}\right\}$$

is a weak $(3,1)$-sum-free set in $\mathbb{Z}_n$ of size $n/4+2$. (See a generalization of this in Proposition G.73 below.) It is tempting to conjecture that

$$\mu\hat{}(\mathbb{Z}_n, \{3,1\}) \geq \lfloor n/4 \rfloor + 2$$

always holds; however, as the computational data of Hallfors (see below) indicates, this is false for $n = 40$. (Note that $n = 24$ has a prime divisor that is congruent to 3 mod 4.)

We pose the following intriguing problem:

Problem G.68 *Suppose that n is congruent to 8 mod 16 and has no prime divisors congruent to 3 mod 4. When is $\mu\hat{}(\mathbb{Z}_n, \{3,1\})$ equal to $n/4+2$ and when is it $n/4+1$?*

As we have discussed above, $\mu\hat{}(\mathbb{Z}_8, \{3,1\}) = 8/4+2$, but $\mu\hat{}(\mathbb{Z}_{40}, \{3,1\}) = 40/4+1$. The next value of n in question is $n = 104$.

Some more general questions:

Problem G.69 *Find $\mu\hat{}(\mathbb{Z}_n, \{3,1\})$ for all values of n.*

Problem G.70 *Find $\mu\hat{}(\mathbb{Z}_n, \{k,1\})$ for other values of k.*

Problem G.71 *Find $\mu\hat{}(\mathbb{Z}_n, \{k,l\})$ for other values of k and l.*

Problem G.72 *Find $\mu\hat{}(G, \{k,l\})$ in noncyclic groups G, in particular, for $G = \mathbb{Z}_k^r$.*

We are not aware of any general exact results besides the ones already discussed. However, there are a variety of constructions yielding lower bounds for $\mu\hat{}(\mathbb{Z}_n, \{k,l\})$, as we present next.

One such result was discovered (although for $l = 1$ only) by Hallfors in [111]:

Proposition G.73 (Hallfors; cf. [111]) *Suppose that n, k, and l are positive integers so that $l < k$ and n is divisible by $(k^2 - l^2)(k-1)$. Let H be the subgroup of order $k-1$ in $\mathbb{Z}_n$, and set*

$$A = \{a, a+1, \dots, a+c\} + H,$$

where

$$a = \frac{ln}{(k^2 - l^2)(k-1)} \quad \text{and} \quad c = \frac{n}{(k+l)(k-1)}.$$

Then A is a weak sum-free set in $\mathbb{Z}_n$, and thus

$$\mu\hat{}(\mathbb{Z}_n, \{k,l\}) \geq \frac{n}{k+l} + k - 1.$$

For example, if n is divisible by 3, then

$$\left\{\frac{n}{3}, \frac{n}{3}+1, \dots, \frac{2n}{3}\right\}$$

is a weak $(2,1)$-sum-free set in $\mathbb{Z}_n$ of size $n/3+1$, and if n is divisible by 16, then we get the weak $(3,1)$-sum-free set

$$\left\{\frac{n}{16}, \frac{n}{16}+1, \dots, \frac{3n}{16}\right\} \cup \left\{\frac{9n}{16}, \frac{9n}{16}+1, \dots, \frac{11n}{16}\right\}$$

of size $n/4+2$ mentioned above. The proof of Proposition G.73 is on page 373.

Another construction of Hallfors from [111] (though only presented there for the special case of $d = 4$) is as follows:

Proposition G.74 (Hallfors; cf. [111]) *Suppose that n, k, and l are positive integers so that $l < k$. Let $d \in D(n)$ be even, and suppose that $n \geq d(d/2-1)$ and that the remainders of k and l when divided by d differ by $d/2$. Let H be the subgroup of order n/d in $\mathbb{Z}_n$, and set*

$$A = (1+H) \cup H_d,$$

where

$$H_d = \{0, d, 2d, \dots, (d/2-2)d\} \subseteq H$$

(with $H_2 = \emptyset$). Then A is a weak sum-free set in $\mathbb{Z}_n$, and thus

$$\mu\hat{}(\mathbb{Z}_n, \{k,l\}) \geq n/d + d/2 - 1.$$

The proof of Proposition G.74 is on page 374.

Observe that Proposition G.74 has the obvious special case for $d = 2$: If n is even, k and l have opposite parity, and $l < k \leq n$, then the set

$$A = \{1, 3, 5, \dots, n-1\}$$

is a weak (k,l)-sum-free set in $\mathbb{Z}_n$, and thus

$$\mu\hat{}(\mathbb{Z}_n, \{k,l\}) \geq n/2.$$

This, however, already follows from Proposition G.63 and Corollary 4.5, since in this case we have:

$$\mu\hat{}(\mathbb{Z}_n, \{k,l\}) \geq \mu(\mathbb{Z}_n, \{k,l\}) \geq v_{k-l}(n, k+l) = n/2.$$

We continue with a set of other (rather easy) constructions.

Proposition G.75 *Suppose that $n = k(k-1)$. Let H be the subgroup of order $k-1$ in $\mathbb{Z}_n$, and set*

$$A = (H \setminus \{0\}) \cup (1 + H).$$

Then A is a weak $(k,1)$-sum-free set in $\mathbb{Z}_n$, so

$$\mu\hat{}(\mathbb{Z}_n, \{k,1\}) \geq 2k - 3.$$

This claim can be verified quickly, as

$$k\hat{}A \subseteq \{2, 3, \ldots, k-1\} + H,$$

so $k\hat{}A$ is disjoint from A. We should point out that Proposition G.75 sometimes supersedes our previous results: for example, it yields $\mu\hat{}(\mathbb{Z}_{20}, \{5,1\}) \geq 7$ (which is the exact value, as it turns out), while $\mu(\mathbb{Z}_{20}, \{5,1\}) = 4$ and $\gamma\hat{}(\mathbb{Z}_{20}, \{5,1\}) = 6$.

In a similar vein:

Proposition G.76 *Suppose that $n = k(k^2-1)$. Let H be the subgroup of order k in $\mathbb{Z}_n$, define*

$$h_0 = \begin{cases} 0 & \text{if } k \text{ is odd,} \\ n/2 & \text{if } k \text{ is even,} \end{cases}$$

and set

$$A = (\{1, 2, \ldots, k-1\} + H) \cup (k + (H \setminus \{h_0\})).$$

Then A is a weak $(k,1)$-sum-free set in $\mathbb{Z}_n$, so

$$\mu\hat{}(\mathbb{Z}_n, \{k,1\}) \geq k^2 - 1 = \frac{n}{k+1} + k - 1.$$

To verify our claim, observe that

$$\begin{aligned} k\hat{}A &= \{k + k(k-1)/2 \cdot (k^2-1)\} \cup (\{k+1, k+2, \ldots, (k-1)k + k - 1\} + H) \\ &= \{k + h_0\} \cup (\{k+1, k+2, \ldots, k^2 - 1\} + H) \\ &= \mathbb{Z}_n \setminus A, \end{aligned}$$

establishing our claim.

Relying on a computer program, Hallfors in [111] exhibited the values of $\mu\hat{}(\mathbb{Z}_n, \{k,l\})$ for $n \leq 40$ and $1 \leq l < k \leq 4$—we provide these data (except for $(k,l) = (2,1)$ for which we have Theorem G.67) in the table below.

n	$\mu\hat{}(\mathbb{Z}_n, \{3,1\})$	$\mu\hat{}(\mathbb{Z}_n, \{4,1\})$	$\mu\hat{}(\mathbb{Z}_n, \{3,2\})$	$\mu\hat{}(\mathbb{Z}_n, \{4,2\})$	$\mu\hat{}(\mathbb{Z}_n, \{4,3\})$
5	3	4	3	3	4
6	3	4	3	4	4
7	3	4	3	4	4
8	4	4	4	4	4
9	4	4	4	4	4
10	4	5	5	4	5
11	4	5	4	4	4
12	5	6	6	5	6
13	5	5	4	5	5
14	5	7	7	5	7
15	5	6	5	5	5
16	6	8	8	5	8
17	6	6	5	5	5
18	6	9	9	6	9
19	6	6	6	6	6
20	7	10	10	6	10
21	7	6	7	7	7
22	7	11	11	6	11
23	7	7	6	6	6
24	8	12	12	8	12
25	8	8	7	7	6
26	8	13	13	7	13
27	9	8	9	9	9
28	9	14	14	8	14
29	9	8	8	7	7
30	10	15	15	10	15
31	9	9	8	8	7
32	10	16	16	9	16
33	11	9	11	11	11
34	10	17	17	8	17
35	10	10	10	8	8
36	12	18	18	12	18
37	11	10	9	9	8
38	11	19	19	9	19
39	13	10	13	13	13
40	11	20	20	11	20

Additionally, Hallfors in [110] also computed the values of $\mu\hat{}(\mathbb{Z}_n, \{k,l\})$ for $n \leq 30$, $l = 1$, and $5 \leq k \leq 9$, which are as follows:

n	$\mu\hat{}(\mathbb{Z}_n,\{5,1\})$	$\mu\hat{}(\mathbb{Z}_n,\{6,1\})$	$\mu\hat{}(\mathbb{Z}_n,\{7,1\})$	$\mu\hat{}(\mathbb{Z}_n,\{8,1\})$	$\mu\hat{}(\mathbb{Z}_n,\{9,1\})$
5	4	5	5	5	5
6	4	5	6	6	6
7	5	6	6	7	7
8	5	6	7	7	8
9	5	6	7	8	8
10	5	6	7	8	8
11	5	6	7	8	9
12	6	6	7	8	9
13	6	6	7	8	9
14	6	7	7	8	9
15	6	7	7	8	9
16	6	8	8	8	9
17	6	7	8	8	9
18	6	9	8	9	9
19	7	7	8	9	9
20	7	10	8	10	10
21	7	8	8	9	10
22	7	11	8	11	10
23	7	8	8	9	10
24	8	12	9	12	10
25	8	8	9	10	10
26	8	13	9	13	10
27	9	9	9	10	10
28	8	14	9	14	10
29	8	9	9	10	10
30	10	15	9	15	11

By carefully analyzing these data, one may be able to generate constructions that are not included in this section:

Problem G.77 *Develop new general constructions for weak (k,l)-sum-free sets in $\mathbb{Z}_n$, and thereby exhibit new lower bounds for $\mu\hat{}(\mathbb{Z}_n,\{k,l\})$.*

It may be interesting to investigate the maximum size of weak (k,l)-sum-free sets in groups from the opposite perspective: Rather than attempting to find $\mu\hat{}(G,\{k,l\})$ for a given group G, we look for all groups G that contain a weak (k,l)-sum-free set of a given size:

Problem G.78 *For each $k,l,m \in \mathbb{N}$ with $l < k$, find all groups G for which $\mu\hat{}(G,\{k,l\}) \geq m$.*

We can formulate two sub-problems of Problem G.78:

Problem G.79 *For each $k,l,m \in \mathbb{N}$ with $l < k$, find the least integer $f(m,\{k,l\})$ for which $\mu\hat{}(\mathbb{Z}_n,\{k,l\}) \geq m$ holds for $n = f(m,\{k,l\})$.*

Problem G.80 *For each $k,l,m \in \mathbb{N}$ with $l < k$, find the least integer $g(m,\{k,l\})$ for which $\mu\hat{}(\mathbb{Z}_n,\{k,l\}) \geq m$ holds for all $n \geq g(m,\{k,l\})$.*

For example, the data above indicate that $f(10,\{3,1\}) = 30$ and $g(10,\{3,1\}) = 32$.
We have the following results:

Proposition G.81 *For all positive integers k, l, and m, with $l < k$, we have*

$$g(m,\{k,l\}) \geq f(m,\{k,l\}) \geq m,$$

with equality if, and only if, $m \leq k-1$.

The fact that both $f(m,\{k,l\})$ and $g(m,\{k,l\})$ must be at least m is obvious, and it is also obvious that equality holds when $m \leq k-1$. The fact that for $m \geq k$ we have

$$g(m,\{k,l\}) \geq f(m,\{k,l\}) \geq m+1$$

follows directly from the Lemma on page 337.

We also have the values of $f(m,\{k,1\})$ and $g(m,\{k,1\})$ for $m = k$ and $m = k+1$:

Proposition G.82 *For every integer $k \geq 2$ we have*

$$f(k,\{k,1\}) = g(k,\{k,1\}) = \begin{cases} k+2 & \text{if } k \equiv 1 \text{ mod } 4, \\ k+1 & \text{otherwise;} \end{cases}$$

and

$$f(k+1,\{k,1\}) = g(k+1,\{k,1\}) = 2k+2.$$

The proof starts on page 374.

Problem G.83 *Find the values of (or at least good bounds for) $f(m,\{k,l\})$ and $g(m,\{k,l\})$; in particular, evaluate $f(k+2,\{k,1\})$ and $g(k+2,\{k,1\})$.*

As we did with (k,l)-sum-free sets in Section G.1.1, we also investigate weak (k,l)-sum-free sets where no elements are "wasted": we say that a weak (k,l)-sum-free set A in G is *complete* if $k\hat{}A$ and $l\hat{}A$ partition G; in other words, not only do we have $k\hat{}A \cap l\hat{}A = \emptyset$ (i.e., A is weakly (k,l)-sum-free), but also $k\hat{}A \cup l\hat{}A = G$.

Looking through our results above, we find the following complete weak (k,l)-sum-free sets:

- The set A of Proposition G.73 is complete when $l \leq k-2$ (see page 373).
- The set A of Proposition G.74 is complete when

  $$d/2 - 1 < l < k < n/d$$

  (see page 374).
- The set A in Proposition G.76 is complete (see page 299).
- The set A in the proof of the second statement of Proposition G.82 is complete when $n = 2k+2$ (see page 374).

Problem G.84 *Find other complete weak (k,l)-sum-free sets in abelian groups.*

G.3.2 Limited number of terms

G.3.3 Arbitrary number of terms

G.4 Restricted signed sumsets

G.4.1 Fixed number of terms

G.4.2 Limited number of terms

G.4.3 Arbitrary number of terms

Part V

Pudding

"The proof is in the pudding!" In this chapter we provide proofs to those of our results in the book that have not been published (and that were not presented where the results are stated). A bit of warning: as is often the case, puddings can be messy ... read at your own risk!

Proof of Proposition 2.2

1. Let us define

$$a'(j,k) := \sum_{i=0}^{k} \binom{k}{i}\binom{j}{i} 2^i.$$

Clearly, $a'(j,0) = a'(0,k) = 1$; below we prove that $a'(j,k)$ also satisfies the recursion.

We have

$$\begin{aligned} a'(j-1,k-1) &= \sum_{i=0}^{k-1} \binom{k-1}{i}\binom{j-1}{i} 2^i \\ &= \sum_{i=0}^{k-2} \binom{k-1}{i}\binom{j-1}{i} 2^i + \binom{j-1}{k-1} 2^{k-1}, \end{aligned}$$

and

$$\begin{aligned} a'(j-1,k) &= \sum_{i=0}^{k} \binom{k}{i}\binom{j-1}{i} 2^i \\ &= \sum_{i=0}^{k-1} \binom{k}{i}\binom{j-1}{i} 2^i + \binom{j-1}{k} 2^k \\ &= \sum_{i=0}^{k-1} \binom{k-1}{i-1}\binom{j-1}{i} 2^i + \sum_{i=0}^{k-2} \binom{k-1}{i}\binom{j-1}{i} 2^i + \binom{j-1}{k-1} 2^{k-1} + \binom{j-1}{k} 2^k. \end{aligned}$$

Next, we add $a'(j-1,k)$ and $a'(j-1,k-1)$. Note that

$$\binom{j-1}{k-1} 2^{k-1} + \binom{j-1}{k-1} 2^{k-1} + \binom{j-1}{k} 2^k = \binom{j}{k} 2^k,$$

and

$$\sum_{i=0}^{k-2} \binom{k-1}{i}\binom{j-1}{i} 2^i + \sum_{i=0}^{k-2} \binom{k-1}{i}\binom{j-1}{i} 2^i = \sum_{i=0}^{k-2} \binom{k-1}{i}\binom{j-1}{i} 2^{i+1},$$

and by replacing i by $i-1$, this sum becomes

$$\sum_{i=0}^{k-1} \binom{k-1}{i-1}\binom{j-1}{i-1} 2^i.$$

Therefore,

$$\begin{aligned} a'(j-1,k) + a'(j-1,k-1) &= \sum_{i=0}^{k-1} \binom{k-1}{i-1}\binom{j-1}{i} 2^i + \sum_{i=0}^{k-1} \binom{k-1}{i-1}\binom{j-1}{i-1} 2^i + \binom{j}{k} 2^k \\ &= \sum_{i=0}^{k-1} \binom{k-1}{i-1}\binom{j}{i} 2^i + \binom{j}{k} 2^k \\ &= \sum_{i=0}^{k} \binom{k-1}{i-1}\binom{j}{i} 2^i \\ &= \sum_{i=0}^{k} \binom{k}{i}\binom{j}{i} 2^i - \sum_{i=0}^{k} \binom{k-1}{i}\binom{j}{i} 2^i \\ &= a'(j,k) - a'(j,k-1). \end{aligned}$$

2. The cases of $j = 0$ or $k = 0$ can be verified easily (see also the comments after the statement of Proposition 2.2). For positive j and k we use Proposition 2.1 and part 1 above:

$$\begin{aligned} c(j,k) &= a(j,k-1) + a(j-1,k-1) \\ &= \sum_{i=0}^{k} \binom{j}{i}\binom{k-1}{i}2^i + \sum_{i=0}^{k} \binom{j-1}{i}\binom{k-1}{i}2^i \\ &= \sum_{i=0}^{k} \left[\binom{j}{i} + \binom{j-1}{i}\right]\binom{k-1}{i-1}2^i, \end{aligned}$$

and

$$\begin{aligned} c(j,k) &= a(j,k) - a(j-1,k) \\ &= \sum_{i=0}^{k} \binom{k}{i}\binom{j}{i}2^i - \sum_{i=0}^{k} \binom{k}{i}\binom{j-1}{i}2^i \\ &= \sum_{i=0}^{k} \binom{k}{i}\left[\binom{j}{i} - \binom{j-1}{i}\right]2^i \\ &= \sum_{i=0}^{k} \binom{k}{i}\binom{j-1}{i-1}2^i. \end{aligned}$$

□

Proof of Proposition 3.1

Suppose that G has invariant factorization

$$\mathbb{Z}_{n_1} \times \cdots \times \mathbb{Z}_{n_r}$$

(so $2 \leq n_1$ and n_i is a divisor of n_{i+1} for $i = 1, \ldots, r-1$). Let p be any prime divisor of n_1, and set

$$H = \langle \{p \cdot g \mid g \in G\} \rangle.$$

(Actually, p would not need to be prime; any divisor $d > 1$ of n_1 would do, including $d = n_1$.) It is easy to see that

$$H \cong \mathbb{Z}_{n_1/p} \times \cdots \times \mathbb{Z}_{n_r/p}$$

and

$$G/H \cong \mathbb{Z}_p^r.$$

Since each nonzero element of $\mathbb{Z}_p^r$ has order p, it cannot be generated by fewer than r elements.

Now suppose that $A = \{a_i \mid i = 1, \ldots, m\} \subseteq G$ has size m and that $\langle A \rangle = G$. One can readily verify that

$$\overline{A} = \{a_i + H \mid i = 1, \ldots, m\}$$

then generates G/H. Thus we have

$$m = |A| \geq |\overline{A}| \geq r,$$

as claimed. □

Proof of Proposition 3.4

As usual, we let $A = \{a_1, \ldots, a_m\}$.

Suppose first that $x \in h(A \cup (-A))$; we then have

$$\begin{aligned} x &= \lambda_1 a_1 + \cdots + \lambda_m a_m + \lambda'_1(-a_1) + \cdots + \lambda'_m(-a_m) \\ &= (\lambda_1 - \lambda'_1)a_1 + \cdots + (\lambda_m - \lambda'_m)a_m \end{aligned}$$

for some nonnegative integers $\lambda_1, \ldots, \lambda_m$ and $\lambda'_1, \ldots, \lambda'_m$ for which

$$\lambda_1 + \cdots + \lambda_m + \lambda'_1 + \cdots + \lambda'_m = h.$$

Divide the set $\{1, 2, \ldots, m\}$ into two subsets, I_1 and I_2, such that $i \in I_1$ when $\lambda_i \geq \lambda'_i$ and $i \in I_2$ when $\lambda_i < \lambda'_i$. We can then write

$$\begin{aligned} \sum_{i=1}^{m} |\lambda_i - \lambda'_i| &= \sum_{i \in I_1} (\lambda_i - \lambda'_i) + \sum_{i \in I_2} (\lambda'_i - \lambda_i) \\ &= \sum_{i=1}^{m} \lambda_i + \sum_{i=1}^{m} \lambda'_i - 2\left(\sum_{i \in I_1} \lambda'_i + \sum_{i \in I_2} \lambda_i\right) \\ &= h - 2h_0 \end{aligned}$$

where

$$\begin{aligned} 2h_0 &= 2\left(\sum_{i \in I_1} \lambda'_i + \sum_{i \in I_2} \lambda_i\right) \\ &= \sum_{i \in I_2} \lambda_i + \sum_{i \in I_1} \lambda'_i + \sum_{i \in I_2} \lambda_i + \sum_{i \in I_1} \lambda'_i \\ &\leq \sum_{i \in I_2} \lambda_i + \sum_{i \in I_1} \lambda'_i + \sum_{i \in I_2} \lambda'_i + \sum_{i \in I_1} \lambda_i \\ &= h. \end{aligned}$$

Thus, $x \in (h - 2h_0)_{\pm}A$.

Conversely, let h_0 be a nonnegative integer (not greater than $\lfloor h/2 \rfloor$), and suppose that $x \in (h - 2h_0)_{\pm}A$; we then have integers $\lambda_1, \ldots, \lambda_m$ with

$$x = \lambda_1 a_1 + \cdots + \lambda_m a_m$$

and

$$|\lambda_1| + \cdots + |\lambda_m| = h - 2h_0.$$

Similar to the above, divide the set $\{1, 2, \ldots, m\}$ into two subsets, I_1 and I_2, such that $i \in I_1$ when $\lambda_i \geq 0$ and $i \in I_2$ when $\lambda_i < 0$. We then have

$$\begin{aligned} x &= \sum_{i \in I_1} \lambda_i a_i + \sum_{i \in I_2} \lambda_i a_i \\ &= \sum_{i \in I_1} \lambda_i a_i + \sum_{i \in I_2} (-\lambda_i)(-a_i) \\ &= \sum_{i \in I_1} \lambda_i a_i + \sum_{i \in I_2} (-\lambda_i)(-a_i) + h_0 a_1 + h_0(-a_1). \end{aligned}$$

This expression of x involves only nonnegative coefficients and thus no cancellation of terms occurs (even when terms can be combined, e.g., when $a_1 = -a_1$), so for the nonnegative coefficients we have

$$\sum_{i \in I_1} \lambda_i + \sum_{i \in I_2} (-\lambda_i) + 2h_0 = |\lambda_1| + \cdots + |\lambda_m| + 2h_0 = h.$$

Thus $x \in h(A \cup (-A))$. $\square$

It may be worthwhile to mention that we can easily prove that

$$h\hat{}(A \cup (-A)) \subseteq h\hat{_\pm}A \cup (h-2)\hat{_\pm}A \cup (h-4)\hat{_\pm}A \cup \cdots.$$

The proof is essentially the same as the corresponding proof above: note that with the additional assumption that each coefficient λ_i and λ_i' is 0 or 1 implies that the coefficients $\lambda_i - \lambda_i'$ in the expression of x above are all in $\{-1, 0, 1\}$, and thus $x \in (h - 2h_0)\hat{_\pm}A$.

Proof of Proposition 3.5

The claim is obvious for $h = 0$ and $h = 1$, since

$$0_\pm A = \{0\} = 0A,$$

and

$$1_\pm A = A \cup -A = A = 1A.$$

For $h \geq 2$, we show that $(h-2)_\pm A \subseteq h_\pm A$, from which our proposition follows by Proposition 3.4.

To show that $(h-2)_\pm A \subseteq h_\pm A$, let $A = \{a_1, \ldots, a_m\}$ and

$$g = \lambda_1 a_1 + \cdots + \lambda_m a_m$$

with coefficients $\lambda_1, \ldots, \lambda_m \in \mathbb{Z}$ for which

$$|\lambda_1| + \cdots + |\lambda_m| = h - 2,$$

so that $g \in (h-2)_\pm A$. Without loss of generality, assume that $\lambda_1 \geq 0$. Since $A = -A$, there is an index $i \in \{1, \ldots, m\}$ so that $-a_1 = a_i$. We consider three cases.

If $i = 1$, then $2a_1 = 0$, and thus

$$g = (\lambda_1 + 2)a_1 + \lambda_2 a_2 + \cdots + \lambda_m a_m,$$

so $g \in h_\pm A$.

If $i \neq 1$, w.l.o.g. $i = 2$, and $\lambda_2 \geq 0$, then $a_1 + a_2 = 0$, and thus

$$g = (\lambda_1 + 1)a_1 + (\lambda_2 + 1)a_2 + \lambda_3 a_3 + \cdots + \lambda_m a_m,$$

so $g \in h_\pm A$.

If $i \neq 1$, w.l.o.g. $i = 2$, and $\lambda_2 < 0$, then

$$g = (\lambda_1 + 1)a_1 + (-\lambda_2 + 1)a_2 + \lambda_3 a_3 + \cdots + \lambda_m a_m,$$

so $g \in h_\pm A$ again. $\square$

Proof of Proposition 4.2

The claim is true for $h = 1$, so we assume that $h \geq 2$.

Since g is not divisible by p, we have $\gcd(p^k, g) = 1$, so g is relatively prime with all divisors of p^r.

Suppose first that $p \equiv 1 \bmod h$. We then have

$$\begin{aligned}
v_g(p^r, h) &= \max\left\{\left(\left\lfloor\frac{p^k-2}{h}\right\rfloor+1\right)\cdot\frac{p^r}{p^k} \mid 1 \leq k \leq r\right\} \\
&= \max\left\{\left(\frac{p^k-1-h}{h}+1\right)\cdot p^{r-k} \mid 1 \leq k \leq r\right\} \\
&= \max\left\{\frac{p^k-1}{h}\cdot p^{r-k} \mid 1 \leq k \leq r\right\} \\
&= \max\left\{\frac{p^r-p^{r-k}}{h} \mid 1 \leq k \leq r\right\} \\
&= \frac{p^r-1}{h}.
\end{aligned}$$

Next, let's assume that $p \equiv 0 \bmod h$; this can only happen if $p = h$. We then have

$$\begin{aligned}
v_g(p^r, p) &= \max\left\{\left(\left\lfloor\frac{p^k-2}{p}\right\rfloor+1\right)\cdot\frac{p^r}{p^k} \mid 1 \leq k \leq r\right\} \\
&= \max\left\{\left(\frac{p^k-p}{p}+1\right)\cdot p^{r-k} \mid 1 \leq k \leq r\right\} \\
&= p^{r-1} \\
&= \left(\left\lfloor\frac{p-2}{p}\right\rfloor+1\right)\cdot p^{r-1}.
\end{aligned}$$

For our main case, assume that $p \equiv i \bmod h$ for some $2 \leq i \leq h-1$. Since $p \in D(p^r)$, we must have

$$v_g(p^r, p) \geq \left(\left\lfloor\frac{p-2}{h}\right\rfloor+1\right)\cdot p^{r-1};$$

so, to prove our claim, we must show that

$$\left(\left\lfloor\frac{p-2}{h}\right\rfloor+1\right)\cdot p^{r-1} \geq \left(\left\lfloor\frac{p^k-2}{h}\right\rfloor+1\right)\cdot p^{r-k}$$

holds for every $2 \leq k \leq r$. (We will, in fact, show that strict inequality holds.) We consider two subcases: when $p < h$ and when $p \geq h$.

If $p < h$, then for every $2 \leq k \leq r$ we have

$$\begin{aligned}
\left(\left\lfloor\frac{p-2}{h}\right\rfloor+1\right)\cdot p^{r-1} &= p^{r-1} \\
&= \frac{p^r+p^{r-2}\cdot p}{p+1} \\
&> \frac{p^r+p^{r-k}\cdot(p-1)}{p+1} \\
&= \left(\frac{p^k-2}{p+1}+1\right)\cdot p^{r-k} \\
&\geq \left(\left\lfloor\frac{p^k-2}{h}\right\rfloor+1\right)\cdot p^{r-k},
\end{aligned}$$

as claimed.

Similarly, if $p \geq h$, then for every $2 \leq k \leq r$ we have

$$\begin{aligned}
\left(\left\lfloor \frac{p-2}{h} \right\rfloor + 1\right) \cdot p^{r-1} &\geq \left(\frac{p-(h-1)}{h} + 1\right) \cdot p^{r-1} \\
&= \frac{p+1}{h} \cdot p^{r-1} \\
&= \frac{p^k + p^{k-1}}{h} \cdot p^{r-k} \\
&> \frac{p^k + h - 2}{h} \cdot p^{r-k} \\
&\geq \left(\left\lfloor \frac{p^k - 2}{h} \right\rfloor + 1\right) \cdot p^{r-k},
\end{aligned}$$

as claimed. □

Proof of Proposition 4.3

The first claim is quite straightforward:

$$\begin{aligned}
v_1(n,1) &= \max\left\{\left(\left\lfloor \frac{d-1-1}{1} \right\rfloor + 1\right) \cdot \frac{n}{d} \mid d \in D(n)\right\} \\
&= \max\left\{n - \frac{n}{d} \mid d \in D(n)\right\} \\
&= n - \frac{n}{n} \\
&= n - 1.
\end{aligned}$$

For $v_1(n,2)$ we have

$$\begin{aligned}
v_1(n,2) &= \max\left\{\left(\left\lfloor \frac{d-1-1}{2} \right\rfloor + 1\right) \cdot \frac{n}{d} \mid d \in D(n)\right\} \\
&= \max\left\{\left\lfloor \frac{d}{2} \right\rfloor \cdot \frac{n}{d} \mid d \in D(n)\right\}.
\end{aligned}$$

Now if n is odd, then all its divisors are odd, and we have

$$\begin{aligned}
v_1(n,2) &= \max\left\{\frac{d-1}{2} \cdot \frac{n}{d} \; d \in D(n)\right\} \\
&= \max\left\{\frac{n}{2} - \frac{n}{2d} \; d \in D(n)\right\} \\
&= \frac{n}{2} - \frac{n}{2n} \\
&= \frac{n-1}{2},
\end{aligned}$$

as claimed.

If n is even, then it has both even and odd divisors. For each even divisor d of n, we get

$$\left\lfloor \frac{d}{2} \right\rfloor \cdot \frac{n}{d} = \frac{n}{2};$$

while for odd divisors d, we have

$$\left\lfloor \frac{d}{2} \right\rfloor \cdot \frac{n}{d} < \frac{n}{2}.$$

Therefore, if n is even, we have

$$v_1(n,2) = \frac{n}{2}.$$

The proof of the third claim is similar. □

Proof of Theorem 4.4

Suppose that d is a positive divisor of n, and let i be the remainder of d when divided by h. We define the function

$$f(d) = \left(\left\lfloor \frac{d-1-\gcd(d,g)}{h} \right\rfloor + 1 \right) \cdot \frac{n}{d}.$$

We first prove the following.

Claim 1: We have

$$f(d) = \begin{cases} \frac{n}{h} \cdot \left(1 + \frac{h-i}{d}\right) & \text{if } \gcd(d,g) < i; \\ \frac{n}{h} \cdot \left(1 - \frac{h}{d}\right) & \text{if } h|d \text{ and } g = h; \\ \frac{n}{h} \cdot \left(1 - \frac{i}{d}\right) & \text{otherwise.} \end{cases}$$

Proof of Claim 1. We start with

$$\left\lfloor \frac{d-1-\gcd(d,g)}{h} \right\rfloor = \frac{d-i}{h} + \left\lfloor \frac{i-1-\gcd(d,g)}{h} \right\rfloor.$$

We investigate the maximum and minimum values of the quantity $\left\lfloor \frac{i-1-\gcd(d,g)}{h} \right\rfloor$.

For the maximum, we have

$$\left\lfloor \frac{i-1-\gcd(d,g)}{h} \right\rfloor \leq \left\lfloor \frac{(h-1)-1-1}{h} \right\rfloor \leq 0,$$

with equality if, and only if, $i-1-\gcd(d,g) \geq 0$; that is, $\gcd(d,g) < i$.

For the minimum, we get

$$\left\lfloor \frac{i-1-\gcd(d,g)}{h} \right\rfloor \geq \left\lfloor \frac{0-1-g}{h} \right\rfloor \geq \left\lfloor \frac{0-1-h}{h} \right\rfloor = -2,$$

with equality if, and only if, $i = 0$, $\gcd(d,g) = g$, and $g = h$; that is, $h|d$ and $g = h$.

The proof of Claim 1 now follows easily. □

Claim 2: Using the notations as above, assume that $\gcd(d,g) \geq i$. Then

$$f(d) \leq \begin{cases} \frac{n}{h} & \text{if } g \neq h; \\ \frac{n-1}{h} & \text{if } g = h. \end{cases}$$

Proof of Claim 2. By Claim 1, we have

$$f(d) = \begin{cases} \frac{n}{h} \cdot \left(1 - \frac{h}{d}\right) & \text{if } h|d \text{ and } g = h; \\ \frac{n}{h} \cdot \left(1 - \frac{i}{d}\right) & \text{otherwise.} \end{cases}$$

Therefore,

$$f(d) \leq \frac{n}{h}.$$

Furthermore, unless $i = 0$ and $g \neq h$, we have

$$f(d) \leq \frac{n}{h} \cdot \left(1 - \frac{1}{d}\right) \leq \frac{n}{h} \cdot \left(1 - \frac{1}{n}\right) = \frac{n-1}{h}.$$

□

Claim 3: For all g, h, and n we have

$$v_g(n,h) \geq \begin{cases} \lfloor \frac{n}{h} \rfloor & \text{if } g \neq h; \\ \lfloor \frac{n-1}{h} \rfloor & \text{if } g = h. \end{cases}$$

Proof of Claim 3. We first note that

$$\begin{aligned} v_g(n,h) &= \max\left\{\left(\left\lfloor \frac{d-1-\gcd(d,g)}{h} \right\rfloor + 1\right) \cdot \frac{n}{d} \mid d \in D(n)\right\} \\ &\geq \left(\left\lfloor \frac{n-1-\gcd(n,g)}{h} \right\rfloor + 1\right) \cdot \frac{n}{n} \\ &= \left\lfloor \frac{n-1-\gcd(n,g)}{h} \right\rfloor + 1 \\ &\geq \left\lfloor \frac{n-1-g}{h} \right\rfloor + 1 \\ &= \left\lfloor \frac{n+(h-g-1)}{h} \right\rfloor. \end{aligned}$$

The claim now follows, since $h - g - 1 \geq 0$, unless $g = h$ in which case $h - g - 1 = -1$. □

We are now ready for the proof of Theorem 4.4.

Proof of Theorem 4.4. If $I = \emptyset$, then by Claims 2 and 3 we have

$$v_g(n,h) = \begin{cases} \lfloor \frac{n}{h} \rfloor & \text{if } g \neq h; \\ \lfloor \frac{n-1}{h} \rfloor & \text{if } g = h \end{cases}$$

Suppose now that $I \neq \emptyset$. Here we call a positive divisor d of n good, if its remainder mod h is larger than $\gcd(d,g)$. Since $I \neq \emptyset$, n has at least one good divisor; we let D_I be the collection of good divisors of n. By Claim 1, we have

$$v_g(n,h) = \max\{f(d) \mid d \in D_I\}.$$

Suppose now that d and d' are two elements of D_I that both leave a remainder of i mod h. If $d' > d$, then by Claim 1, we have

$$f(d) = \frac{n}{h} \cdot \left(1 + \frac{h-i}{d}\right) > \frac{n}{h} \cdot \left(1 + \frac{h-i}{d'}\right) = f(d'),$$

from which our result follows. □

Proof of Proposition 4.9

For a divisor d of n, let

$$g_d(n,h) = \left(2 \cdot \left\lfloor \frac{d-2}{2h} \right\rfloor + 1\right) \cdot \frac{n}{d}.$$

Regarding $h = 1$, we see that

$$g_d(n,1) = \left(2 \cdot \left\lfloor \frac{d-2}{2} \right\rfloor + 1\right) \cdot \frac{n}{d} = (d-\delta) \cdot \frac{n}{d} = \left(1 - \frac{\delta}{d}\right) \cdot n,$$

where $\delta = 1$ when d is even and $\delta = 2$ when d is odd. This quantity is maximized with $d = n$, from which the result for $v_{\pm}(n,1)$ follows.

Now for $h = 2$ and even n, we have $2 \in D(n)$, and thus

$$\frac{n}{2} = g_2(n,2) \leq v_{\pm}(n,2) \leq v_1(n,2) = \frac{n}{2},$$

so $v_{\pm}(n,2) = n/2$. When n is odd, then each $d \in D(n)$ is odd too, so

$$v_{\pm}(n,2) = \max\{g_d(n,2)\} \leq \max\left\{\left(2 \cdot \frac{d-3}{4} + 1\right) \cdot \frac{n}{d}\right\} = \max\left\{\left(1 - \frac{1}{d}\right) \cdot \frac{n}{2}\right\}.$$

Therefore, when $n \equiv 3 \mod 4$, then

$$\frac{n-1}{2} = g_n(n,2) \leq v_{\pm}(n,2) \leq \left(1 - \frac{1}{n}\right) \cdot \frac{n}{2} = \frac{n-1}{2},$$

so equality holds throughout. When $n \equiv 1 \mod 4$, then we get

$$\frac{n-3}{2} = g_n(n,2) \leq v_{\pm}(n,2),$$

and there is no $d \in D(n)$ for which $g_d(n,2)$ is larger than $(n-3)/2$, since if $d \neq n$, then $d \leq n/3$, and thus

$$\left(1 - \frac{1}{d}\right) \cdot \frac{n}{2} \leq \left(1 - \frac{1}{n/3}\right) \cdot \frac{n}{2} = \frac{n-3}{2}.$$

For $h = 3$ and even n, we get $v_{\pm}(n,3) = n/2$ as we did for $h = 2$. When n is odd but divisible by 3, then $3 \in D(n)$, so

$$\frac{n}{3} = g_3(n,3) \leq v_{\pm}(n,3) = \max\{g_d(n,3)\} \leq \max\left\{\left(2 \cdot \frac{d-3}{6} + 1\right) \cdot \frac{n}{d}\right\} = \frac{n}{3},$$

and thus $v_{\pm}(n,3) = n/3$. When n is odd and is not divisible by 3, then each $d \in D(n)$ leaves a remainder of 1 or 5 mod 6, so

$$v_{\pm}(n,3) = \max\{g_d(n,3)\} \leq \max\left\{\left(2 \cdot \frac{d-5}{6} + 1\right) \cdot \frac{n}{d}\right\} = \max\left\{\left(1 - \frac{2}{d}\right) \cdot \frac{n}{3}\right\}.$$

Therefore, when $n \equiv 5 \mod 6$, then

$$\frac{n-2}{3} = g_n(n,3) \leq v_{\pm}(n,3) \leq \left(1 - \frac{2}{n}\right) \cdot \frac{n}{3} = \frac{n-2}{3},$$

so equality holds throughout. When $n \equiv 1 \mod 6$, then we get

$$\frac{n-4}{2} = g_n(n,3) \leq v_{\pm}(n,3),$$

and there is no $d \in D(n)$ for which $g_d(n,3)$ is larger than $(n-4)/3$, since if $d \neq n$, then $d \leq n/5$, and thus

$$\left(1-\frac{2}{d}\right)\cdot\frac{n}{3} \leq \left(1-\frac{2}{n/5}\right)\cdot\frac{n}{3} = \frac{n-10}{3}.$$

This completes our proof. □

Proof of Proposition 4.10

For a divisor d of n, let

$$g_d(n,4) = \left(2\cdot\left\lfloor\frac{d-2}{8}\right\rfloor+1\right)\cdot\frac{n}{d}.$$

As before, for even n, we have $2 \in D(n)$, and thus

$$\frac{n}{2} = g_2(n,4) \leq v_\pm(n,4) \leq v_1(n,4) = \frac{n}{2},$$

so $v_\pm(n,4) = n/2$.

Suppose that n is odd but has one or more divisors d congruent to 3 mod 8, then for all such d we have

$$g_d(n,4) = \left(2\cdot\frac{d-3}{8}+1\right)\cdot\frac{n}{d} = \left(1+\frac{1}{d}\right)\frac{n}{4},$$

which is maximal when d is minimal. For any other $d \in D(n)$ we have

$$g_d(n,4) \leq \left(2\cdot\frac{d-5}{8}+1\right)\cdot\frac{n}{d} < \frac{n}{4},$$

so our claim holds in this case.

Suppose now that n is odd and has no divisors congruent to 3 mod 8; in particular, $3 \notin D(n)$ and thus $n/3 \notin D(n)$, so the second largest divisor of n is at most $n/5$. We have

$$g_d(n,4) \leq \left(2\cdot\frac{d-5}{8}+1\right)\cdot\frac{n}{d} = \left(1-\frac{1}{d}\right)\cdot\frac{n}{4}.$$

We consider the cases of $n \equiv 5$ mod 8, $n \equiv 7$ mod 8, and $n \equiv 1$ mod 8 separately. Let $d_0 \in D(n)$ be such that $v_\pm(n,4) = g_{d_0}(n,4)$.

If $n \equiv 5$ mod 8, then

$$\frac{n-1}{4} = 2\cdot\left\lfloor\frac{n-2}{8}\right\rfloor+1 = g_n(n,4) \leq g_{d_0}(n,4) \leq \left(1-\frac{1}{d_0}\right)\cdot\frac{n}{4} \leq \left(1-\frac{1}{n}\right)\cdot\frac{n}{4} = \frac{n-1}{4},$$

so equality holds throughout.

If $n \equiv 7$ mod 8, then $d_0 = n$ and thus

$$v_\pm(n,4) = g_n(n,4) = 2\cdot\left\lfloor\frac{n-2}{8}\right\rfloor+1 = \frac{n-3}{4},$$

since if $d_0 \neq n$, then $d_0 \leq n/5$, so

$$g_{d_0}(n,4) \leq \left(1-\frac{1}{d_0}\right)\cdot\frac{n}{4} \leq \left(1-\frac{1}{n/5}\right)\cdot\frac{n}{4} = \frac{n-5}{4}.$$

Suppose, finally, that $n \equiv 1$ mod 8. If $d_0 = n$, we get

$$v_\pm(n,4) = g_n(n,4) = 2\cdot\left\lfloor\frac{n-2}{8}\right\rfloor+1 = \frac{n-5}{4}.$$

If $d_0 \neq n$, then $d_0 \leq n/5$. If $d_0 = n/5$, we get

$$v_{\pm}(n,4) = g_{n/5}(n,4) \leq \left(1 - \frac{1}{n/5}\right) \cdot \frac{n}{4} = \frac{n-5}{4}.$$

If $d_0 \leq n/7$, then

$$v_{\pm}(n,4) = g_{d_0}(n,4) \leq \left(1 - \frac{1}{d_0}\right) \cdot \frac{n}{4} \leq \left(1 - \frac{1}{n/7}\right) \cdot \frac{n}{4} = \frac{n-7}{4}.$$

Therefore,

$$v_{\pm}(n,4) = \frac{n-5}{4} = 2 \cdot \left\lfloor \frac{n-2}{8} \right\rfloor + 1.$$

That completes our proof. □

Proof of Theorem 4.17

We need to prove that, for $m = v_1(n,h)$, we have $u(n,m,h) < n$ but $u(n,m+1,h) \geq n$. Let $d_0 \in D(n)$ be such that

$$v_1(n,h) = \max\left\{\left(\left\lfloor \frac{d-2}{h} \right\rfloor + 1\right) \cdot \frac{n}{d} \mid d \in D(n)\right\} = \left(\left\lfloor \frac{d_0-2}{h} \right\rfloor + 1\right) \cdot \frac{n}{d_0}.$$

To establish the first inequality, simply note that $u(n,m,h) \leq f_{n/d_0}(m,h)$ where

$$f_{n/d_0}(m,h) = \left(h \cdot \left(\left\lfloor \frac{d_0-2}{h} \right\rfloor + 1\right) - h + 1\right) \cdot \frac{n}{d_0} = \left(h \cdot \left\lfloor \frac{d_0-2}{h} \right\rfloor + 1\right) \cdot \frac{n}{d_0} \leq (d_0 - 1) \cdot \frac{n}{d_0},$$

which is less than n.

For the second inequality, we must prove that, for any $d \in D(n)$, we have $f_d(m+1,h) \geq n$; that is,

$$h \cdot \left\lceil \frac{(\lfloor \frac{d_0-2}{h} \rfloor + 1) \cdot \frac{n}{d_0} + 1}{d} \right\rceil - h + 1 \geq \frac{n}{d}.$$

But $n/d \in D(n)$, so by the choice of d_0, we have

$$\left(\left\lfloor \frac{d_0-2}{h} \right\rfloor + 1\right) \cdot \frac{n}{d_0} \geq \left(\left\lfloor \frac{n/d-2}{h} \right\rfloor + 1\right) \cdot \frac{n}{n/d},$$

and thus

$$\begin{aligned} h \cdot \left\lceil \frac{(\lfloor \frac{d_0-2}{h} \rfloor + 1) \cdot \frac{n}{d_0} + 1}{d} \right\rceil - h + 1 &\geq h \cdot \left\lceil \left(\left\lfloor \frac{n/d-2}{h} \right\rfloor + 1\right) + \frac{1}{d} \right\rceil - h + 1 \\ &= h \cdot \left(\left\lfloor \frac{n/d-2}{h} \right\rfloor + 2\right) - h + 1 \\ &\geq h \cdot \left(\frac{n/d - 2 - (h-1)}{h} + 2\right) - h + 1 \\ &= \frac{n}{d}. \end{aligned}$$

Our proof is complete. □

Proof of Proposition 4.22

From the formulae on page 87 we already have the following values:

$$f_{\hat{1}}(n,m,2) = \min\{n, 2m-3\};$$

and

$$f_{\hat{2}}(n,m,2) = \begin{cases} \min\{n, 2m-4\} & \text{if } m \text{ is even,} \\ \min\{n, 2m-3\} & \text{if } m \text{ is odd.} \end{cases}$$

For $d \geq 3$, we have

$$f_{\hat{d}}(n,m,2) = \begin{cases} \min\{n, f_d, 2m-3\} & \text{if } d \not| m-1, \\ \min\{n, 2m-2\} & \text{otherwise.} \end{cases}$$

To prove Proposition 4.22, we suppose first that m is odd. Then

$$f_{\hat{1}}(n,m,2) = f_{\hat{2}}(n,m,2) = \min\{n, 2m-3\}.$$

We consider three subcases: when $u(n,m,2) = n$, when $2m-3 \leq u(n,m,2) < n$, and when $u(n,m,2) < 2m-3$. (Note that we always have $u(n,m,2) \leq n$.)

If $u(n,m,2) = n$, then for every $d \in D(n)$ we have $f_d \geq n$, and thus for $d \geq 3$ we have

$$f_{\hat{d}}(n,m,2) = \begin{cases} \min\{n, 2m-3\} & \text{if } d \not| m-1, \\ \min\{n, 2m-2\} & \text{otherwise.} \end{cases}$$

Therefore,

$$u\hat{\ }(n,m,2) = \min\{f_{\hat{d}}(n,m,h) \mid d \in D(n)\} = \min\{n, 2m-3\} = \min\{u(n,m,2), 2m-3\},$$

as claimed.

Similarly, if $2m-3 \leq u(n,m,2) < n$, then for every $d \in D(n)$ we have $f_d \geq 2m-3$, and thus for $d \geq 3$ we have

$$f_{\hat{d}}(n,m,2) = \begin{cases} \min\{n, 2m-3\} & \text{if } d \not| m-1, \\ \min\{n, 2m-2\} & \text{otherwise.} \end{cases}$$

Therefore,

$$u\hat{\ }(n,m,2) = \min\{n, 2m-3\} = 2m-3 = \min\{u(n,m,2), 2m-3\},$$

as claimed.

Finally, if $u(n,m,2) < 2m-3$, then let us choose a $d_0 \in D(n)$ for which $u(n,m,2) = f_{d_0}$. Note that $f_{d_0} < 2m-3$ but, as a quick computation shows, $f_1 = 2m-1$ and $f_2 = 2m$, so $d_0 \geq 3$. Furthermore, whenever $d | m-1$, we get $f_d = 2m+d-2 > 2m-3$, so $d_0 \not| m-1$, and thus $f_{\hat{d_0}} = f_{d_0}$.

We will verify that $u\hat{\ }(n,m,2) = f_{d_0}$. Clearly, $u\hat{\ }(n,m,2) \leq f_{\hat{d_0}} = f_{d_0}$, so we only need to show that there is no $d \in D(n)$ for which $f_{\hat{d}} < f_{d_0}$. Since $f_{d_0} < 2m-3$, this could only happen if $f_{\hat{d}} = f_d$, but that is not possible as $f_d \geq f_{d_0}$. Thus, again we get,

$$u\hat{\ }(n,m,2) = f_{d_0} = u(n,m,2) = \min\{u(n,m,2), 2m-3\},$$

as claimed.

The case when m is even can be analyzed similarly. □

Proof of Proposition 4.23

We can carry out a similar analysis for the case of $h = 3$, where we may assume that $4 \le m \le n$. If $d \ge 4$ and $k \ge 3$, then

$$f_{\hat{d}}(n, m, 3) = \min\{n, f_d, 3m - 8\}.$$

For the remaining cases we have the following.

d	k	r	δ_d	$f_{\hat{d}}(n, m, 3)$
1	1	1	0	$\min\{n, 3m - 8\}$
2	1	1	0	$\min\{n, 3m - 8\}$
2	2	1	0	$\min\{n, 3m - 8\}$
3	1	3	0	$\min\{n, 3m - 8\}$
3	2	3	0	$\min\{n, 3m - 8\}$
3	3	3	2	$\min\{n, 3m - 10\}$
≥ 4	1	3	$d - 5$	$\min\{n, 3m - 3 - d\}$
≥ 4	2	3	-2	$\min\{n, 3m - 6\}$

The case of $d \ge 4$ and $k = 1$ is interesting: it implies that $m - 1$ is divisible by d; thus, to minimize $3m - 3 - d$, we select $d = \gcd(n, m - 1)$. Note also that when $3 | m$ then $\gcd(n, m - 1) \ne 6$. Therefore,

$$u_{\hat{}}(n, m, 3) = \begin{cases} \min\{u(n, m, 3), 3m - 3 - \gcd(n, m - 1)\} & \text{if } \gcd(n, m - 1) \ge 8; \\ \min\{u(n, m, 3), 3m - 10\} & \text{if } \gcd(n, m - 1) = 7, \text{ or} \\ & \gcd(n, m - 1) \le 5, \; 3|n, \text{ and } 3|m; \\ \min\{u(n, m, 3), 3m - 9\} & \text{if } \gcd(n, m - 1) = 6; \\ \min\{u(n, m, 3), 3m - 8\} & \text{otherwise.} \end{cases}$$

Proof of Proposition 4.26

Choose $d_0 \in D(n)$ so that $f_{d_0} = u(n, m, h)$. We will show that then $f_{\hat{d_0}} \le f_{d_0}$; this will then imply our claim since

$$u_{\hat{}}(n, m, h) = \min\{f_{\hat{d}} \mid d \in D(n)\} \le f_{\hat{d_0}} \le f_{d_0} = u(n, m, h).$$

Suppose, indirectly, that $f_{\hat{d_0}} > f_{d_0}$; we then have $h > \min\{d_0 - 1, k\}$ and

$$f_{\hat{d_0}} = \min\{n, hm - h^2 + 1 - \delta_{d_0}\},$$

otherwise we would have

$$f_{\hat{d_0}} = \min\{n, f_{d_0}, hm - h^2 + 1\} \le f_{d_0},$$

a contradiction.

Therefore,

$$f_{\hat{d_0}} \le hm - h^2 + 1 - \delta_{d_0}.$$

Assume first that $h < d_0$. Then $k < h = r < d_0$, so $\delta_{d_0} = (d_0 - r)(r - k) - (d_0 - 1)$, and thus

$$\begin{aligned}
hm - hk + d_0 &= \left(h \cdot \frac{m - k + d_0}{d_0} - h + 1\right) \cdot d_0 \\
&= \left(h \left\lceil \frac{m}{d_0} \right\rceil - h + 1\right) \cdot d_0 \\
&= f_{d_0} \\
&< f_{\hat{d_0}} \\
&\leq hm - h^2 + 1 - \delta_{d_0} \\
&= hm - h^2 - (d_0 - r)(r - k) + d_0 \\
&= hm - h^2 - (d_0 - h)(h - k) + d_0 \\
&= hm - hk + d_0 - d_0(h - k) \\
&< hm - hk + d_0,
\end{aligned}$$

which is a contradiction.

Assume now that $h \geq d_0$. We then have $h \geq k$ since $k \leq d_0$. Furthermore, we see from the definition of δ that $\delta \geq 1 - (d - 1)$ in all cases; therefore, $-\delta_{d_0} \leq d_0 - 2$. We now see that

$$\begin{aligned}
hm - hk + d_0 &= \left(h \cdot \frac{m - k + d_0}{d_0} - h + 1\right) \cdot d_0 \\
&= \left(h \left\lceil \frac{m}{d_0} \right\rceil - h + 1\right) \cdot d_0 \\
&= f_{d_0} \\
&< f_{\hat{d_0}} \\
&\leq hm - h^2 + 1 - \delta_{d_0} \\
&\leq hm - h^2 + d_0 - 1 \\
&\leq hm - hk + d_0 - 1
\end{aligned}$$

which is a contradiction. □

Proof of Proposition 4.29

Let $d \in D(n)$. If $h \leq \min\{d - 1, k\}$, then

$$f_{\hat{d}} = \min\{n, f_d, hm - h^2 + 1\}.$$

Clearly, $n \geq m$, and we can easily see that $hm - h^2 + 1 \geq m$ holds as well, with equality if, and only if, $h = 1$ or $h = m - 1$. Furthermore, from the argument for Proposition 4.15 we see that we also have $f_d \geq m$ with equality if, and only if, $h = 1$ or $m = d$. Therefore, when $h \leq \min\{d - 1, k\}$, we have $f_{\hat{d}} \geq m$ with equality if, and only if, $h = 1$, $h = m - 1$, or $m = d$.

Suppose now that $h > \min\{d - 1, k\}$, in which case

$$f_{\hat{d}} = \min\{n, hm - h^2 + 1 - \delta_d\}.$$

If $\delta_d = 0$, then $f_{\hat{d}} \geq m$ with equality if, and only if, $h = 1$ or $h = m - 1$. This leaves three cases, according to the definition of δ_d.

Assume first that $r = k = d$. Note that, if $d = 1$, then $\delta_d = 0$, so we assume that $d \geq 2$; this then also implies that $h \geq 2$. Furthermore, $r = d$ implies that $h \geq d$ and $k = d$ implies that $m - h \geq d$. Therefore, we have

$$\begin{aligned} hm - h^2 + 1 - \delta_d - m &= (h-1)(m-h-1) - (d-1) \\ &\geq (h-1)(m-h-1) - (h-1) \\ &= (h-1)(m-h-2) \\ &\geq (h-1)(d-2) \\ &\geq 0, \end{aligned}$$

with equality if, and only if, $h = d$, $m - h = d$, and $d = 2$; that is, $h = 2$, $m = 4$, and $d = 2$.

Next, assume that $r < k$; note that this then implies that $r \leq d - 1$. We now have

$$hm - h^2 + 1 - \delta_d - m = (h-1)(m-h-1) - (k-r)r + (d-1) = (h-1)(m-h) - (k-r)r - (h-d).$$

Here

$$m - h = (cd + k) - (qd + r) = (c - q)d + (k - r) \geq k - r,$$

so

$$\begin{aligned} hm - h^2 + 1 - \delta_d - m &\geq (h-1)(k-r) - (k-r)r - (h-d) \\ &= (h-r-1)(k-r) - (h-d) \\ &\geq (h-r-1) - (h-d) \\ &= d-1-r \\ &\geq 0. \end{aligned}$$

We see that equality holds if, and only if, $m - h = k - r$, $k - r = 1$, and $r = d - 1$; that is, $h = qd + d - 1$, and $m = h + 1 = qd + d$.

In our last case, we have $k < r < d$. In this case, we must have $d \geq 3$ and

$$1 \leq m - h = (cd + k) - (qd + r) = d(c - q) + (k - r) < d(c - q),$$

so $c - q \geq 1$ and thus

$$m - h = d(c - q - 1) + (d + k - r) \geq d + k - r.$$

Thus,

$$\begin{aligned} hm - h^2 + 1 - \delta_d - m &= (h-1)(m-h-1) - (d-r)(r-k) + (d-1) \\ &\geq (h-1)(d+k-r) - (d-r)(r-k) + (d-1) \\ &= (d-r)(h-r+k-1) + k(h-1) + (d-1). \end{aligned}$$

Here each term is nonnegative, but $d - 1 > 0$.

Therefore, we have proved that $f_{\hat{d}} \geq m$; since $d \in D(n)$ was arbitrary, this implies that $u\hat{}(n, m, h) \geq m$.

If $h = 1$ or $h = m - 1$, then with $d = n$ we get $h \leq \min\{d-1, k\}$, and thus $u\hat{}(n, m, h) \leq f_{\hat{d}} = m$.

If m is a divisor of n, then we may choose $d = m$, with which again $h \leq \min\{d-1, k\}$, and thus $u\hat{}(n, m, h) \leq f_{\hat{d}} = m$.

If $h = 2$, $m = 4$, and n is even, we may pick $d = 2$, with which $h > \min\{d-1, k\}$, and so $u\hat{}(n, m, h) \leq f_{\hat{d}} = m$ holds again.

If none of these occur, then $f_{\hat{d}} > m$ for all $d \in D(n)$ and thus $u\hat{}(n, m, h) > m$. □

Proof of Proposition A.42

We first show that we can write every odd integer between -2^m and 2^m in the form $\Sigma_{i=0}^{m-1} \pm 2^i$. There are exactly 2^m such odd integers, so we just need to verify that no two of the 2^m signed sums yield the same integer. Suppose we have

$$\Sigma_{i=0}^{m-1} \lambda_i \cdot 2^i = \Sigma_{i=0}^{m-1} \lambda_i' \cdot 2^i$$

with coefficients λ_i and λ_i' all from the set $\{-1, 1\}$, and assume that j is the largest index for which $\lambda_i \neq \lambda_i'$; without loss of generality, let $\lambda_j = 1$ and $\lambda_j' = -1$. But then

$$\Sigma_{i=0}^{m-1} \lambda_i \cdot 2^i \geq 1 + \Sigma_{i=j+1}^{m-1} \lambda_i \cdot 2^i,$$

and

$$\Sigma_{i=0}^{m-1} \lambda_i' \cdot 2^i \leq -1 + \Sigma_{i=j+1}^{m-1} \lambda_i \cdot 2^i,$$

which is a contradiction.

Now if n is odd, then our 2^m odd integers yield exactly $\min\{n, 2^m\}$ elements in $\mathbb{Z}_n$; since by Proposition A.40 we cannot hope for more, we have

$$\nu_{\pm}^{\hat{}}(\mathbb{Z}_n, m, h) = \min\{n, 2^m\}.$$

If n is even, our integers yield only $\min\{n/2, 2^m\}$ elements in $\mathbb{Z}_n$, which is less than the upper bound in Proposition A.40 when $n < 2^{m+1}$. But note that, whenever n is even, no m-subset $A = \{a_1, \ldots, a_m\}$ of $\mathbb{Z}_n$ has a restricted m-fold signed sumset of size more than $n/2$: indeed, the signed sums in

$$m_{\pm}^{\hat{}}A = \{\pm a_1 \pm \cdots \pm a_m\}$$

all have the same parity. □

Proof of Theorem B.8

We consider the cases of $s = 2$ and $s = 3$ separately.

Suppose that A is a perfect 2-basis of size m in G. Then we must have

$$n = \binom{m+2}{2}.$$

Clearly, $0 \in A - A$; we show that $A - A$ has size $m(m-1) + 1$: in other words, if a_1 and a_2 are distinct elements of A and a_3 and a_4 are distinct elements of A so that

$$a_1 - a_2 = a_3 - a_4,$$

then $a_1 = a_3$ and $a_2 = a_4$. Indeed: our hypothesis implies that

$$a_1 + a_4 = a_2 + a_3,$$

so if A is a perfect 2-basis, then $a_1 \in \{a_2, a_3\}$ and $a_4 \in \{a_2, a_3\}$, from which our claim follows since $a_1 \neq a_2$ and $a_3 \neq a_4$.

Furthermore, $A - A$ and A are disjoint, since if we were to have $a_1, a_2, a_3 \in A$ for which

$$a_1 - a_2 = a_3,$$

then

$$a_1 = a_2 + a_3,$$

contradicting the assumption that A is a perfect 2-basis.

Therefore,

$$|(A - A) \cup A| = m(m-1) + 1 + m \le n = \binom{m+2}{2},$$

from which $m \le 3$ follows. We will rule out $m = 2$ and $m = 3$ as follows.

If $m = 2$, then $n = \binom{m+2}{2} = 6$, so $G = \mathbb{Z}_6$. Suppose that $A = \{a, b\} \subseteq \mathbb{Z}_6$. Then

$$[0, 2]A = \{0, a, b, 2a, 2b, a + b\} = \mathbb{Z}_6,$$

so a, b, and $a + b$ must all be odd, which is impossible.

Similarly, if $m = 3$, then $n = \binom{m+2}{2} = 10$, so $G = \mathbb{Z}_{10}$. Suppose that $A = \{a, b\} \subseteq \mathbb{Z}_{10}$. Then

$$[0, 2]A = \{0, a, b, c, 2a, 2b, 2c, a + b, a + c, b + c\} = \mathbb{Z}_{10},$$

so exactly one of a, b, c, $a + b$, $a + c$, or $b + c$ is even, which is impossible.

Now we turn to $s = 3$. Suppose that A is a perfect 3-basis of size m in G. Then we must have

$$n = \binom{m+3}{3}.$$

(Our argument here is similar to the proof of Theorem 3 in [43].) Note that the set $2A - A$ is the disjoint union of the sets

$$B = \{a_1 + a_2 - a_3 \mid a_1, a_2, a_3 \in A; a_1 \neq a_2 \neq a_3 \neq a_1\},$$

$$C = \{2a_1 - a_2 \mid a_1, a_2 \in A, a_1 \neq a_2\},$$

and A since, for example,

$$a_1 + a_2 - a_3 = 2a_1' - a_2'$$

would imply

$$a_1 + a_2 + a_2' = 2a_1' + a_3,$$

from which $a_3 \in \{a_1, a_2, a_2'\}$ so $a_3 = a_2'$, and thus $a_1 = a_2$, which is a contradiction. Furthermore, B and C each have maximum cardinality; that is, if

$$a_1 + a_2 - a_3 = a_1' + a_2' - a_3',$$

then $\{a_1, a_2\} = \{a_1', a_2'\}$ and $a_3 = a_3'$; similarly for C. Therefore,

$$|2A - A| = |B| + |C| + |A| = m(m-1)(m-2)/2 + m(m-1) + m.$$

We can similarly see that $2A - A$ must also be disjoint from $A - A$, and, as above, we have

$$|A - A| = m(m-1) + 1.$$

Therefore, we have

$$m(m-1)(m-2)/2 + m(m-1) + m + m(m-1) + 1 \le n = \binom{m+3}{3},$$

from which $m \le 3$.

We can easily rule out $m = 2$ by hand: If $A = \{a, b\}$ were to be a perfect 3-basis in G, then $n = 10$ and thus $G = \mathbb{Z}_{10}$; with

$$[0,3]A = \{0, a, b, 2a, 2b, a+b, 3a, 3b, 2a+b, a+2b\}$$

we see that (i) if both a and b are odd, then 6 elements of $[0,3]A$ are odd; (ii) if exactly one of a or b is odd, then 4 elements of $[0,3]A$ are odd; and (iii) if both a and b are even, then no element of $[0,3]A$ is odd. The case of $m = 3$ leads to $G = \mathbb{Z}_{20}$ or $G = \mathbb{Z}_2 \times \mathbb{Z}_{10}$; we ruled out these by the computer program [120]. □

Proof of Proposition B.28

We will prove that for each $2 \leq n \leq \lfloor \frac{f^2+6f+5}{4} \rfloor$, the set $A = \{1, a\}$ with $a = \lfloor \frac{f+3}{2} \rfloor$ has folding number at most f in $\mathbb{Z}_n$.

What we need to verify is that for every $g \in \mathbb{Z}_n$, we have $g \in \cup_{h=0}^{f} hA$. We may assume that g is a nonnegative integer which is less than n.

By the Division Theorem, there exist (unique) integers q and r with $0 \leq r \leq a-1$ for which $g = q \cdot a + r \cdot 1$. We will prove that $q + r \leq f$, from which our claim follows.

Case 1: f is even. In this case, we have $a = \frac{f+2}{2}$ and $n \leq \frac{f^2+6f+4}{4}$. We separate several subcases.

Subcase 1.1: $q \leq a-1$. Then $q + r \leq 2a - 2 \leq f$, as claimed.

Subcase 1.2: $q = a$ and $r \leq a-2$. Again, $q + r \leq 2a - 2 \leq f$, as claimed.

Subcase 1.3: $q = a$ and $r = a-1$. Then we have

$$g = a^2 + a - 1 = \frac{f^2 + 6f + 4}{4} \geq n,$$

which cannot happen, since we assumed that g is less than n.

Subcase 1.4: $q \geq a+1$. Then we have $g \geq a^2 + a$, which is a contradiction as in Subcase 1.3.

Case 2: f is odd. In this case, we have $a = \frac{f+3}{2}$ and $n \leq \frac{f^2+6f+5}{4}$. We again separate several subcases.

Subcase 2.1: $q \leq a-2$. Then $q + r \leq 2a - 3 \leq f$, as claimed.

Subcase 2.2: $q = a-1$ and $r \leq a-2$. Again, $q + r \leq 2a - 3 \leq f$, as claimed.

Subcase 2.3: $q = a-1$ and $r = a-1$. Then we have

$$g = a(a-1) + a - 1 = \frac{f^2 + 6f + 5}{4} \geq n,$$

which cannot happen, since we assumed that g is less than n.

Subcase 2.4: $q \geq a$. Then we have $g \geq a^2$, which is a contradiction as in Subcase 2.3. □

Proof of Proposition B.46

As we noted below the statement of Proposition B.46, here we only need to prove that the set $\{1, 2s+1\}$ is a perfect s-spanning set in $\mathbb{Z}_n$ for $n = 2s^2 + 2s + 1$.

Let $n = 2s^2 + 2s + 1$. We will prove that for every integer g between $-s$ and $n - s - 1$, inclusive, one can find integers λ_1 and λ_2 so that $|\lambda_1| + |\lambda_2| \leq s$ and

$$g \equiv \lambda_1 \cdot 1 + \lambda_2 \cdot (2s+1) \bmod n.$$

Let k be the largest odd integer for which $g > k(s+1)$; k is then uniquely defined by the inequalities

$$k(s+1)+1 \leq g \leq (k+2)(s+1).$$

Note that $g \geq -s$ implies that $k \geq -1$, and $g \leq 2s^2+s$ implies that $k \leq 2s-1$, so we have

$$-1 \leq k \leq 2s-1.$$

We consider two cases.

Case 1: Suppose that

$$k(s+1)+1 \leq g \leq k(s+1)+(2s-k) = (k+2)s.$$

(Note that $2s-k \geq 1$.) Let

$$\lambda_1 = g-(k+1)s-\frac{k+1}{2},$$

and

$$\lambda_2 = \frac{k+1}{2}.$$

We can readily verify that

$$g = \lambda_1 \cdot 1 + \lambda_2 \cdot (2s+1);$$

we need to prove that $|\lambda_1|+|\lambda_2| \leq s$ holds as well.

Note that $k \geq -1$, so

$$|\lambda_2| = \frac{k+1}{2}.$$

Since λ_1 is a linear function of g, $|\lambda_1|$ achieves its maximum value over the interval for g at one (or both) of the endpoints; therefore, we need to evaluate

$$\left| k(s+1)+1-(k+1)s-\frac{k+1}{2} \right|$$

and

$$\left| (k+2)s-(k+1)s-\frac{k+1}{2} \right|;$$

since $s \geq \frac{k+1}{2}$, both expressions equal $s-\frac{k+1}{2}$. Therefore, $|\lambda_1|+|\lambda_2| \leq s$, as claimed.

Case 2: Suppose now that

$$(k+2)s+1 \leq g \leq (k+2)(s+1) = (k+2)s+(k+2).$$

(Note that $k+2 \geq 1$.) Let

$$\lambda_1 = g-(k+2)s-\frac{k+3}{2},$$

and

$$\lambda_2 = \frac{k+1}{2}-s.$$

We can verify that

$$\lambda_1 \cdot 1 + \lambda_2 \cdot (2s+1) = g-(2s^2+2s+1),$$

and thus $\lambda_1 \cdot 1 + \lambda_2 \cdot (2s+1) \equiv g \bmod n$. Next, we prove that $|\lambda_1|+|\lambda_2| \leq s$ holds as well.

First, since $s \geq \frac{k+1}{2}$, we have

$$|\lambda_2| = s-\frac{k+1}{2}.$$

Since λ_1 is a linear function of g, $|\lambda_1|$ achieves its maximum value over the interval for g at one (or both) of the endpoints; therefore, we need to evaluate

$$\left|(k+2)s+1-(k+2)s-\frac{k+3}{2}\right|$$

and

$$\left|(k+2)(s+1)-(k+2)s-\frac{k+3}{2}\right|;$$

since $k \geq -1$, both expressions equal $\frac{k+1}{2}$. Therefore, we again have $|\lambda_1|+|\lambda_2| \leq s$, as claimed. □

Proof of Proposition B.54

We verify that for every $n \leq 2s^2+2s+1$, we have

$$[0,s]_{\pm}\{s,s+1\} = \{\lambda_1 s + \lambda_2(s+1) \in \mathbb{Z}_n \mid \lambda_1,\lambda_2 \in \mathbb{Z}, |\lambda_1|+|\lambda_2| \leq s\} = \mathbb{Z}_n.$$

Considering the elements of $[0,s]_{\pm}\{s,s+1\}$ in $\mathbb{Z}$ (rather than $\mathbb{Z}_n$), we see that the elements of

$$\Sigma = \{\lambda_1 s + \lambda_2(s+1) \in \mathbb{Z} \mid \lambda_1,\lambda_2 \in \mathbb{Z}, |\lambda_1|+|\lambda_2| \leq s\}$$

lie in the interval $[-(s^2+s),(s^2+s)]$. Since the index set

$$I_{\pm}(2,[0,s]) = \{(\lambda_1,\lambda_2) \mid \lambda_1,\lambda_2 \in \mathbb{Z}, |\lambda_1|+|\lambda_2| \leq s\}$$

contains exactly $2s^2+2s+1$ elements, it suffices to prove that no integer in $[-(s^2+s),(s^2+s)]$ can be written as an element of Σ in two different ways.

For that, suppose that

$$\lambda_1 s + \lambda_2(s+1) = \lambda_1' s + \lambda_2'(s+1)$$

for some $(\lambda_1,\lambda_2) \in I_{\pm}(2,[0,s])$ and $(\lambda_1',\lambda_2') \in I_{\pm}(2,[0,s])$; w.l.o.g., we can assume that $\lambda_2 \geq \lambda_2'$. Our equation implies that $\lambda_2-\lambda_2'$ is divisible by s and is, therefore, equal to 0, s, or $2s$.

If $\lambda_2-\lambda_2' = 2s$, then $\lambda_2 = s$ and $\lambda_2' = -s$, which can only happen if $\lambda_1 = \lambda_1' = 0$; this case leads to a contradiction with our equation. Assume, next, that $\lambda_2-\lambda_2' = s$. Our equation then yields $\lambda_1-\lambda_1' = -s-1$. In this case, we have $\lambda_1 \leq 0$, $\lambda_1' \geq 0$, $\lambda_2 \geq 0$, and $\lambda_2' \leq 0$, and we see that

$$|\lambda_1|+|\lambda_2| = -\lambda_1+\lambda_2 = (s+1-\lambda_1')+(s+\lambda_2') = 2s+1-(|\lambda_1'|+|\lambda_2'|) \geq s+1,$$

but that is a contradiction.

This leaves us with the case that $\lambda_2 = \lambda_2'$, which implies $\lambda_1 = \lambda_1'$, as claimed, and therefore the set $\{s,s+1\}$ is s-spanning in $\mathbb{Z}_n$. □

Proof of Proposition B.57

Suppose first that $s \geq 2$, $k \leq 3s-4$, and $k \neq 2s-1$. We will show that with

$$A = \{(0,1),(1,s-1)\},$$

we have $[0,s]_{\pm}A = \mathbb{Z}_2 \times \mathbb{Z}_{2k}$.

Let x and y be integers. Observe that

(i) if $0 \le x \le s$, then

$$(0, x) = x \cdot (0, 1) \in [0, s]_{\pm}A;$$

(ii) if $s \le x \le 3s - 4$, then

$$(0, x) = 2 \cdot (1, s - 1) + (x - 2s + 2) \cdot (0, 1) \in [0, s]_{\pm}A;$$

(iii) if $0 \le y \le 2s - 2$, then

$$(1, y) = 1 \cdot (1, s - 1) + (y - s + 1) \cdot (0, 1) \in [0, s]_{\pm}A;$$

and

(iv) if $2s \le y \le 4s - 6$, then

$$(1, y) = 3 \cdot (1, s - 1) + (y - 3s + 3) \cdot (0, 1) \in [0, s]_{\pm}A.$$

Therefore, by (i) and (ii), $(0, x) \in [0, s]_{\pm}A$ for all $0 \le x \le 3s - 4$, thus, since $2k \le 6s - 8$, by symmetry, $[0, s]_{\pm}A$ contains all group elements whose first component is 0.

If $k \le 2s - 2$, then by (iii) and by symmetry, $[0, s]_{\pm}A$ contains all group elements whose first component is 1, and we are done.

Suppose now that $2s \le k \le 3s - 4$. By (iii), $(1, y) \in [0, s]_{\pm}A$ for all $0 \le y \le 2s - 2$, so, by symmetry, $(1, y) \in [0, s]_{\pm}A$ for all $2k - 2s + 2 \le y \le 2k$. Note that

$$2k - 2s + 2 \le 4s - 6,$$

so combining this with (iv), we get that $(1, y) \in [0, s]_{\pm}A$ for all $2s \le y \le 2k$. We are under the assumption that $2s \le k$, however, so by symmetry, again $[0, s]_{\pm}A$ contains all group elements whose first component is 1, and $[0, s]_{\pm}A = \mathbb{Z}_2 \times \mathbb{Z}_{2k}$

This completes all cases of Proposition B.57, except for $k = 2s - 1$ with $s \ge 5$. We will show that, in this case, for

$$A = \{(0, 1), (1, s + 1)\},$$

we have $[0, s]_{\pm}A = \mathbb{Z}_2 \times \mathbb{Z}_{2k}$.

Let x and y be integers. Observe that

(i) if $0 \le x \le s$, then

$$(0, x) = x \cdot (0, 1) \in [0, s]_{\pm}A;$$

(ii) if $s + 1 \le x \le s + 3$, then

$$(0, x) = (-2) \cdot (1, s + 1) + (x - 2s + 4) \cdot (0, 1) \in [0, s]_{\pm}A;$$

(iii) if $s + 4 \le x \le 3s$, then

$$(0, x) = 2 \cdot (1, s + 1) + (x - 2s - 2) \cdot (0, 1) \in [0, s]_{\pm}A;$$

(iv) if $0 \le y \le 1$, then

$$(1, y) = (-3) \cdot (1, s + 1) + (y - s + 5) \cdot (0, 1) \in [0, s]_{\pm}A;$$

and

(v) if $2 \le y \le 2s$, then

$$(1, y) = 1 \cdot (1, s + 1) + (y - s - 1) \cdot (0, 1) \in [0, s]_{\pm}A.$$

As an explanation for (ii), we should add that, since $k = 2s - 1$, we have

$$-2(s+1) + (x - 2s + 4) = -2k + x \equiv x \bmod 2k,$$

and, since $s \geq 5$,

$$|-2| + |x - 2s + 4| \leq s$$

always holds when $s + 1 \leq x \leq s + 3$. The explanation for (iv) is similar.

The result now follows by symmetry, as above. □

Proof of Proposition C.36

We will prove a slightly more general result.

For positive integers n, m, h, a, and b, consider the geometric progression

$$A = \{a, ab, ab^2, \ldots, ab^{m-1}\}$$

in the cyclic group $\mathbb{Z}_n$. (By this we mean that each integer in the set is considered mod n; we will talk about integers and their values mod n interchangeably.) Suppose that $b \geq h \geq 2$ and

$$n \geq hab^{m-1}.$$

We first note that with these conditions, A has size m: indeed, we have

$$1 \leq a < ab < \cdots < ab^{m-1} < n.$$

Furthermore, we show that two h-fold sums of A can be equal only if they correspond to the same element of the index set

$$\mathbb{N}_0^m(h) = \{(\lambda_1, \ldots, \lambda_m) \in \mathbb{N}_0^m \mid \lambda_1 + \cdots + \lambda_m = h\}.$$

First, observe that, for nonnegative integer coefficients $\lambda_1, \ldots, \lambda_m$ with

$$\lambda_1 + \cdots + \lambda_m = h,$$

we have

$$\lambda_1 a + \lambda_2 ab + \cdots + \lambda_m ab^{m-1} \geq \lambda_1 a + \lambda_2 a + \cdots + \lambda_m a = ha > 0$$

and

$$\lambda_1 a + \lambda_2 ab + \cdots + \lambda_m ab^{m-1} \leq \lambda_1 ab^{m-1} + \lambda_2 ab^{m-1} + \cdots + \lambda_m ab^{m-1} = hab^{m-1} \leq n,$$

so linear combinations

$$\lambda_1 a + \lambda_2 ab + \cdots + \lambda_m ab^{m-1}$$

and

$$\lambda_1' a + \lambda_2' ab + \cdots + \lambda_m' ab^{m-1}$$

in hA can only be equal in $\mathbb{Z}_n$ if they are also equal in $\mathbb{Z}$. But

$$\lambda_1 a + \lambda_2 ab + \cdots + \lambda_m ab^{m-1} = \lambda_1' a + \lambda_2' ab + \cdots + \lambda_m' ab^{m-1}$$

in $\mathbb{Z}$ implies that $\lambda_1 - \lambda_1'$ is divisible by b. Assume, without loss of generality, that $\lambda_1 \geq \lambda_1'$; then

$$0 \leq \lambda_1 - \lambda_1' \leq h.$$

In fact, we cannot have $\lambda_1 - \lambda_1' = h$, since that could only happen if $\lambda_1 = h$ and $\lambda_1' = 0$, but that would imply that

$$ha = \lambda_2' ab + \cdots + \lambda_m' ab^{m-1},$$

which is impossible since

$$\lambda_2' ab + \cdots + \lambda_m' ab^{m-1} \geq \lambda_2' ab + \cdots + \lambda_m' ab = hab > ha.$$

Therefore,

$$0 \leq \lambda_1 - \lambda_1' < h \leq b,$$

so the only way for $\lambda_1 - \lambda_1'$ to be divisible by b is if $\lambda_1 = \lambda_1'$.

This, in turn, implies that $\lambda_i = \lambda_i'$ for every $i = 2, 3, \ldots, m$ as well; therefore, each element of the index set yields a different element. Thus we have shown that, under the conditions of $b \geq h \geq 2$ and

$$n \geq hab^{m-1},$$

the geometric progression

$$A = \{a, ab, ab^2, \ldots, ab^{m-1}\}$$

has size m and is a B_h set in the cyclic group $\mathbb{Z}_n$. (We should note that there is nothing special about A being a geometric progression: all that's needed is that each element in the progression is much larger than the previous one.)

By choosing $a = 1$ and $b = h$, we arrive at Proposition C.36. □

Proof of Proposition C.50

Again we will prove a slightly more general result.

We consider, for positive integers n, m, h, a, and b, the geometric progression

$$A = \{a, ab, ab^2, \ldots, ab^{m-1}\}$$

in the cyclic group $\mathbb{Z}_n$. Here we assume that $b \geq 2h \geq 2$ and

$$n > 2hab^{m-1}.$$

Again we see that, with these conditions, A has size m. Similarly as before, we see that for integer coefficients $\lambda_1, \ldots, \lambda_m$ with

$$|\lambda_1| + \cdots + |\lambda_m| = h,$$

we have

$$\lambda_1 a + \lambda_2 ab + \cdots + \lambda_m ab^{m-1} \geq -|\lambda_1| ab^{m-1} - |\lambda_2| ab^{m-1} - \cdots - |\lambda_m| ab^{m-1} = -hab^{m-1},$$

and

$$\lambda_1 a + \lambda_2 ab + \cdots + \lambda_m ab^{m-1} \leq |\lambda_1| ab^{m-1} + |\lambda_2| ab^{m-1} + \cdots + |\lambda_m| ab^{m-1} = hab^{m-1},$$

so linear combinations

$$\lambda_1 a + \lambda_2 ab + \cdots + \lambda_m ab^{m-1}$$

and

$$\lambda_1' a + \lambda_2' ab + \cdots + \lambda_m' ab^{m-1}$$

in $h_{\pm} A$ can only be equal in $\mathbb{Z}_n$ if they are also equal in $\mathbb{Z}$.

But

$$\lambda_1 a + \lambda_2 ab + \cdots + \lambda_m ab^{m-1} = \lambda_1' a + \lambda_2' ab + \cdots + \lambda_m' ab^{m-1}$$

in $\mathbb{Z}$ implies that $\lambda_1 - \lambda_1'$ is divisible by b. Assume, without loss of generality, that $\lambda_1 \geq \lambda_1'$; then

$$0 \leq \lambda_1 - \lambda_1' \leq 2h.$$

In fact, we cannot have $\lambda_1 - \lambda_1' = 2h$, since that could only happen if $\lambda_1 = h$ and $\lambda_1' = -h$, but that would imply that

$$ha = -ha,$$

which is impossible. Therefore,

$$0 \leq \lambda_1 - \lambda_1' < 2h \leq b,$$

so the only way for $\lambda_1 - \lambda_1'$ to be divisible by b is if $\lambda_1 = \lambda_1'$.

This, in turn, implies that $\lambda_i = \lambda_i'$ for every $i = 2, 3, \ldots, m$ as well; therefore, each element of the index set yields a different element in the h-fold signed sumset, and

$$|h_{\pm}A| = |\mathbb{Z}^m(h)| = c(h,m).$$

In summary, we have shown that, under the conditions $b \geq 2h \geq 2$ and

$$n > 2hab^{m-1},$$

the geometric progression

$$A = \{a, ab, ab^2, \ldots, ab^{m-1}\}$$

in the cyclic group $\mathbb{Z}_n$ has size m, and its h-fold signed sumset $h_{\pm}A$ has size

$$|\mathbb{Z}^m(h)| = c(h,m)$$

and thus it is a B_h set over $\mathbb{Z}$ in $\mathbb{Z}_n$. (We should note that there is nothing special about A being a geometric progression: all that's needed is that each element in the progression is much larger than the previous one.)

By choosing $a = 1$ and $b = 2h$, we arrive at Proposition C.50. □

Proof of Proposition C.51

We first provide a proof for the case when h is even: we will show that, in this case, $A = \{1, h\}$ has an h-fold signed sumset of size exactly $4h$.

We start by observing that the relevant index set is $\mathbb{Z}^2(h)$, which consists of the $4h$ points

$$\{(0, \pm h), (\pm 1, \pm(h-1)), (\pm 2, \pm(h-2)), \ldots, (\pm(h-1), \pm 1), (\pm h, 0)\};$$

we may rewrite this layer of the integer lattice as

$$\mathbb{Z}^2(h) = I \cup J \cup (-I) \cup (-J)$$

where

$$I = \{(i, h-i) \mid i = 0, 1, 2, \ldots, h\}$$

and

$$J = \{(-j, h-j) \mid j = 1, 2, \ldots, h-1\}.$$

(Note that $|I| + |J| + |-I| + |-J| = (h+1) + (h-1) + (h+1) + (h-1) = 4h$.)

Therefore, we get $h_{\pm}A = S \cup T \cup (-S) \cup (-T)$, where

$$S = \{i \cdot 1 + (h-i) \cdot h \mid i = 0, 1, 2, \dots, h\}$$

and

$$T = \{-j \cdot 1 + (h-j) \cdot h \mid j = 1, 2, \dots, h-1\}.$$

We need to prove that the four sets S, T, $-S$, and $-T$ are pairwise disjoint in $\mathbb{Z}_n$. First, we prove that S and T are disjoint. Suppose, indirectly, that this is not so and we have indices $i \in \{0, 1, 2, \dots, h\}$ and $j \in \{1, 2, \dots, h-1\}$ for which

$$i \cdot 1 + (h-i) \cdot h = -j \cdot 1 + (h-j) \cdot h.$$

Rearranging yields that

$$j \cdot (h+1) - i \cdot (h-1) = 0$$

in $\mathbb{Z}_n$. Now the integer $j \cdot (h+1) - i \cdot (h-1)$ is at least

$$1 \cdot (h+1) - h \cdot (h-1) = -h^2 + 2h + 1 > -n$$

and at most

$$(h-1) \cdot (h+1) - 0 \cdot (h-1) = h^2 - 1 < n,$$

so if

$$j \cdot (h+1) - i \cdot (h-1) = 0$$

in $\mathbb{Z}_n$, then

$$j \cdot (h+1) - i \cdot (h-1) = 0$$

in $\mathbb{Z}$. Rearranging again, we get

$$i + j = (i-j) \cdot h,$$

which implies that $i + j$ is divisible by h; since $1 \leq i + j \leq 2h - 1$, this can only occur if $i + j = h$. Our equation then yields $i - j = 1$, adding these two equations results in $2i = h + 1$, which is a contradiction as the left-hand side is even and the right-hand side is odd. Therefore, we proved that $S \cap T = \emptyset$.

Observe that the elements of S are between h and h^2 (inclusive), and the elements of T are between 1 and $h^2 - h - 1$ (inclusive); in particular, all of them are between 1 and h^2. Therefore, if S and T are disjoint in $\mathbb{Z}_n$ and $n > 2h^2$, then the four sets S, T, $-S$, and $-T$ must be pairwise disjoint. This completes our proof for the case when h is even.

Next, we turn to the case when h is odd, and show that $A = \{2, h\}$ has an h-fold signed sumset of size exactly $4h$. Again, we write $h_{\pm}A$ as $S \cup T \cup (-S) \cup (-T)$, but this time with

$$S = \{i \cdot 2 + (h-i) \cdot h \mid i = 0, 1, 2, \dots, h\}$$

and

$$T = \{-j \cdot 2 + (h-j) \cdot h \mid j = 1, 2, \dots, h-1\}.$$

As above, we can show that the four sets S, T, $-S$, and $-T$ must be pairwise disjoint. □

Proof of Proposition D.6

Since the claim is obviously true for $h = 1$, we will assume that $h \geq 2$.

Suppose first that $A = a + H$ for some $a \in G$ and $H \leq G$; we then have $|A| = |H| = m$. Note that $hA = ha + H$, since

$$a_1 + \cdots + a_h = (a + h_1) + \cdots + (a + h_h) \in ha + H$$

and

$$ha + h_0 = (a + 0) + \cdots + (a + 0) + (a + h_0) \in hA.$$

Therefore,

$$|hA| = |ha + H| = |H| = |A|.$$

Conversely, suppose that $|hA| = |A|$. Let H be the stabilizer of $(h-1)A$; that is,

$$H = \{g \in G \mid g + (h-1)A = (h-1)A\}.$$

Then $H \leq G$. Choose any $a \in A$; we will show that $A = a + H$.

Consider the set $A' = A - a$. Then $|A'| = m$ and $0 \in A'$, and therefore

$$(h-1)A = \{0\} + (h-1)A \subseteq A' + (h-1)A.$$

But then

$$|hA| = |hA - a| = |A' + (h-1)A| \geq |(h-1)A| \geq |(h-2) \cdot a + A| = |A|;$$

since we assumed $|hA| = |A|$, equality must hold throughout, and thus

$$A' + (h-1)A = (h-1)A,$$

so $A' \subseteq H$ by definition. This means that $A \subseteq a + H$. This implies that

$$|a + H| \geq |A| = |hA| = |(h-1)A + A| \geq |(h-1)A| = |H + (h-1)A| \geq |H| = |a + H|.$$

Therefore, equality holds throughout and $|a + H| = |A|$ and thus $a + H = A$. □

Proof of Theorem D.8

Suppose, first, that A is an arithmetic progression (AP) of length m; that is,

$$A = \{a + ig \mid i = 0, 1, \ldots, m-1\}$$

for some $a, g \in \mathbb{Z}_p$. Then

$$hA = \{ha + ig \mid i = 0, 1, \ldots, hm - h\},$$

and thus

$$|hA| = \min\{p, hm - h + 1\} = hm - h + 1,$$

by assumption.

For the other direction, we will use the Cauchy–Davenport Theorem (see [52] and [59]) and Vosper's Theorem (see [197] or Theorem 2.7 in [161]). They both refer to the sum

$$A + B = \{a + b \mid a \in A, b \in B\}$$

of subsets A and B in $\mathbb{Z}_p$, and can be stated as follows.

Theorem (Cauchy–Davenport Theorem) *Suppose that p is a prime, and let $A, B \subseteq \mathbb{Z}_p$ for which $|A+B| < p$. Then $|A+B| \geq |A| + |B| - 1$.*

Theorem (Vosper's Theorem) *Suppose that p is a prime, and let $A, B \subseteq \mathbb{Z}_p$ for which $|A+B| < p$. Then $|A+B| = |A| + |B| - 1$ if, and only if, at least one of the following three conditions holds:*

i $|A| = 1$ *or* $|B| = 1$,

ii there is an element $c \in \mathbb{Z}_p$ for which $(A+B) \cup \{c\} = \mathbb{Z}_p$ and $\mathbb{Z}_p \setminus B = \{c\} - A$, or

iii A and B are APs with a common difference.

Suppose now that
$$|hA| = hm - h + 1 < p.$$
We first observe that for any positive integer k and any $a \in A$, we have
$$(k-1)A + \{a\} \subseteq kA,$$
and thus $|(k-1)A| \leq |kA|$. Therefore, $|kA| < p$ for $k = 1, 2, \ldots, h$, and we can repeatedly apply the Cauchy–Davenport Theorem to get
$$\begin{aligned} hm - h + 1 &= |hA| \\ &= |A + (h-1)A| \\ &\geq |A| + |(h-1)A| - 1 \\ &\geq 2|A| + |(h-2)A| - 2 \\ &\vdots \\ &\geq (h-1)|A| + |A| - (h-1) \\ &= hm - h + 1. \end{aligned}$$
Therefore, we must have equality throughout; in particular, $|kA| = km - k + 1$ for $k = 1, 2, \ldots, h$.

In fact, to complete our proof, we will only need the fact that $|2A| = 2m - 1 < p$. If $m = 1$, then we are done since if $A = \{a\}$ for some $a \in \mathbb{Z}_p$, then A is an AP (of length 1). Assume then that $m \geq 2$. Using the "only if" part of Vosper's Theorem for $A = B$ (which we are allowed as the hypotheses hold), we see that conditions i and ii cannot occur as condition ii would imply that $p = 2m$ and thus $m = 1$. Thus condition iii holds, proving our claim. □

Proof of Theorem D.9

Recall that the stabilizer of a nonempty subset S in G is the set
$$H = \{g \in G \mid g + S = S\}.$$
It is easy to verify that the stabilizer of S is a subgroup of G, and we have
$$H + S = S;$$
that is, S is the union of some cosets of H.

We shall use the following famous result of Kneser (cf. [133] or Theorem 4.5 in [161]):

Theorem (Kneser's Theorem) *Suppose that A is an m-subset of a finite abelian group G, h is a positive integer, and H is the stabilizer subgroup of hA in G. We then have*

$$|hA| \geq h|A+H| - (h-1)|H|.$$

Suppose now that A is an m-subset of G and that hA has size p, where p is the smallest prime divisor of the order n of G. We also assume that $m \leq p < hm - h + 1$. Note that if the stabilizer H of hA were to have order 1, then by Kneser's Theorem, we would get

$$p = |hA| \geq h|A+H| - (h-1)|H| \geq hm - (h-1),$$

contradicting our assumption that $p < hm - h + 1$. Therefore, $|H| > 1$, and by the definition of p, we have $|H| \geq p$. Since H is the stabilizer of hA, hA is the union of some cosets of H; with $|H| \geq p$ and $|hA| = p$ this is only possible if $|H| = p$ and hA equals a coset of H.

Note that if A would not lie in a single coset of H, then it would have elements a_1 and a_2 for which the cosets $a_1 + H$ and $a_2 + H$ are different, but then the cosets $(h-1)a_1 + a_2 + H$ and $ha_1 + H$ would be different too, so hA could not lie in a single coset of H, which is a contradiction. □

Proof of Theorem D.10

Suppose, first, that A is an arithmetic progression (AP) of length m; that is,

$$A = \{a + ig \mid i = 0, 1, \ldots, m-1\}$$

for some $a, g \in G$. Then

$$hA = \{ha + ig \mid i = 0, 1, \ldots, hm - h\},$$

and thus

$$|hA| = \min\{p, hm - h + 1\} = hm - h + 1,$$

by assumption.

For the other direction, we will use the Corollary on page 74 of Kemperman's work [129]:

Theorem (Kemperman; cf. [129]) *Let p be the minimum prime divisor of n. Suppose that $m \geq 2$ and*

$$|2A| \leq \min\{p-2, 2m-1\}$$

for some $A \subseteq G$. Then A is an AP.

Note that our assumption of $hm - h + 1 < p$ implies that $2m - 1 < p$ and, since p is odd, we then have $2m + 1 \leq p$ or $p - 2 \geq 2m - 1$. Hence, the conditions of Kemperman's Theorem above are met. The proof of Theorem D.10 then follows as the proof of Theorem D.8 above.

Proof of Theorem D.40

We can evaluate $u_{\pm}(p, m, [0, s])$ for odd prime values of p: we have

$$D_1(p) = \begin{cases} \{1, p\} & \text{if} \quad m \text{ is odd,} \\ \{p\} & \text{if} \quad m \text{ is even;} \end{cases}$$

and

$$D_2(p) = \begin{cases} \emptyset & \text{if } \ m \text{ is odd,} \\ \{1\} & \text{if } \ m \text{ is even.} \end{cases}$$

Therefore, when m is odd, we get

$$u_{\pm}(p, m, [0, s]) = \min\{f_1(m, s), f_p(m, s)\} = \min\{p, sm - s + 1\} = \min\{p, 2s\lfloor m/2 \rfloor + 1\},$$

and when m is even, we get

$$u_{\pm}(p, m, [0, s]) = \min\{f_1(m + 1, s), f_p(m, s)\} = \min\{p, sm + 1\} = \min\{p, 2s\lfloor m/2 \rfloor + 1\}.$$

This means that, by Theorem D.37, we have

$$\rho_{\pm}(\mathbb{Z}_p, m, [0, s]) \leq \min\{p, 2s\lfloor m/2 \rfloor + 1\}.$$

Now we prove that for every m-subset A of $\mathbb{Z}_p$, we have

$$|[0, s]_{\pm}A| \geq \min\{p, 2s\lfloor m/2 \rfloor + 1\}.$$

When m is odd, this follows immediately from the Cauchy–Davenport Theorem, since

$$|[0, s]_{\pm}A| \geq |sA| \geq \min\{p, sm - s + 1\} = \min\{p, 2s\lfloor m/2 \rfloor + 1\}.$$

Observe that when m is even, then A is a proper subset of $A \cup (-A) \cup \{0\}$; therefore, by Propositions 3.3 and 3.4, we get

$$|[0, s]_{\pm}A| = |s(A \cup (-A) \cup \{0\})| \geq \min \rho(\mathbb{Z}_p, m + 1, s).$$

Our claim again follows from the Cauchy–Davenport Theorem, since

$$\rho(\mathbb{Z}_p, m + 1, s) = \min\{p, s(m + 1) - s + 1\} = \min\{p, 2s\lfloor m/2 \rfloor + 1\}.$$

Our proof is now complete. □

Proof of Proposition D.42

Let us write an arbitrary 4-subset A of G in the form

$$A = \{a, a + d_1, a + d_2, a + d_3\}.$$

Here d_1, d_2, and d_3 are distinct nonzero elements of G.

We first prove the following.

Proposition *Suppose that d_1, d_2, and d_3 are distinct nonzero elements of G, and let $A = \{a, a + d_1, a + d_2, a + d_3\}$.*

1. *If none of d_1, d_2, or d_3 equals the sum of the other two, then $|2\hat{\ }A| = 6$.*
2. *Suppose (w.l.o.g.) that $d_3 = d_1 + d_2$.*
   - *(a) If both d_1 and d_2 have order 2, then $|2\hat{\ }A| = 3$;*
   - *(b) if exactly one of d_1 or d_2 have order 2, then $|2\hat{\ }A| = 4$;*
   - *(c) if neither of d_1 or d_2 have order 2, then $|2\hat{\ }A| = 5$.*

Proof of Proposition: Since $2a + d_1$, $2a + d_2$, and $2a + d_3$ are distinct,

$$2\hat{}A = \{2a + d_1, 2a + d_2, 2a + d_3, 2a + d_1 + d_2, 2a + d_1 + d_3, 2a + d_2 + d_3\}$$

has size 3, 4, 5, or 6, exactly when three, two, one, or none of the equations

$$d_3 = d_1 + d_2, \quad d_2 = d_1 + d_3, \quad \text{or } d_1 = d_2 + d_3$$

hold, respectively. This proves 1.

Assume now that $d_3 = d_1 + d_2$, and that both d_1 and d_2 have order 2. Then

$$d_1 + d_3 = d_1 + (d_1 + d_2) = d_2;$$

similarly, $d_2 + d_3 = d_1$. This proves 2 (a).

If $d_3 = d_1 + d_2$, and (w.l.o.g.) d_1 has order 2 but d_2 does not, then we still have $d_1 + d_3 = d_2$, but

$$d_2 + d_3 = d_2 + (d_1 + d_2) \neq d_1.$$

Finally, if neither of d_1 or d_2 have order 2, then $d_2 \neq d_1 + d_3$ and $d_1 \neq d_2 + d_3$. □

Proof of Proposition D.42. By our proposition above, it suffices to prove that

$$\rho\hat{}(G, 4, 2) \leq \begin{cases} 3 & \text{if } |\mathrm{Ord}(G, 2)| \geq 2, \\ 4 & \text{if } |\mathrm{Ord}(G, 2)| = 1, \\ 5 & \text{if } |\mathrm{Ord}(G, 2)| = 0. \end{cases}$$

Suppose, first, that $|\mathrm{Ord}(G, 2)| \geq 2$, and let d_1 and d_2 be two distinct elements of $\mathrm{Ord}(G, 2)$. Then $d_3 = d_1 + d_2$ (which is also of order 2) is nonzero, and distinct from d_1 or d_2. Thus, by our proposition above, the set $A = \{0, d_1, d_2, d_3\}$ has $|2\hat{}A| = 3$.

Suppose now that d_1 is the unique element of G of order 2; let d_2 be any other nonzero element of G (exists since $n \geq m = 4$). Again, $d_3 = d_1 + d_2$ is nonzero, and is distinct from d_1 or d_2, and thus, by our proposition above, the set $A = \{0, d_1, d_2, d_3\}$ has $|2\hat{}A| = 4$.

Finally, assume that $\mathrm{Ord}(G, 2) = \emptyset$, and let d_1 and d_2 be distinct nonzero elements of G; assume further that $d_2 \neq -d_1$ (this is possible since $n \geq m = 4$). As before, $d_3 = d_1 + d_2$ is nonzero, and is distinct from d_1 or d_2, and thus, by our proposition above, the set $A = \{0, d_1, d_2, d_3\}$ has $|2\hat{}A| = 5$. □

Proof of Theorem D.47

We first state and prove the following lemma.

Lemma *Suppose that d and t are positive integers with $t \leq d - 1$, and let $j \in \mathbb{Z}_d$. Then there is a t-subset $J = \{j_1, \ldots, j_t\}$ of $\mathbb{Z}_d$ for which*

$$j_1 + j_2 + \cdots + j_t = j.$$

Note that the restriction of $t \leq d - 1$ is necessary: for $t = d$ the only $j \in \mathbb{Z}_d$ for which such a set exists is, of course,

$$j = 0 + 1 + \cdots + (d - 1) = \frac{d(d-1)}{2} = \begin{cases} \frac{d}{2} & \text{if } d \text{ is even;} \\ 0 & \text{if } d \text{ is odd.} \end{cases}$$

Proof of lemma: Write j as

$$j = \frac{t^2 - t}{2} + j_0 \bmod d$$

where $j_0 = 0, 1, \ldots, d-1$. Note that $1 \leq t \leq d-1$. We will separate two cases: when j_0 is less than t and when it is not.

If $0 \leq j_0 \leq t-1$, let

$$J = \{0, 1, \ldots, t-1, t\} \setminus \{t - j_0\}.$$

Then $|J| = t$, and the elements of J add up to

$$\frac{t^2+t}{2} - (t - j_0) = j.$$

If $t \leq j_0 \leq d-1$, take

$$J = \{1, \ldots, t-1\} \cup \{j_0\}.$$

Again, $|J| = t$, and the elements of J add up to j. □

Before we turn to the proof of Proposition D.47, let us recall our notations and perform some computations. We write

$$m = dc + k \text{ with } c = \left\lceil \frac{m}{d} \right\rceil - 1,$$

and

$$h = dq + r \text{ with } q = \left\lceil \frac{h}{d} \right\rceil - 1.$$

Note that $c \geq 0$, $q \geq 0$, $1 \leq k \leq d$, and $1 \leq r \leq d$.

Recall that we have set

$$A = A_d(n,m) = \bigcup_{i=0}^{c-1} (i + H) \cup \left\{ c + j \cdot \frac{n}{d} \mid j = 0, 1, 2, \ldots, k-1 \right\}.$$

Here

$$\bigcup_{i=0}^{c-1} (i + H) = \emptyset$$

when $c = 0$ (that is, when $m \leq d$), but

$$\left\{ c + j \cdot \frac{n}{d} \mid j = 0, 1, 2, \ldots, k-1 \right\} \neq \emptyset.$$

Note that every element of $h\hat{\ }A$ is of the form

$$(i_1 + i_2 + \cdots + i_h) + (j_1 + j_2 + \cdots + j_h) \cdot \frac{n}{d}$$

with $i_1, \ldots, i_h \in \{0, 1, \ldots, c\}$ and $j_1, \ldots, j_h \in \{0, 1, \ldots, d-1\}$, with the added conditions that when any of the i-indices equals c, the corresponding j-index is at most $k-1$, and that when two i-indices are equal, the corresponding j-indices are distinct.

Clearly, the least value of $i_1 + \cdots + i_h$ is

$$i_{\min} = d(0 + 1 + \cdots + (q-1)) + rq = q \cdot \frac{h + r - d}{2}.$$

To compute the largest value $i_{\max}$ of $i_1 + \cdots + i_h$, we consider four cases depending on whether $r > k$ or not and whether $q = 0$ or not.

First, when $q = 0$ and $r > k$, then $r = h$ and $1 \leq h - k = r - k < r \leq d$, so it is easy to see that

$$i_{\max} = kc + (h-k)(c-1) = h(c-1) + k.$$

In the case when $r > k$ and $q \geq 1$, we write h as $h = k + dq + (r-k)$; thus

$$\begin{aligned} i_{\max} &= kc + [(c-1)+(c-2)+\cdots+(c-q)]d + (r-k)(c-q-1) \\ &= kc + qcd - \frac{q(q+1)}{2}d + rc - rq - r - kc + kq + k \\ &= qcd + rc - q \cdot \frac{dq+d+2r}{2} - r + kq + k \\ &= hc - q \cdot \frac{h+r-d}{2} - dq - r + kq + k \\ &= h(c-1) - q \cdot \frac{h+r-d}{2} + kq + k. \end{aligned}$$

Next, when $q = 0$ and $r \leq k$, then $r = h$ and $h = r \leq k$, so we have

$$i_{\max} = hc = h(c-1) + r.$$

Finally, in the case when $q \geq 1$ and $r \leq k$, then $0 < d - k + r \leq d$; we write h as $h = k + d(q-1) + (d-k+r)$ and thus (using our result from the first case above)

$$\begin{aligned} i_{\max} &= kc + [(c-1)+(c-2)+\cdots+(c-q+1)]d + (d+r-k)(c-q) \\ &= kc + [(c-1)+(c-2)+\cdots+(c-q+1)+(c-q)]d + (r-k)(c-q-1) + (r-k) \\ &= h(c-1) - q \cdot \frac{h+r-d}{2} + kq + r. \end{aligned}$$

All four cases can be summarized by the formula

$$i_{\max} = h(c-1) - q \cdot \frac{h+r-d}{2} + kq + \min\{r,k\}.$$

Clearly, when $m \leq d$, then $i_{\min} = i_{\max} = 0$. But when $m > d$, we can verify that $i_{\max} > i_{\min}$ as follows. We have

$$\begin{aligned} i_{\max} - i_{\min} &= h(c-1) - q \cdot (h+r-d) + kq + \min\{r,k\} \\ &= h \cdot (c-q-1) + q(d-r+k) + \min\{r,k\}; \end{aligned}$$

this quantity is positive when $c \geq q+1$. Note that we must have $c - q = \lceil \frac{m}{d} \rceil - \lceil \frac{h}{d} \rceil \geq 0$, thus the only remaining case is when $c = q$, in which case $k - r = m - h > 0$ and $q > 0$ (since $q = c$ and $m > d$), so we now have

$$i_{\max} - i_{\min} = -h + q(d-r+k) + \min\{r,k\} = -h + qd + q(k-r) + r = q(k-r) > 0.$$

Obviously, $i = i_1 + i_2 + \cdots + i_h$ can assume the value of any integer between these two bounds, and thus $h\hat{}A$ lies in exactly

$$\min\left\{\frac{n}{d}, i_{\max} - i_{\min} + 1\right\}$$

cosets of H.

Proof of Proposition D.47. We can easily check that the result holds for $h = 1$, so below we assume that $2 \leq h \leq m-1$. We will separate the rest of the proof into several cases. In the first two cases, we have $h \leq k \leq d$, and thus $q = 0$ and $h = r$, so we have $i_{\min} = 0$ and $i_{\max} = hc$.

Claim 1: If $h \leq k$ and $h < d$, then $|h\hat{}A| = \min\{n, hcd + d, hm - h^2 + 1\}$.

We here recognize the quantity $f_d = hcd + d$.

Proof of Claim 1: Note that the assumptions, using our lemma above, imply that

$$h\hat{\ }A = \bigcup_{i=0}^{hc-1} (i+H) \cup \left\{ hc + j \cdot \frac{n}{d} \mid j = \frac{h(h-1)}{2}, \dots, h(k-1) - \frac{h(h-1)}{2} \right\}.$$

We first consider the case when $c = 0$ or, equivalently, when $m \leq d$, i.e., when $m = k$. In this case

$$h\hat{\ }A = \left\{ j \cdot \frac{n}{d} \mid j = \frac{h(h-1)}{2}, \dots, h(m-1) - \frac{h(h-1)}{2} \right\},$$

and therefore

$$|h\hat{\ }A| = \min\{d, hm - h^2 + 1\} = \min\{n, d, hm - h^2 + 1\} = \min\{n, hcd + d, hm - h^2 + 1\},$$

as claimed.

Assume now that $c \geq 1$ (iff $m > d$, iff $m > k$). If we also have

$$hc - 1 \geq \frac{n}{d} - 1,$$

then

$$|h\hat{\ }A| = n.$$

But note that, when $k \geq h$ and

$$hc - 1 \geq \frac{n}{d} - 1,$$

then we also have

$$hm - h^2 + 1 = hcd + hk - h^2 + 1 \geq hcd + 1 > n$$

and

$$hcd + d \geq n + d > n,$$

and thus

$$|h\hat{\ }A| = n = \min\{n, hcd + d, hm - h^2 + 1\},$$

as claimed.

If, on the other hand, $c \geq 1$ and

$$hc - 1 \leq \frac{n}{d} - 2,$$

then

$$|h\hat{\ }A| = hcd + \min\{d, hk - h^2 + 1\} = \min\{hcd + d, hcd + hk - h^2 + 1\} = \min\{hcd + d, hm - h^2 + 1\}.$$

Note that, under the assumption that

$$hc - 1 \leq \frac{n}{d} - 2,$$

we have

$$\min\{hcd + d, hm - h^2 + 1\} \leq hcd + d \leq n,$$

and therefore again we get

$$|h\hat{\ }A| = n = \min\{n, hcd + d, hm - h^2 + 1\},$$

as claimed. This completes the proof of Claim 1.

Claim 2: If $h = k = d$, then $|h\hat{}A| = \min\{n, hm - h^2 - h + 2\}$.

Proof of Claim 2: First, we note that $h < m = dc + k = dc + h$, so $c \geq 1$. In the case of $h = k = d$, our lemma above cannot be used for the coset $i + H$ when $i = 0$; we in fact now have

$$h\hat{}A = \left\{\frac{d(d-1)}{2} \cdot \frac{n}{d}\right\} \cup \bigcup_{i=1}^{dc-1} (i + H) \cup \left\{dc + \frac{d(d-1)}{2} \cdot \frac{n}{d}\right\}.$$

If we have

$$dc - 1 \geq \frac{n}{d},$$

then clearly $|h\hat{}A| = n$, but we also have

$$hm - h^2 - h + 2 = h(dc + h) - h^2 - h + 2 = h(dc - 1) + 2 \geq n + 2 > n,$$

so

$$|h\hat{}A| = \min\{n, hm - h^2 - h + 2\},$$

as claimed.

Assume now that

$$dc - 1 \leq \frac{n}{d} - 2.$$

In this case

$$|h\hat{}A| = 1 + (dc - 1)d + 1 = h(cd + h) - h^2 - h + 2 = hm - h^2 - h + 2;$$

furthermore,

$$1 + (dc - 1)d + 1 \leq n - 2d + 2 \leq n,$$

thus

$$|h\hat{}A| = \min\{n, hm - h^2 - h + 2\},$$

as claimed.

This leaves us with the case of

$$dc - 1 = \frac{n}{d} - 1,$$

when we have

$$\begin{aligned} h\hat{}A &= \left\{\frac{d(d-1)}{2} \cdot \frac{n}{d}\right\} \cup \bigcup_{i=1}^{dc-1} (i + H) \cup \left\{dc + \frac{d(d-1)}{2} \cdot \frac{n}{d}\right\} \\ &= \left\{\frac{d(d-1)}{2} \cdot \frac{n}{d}\right\} \cup \bigcup_{i=1}^{\frac{n}{d}-1} (i + H) \cup \left\{\frac{n}{d} + \frac{d(d-1)}{2} \cdot \frac{n}{d}\right\} \\ &= \left\{j \cdot \frac{n}{d} \mid j = \frac{d(d-1)}{2}, \frac{d(d-1)}{2} + 1\right\} \cup \bigcup_{i=1}^{\frac{n}{d}-1} (i + H), \end{aligned}$$

and $d = h \geq 2$, so

$$|h\hat{}A| = 2 + \left(\frac{n}{d} - 1\right) \cdot d = 2 + (dc-1)d = cdh - h + 2 = (cd+h)h - h^2 - h + 2 = hm - h^2 - h + 2.$$

Furthermore, $2 + \left(\frac{n}{d} - 1\right) \cdot d = n - (d - 2) \leq n$, so again we have

$$|h\hat{}A| = \min\{n, hm - h^2 - h + 2\},$$

completing the proof of Claim 2.

We can also observe that, since we always have $k \leq d$, Claims 1 and 2 cover all possibilities under the assumption $h \leq k$.

Claim 3: If $h > k$, $r \neq d$, and $r \neq k$, then

$$|h\hat{\ }A| = \min\{n, hcd + d - (h-r)(h+r-k) - d \cdot \max\{0, r-k\}\}.$$

Proof of Claim 3: First, observe that by our lemma above, the three conditions imply that

$$h\hat{\ }A = \bigcup_{i=i_{\min}}^{i_{\max}} (i + H).$$

Therefore, we just need to prove that

$$(i_{\max} - i_{\min} + 1) \cdot d = hcd + d - (h-r)(h+r-k) - d \cdot \max\{0, r-k\}.$$

Indeed, from our computations above we get

$$\begin{aligned} i_{\max} - i_{\min} + 1 &= h(c-1) - q \cdot (h+r-d) + kq + \min\{r,k\} + 1 \\ &= hc - h + qd - q \cdot (h+r-k) + \min\{r,k\} + 1 \\ &= hc - r - q \cdot (h+r-k) + \min\{r,k\} + 1 \\ &= hc + 1 - \frac{h-r}{d} \cdot (h+r-k) + \min\{0, k-r\}, \end{aligned}$$

from which our result follows.

Claim 4: If $h > k$, $r = d$, and $r \neq k$, then

$$|h\hat{\ }A| = \min\{n, hcd + d - (h-r)(h+r-k) - d \cdot \max\{0, r-k\} - (d-1)\}.$$

Proof of Claim 4: The only difference between the conditions here and the conditions for Claim 3 is that this time we have

$$h\hat{\ }A = \{x_{\min}\} \cup \bigcup_{i=i_{\min}+1}^{i_{\max}} (i + H)$$

where $x_{\min}$ equals the sum of the h elements of the set

$$\bigcup_{i=0}^{q} (i + H).$$

Therefore,

$$|h\hat{\ }A| = \begin{cases} n & \text{if } i_{\max} - i_{\min} \geq \frac{n}{d}, \\ (i_{\max} - i_{\min}) \cdot d + 1 & \text{if } i_{\max} - i_{\min} \leq \frac{n}{d} - 1 \end{cases}$$

or, equivalently,

$$|h\hat{\ }A| = \min\{n, (i_{\max} - i_{\min}) \cdot d + 1\}.$$

Our claim then follows, since, as can be seen from the proof of Claim 3,

$$\begin{aligned} (i_{\max} - i_{\min}) \cdot d + 1 &= (i_{\max} - i_{\min} + 1) \cdot d - (d-1) \\ &= hcd + d - (h-r)(h+r-k) - d \cdot \max\{0, r-k\} - (d-1). \end{aligned}$$

Claim 5: If $h > k$, $r \neq d$, and $r = k$, then

$$|h\hat{\ }A| = \min\{n, hcd + d - (h-r)(h+r-k) - d \cdot \max\{0, r-k\} - (d-1)\}.$$

Proof of Claim 5: The only difference between the conditions here and the conditions for Claim 3 is that this time we have

$$h\hat{\ }A = \bigcup_{i=i_{\min}}^{i_{\max}-1} (i+H) \cup \{x_{\max}\}$$

where $x_{\max}$ equals the sum of the h elements of the set

$$\bigcup_{i=c-q}^{c-1} (i+H) \cup \left\{c + j \cdot \frac{n}{d} \mid j = 0, 1, 2, \ldots, k-1\right\}.$$

Our claim then follows as in Claim 5.

Claim 6: If $h > k$, $r = d$, and $r = k$, then

$$|h\hat{\ }A| = \min\{n, n - (d-2)\}$$

if $m = \frac{n}{h} + h$, and

$$|h\hat{\ }A| = \min\{n, hcd + d - (h-r)(h+r-k) - d \cdot \max\{0, r-k\} - 2(d-1)\}$$

otherwise.

Proof of Claim 6: This time we get

$$h\hat{\ }A = \{x_{\min}\} \cup \bigcup_{i=i_{\min}+1}^{i_{\max}-1} (i+H) \cup \{x_{\max}\}$$

where $x_{\min}$ and $x_{\max}$ were defined in Claims 4 and 5, respectively.

With $r = k = d$, we can simplify the expression we got for $i_{\max} - i_{\min} + 1$ in the proof of Claim 3:

$$\begin{aligned} i_{\max} - i_{\min} + 1 &= hc + 1 - \frac{h-r}{d} \cdot (h+r-k) + \min\{0, k-r\} \\ &= hc + 1 - \frac{(h-d)h}{d} \\ &= \frac{h(cd+d) - h^2 + d}{d} \\ &= \frac{hm - h^2 + d}{d}. \end{aligned}$$

Thus we see that, if $m \neq \frac{n}{h} + h$, then $i_{\max} - i_{\min} - 1 \neq \frac{n}{d} - 1$. Moreover, if $m > \frac{n}{h} + h$, then $i_{\max} - i_{\min} - 1 \geq \frac{n}{d}$, and thus

$$\begin{aligned} |h\hat{\ }A| &= n \\ &= \min\{n, (i_{\max} - i_{\min} - 1) \cdot d + 2\} \\ &= \min\{n, (i_{\max} - i_{\min} + 1) \cdot d - 2(d-1)\} \\ &= \min\{n, hcd + d - (h-r)(h+r-k) - d \cdot \max\{0, r-k\} - 2(d-1)\}. \end{aligned}$$

If, on the other hand, $m < \frac{n}{h} + h$, then $i_{\max} - i_{\min} - 1 \leq \frac{n}{d} - 2$, and thus

$$\begin{aligned}
|h\hat{\ }A| &= (i_{\max} - i_{\min} - 1) \cdot d + 2 \\
&= \min\{n, (i_{\max} - i_{\min} - 1) \cdot d + 2\} \\
&= \min\{n, (i_{\max} - i_{\min} + 1) \cdot d - 2(d-1)\} \\
&= \min\{n, hcd + d - (h-r)(h+r-k) - d \cdot \max\{0, r-k\} - 2(d-1)\}.
\end{aligned}$$

This leaves us with the case of $m = \frac{n}{h} + h$. In this case

$$\begin{aligned}
h\hat{\ }A &= \{x_{\min}\} \cup \bigcup_{i=i_{\min}+1}^{i_{\max}-1} (i+H) \cup \{x_{\max}\} \\
&= \bigcup_{i=i_{\min}+1}^{i_{\min}+\frac{n}{d}-1} (i+H) \cup \{x_{\min}, x_{\max}\}.
\end{aligned}$$

A simple computation shows that, denoting the sum of the elements in a subset S of $\mathbb{Z}_n$ by $\sum S$, we have

$$x_{\min} = \sum \bigcup_{i=0}^{q} (i+H) = \frac{dq(q+1)}{2} + \frac{d(d-1)(q+1)}{2} \cdot \frac{n}{d}$$

and

$$x_{\max} = \sum \bigcup_{i=c-q}^{c} (i+H) = cd(q+1) - \frac{dq(q+1)}{2} + \frac{d(d-1)(q+1)}{2} \cdot \frac{n}{d}.$$

But

$$cd = m - d = \frac{n}{h} + h - d = \frac{n}{d(q+1)} + dq,$$

thus

$$x_{\max} = \frac{dq(q+1)}{2} + \left(\frac{d(d-1)(q+1)}{2} + 1\right) \cdot \frac{n}{d},$$

showing that $x_{\min} = x_{\max}$ if, and only if, $d = 1$. Therefore, when $d \geq 2$, we get

$$|h\hat{\ }A| = \left(\frac{n}{d} - 1\right) d + 2 = n - d + 2,$$

and when $d = 1$ we get

$$|h\hat{\ }A| = \left(\frac{n}{d} - 1\right) d + 1 = n.$$

This completes the proof of Claim 6.

Now it is an easy exercise to verify that in all cases, $|h\hat{\ }A|$ is as claimed in the statement of Theorem D.47.

Proof of Proposition D.59

Our task is to prove that, when n is not a power of 2, then

$$\rho\hat{\ }(G, m_0, 2) \leq n - 2,$$

where

$$m_0 = \frac{n + |\mathrm{Ord}(G,2)| + 1}{2}.$$

We use induction on the rank of G. When G is cyclic, the claim follows easily from Corollary D.52; for the sake of completeness—and since the rest of our proof is constructive—we exhibit a subset A of $\mathbb{Z}_n$ of size m_0 for which $2\hat{}A$ is of size $n-2$.

When n is odd, we have $m_0 = (n+1)/2$, and the set

$$A = \{0, 1, 2, \dots, (n-1)/2\}$$

has restricted sumset

$$2\hat{}A = \{1, 2, \dots, n-2\},$$

as claimed.

Suppose now that n is even, and set $n = 2^k \cdot d$ where $k \geq 1$ and $d \geq 3$ is odd. In this case,

$$m_0 = n/2 + 1 = 2^{k-1} \cdot d + 1.$$

Consider the set

$$A = B_d(n, m_0; (d+1)/2, (d+1)/2, 1, (d-1)/2),$$

defined on page 168: in our case, this set becomes

$$A = B' \cup \bigcup_{i=1}^{2^{k-1}-1} \{i + j \cdot 2^k \mid j = 0, 1, \dots, d-1\} \cup B'',$$

with

$$B' = \{j \cdot 2^k \mid j = 0, 1, \dots, (d-1)/2\},$$

and

$$B'' = \{2^{k-1} + j \cdot 2^k \mid j = (d-1)/2, (d+1)/2, \dots, d-1\}.$$

Then

$$|A| = (d+1)/2 + (2^{k-1} - 1) \cdot d + (d+1)/2 = 2^{k-1} \cdot d + 1 = m_0,$$

and

$$2\hat{}A = \mathbb{Z}_n \setminus \{0, (d-1) \cdot 2^k\},$$

since

$$2\hat{}B' = \{j \cdot 2^k \mid j = 1, 2, \dots, d-2\}$$

and

$$2\hat{}B'' = \{2^k + j \cdot 2^k \mid j = d, d+1, \dots, 2d-3\} = 2\hat{}B'.$$

This completes the case when G is cyclic.

Assume now that our claim holds for groups of rank $r-1$, and let G be a group of rank $r \geq 2$. Suppose further that G is of type $(n_1, \dots, n_r)$; we write

$$G = \mathbb{Z}_{n_1} \times G_2$$

where G_2 has order n/n_1 and rank $r-1$. Note that n_1 is a divisor of n/n_1, so if n is not a power of 2, then n/n_1 is not a power of 2 either. Therefore, by our inductive assumption, G_2 contains a subset A_2 of size

$$|A_2| = \frac{n/n_1 + |\mathrm{Ord}(G_2, 2)| + 1}{2}$$

for which

$$2\hat{}A_2 = G_2 \setminus X$$

for some $X \subseteq G_2$ of size $|X| \geq 2$.

We separate two cases depending on the parity of n_1.

Observe that when n_1 is odd, then

$$|\mathrm{Ord}(G,2)| = |\mathrm{Ord}(G_2,2)|,$$

so

$$m_0 = \frac{n + |\mathrm{Ord}(G_2,2)| + 1}{2}.$$

Set

$$A = (\{0\} \times A_2) \cup (\{1, 2, \ldots, (n_1-1)/2\} \times G_2).$$

The size of A is then

$$|A| = |A_2| + (n_1-1)/2 \cdot |G_2| = \frac{n/n_1 + |\mathrm{Ord}(G_2,2)| + 1}{2} + \frac{(n_1-1) \cdot n/n_1}{2} = m_0.$$

We can also see that

$$2\hat{}A \subseteq (\mathbb{Z}_{n_1} \times G_2) \setminus (\{0\} \times X),$$

and hence

$$|2\hat{}A| \leq n - |X| \leq n-2,$$

as claimed.

Suppose now that n_1 is even. In this case, we can easily see that

$$|\mathrm{Ord}(G,2)| = 2 \cdot |\mathrm{Ord}(G_2,2)| + 1,$$

so

$$m_0 = \frac{n + 2 \cdot |\mathrm{Ord}(G_2,2)| + 2}{2} = \frac{n}{2} + |\mathrm{Ord}(G_2,2)| + 1.$$

This time, we set

$$A = (\{0, n_1/2\} \times A_2) \cup (\{1, 2, \ldots, n_1/2 - 1\} \times G_2).$$

We again have

$$|A| = 2 \cdot |A_2| + (n_1/2 - 1) \cdot |G_2| = (n/n_1 + |\mathrm{Ord}(G_2,2)| + 1) + (n/2 - n/n_1) = m_0$$

and

$$2\hat{}A \subseteq (\mathbb{Z}_{n_1} \times G_2) \setminus (\{0\} \times X),$$

and thus

$$|2\hat{}A| \leq n - |X| \leq n-2,$$

which completes our proof. □

Proof of Theorem D.72

Our claim is already established for cyclic groups by, for example, Proposition D.46.

Turning to noncyclic groups, let us first consider the elementary abelian 2-group $\mathbb{Z}_2^r$, which we write as $G_1 \times \mathbb{Z}_2$ with $G_1 = \mathbb{Z}_2^{r-1}$. The result for $h = 2$ has been delivered in Proposition D.64, so assume that $h \geq 3$. Let $g_1 \in G_1$ and $g_2 \in \mathbb{Z}_2$ be arbitrary; we need to prove that there are h pairwise distinct elements in $G_1 \times \mathbb{Z}_2$ that add to (g_1, g_2).

Noting that we have $h \leq 2^{r-1}$, we choose $h-1$ arbitrary elements $a_1, \ldots, a_{h-1}$ in G_1. Set

$$a = g_1 - (a_1 + \cdots + a_{h-1}).$$

(To make the proof more transparent, we used subtraction here, though of course in G_1 it is equivalent to addition.) If a is distinct from a_i for all $1 \leq i \leq h-1$, then

$$(a, g_2), (a_1, 0), \ldots, (a_{h-1}, 0)$$

are h distinct elements of $G_1 \times \mathbb{Z}_2$ that add to (g_1, g_2).

In the case when a equals one of $a_1, \ldots, a_{h-1}$, say $a = a_1$, then, noting also that $h-1 \geq 2$, we see that

$$(a_1, 0), (a_1, 1), (a_2, g_2 - 1), (a_3, 0), \ldots, (a_{h-1}, 0)$$

are h distinct elements of $G_1 \times \mathbb{Z}_2$ that add to (g_1, g_2).

Next, we consider the group $\mathbb{Z}_2 \times \mathbb{Z}_\kappa$ with $\kappa \geq 4$ even. Let $g_1 \in \mathbb{Z}_2$ and $g_2 \in \mathbb{Z}_\kappa$; we need to find h distinct elements of $\mathbb{Z}_2 \times \mathbb{Z}_\kappa$ that add to (g_1, g_2). If $h < \kappa$, then this follows easily from the result for cyclic groups. Given that $h \leq n/2$, the remaining case is that $h = \kappa$, which needs more attention. We separate two cases depending on κ mod 4.

Suppose first that κ is divisible by 4, and let $c = \kappa/4$. Since

$$c \leq \kappa/2 - 1,$$

we have pairwise distinct integers $a_1, \ldots, a_c$ that are all between 1 and $\kappa/2 - 1$, inclusive. We consider two subcases as follows: if $g_2 = 0$, let

$$A = \{(0, a_i), (0, -a_i), (1, a_i), (1, -a_i) \mid i = 1, \ldots, c-1\} \cup \{(1, a_c), (g_1, -a_c), (0, 0), (1, 0)\}.$$

If $g_2 \neq 0$, then one of the integers $a_1, \ldots, a_c$ may equal g_2 or $-g_2$; we will assume that none of $a_1, \ldots, a_{c-1}$ equals $\pm g_2$, and we let

$$A = \{(0, a_i), (0, -a_i), (1, a_i), (1, -a_i) \mid i = 1, \ldots, c-1\} \cup \{(0, a_c), (0, -a_c), (1, g_2), (g_1 - 1, 0)\}.$$

In both subcases, A consists of $h = \kappa$ pairwise distinct elements that add to (g_1, g_2).

Now suppose that $\kappa \equiv 2$ mod 4, and let $c = (\kappa - 2)/4$. This time,

$$c \leq \kappa/2 - 2,$$

so we have pairwise distinct integers $a_1, \ldots, a_{c+1}$ that are all between 1 and $\kappa/2 - 1$, inclusive. Again we consider two subcases: if $g_2 = 0$, let

$$A = \{(0, a_i), (0, -a_i), (1, a_i), (1, -a_i) \mid i = 1, \ldots, c\} \cup \{(0, a_{c+1}), (g_1, -a_{c+1})\}.$$

If $g_2 \neq 0$, then one of the $c+1$ integers may equal g_2 or $-g_2$; we will assume that none of $a_1, \ldots, a_c$ equals $\pm g_2$, and we let

$$A = \{(0, a_i), (0, -a_i), (1, a_i), (1, -a_i) \mid i = 1, \ldots, c\} \cup \{(0, 0), (g_1, g_2)\}.$$

In both subcases, A consists of $h = \kappa$ pairwise distinct elements that add to (g_1, g_2).

This completes the cases when G is cyclic, the elementary abelian 2-group, or is of the form $\mathbb{Z}_2 \times \mathbb{Z}_\kappa$, so we assume that $G \cong G_1 \times \mathbb{Z}_\kappa$, where G_1 is of order at least three, κ is the exponent of G, and $\kappa \geq 3$. We will also write

$$h = c\kappa + b$$

where

$$0 \leq b \leq \kappa - 1.$$

Furthermore, since $|G_1| \geq 3$ and $h \leq |G_1| \cdot \kappa/2$, we also have

$$c \leq |G_1| - 2.$$

Let $g_1 \in G_1$ and $g_2 \in \mathbb{Z}_\kappa$; we need to find h distinct elements of $G_1 \times \mathbb{Z}_\kappa$ that add to (g_1, g_2). We consider two cases depending on whether b is positive or not.

If $b \geq 1$, we let A_1 be any c distinct elements of $G_1 \setminus \{0, g_1\}$ (possible since $c \leq |G_1| - 2$), and let A_2 be any b distinct elements of $\mathbb{Z}_\kappa$ that add to

$$g_2 - \kappa(\kappa - 1)c/2$$

(possible since $1 \leq b \leq \kappa - 1$). Let a be any element of A_2. Then

$$(A_1 \times \mathbb{Z}_\kappa) \cup (\{0\} \times (A_2 \setminus \{a\})) \cup \{(g_1, a)\}$$

is a set of h distinct elements of G, and its elements add to (g_1, g_2). (Note that, since κ is the exponent of G, $\kappa a_1 = 0$ for any $a_1 \in A_1$.)

Suppose now that $b = 0$; we need to separate several subcases. If $g_1 \neq 0$, we let $A_1 \subset G_1$ be any $c - 1$ distinct elements of $G_1 \setminus \{0, g_1\}$, and let A_2 be the $\kappa - 1$ distinct elements of $\mathbb{Z}_\kappa$ that add to

$$g_2 - \kappa(\kappa - 1)(c - 1)/2.$$

Then

$$(A_1 \times \mathbb{Z}_\kappa) \cup (\{0\} \times A_2) \cup \{(g_1, 0)\}$$

is a set of h distinct elements of G and its elements add to (g_1, g_2).

Next, suppose that $b = 0$, $g_1 = 0$, and κ is odd. Let a be any nonzero element of G_1, and choose A_1 to be any $c - 1$ distinct elements of $G_1 \setminus \{0, a, -a\}$ (note that this is possible even when $a \neq -a$, since $c - 1 \leq |G_1| - 3$); furthermore, let

$$A_2 = \{1, 2, 3, \ldots, (\kappa - 1)/2\}.$$

Then

$$(A_1 \times \mathbb{Z}_\kappa) \cup (\{a\} \times A_2) \cup (\{-a\} \times (-A_2)) \cup \{(0, g_2)\}$$

is a set of h distinct elements of G and its elements add to $(0, g_2)$. (Note that, since $\kappa - 1$ is even, $\kappa(\kappa - 1)(c - 1)/2)$ equals zero in $\mathbb{Z}_\kappa$.)

The case when $b = 0$, $g_1 = 0$, κ is even, and $g_2 \neq t$ where

$$t = \kappa(\kappa - 1)(c - 1)/2$$

is very similar: again we let a be any nonzero element of G_1, and choose A_1 to be any $c - 1$ distinct elements of $G_1 \setminus \{0, a, -a\}$, but now we set

$$A_2 = \{1, 2, \ldots, \kappa/2 - 1\}.$$

Then

$$(A_1 \times \mathbb{Z}_\kappa) \cup (\{0\} \times A_2) \cup (\{0\} \times (-A_2)) \cup \{(a, 0), (-a, g_2 - t)\}$$

is a set of h distinct elements of G and its elements add to $(0, g_2)$. This very construction works even if $g_2 = t$ as long as a has order at least three in G_1.

This leaves us with the case when $b = 0$, $g_1 = 0$, κ is even, $g_2 = t$, and G_1 is an elementary abelian 2-group. Our construction is again similar, but we start with two distinct nonzero elements a_1 and a_2 of G_1. We then let A_1 be any $c - 1$ distinct elements of $G_1 \setminus \{0, a_1, a_2\}$, and set

$$A_2 = \{1, 2, \ldots, \kappa/2 - 2\}.$$

Then

$$(A_1 \times \mathbb{Z}_\kappa) \cup (\{0\} \times A_2) \cup (\{0\} \times (-A_2)) \cup \{(a_1, 0), (a_2, 0), (a_1, \kappa/2), (a_2, \kappa/2\}$$

is a set of h distinct elements of G and its elements add to $(0, g_2)$. □

Proof of Proposition D.128

Let $A = \{a, b, c\}$ be a weakly zero-sum-free 3-subset of G; we then have

$$\Sigma^* A = \{a, b, c, a+b, a+c, b+c, a+b+c\}.$$

We can observe right away that the elements a, b, c, and $a+b+c$ must be pairwise distinct (otherwise one of $a+b$, $a+c$, or $b+c$ would be 0). Furthermore, none of $a+b$, $a+c$, or $b+c$ can equal $a+b+c$. Therefore, the size of $\Sigma^* A$ is 4, 5, 6, or 7, depending on how many of the equations $a+b=c$, $a+c=b$, and $b+c=a$ hold. Note that if all three of them hold, then $a+b+c=0$. Therefore, the size of $\Sigma^* A$ is 5, 6, or 7.

Suppose first that n is odd. In this case, $|\Sigma^* A| \geq 6$, since otherwise we would have, wlog, $a+b=c$ and $a+c=b$, which would imply that $2a=0$, and that is impossible if n is odd. On the other hand, if $c=a+b$, then clearly $|\Sigma^* A| = 6$.

Now if $n \in \{1, 3, 5\}$, then G has no weakly zero-sum-free 3-subsets. In the case when $n \geq 7$ and odd, we show that we can always find elements $a \in G$ and $b \in G$ so that $A = \{a, b, a+b\}$ is weakly zero-sum-free. If G is isomorphic to $G_1 \times G_2$, then for any $g_1 \in G_1 \setminus \{0\}$ and $g_2 \in G_2 \setminus \{0\}$, we can take $a = (g_1, 0)$ and $b = (0, g_2)$, since then

$$0 \notin \Sigma^* A = \{(g_1, 0), (0, g_2), (g_1, g_2), (2g_1, g_2), (g_1, 2g_2), (2g_1, 2g_2)\}.$$

This leaves us with cyclic groups of (prime) order at least 7, in which case we take $a=1$ and $b=2$, for which

$$0 \notin \Sigma^* A = \{1, 2, 3, 4, 5, 6\}.$$

Suppose next that n is even and that the exponent κ of G is at least 5. In this case, let a be an order 2 element of G and b to be an order κ element of G; we also set $c = a+b$. In this case,

$$\Sigma^* A = \{a, b, a+b, 2b, a+2b\}.$$

None of these elements are 0; for example, $a+2b=0$ would imply that $4b=-2a=0$, contradicting the fact that b has order at least 5.

Suppose now that G has exponent 4 and order at least 8. In this case, we can write G as $G_1 \times \mathbb{Z}_4$; let g_1 be an element of G_1 of order 2. Choosing $a = (g_1, 0)$, $b = (0, 1)$ and $c = (g_1, 1)$ gives us

$$\Sigma^* A = \{(g_1, 0), (0, 1), (g_1, 1), (0, 2), (g_1, 2)\},$$

so A is weakly zero-sum-free.

We are left with the group $\mathbb{Z}_2^r$, for which we prove that $\Sigma^* A$ has size 7 for every weakly zero-sum-free subset $A = \{a, b, c\}$ of size 3. Indeed, none of the equations $a+b=c$, $a+c=b$, and $b+c=a$ hold: for example, $a+b=c$ implies that $a+b+c=2c=0$. Finally, observe that when $r \geq 3$, then letting a, b, and c denote the elements of $\mathbb{Z}_2^r$ that consist of $r-1$ zero components, with the 1 component being in three different places, the set $A = \{a, , c\}$ is weakly zero-sum-free. This completes our proof. □

Proof of Theorem E.15

We will use the notations of [132]. First, note that

$$\hat{\chi}(G, [0, s]) = s_{s+1}^{+}(G) + 1.$$

Therefore, we need to prove that

$$s_\rho^{+}(G) \leq \hat{v}(n, \rho-1) = \max\left\{\left(\left\lfloor \frac{d-2}{\rho-1} \right\rfloor + 1\right) \cdot \frac{n}{d} \mid d \in D(n), d \geq \rho+1\right\}$$

for every integer $\rho \geq 2$.

By Lemma 2.3 in [132], we have

$$s_\rho^+(G) = \max\{|H| \cdot t_\rho^+(G/H) \mid H \leq G, H \neq G\}.$$

Let $H \leq G, H \neq G$ be such that

$$s_\rho^+(G) = |H| \cdot t_\rho^+(G/H),$$

and suppose that $|G/H| = d$ and that G/H has invariant factorization

$$G/H \cong \mathbb{Z}_{d_1} \times \cdots \times \mathbb{Z}_{d_r}.$$

By Theorem 2.1 in [132], we have

$$\text{diam}^+(G/H) = \Sigma_{i=1}^r (d_i - 1).$$

Since the number of elements in $\mathbb{Z}_{d_1} \times \cdots \times \mathbb{Z}_{d_r}$ with exactly one nonzero coordinate equals $\Sigma_{i=1}^r (d_i - 1)$ and this number is clearly at most the number of nonzero elements in G/H, we have

$$\text{diam}^+(G/H) \leq |G/H| - 1 = d - 1.$$

In particular, when $d \leq \rho$, we have $\text{diam}^+(G/H) < \rho$. According to page 27 in [132], we then have $t_\rho^+(G/H) = 0$, so $s_\rho^+(G) = 0$, from which our claim trivially follows.

Assume now that $d \geq \rho + 1$. By Proposition 2.8 in [132], we have

$$t_\rho^+(G/H) \leq \left\lfloor \frac{d-2}{\rho - 1} \right\rfloor + 1,$$

and so

$$\begin{aligned} s_\rho^+(G) &\leq \left(\left\lfloor \frac{d-2}{\rho - 1} \right\rfloor + 1 \right) \cdot \frac{n}{d} \\ &\leq \widehat{v}(n, \rho - 1), \end{aligned}$$

as claimed. □

Proof of Proposition E.76

Clearly,

$$\chi\hat{\ }(G^*, h) \leq \chi\hat{\ }(G, h),$$

since if there is a subset A of $G \setminus \{0\}$ with $h\hat{\ }A \neq G$, then $|A| + 1$ provides a lower bound for $\chi\hat{\ }(G, h)$.

For the other direction, let B be a subset of G of size $\chi\hat{\ }(G, h) - 1$ for which $h\hat{\ }B \neq G$. Since $\chi\hat{\ }(G, h) \leq n$, we have $|B| \leq n - 1$, and so $|-B| \leq n - 1$ as well. Let $g \in G \setminus (-B)$. Then $A = g + B$ has size $\chi\hat{\ }(G, h) - 1$, and $A \subseteq G \setminus \{0\}$, since $0 \in A = g + B$ would imply that $0 = g + b$ for some $b \in B$, and thus $g = -b \in -B$, a contradiction. But $h\hat{\ }A$ and $h\hat{\ }B$ have the same size, so we conclude that $h\hat{\ }A \neq G$, from which

$$\chi\hat{\ }(G^*, h) \geq \chi\hat{\ }(G, h)$$

follows. □

Proof of Lemma E.87

Suppose first that k is odd, in which case $h = (k+1)/2$, and our desired inequality becomes

$$\lfloor (2n-4)/(k+1) \rfloor + (k+1)/2 \leq k$$

or

$$\lfloor (2n-4)/(k+1) \rfloor \leq (k-1)/2.$$

Since $(k-1)/2$ is an integer, it suffices to prove that

$$(2n-4)/(k+1) < (k-1)/2 + 1$$

or, equivalently, that

$$4n - 8 < (k+1)^2.$$

But

$$(k+1)^2 = (\lfloor 2\sqrt{n-2} \rfloor + 1)^2 > (2\sqrt{n-2})^2 = 4n - 8,$$

as claimed.

Assume now that k is even, in which case $h = k/2$, and our desired inequality becomes

$$\lfloor (2n-4)/k \rfloor + k/2 \leq k$$

or

$$\lfloor (2n-4)/k \rfloor \leq k/2.$$

Since $k/2$ is an integer, it suffices to prove that

$$(2n-4)/k < k/2 + 1$$

or, equivalently, that

$$4n - 7 < (k+1)^2.$$

As above, we see that

$$(k+1)^2 > 4n - 8,$$

so we just need to rule out the possibility that

$$(k+1)^2 = 4n - 7.$$

This is indeed impossible, as the left-hand side is the square of an odd integer and thus congruent to 1 mod 8, while the right-hand side, since n is odd, is congruent to 5 mod 8. □

Proof of Theorem E.100

Clearly, any P1-set is a P3-set, and any P2-set is a P4-set.

Suppose now that A is a P3-set; we will prove that $0 \in A$ and $A \setminus \{0\}$ is a P2-set as follows. If we were to have $0 \notin A$, then, since by Theorem E.98 A has size

$$|A| = \chi\hat{\ }(G, \mathbb{N}) - 1 = \chi\hat{\ }(G^*, \mathbb{N}),$$

we have $\Sigma^* A = G$, but that contradicts our assumption that A is a P3-set. Therefore, $0 \in A$, and, using Theorem E.98 again,

$$|A \setminus \{0\}| = |A| - 1 = \chi\hat{\ }(G, \mathbb{N}) - 2 = \chi\hat{\ }(G^*, \mathbb{N}_0) - 1.$$

Furthermore, since $0 \in A$, we have $\Sigma^* A = \Sigma A$, but then $\Sigma(A \setminus \{0\}) \neq G$, since $\Sigma^* A \neq G$. This proves that $A \setminus \{0\}$ is a P2-set.

Finally, Theorem E.98 immediately implies that if $0 \in A$ and $A \setminus \{0\}$ is a P2-set, then A is a P1-set. This completes our proof. □

Proof of Theorem E.108

Our proof will follow that of Theorem 3.1 in [94] for the case of $|S| = n/2 - 1$. We will need two lemmas, also found (though one stated with a crucial condition missing) in [94]:

Lemma 1 (Gao, Hamidoune, Lladó, and Serra; cf. [94] Lemma 2.7) *Let* $S \subseteq G \setminus \{0\}$, $\langle S \rangle = G$, *and* $|S| \geq 14$. *Suppose further that there is no proper subgroup* H *of* G *for which*

$$|S \cap H| \geq |S| - 1.$$

Then

$$|\Sigma S| \geq \min\{n - 3, 3|S| - 3\}.$$

Lemma 2 (Gao, Hamidoune, Lladó, and Serra; cf. [94] Lemma 2.9) *Let* $S \subseteq G \setminus \{0\}$. *Suppose further that* H *is a subgroup of* G *of prime index* p, *for which*

$$|S \cap H| \geq p - 1.$$

Then

$$\Sigma(S \setminus H) + H = G.$$

Proof of Theorem E.108: If A is a subgroup of order $n/2$, then clearly $\Sigma A \neq G$, so we only need to prove the converse. We can check the claim for all groups of order 16 using the computer program [120], so we will assume that $n \geq 18$.

Let us assume, indirectly, that there is a subset A of G of size $|A| = n/2$ that is not a subgroup of G but for which $\Sigma A \neq G$. Since A is not a subgroup of G, $A \subset \langle A \rangle$, which implies that $\langle A \rangle = G$.

Observe that by Theorem E.90, if $|A| = n/2$ and $\Sigma A \neq G$, then $0 \in A$. Let $S = A \setminus \{0\}$. Then

$$\langle S \rangle = \langle S \cup \{0\} \rangle = \langle A \rangle = G.$$

Since $\Sigma A \neq G$ implies that $\Sigma S \neq G$, we may choose an element $g \in G \setminus \Sigma S$.

Claim: For every proper subgroup H of G we have

$$|S \cap H| \leq n/4.$$

Proof of Claim: Suppose that our claim is false; we then have a proper subgroup H of G for which

$$|S \cap H| \geq \lceil (n+2)/4 \rceil.$$

The index of H in G then is either 2 or 3. Since $\langle S \rangle = G$, S cannot be a subset of H; furthermore, if the index of H in G is 3, then $S \setminus H$ cannot consist of a single element. Therefore, H is a subgroup of G of order $p \in \{2, 3\}$ so that

$$|S \setminus H| \geq p - 1.$$

Thus, by Lemma 2 above,

$$\Sigma(S \setminus H) + H = G.$$

But then

$$\Sigma(S \cap H) \neq H,$$

otherwise we would have

$$G = \Sigma(S \setminus H) + \Sigma(S \cap H) = \Sigma S,$$

contradicting our assumption. We fix an element $h \in H \setminus \Sigma(S \cap H)$.

Since for $n \geq 16$, $|S \cap H| \geq \lceil (n+2)/4 \rceil \geq 5$, we see that we can find an element $s \in S \cap H$ so that $S_1 = (S \cap H) \setminus \{s\}$ either has size at least 5, or has size 4 but is not of the form $\{\pm a, \pm 2a\}$ for any $a \in S$.

Now
$$|S_1 \cup \{0\}| = |S \cap H| > n/4 \geq |H|/2,$$
so
$$\langle S_1 \rangle = \langle S_1 \cup \{0\} \rangle = H,$$
and, therefore, by Theorem D.123,
$$|\Sigma S_1| \geq \min\{|H| - 1, 2|S_1|\}.$$
Here
$$|H| - 1 \leq n/2 - 1 = 2 \cdot ((n+2)/4 - 1) \leq 2 \cdot (|S \cap H| - 1) = 2 \cdot |S_1|,$$
so
$$|\Sigma S_1| \geq |H| - 1.$$
This is impossible, however, since we know of at least two distinct elements of H that are not in ΣS_1: h and $h - s$. Indeed, $h \in H \setminus \Sigma(S \cap H)$, so $h \in H \setminus \Sigma S_1$; for the same reason, if $h - s \in \Sigma S_1$, then $h = s + (h - s) \in \Sigma(S \cap H)$, a contradiction. This proves our claim.

Since $|S| = n/2 - 1 > 2$, we can select distinct elements $s_1, s_2 \in S$ so that $s_1 + s_2 \neq 0$. Set $S_2 = S \setminus \{s_1, s_2\}$. We find that $\langle S_2 \rangle = G$, otherwise $\langle S_2 \rangle$ would be a proper subgroup of G for which
$$|S \cap \langle S_2 \rangle| \geq |S_2| = n/2 - 3 > n/4,$$
contradicting our claim above. By the same claim, when $n \geq 18$, then for each proper subgroup H of G, we have
$$|S_2 \cap H| \leq |S \cap H| \leq \lfloor n/4 \rfloor \leq n/2 - 5 = |S_2| - 2.$$
We can then apply Lemma 1 above, and get
$$|\Sigma S_2| \geq \min\{n - 3, 3|S_2| - 3\} = \min\{n - 3, 3(n/2 - 3) - 3\} = n - 3.$$
This is impossible; recall that $g \notin \Sigma S$, so none of the four pairwise distinct elements $g, g - s_1, g - s_2$, or $g - s_1 - s_2$ are in ΣS_2. □

Proof of Theorem E.109

Our proof will follow that of Theorem 3.2 in [94] for the case of $|S| = n/3 + 1$. Besides the two lemmas in the proof of Theorem E.108 above, we will also need the following from [94]:

Lemma (Gao, Hamidoune, Lladó, and Serra; cf. [94] Lemma 2.8) *Let X be a generating subset of G that is asymmetric (i.e. $X \cap -X = \emptyset$), and suppose that $|\Sigma X| \leq n/2$. Then there is a proper subset V of X so that*
$$|\Sigma X| \geq 4|V| + \frac{(|X| + |V| + 5)(|X| - |V| - 1) - 2}{4}.$$

Proof of Theorem E.109: If A is of the form specified, then clearly $\Sigma A \neq G$, so we only need to prove the converse. Since 27 and 45 are the only odd values of n up to 62 for which $n/3$ is a composite integer, we may assume that $n \geq 63$.

Let us assume that A is a subset of G so that $|A| = n/3+1$ and $\Sigma A \neq G$. Then $\langle A \rangle = G$; furthermore, by Theorem E.92, we see that $0 \in A$. Let $S = A \setminus \{0\}$; then S has size $n/3$.

Claim 1: There is an asymmetric subset X of S of size $(n-3)/6$ so that $|\Sigma X| \leq (n-1)/2$.

Proof of Claim 1: Let s be an arbitrary element of S, and consider $S \setminus \{s\}$. Observe that $S \setminus \{s\}$ can be partitioned as $X_1 \cup X_2$ so that X_1 and X_2 are both asymmetric and have size $(n-3)/6$. We just have to prove that at least one of ΣX_1 or ΣX_2 has size at most $(n-1)/2$. If this were not the case, then for any $g \in G \setminus \Sigma S$, we would have

$$|\Sigma X_1| + |g - \Sigma X_2| = |\Sigma X_1| + |\Sigma X_2| \geq n+1,$$

so ΣX_1 and $g - \Sigma X_2$ cannot be disjoint and thus

$$g \in \Sigma(X_1 \cup X_2) \subseteq \Sigma S,$$

a contradiction. This proves Claim 1.

Let X be a set specified by Claim 1, and set $H = \langle X \rangle$.

Claim 2: $|H| = n/3$.

Proof of Claim 2: Since X is asymmetric, we must have

$$|H| \geq 2|X| + 1 = n/3;$$

since n is odd, we only need to prove that $H \neq G$.

If $H = G$, then by the lemma above, there is a proper subset V of X so that

$$|\Sigma X| \geq 4|V| + \frac{(|X| + |V| + 5)(|X| - |V| - 1) - 2}{4}$$

and thus, by Claim 1,

$$\frac{n-1}{2} \geq 4|V| + \frac{((n-3)/6 + |V| + 5)((n-3)/6 - |V| - 1) - 2}{4}.$$

However, we find that, when $n \geq 63$, there is no value of $0 \leq |V| \leq (n-3)/6 - 1$ for which this inequality holds. (The minimum value of the right-hand side must occur either at $|V| = 0$ or at $|V| = (n-3)/6 - 1$, but both possibilities yield values more than $(n-1)/2$.) This proves our claim.

Proof of Theorem E.109: Let p be the smallest prime divisor of $n/3$. Since

$$|X| = \frac{n-3}{6} \geq \frac{n}{9} + 2 = \frac{3n + 9p^2 - 9p + (p-3)(n-9p)}{9p} \geq \frac{n/3}{p} + p - 1,$$

by Theorems E.91–E.95, we get $\Sigma X = H$. Therefore,

$$\Sigma(S \setminus H) + H = \Sigma(S \setminus \Sigma X) + \Sigma X \subseteq \Sigma(S \setminus X) + \Sigma X = \Sigma S \neq G,$$

so by Lemma 2 from the proof of Theorem E.108, $|S \setminus H| \leq 1$. But $|S| = |H| = n/3$ and $0 \in H \setminus S$, so $|S \setminus H| = 1$, and our claim follows. □

Remark: We did not use in this proof that $n/3$ is composite, so as long as $n \geq 63$, our claim holds.

Proof of Theorem F.6

If $n|h$, then $\tau(\mathbb{Z}_n, h) = 0$ and, since for each $d \in D(n)$ we have $\gcd(d, h) = d$,

$$v_h(n, h) = \max\left\{\left(\left\lfloor \frac{d - 1 - \gcd(d, h)}{h} \right\rfloor + 1\right) \cdot \frac{n}{d} \mid d \in D(n)\right\} = 0$$

as well.

Suppose now that h is not divisible by n, and let $d \in D(n)$. If $d|h$, then

$$\left(\left\lfloor \frac{d-1-\gcd(d,h)}{h} \right\rfloor + 1 \right) \cdot \frac{n}{d} = 0;$$

otherwise (and there is at least one such d)

$$d - 1 - \gcd(d,h) \geq 0.$$

Let

$$c = \left\lfloor \frac{d-1-\gcd(d,h)}{h} \right\rfloor .$$

Note that

$$1 \leq \gcd(d,h) \leq \gcd(d,h) + hc \leq d-1.$$

Choose integers a and b for which

$$\gcd(d,h) = ha + db.$$

Let H be the subgroup of $\mathbb{Z}_n$ of index d, and set

$$A = \bigcup_{i=a}^{a+c} (i+H);$$

then A has size $(c+1)n/d$. We also see that A is zero-h-sum-free in $\mathbb{Z}_n$, since

$$hA = \bigcup_{i=ha}^{ha+hc} (i+H) = \bigcup_{i=\gcd(d,h)-db}^{\gcd(d,h)-db+hc} (i+H) = \bigcup_{i=\gcd(d,h)}^{\gcd(d,h)+hc} (i+H) \subseteq \bigcup_{i=1}^{d-1} (i+H).$$

Thus we have constructed a zero-h-sum-free set of size $(c+1)n/d$ in $\mathbb{Z}_n$ for each $d \in D(n)$, which proves our lower bound. □

Proof of Proposition F.27

Suppose that $A = \{a_1, \ldots, a_m\}$ is a B_h set over $\mathbb{Z}$ in G. Furthermore, suppose that $\lambda_1, \ldots, \lambda_m$ are integers for which

$$0 = \lambda_1 a_1 + \cdots + \lambda_m a_m$$

and

$$|\lambda_1| + \cdots + |\lambda_m| = 2h.$$

Let k be the smallest index for which

$$h < |\lambda_1| + \cdots + |\lambda_k|;$$

w.l.o.g., we may assume that $\lambda_k > 0$. Let us write $\lambda = |\lambda_1| + \cdots + |\lambda_{k-1}|$ (if $k = 1$, then simply let $\lambda = 0$).

We may then write our original equation as

$$\lambda_1 a_1 + \cdots + \lambda_{k-1} a_{k-1} + (h - \lambda) a_k = (h - \lambda - \lambda_k) a_k - \lambda_{k+1} a_{k+1} - \cdots - \lambda_m a_m$$

(with the understanding that if $k = 1$ or $k = m$, then the terms before or after the one with a_k in it vanish).

We first note that the left-hand side above is a signed sum of exactly h terms, since

$$|\lambda_1| + \cdots + |\lambda_{k-1}| + (h - \lambda) = h.$$

Before calculating the number of terms on the right-hand side, note that

$$h - \lambda - \lambda_k = h - \lambda - |\lambda_k| = h - (|\lambda_1| + \cdots + |\lambda_k|) < 0,$$

so the right-hand side consists of

$$\begin{aligned} |h - \lambda - \lambda_k| + |\lambda_{k+1}| + \cdots + |\lambda_m| &= |\lambda_1| + \cdots + |\lambda_k| - h + |\lambda_{k+1}| + \cdots + |\lambda_m| \\ &= |\lambda_1| + \cdots + |\lambda_m| - h \\ &= 2h - h = h \end{aligned}$$

terms. Thus, the equation

$$\lambda_1 a_1 + \cdots + \lambda_{k-1} a_{k-1} + (h - \lambda) a_k = (h - \lambda - \lambda_k) a_k - \lambda_{k+1} a_{k+1} - \cdots - \lambda_m a_m$$

has exactly h terms on each side; since A is a B_h set over $\mathbb{Z}$, we get that $\lambda_i = 0$ for each $i = 1, \ldots, m$, and therefore A is a zero-$2h$-sum-free set over $\mathbb{Z}$ in G. □

Proof of Proposition F.28

Let $A = \{a_1, \ldots, a_m\}$ be a zero-h-sum-free set over $\mathbb{Z}$ in G for some positive integer h which is divisible by 4, and suppose that

$$\lambda_1 a_1 + \cdots + \lambda_m a_m = \lambda_1' a_1 + \cdots + \lambda_m' a_m,$$

where the coefficients are integers and

$$|\lambda_1| + \cdots + |\lambda_m| = |\lambda_1'| + \cdots + |\lambda_m'| = 2.$$

We will prove that we must have $\lambda_i = \lambda_i'$ for every $1 \leq i \leq m$.

It doesn't take long to verify that the only possible values of

$$|\lambda_1 - \lambda_1'| + \cdots + |\lambda_m - \lambda_m'|$$

are 0, 2, or 4.

If it is 0, then we are done. If it is 2 or 4, then consider the sum

$$\tfrac{h}{2}(\lambda_1 - \lambda_1')a_1 + \cdots + \tfrac{h}{2}(\lambda_m - \lambda_m')a_m$$

or

$$\tfrac{h}{4}(\lambda_1 - \lambda_1')a_1 + \cdots + \tfrac{h}{4}(\lambda_m - \lambda_m')a_m,$$

respectively. Since the sum of the absolute values of the coefficients equals h in both cases, the expressions cannot be zero, and therefore we cannot have

$$\lambda_1 a_1 + \cdots + \lambda_m a_m = \lambda_1' a_1 + \cdots + \lambda_m' a_m,$$

which is a contradiction. □

Proof of Proposition F.32

First, we note that, for any $k \in \mathbb{N}$, the equation $k \cdot x = 0$ has exactly $\gcd(k,n)$ solutions in $\mathbb{Z}_n$: indeed, the equation is equivalent to kx being divisible by n and thus to $(k/\gcd(k,n)) \cdot x$ being divisible by $n/\gcd(k,n)$, which happens if, and only if, x itself is divisible by $n/\gcd(k,n)$. Consequently, the equation $k \cdot x = 0$ has at most k solutions in $\mathbb{Z}_n$.

We use the sequences $a(j,k)$ and $c(j,k)$ defined in Section 2.4.

Suppose that

$$n > a(0,m-1) + a(1,m-1) + \cdots + a(h-1,m-1).$$

We construct the set $A = \{a_1, \dots, a_m\}$ recursively as follows. Since $a(j,k) \geq 1$ for all j,k implies $n > h$, we can find an element $a_1 \in \mathbb{Z}_n$ for which $h \cdot a_1 \neq 0$.

Suppose now that we have already found the $(m-1)$-subset $A_{k-1} = \{a_1, \dots, a_{k-1}\}$ in $\mathbb{Z}_n$ so that A_{k-1} is zero-h-sum-free over $\mathbb{Z}$ in $\mathbb{Z}_n$. To find an element $a_k \in \mathbb{Z}_n$ so that $A_k = A_{k-1} \cup \{a_k\}$ is zero-h-sum-free over $\mathbb{Z}$ in $\mathbb{Z}_n$, we must have that none of

$$h \cdot a_k, \ (h-1) \cdot a_k + g_1, \ (h-2) \cdot a_k + g_2, \ \dots, \ 1 \cdot a_k + g_{h-1}$$

equals zero for any $g_1 \in 1_{\pm}A_{k-1}, g_2 \in 2_{\pm}A_{k-1}, \dots, g_{h-1} \in (h-1)_{\pm}A_{k-1}$. This rules out at most

$$h + (h-1) \cdot |1_{\pm}A_{k-1}| + (h-2) \cdot |2_{\pm}A_{k-1}| + \cdots + 1 \cdot |(h-1)_{\pm}A_{k-1}|$$

elements. This quantity is at most

$$h \cdot c(0,k-1) + (h-1) \cdot c(1,k-1) + (h-2) \cdot c(2,k-1) + \cdots + 1 \cdot c(h-1,k-1)$$

which, using Proposition 2.1, can be rewritten as

$$a(0,k-1) + a(1,k-1) + \cdots + a(h-1,k-1).$$

Therefore, when $k \leq m$, we can find the desired element a_k.

In particular, for $h = 4$, it is sufficient that

$$\begin{aligned} n &> a(0,m-1) + a(1,m-1) + a(2,m-1) + a(3,m-1) \\ &= 4 + 12(m-1) + 16\binom{m-1}{2} + 8\binom{m-1}{3} \\ &= \frac{4}{3}m^3 + \frac{8}{3}m. \end{aligned}$$

□

Proof of Proposition F.35

Let A be the set of integers (viewed as elements of $G = \mathbb{Z}_n$) that are strictly between $\frac{h-1}{2}\frac{n}{h}$ and $\frac{h+1}{2}\frac{n}{h}$. In particular, when n is divisible by h, we take

$$A = \left\{ \frac{h-1}{2}\frac{n}{h} + 1, \frac{h-1}{2}\frac{n}{h} + 2, \dots, \frac{h+1}{2}\frac{n}{h} - 1 \right\}.$$

When n is not divisible by h, then neither are $\frac{h-1}{2}n$ and $\frac{h+1}{2}n$, since h is relatively prime to $\frac{h-1}{2}$ and $\frac{h+1}{2}$; so, if n is not divisible by h, we have

$$A = \left\{ \left\lfloor \frac{h-1}{2}\frac{n}{h} \right\rfloor + 1, \left\lfloor \frac{h-1}{2}\frac{n}{h} \right\rfloor + 2, \dots, \left\lfloor \frac{h+1}{2}\frac{n}{h} \right\rfloor \right\}.$$

We prove that $0 \notin h_{\pm}A$ by showing that, for every $k = 0, 1, \ldots, h$, the set $kA - (h-k)A$ does not contain 0. Indeed, the smallest integer in $kA - (h-k)A$ is greater than

$$k \cdot \frac{h-1}{2}\frac{n}{h} - (h-k) \cdot \frac{h+1}{2}\frac{n}{h} = \left(k - \frac{h+1}{2}\right) n$$

and less than

$$k \cdot \frac{h+1}{2}\frac{n}{h} - (h-k) \cdot \frac{h-1}{2}\frac{n}{h} = \left(k - \frac{h-1}{2}\right) n,$$

or two consecutive multiples of n, and therefore $0 \notin kA - (h-k)A$.

It remains to be verified that the size of A equals $2\left\lfloor \frac{n+h-2}{2h} \right\rfloor$.

This is easy to see if n is divisible by h, since then the size of A is clearly

$$\frac{h+1}{2}\frac{n}{h} - \frac{h-1}{2}\frac{n}{h} - 1 = 2\left(\frac{n+h}{2h} - 1\right) = 2\left(\left\lfloor \frac{n+h}{2h} \right\rfloor - 1\right) = 2\left\lfloor \frac{n+h-2}{2h} \right\rfloor.$$

When n is not divisible by h, then

$$\begin{aligned} |A| &= \left\lfloor \frac{h+1}{2}\frac{n}{h} \right\rfloor - \left\lfloor \frac{h-1}{2}\frac{n}{h} \right\rfloor \\ &= \left(\frac{n+1}{2} + \left\lfloor \frac{n-h}{2h} \right\rfloor\right) - \left(\frac{n-1}{2} - \left\lfloor \frac{n-h}{2h} \right\rfloor - 1\right) \\ &= 2\left(\left\lfloor \frac{n-h}{2h} \right\rfloor + 1\right) \\ &= 2\left\lfloor \frac{n+h}{2h} \right\rfloor; \end{aligned}$$

since neither $n+h$ nor $n+h-1$ is divisible by $2h$ (the first would imply that $h|n$ and the second that $n+h$ is odd), this equals $2\left\lfloor \frac{n+h-2}{2h} \right\rfloor$, as claimed. □

Proof of Proposition F.46

For statement 2 (a), we prove that for $t = 2s$, the set $\{s, s+1\}$ is $2s$-independent in $\mathbb{Z}_n$ for all $n \geq 2s^2 + 2s + 1$. Suppose that for the integer

$$a = \lambda_1 \cdot s + \lambda_2 \cdot (s+1)$$

we have $a = 0$ in $\mathbb{Z}_n$ for some integer coefficients with $|\lambda_1| + |\lambda_2| \leq 2s$; w.l.o.g., assume that $\lambda_2 \geq 0$. Then

$$-2s^2 \leq a \leq 2s^2 + 2s,$$

so $a = 0$ in $\mathbb{Z}_n$ means that $a = 0$ in $\mathbb{Z}$. Therefore, λ_2 is divisible by s, hence $\lambda_2 = 0$, $\lambda_2 = s$, or $\lambda_2 = 2s$. In the first and last cases, we have $\lambda_1 = 0$, and we are done. The second case cannot happen, since

$$\lambda_1 \cdot s + s \cdot (s+1) = 0$$

implies that $\lambda_1 = -(s+1)$, contradicting $|\lambda_1| + |\lambda_2| \leq 2s$.

Similarly, for statement 2 (b) (ii), we show (somewhat clarifying Miller's proof) that the set $\{s, s+1\}$ is $(2s-1)$-independent in $\mathbb{Z}_n$ for $n \geq 2s^2 + s$. Suppose that for the integer

$$a = \lambda_1 \cdot s + \lambda_2 \cdot (s+1)$$

we have $a = 0$ in $\mathbb{Z}_n$ for some integer coefficients with $|\lambda_1| + |\lambda_2| \leq 2s - 1$; w.l.o.g., assume that $\lambda_2 \geq 0$. Then

$$-(2s^2 - s) \leq a \leq 2s^2 + s - 1,$$

so $a = 0$ in $\mathbb{Z}_n$ means that $a = 0$ in $\mathbb{Z}$. Therefore, λ_2 is divisible by s, hence $\lambda_2 = 0$ or $\lambda_2 = s$. In the first case, we have $\lambda_1 = 0$, and we are done. The second case cannot happen, since

$$\lambda_1 \cdot s + s \cdot (s + 1) = 0$$

implies that $\lambda_1 = -(s + 1)$, contradicting $|\lambda_1| + |\lambda_2| \leq 2s - 1$.

Next, we consider 2 (b) (iii), and prove that $\{1, 2s - 1\}$ is $(2s - 1)$-independent in $\mathbb{Z}_n$ for $n = 2s^2$. Assume that

$$a = \lambda_1 \cdot 1 + \lambda_2 \cdot (2s - 1) = 0$$

in $\mathbb{Z}_n$ for some integer coefficients with $|\lambda_1| + |\lambda_2| \leq 2s - 1$; w.l.o.g., assume that $\lambda_2 \geq 0$. Then for the integer a we have

$$-(2s - 1) \leq a \leq 4s^2 - 4s + 1,$$

so $a = 0$ in $\mathbb{Z}_n$ means that either $a = 0$ in $\mathbb{Z}$ or $a = 2s^2$ in $\mathbb{Z}$. If $a = 0$, then λ_1 is divisible by $2s - 1$, hence $\lambda_1 = 0$, which implies that $\lambda_2 = 0$ and we are done, or $|\lambda_1| = 2s - 1$, which again implies that $\lambda_2 = 0$, which can only happen if $\lambda_1 = 0$ as well, a contradiction.

If $a = 2s^2$ in $\mathbb{Z}$, then $\lambda_1 - \lambda_2$ must be divisible by $2s$, which can only happen if it is 0, since

$$|\lambda_1 - \lambda_2| \leq |\lambda_1| + |\lambda_2| \leq 2s - 1.$$

But solving the system

$$\lambda_1 \cdot 1 + \lambda_2 \cdot (2s - 1) = 2s^2$$

and

$$\lambda_1 - \lambda_2 = 0$$

yields $\lambda_1 = \lambda_2 = s$, contradicting $|\lambda_1| + |\lambda_2| \leq 2s - 1$.

For statement 2 (b) (iv), we assume that s and n are both even, $n \geq 2s^2$, and prove that the set $\{s - 1, s + 1\}$ is $(2s - 1)$-independent in $\mathbb{Z}_n$. Suppose that

$$a = \lambda_1 \cdot (s - 1) + \lambda_2 \cdot (s + 1) = 0$$

in $\mathbb{Z}_n$ for some integer coefficients with $|\lambda_1| + |\lambda_2| \leq 2s - 1$; w.l.o.g., assume that $\lambda_2 \geq 0$.

Note that every integer has the same parity as its absolute value. Therefore,

$$\lambda_1 \cdot (s - 1) + \lambda_2 \cdot (s + 1) \equiv |\lambda_1| \cdot (s - 1) + |\lambda_2| \cdot (s + 1) \equiv (|\lambda_1| + |\lambda_2|) \cdot (s + 1) \bmod 2.$$

So if $|\lambda_1| + |\lambda_2|$ is odd and s is even, then a is odd and, since n is even, $a \neq 0$ in $\mathbb{Z}_n$. In particular, $0 \notin (2s-1)_{\pm}\{s-1, s+1\}$. It remains to be shown that $0 \notin [1, 2s-2]_{\pm}\{s-1, s+1\}$.

For $|\lambda_1| + |\lambda_2| \leq 2s - 2$, we have

$$-(2s^2 - 4s + 2) \leq a \leq 2s^2 - 2,$$

so $a = 0$ in $\mathbb{Z}_n$ means that $a = 0$ in $\mathbb{Z}$. Therefore, λ_1 is divisible by $s + 1$ and λ_2 is divisible by $s - 1$, hence either one of them equals 0, in which case they both do, or

$$|\lambda_1| + |\lambda_2| \geq (s + 1) + (s - 1) = 2s,$$

a contradiction.

For statement 2 (b) (v), we assume that s is odd, n is congruent to 2 mod 4, $n \geq 2s^2$, and prove that the set $\{s-2, s+2\}$ is $(2s-1)$-independent in $\mathbb{Z}_n$. Suppose that

$$a = \lambda_1 \cdot (s-2) + \lambda_2 \cdot (s+2) = 0$$

in $\mathbb{Z}_n$ for some integer coefficients with $|\lambda_1| + |\lambda_2| \leq 2s-1$; w.l.o.g., assume that $\lambda_2 \geq 0$.

Like in the previous case, we can show that if $|\lambda_1| + |\lambda_2|$ is odd and s is odd, then a is odd and, since n is even, $a \neq 0$ in $\mathbb{Z}_n$. Therefore, $0 \not\in (2s-1)_{\pm}\{s-2, s+2\}$ and $0 \not\in (2s-3)_{\pm}\{s-2, s+2\}$. We will now show that $0 \not\in (2s-2)_{\pm}\{s-2, s+2\}$.

Assume first that $\lambda_1 \geq 0$. In this case,

$$0 < (2s-2)(s-2) \leq a \leq (2s-2)(s+2) < 2n,$$

so $a = 0$ in $\mathbb{Z}_n$ implies that $a = n$. But then

$$n = \lambda_1 \cdot (s-2) + \lambda_2 \cdot (s+2) = (\lambda_1 + \lambda_2) \cdot (s-2) + 4\lambda_2 = (2s-2) \cdot (s-2) + 4\lambda_2 \equiv 0 \text{ mod } 4,$$

a contradiction, since n is congruent to 2 mod 4.

If $\lambda_1 < 0$, then

$$-n < -(2s-2)(s-2) \leq a \leq (-1)(s-2) + (2s-3)(s+2) < n.$$

Therefore, $a = 0$ in $\mathbb{Z}$, and since $s-2$ and $s+2$ are relatively prime, this implies that λ_1 is divisible by $s+2$ and λ_2 is divisible by $s-2$, hence either one of them equals 0, in which case they both do, or

$$|\lambda_1| + |\lambda_2| \geq (s+2) + (s-2) = 2s,$$

a contradiction.

The proof that $0 \not\in [1, 2s-4]_{\pm}\{s-2, s+2\}$ is similar: for $|\lambda_1| + |\lambda_2| \leq 2s-4$, we have

$$-n < -(2s-4)(s-2) \leq a \leq (2s-4)(s+2) < n$$

so $a = 0$ in $\mathbb{Z}$, and we can complete the proof as above. □

Proof of Proposition F.80

Let n and h be fixed positive integers; note that

$$\overline{h} = \frac{h^2 - h - 2}{2}$$

is an integer. Let $d = \gcd(n, h)$, and set r equal to the remainder of $\overline{h}$ when divided by d. Furthermore, let

$$c = \left\lfloor \frac{n + h^2 - r - 2}{h} \right\rfloor.$$

Note that $h \leq n-1$, so

$$c \leq (n + h^2 - 2)/h < (n + h^2 - 1)/h \leq (n + (h-1)n)/h = n;$$

we need to prove that $\tau\hat{\ }(\mathbb{Z}_n, h) \geq c$.

Since d is a divisor of $\overline{h} - r$, we can find integers a and b for which

$$a \cdot h - b \cdot n = -(\overline{h} - r);$$

furthermore, we may assume w.l.o.g. that $b \geq 0$.

Now define the set A as

$$A = \{a, a+1, \dots, a+c-1\}.$$

(Throughout this proof, we will talk about integers and elements of the group $\mathbb{Z}_n$ interchangeably: when an integer is considered as an element of $\mathbb{Z}_n$, we regard it mod n; on the other hand, an element k of $\mathbb{Z}_n$ is treated as the integer k in $\mathbb{Z}$.)

Clearly, $|A| = c$. We will show that A is a weak zero-h-sum-free set in $\mathbb{Z}_n$ by showing that every element of $h\hat{}A$ is strictly between bn and $(b+1)n$.

The smallest element of $h\hat{}A$ is

$$ha + \frac{h(h-1)}{2} = ha + \overline{h} + 1 = bn - (\overline{h} - r) + \overline{h} + 1 = bn + 1 + r \geq bn + 1.$$

For the largest element of $h\hat{}A$ we have

$$\begin{aligned} ha + hc - \frac{h(h+1)}{2} &= (ha - bn) + bn + hc - \frac{h(h+1)}{2} \\ &= -(\overline{h} - r) + bn + hc - \frac{h(h+1)}{2} \\ &= bn + hc - h^2 + r + 1 \\ &\leq bn + (n + h^2 - r - 2) - h^2 + r + 1 \\ &= (b+1)n - 1, \end{aligned}$$

as claimed. □

Proof of Proposition F.83

Let n and h be fixed positive integers, and assume that $h^2 | n$. We let ϵ equal 0 if h is even and 1 if h is odd. We define the set

$$A = \bigcup_{i=0}^{h-1} \left\{ (ih + \epsilon) \cdot \frac{n}{h^2} + j \mid j = 0, 1, \dots, \frac{n}{h^2} \right\}.$$

We will prove that A is zero-h-sum-free in $\mathbb{Z}_n$.

First, observe that every element of $h\hat{}A$ can be written as

$$\begin{aligned} b &= ((i_1 + \cdots + i_h)h + \epsilon h) \cdot \frac{n}{h^2} + (j_1 + \cdots + j_h) \\ &= ((i_1 + \cdots + i_h) + \epsilon) \cdot \frac{n}{h} + (j_1 + \cdots + j_h) \end{aligned}$$

where

$$\{i_1, \dots, i_h\} \subseteq \{0, 1, \dots, h-1\}$$

and

$$\{j_1, \dots, j_h\} \subseteq \left\{0, 1, \dots, \frac{n}{h^2}\right\}.$$

If b were to be divisible by n (that is, $b = 0$ in $\mathbb{Z}_n$), then it would need to be divisible by $\frac{n}{h}$. Since the first term above is always divisible by $\frac{n}{h}$, the second term must be as well. However, since we have

$$0 \leq j_1 + \cdots + j_h \leq h \cdot \frac{n}{h^2} = \frac{n}{h},$$

this can only happen if either $j_1 = \cdots = j_h = 0$ or $j_1 = \cdots = j_h = \frac{n}{h^2}$. In either case, all j values are equal.

Therefore, since $b \in h\hat{}A$, all i values must be distinct, and thus

$$\{i_1, \ldots, i_h\} = \{0, 1, \ldots, h-1\}.$$

Thus we must have

$$\begin{aligned} b &= ((i_1 + \cdots + i_h)h + \epsilon h) \cdot \frac{n}{h^2} + (j_1 + \cdots + j_h) \\ &= ((0 + 1 + \cdots + (h-1)) + \epsilon) \cdot \frac{n}{h} + (j_1 + \cdots + j_h) \\ &= \left(\frac{h(h-1)}{2} + \epsilon\right) \cdot \frac{n}{h} + (j_1 + \cdots + j_h). \end{aligned}$$

We will show that b is not divisible by n. We have four cases.

Case 1: h is even and $j_1 = \cdots = j_h = 0$. In this case we get

$$b = \frac{h(h-1)}{2} \cdot \frac{n}{h} = \frac{h-1}{2} \cdot n,$$

and this is not divisible by n since $\frac{h-1}{2}$ is not an integer.

Case 2: h is odd and $j_1 = \cdots = j_h = 0$. In this case we get

$$b = \left(\frac{h(h-1)}{2} + 1\right) \cdot \frac{n}{h} = \left(\frac{h-1}{2} + \frac{1}{h}\right) \cdot n,$$

and this is not divisible by n since $\frac{h-1}{2}$ is an integer but $\frac{1}{h}$ is not (if $h > 1$).

Case 3: h is even and $j_1 = \cdots = j_h = \frac{n}{h^2}$. In this case we get

$$b = \frac{h(h-1)}{2} \cdot \frac{n}{h} + \frac{n}{h} = \left(\frac{h}{2} - \frac{h-2}{2h}\right) \cdot n,$$

and this is not divisible by n since $\frac{h}{2}$ is an integer, but $\frac{h-2}{2h}$ is not an integer as it is strictly between 0 and $\frac{1}{2}$ (if $h > 2$).

Case 4: h is odd and $j_1 = \cdots = j_h = \frac{n}{h^2}$. In this case we get

$$b = \left(\frac{h(h-1)}{2} + 1\right) \cdot \frac{n}{h} + \frac{n}{h} = \left(\frac{h-1}{2} + \frac{2}{h}\right) \cdot n,$$

and this is not divisible by n since $\frac{h-1}{2}$ is an integer but $\frac{2}{h}$ is not (if $h > 2$). □

Proof of Theorem F.88

By Proposition F.69 and Theorem E.60 we have

$$\tau\hat{}(G,h) \leq \chi\hat{}(G,h) - 1 = \begin{cases} n-1 & \text{if } h = 1; \\ n/2 & \text{if } 3 \leq h \leq n/2 - 2; \\ n/2 + 1 & \text{if } h = n/2 - 1; \\ h+1 & \text{if } n/2 \leq h \leq n-2; \\ n-1 & \text{if } h = n-1. \end{cases}$$

Therefore, it suffices to prove that a weakly zero-h-sum-free set of the desired size exists.

For $h = 1$ we take, of course, all nonzero elements of $\mathbb{Z}_n$, and for $3 \leq h \leq n/2 - 2$ (indeed, for any odd h) we take all odd elements of $\mathbb{Z}_n$.

When $h = n/2 - 1$, then, since h is odd, $\gcd(n, h) = 1$, so by Corollary F.81, we get

$$\tau\hat{}(\mathbb{Z}_n, h) \geq \left\lfloor \frac{n + h^2 - 2}{h} \right\rfloor = \left\lfloor \frac{2h + 2 + h^2 - 2}{h} \right\rfloor = h + 2 = \frac{n}{2} + 1.$$

The result for $n/2 \leq h \leq n - 2$ follows from Proposition F.84.

Finally, when $h = n - 1$, we again have $\gcd(n, h) = 1$, with which Corollary F.81 yields

$$\tau\hat{}(\mathbb{Z}_n, h) \geq \left\lfloor \frac{n + h^2 - 2}{h} \right\rfloor = \left\lfloor \frac{h + 1 + h^2 - 2}{h} \right\rfloor = h = n - 1,$$

which completes our proof. □

We should also note that some of our cases carry through for even h as well.

Proof of Proposition F.116

Since no two distinct elements of $\mathbb{Z}_2^r$ add to zero, $\{1\} \times \mathbb{Z}_2^{r-1}$ is a weak zero-$[1,3]$-free set in $\mathbb{Z}_2^r$, and thus

$$\tau\hat{}(\mathbb{Z}_2^r, [1, 3]) \geq 2^{r-1}.$$

To prove the reverse inequality, suppose that A is a weak zero-$[1,3]$-free set in $\mathbb{Z}_2^r$. Let $a \in A$ be arbitrary, and let

$$B = a + (A \setminus \{a\}).$$

Then $|B| = |A| - 1$; furthermore, $0 \notin A$, and $0 \notin B$.

Furthermore, A and B are disjoint, since if we had an $a' \in A \setminus \{a\}$ for which $a + a' = a''$ for some $a'' \in A$, then $a + a' + a''$ would be 0, which can only happen if $a'' \in \{a, a'\}$; however, this cannot be, as $0 \notin A$.

Therefore, A and B are disjoint subsets of $\mathbb{Z}_2^r \setminus \{0\}$, from which

$$|A| + |B| = |A| + |A| - 1 \leq 2^r - 1,$$

and thus $|A| \leq 2^{r-1}$, as claimed. □

Proof of Proposition F.156

Note that

$$\left\lfloor \frac{2n + h^2 - 3}{2h} \right\rfloor \leq \left\lfloor \frac{2n + h^2}{2h} \right\rfloor \leq \left\lfloor \frac{2n + hn}{2h} \right\rfloor \leq \left\lfloor \frac{2hn}{2h} \right\rfloor = n.$$

Let

$$2n + h^2 - 3 = 4hq + r$$

with

$$0 \leq r < 4h.$$

Then

$$\left\lfloor \frac{2n + h^2 - 3}{2h} \right\rfloor = \begin{cases} 2q & \text{if } 0 \leq r < 2h, \\ 2q + 1 & \text{if } 2h \leq r < 4h. \end{cases}$$

We treat the cases of $0 \leq r < 2h$ and $2h \leq r < 4h$ separately.

Suppose first that $0 \leq r < 2h$; set

$$A = \{(n+1)/2 - q, (n+1)/2 - q + 1, \ldots, (n+1)/2 + q - 1\}.$$

As we showed above, $2q \leq n$, so we have $|A| = 2q$.

For each integer $0 \leq k \leq h$, let us denote by $h(k)\hat{\pm}A$ the collection of all signed sums of h distinct terms of A in which exactly k terms are added and $h-k$ are subtracted. (For example, $h(h)\hat{}A = h\hat{}A$, and $h(0)\hat{\pm}A = -h\hat{}A$.) Considering the elements of A as integers (rather than elements of $\mathbb{Z}_n$), we will compute $a_{\min}(k)$ and $a_{\max}(k)$, the smallest and largest elements of $h(k)\hat{\pm}A$, respectively.

For $a_{\min}(k)$ we get

$$\begin{aligned}
a_{\min}(k) &= (k \cdot ((n+1)/2 - q) + k(k-1)/2) \\
&\quad - ((h-k) \cdot ((n+1)/2 + q) - (h-k)(h-k+1)/2) \\
&= k(n+1) - h(n+1)/2 - hq + \tfrac{k(k-1)+(h-k)(h-k+1)}{2} \\
&\geq k(n+1) - h(n+1)/2 - \tfrac{2n+h^2-3}{4} + \tfrac{k(k-1)+(h-k)(h-k+1)}{2} \\
&= (k-(h+1)/2)n + ((h-2k)^2+3)/4 \\
&> (k-(h+1)/2)n.
\end{aligned}$$

Similarly,

$$\begin{aligned}
a_{\max}(k) &= (k \cdot ((n+1)/2 + q) - k(k+1)/2) \\
&\quad - ((h-k) \cdot ((n+1)/2 - q) + (h-k)(h-k-1)/2) \\
&= k(n+1) - h(n+1)/2 + hq - \tfrac{k(k+1)+(h-k)(h-k-1)}{2} \\
&\leq k(n+1) - h(n+1)/2 + \tfrac{2n+h^2-3}{4} - \tfrac{k(k+1)+(h-k)(h-k-1)}{2} \\
&= (k-(h+1)/2+1)n - ((h-2k)^2+3)/4 \\
&< (k-(h+1)/2+1)n.
\end{aligned}$$

Therefore, all elements of $h(k)\hat{\pm}A$ lie strictly between two consecutive multiples of n, thus $0 \not\in h(k)\hat{\pm}A$. Since this holds for every $0 \leq k \leq h$ and

$$h\hat{\pm}A = \cup_{k=0}^{h} h(k)\hat{\pm}A,$$

we have $0 \not\in h\hat{\pm}A$.

Suppose now that $2h \leq r < 4h$; set

$$B = \{(n+1)/2 - q, (n+1)/2 - q + 1, \ldots, (n+1)/2 + q\}.$$

As we showed above, $2q+1 \leq n$, so we have $|B| = 2q+1$.

Since $B = A \cup \{(n+1)/2 + q\}$, we have

$$\begin{aligned}
b_{\min}(k) &= a_{\max}(k) - k \\
&= (k \cdot ((n+1)/2 - q) + k(k-1)/2) \\
&\quad - ((h-k) \cdot ((n+1)/2 + q) - (h-k)(h-k+1)/2) - k \\
&= k(n+1) - h(n+1)/2 - hq + \tfrac{k(k-1)+(h-k)(h-k+1)}{2} - k \\
&\geq k(n+1) - h(n+1)/2 - \tfrac{2n+h^2-3-2h}{4} + \tfrac{k(k-1)+(h-k)(h-k+1)}{2} - k \\
&= (k-(h+1)/2)n + ((h-2k+1)^2+2)/4 \\
&> (k-(h+1)/2)n.
\end{aligned}$$

Similarly,

$$\begin{aligned}
a_{\max}(k) &= a_{\max}(k) + k \\
&= (k \cdot ((n+1)/2 + q) - k(k+1)/2) \\
&\quad - ((h-k) \cdot ((n+1)/2 - q) + (h-k)(h-k-1)/2) + k \\
&= k(n+1) - h(n+1)/2 + hq - \tfrac{k(k+1)+(h-k)(h-k-1)}{2} + k \\
&\leq k(n+1) - h(n+1)/2 + \tfrac{2n+h^2-3-2h}{4} - \tfrac{k(k+1)+(h-k)(h-k-1)}{2} + k \\
&= (k - (h+1)/2 + 1)n - ((h-2k+1)^2 + 2)/4 \\
&< (k - (h+1)/2 + 1)n.
\end{aligned}$$

Again, all elements of $h(k)\hat{}_{\pm}B$ lie strictly between two consecutive multiples of n, thus $0 \not\in h(k)\hat{}_{\pm}B$, and, therefore, $0 \not\in h\hat{}_{\pm}B$. This completes our proof. □

Proof of Proposition F.179

First, we prove that if $A = \{a_1, \ldots, a_t\}$ is a t-subset of $\mathbb{Z}_n$, then $0 \in [1,t]\hat{}_{\pm}A$. Observe that since $|\Sigma A| \leq n < 2^t$, we must have

$$\epsilon_1, \ldots, \epsilon_t, \delta_1, \ldots, \delta_t \in \{0, 1\}$$

for which

$$\epsilon_1 a_1 + \cdots + \epsilon_t a_t = \delta_1 a_1 + \cdots + \delta_t a_t,$$

but for at least one index $1 \leq i \leq t$, $\epsilon_i \neq \delta_i$. But then

$$1 \leq |\epsilon_1 - \delta_1| + \cdots + |\epsilon_t - \delta_t| \leq t,$$

and

$$0 = (\epsilon_1 - \delta_1)a_1 + \cdots + (\epsilon_t - \delta_t)a_t \in [1,t]\hat{}_{\pm}A,$$

as claimed.

Second, we prove that for the set

$$A = \{1, 2, \ldots, 2^{t-2}\} \subseteq \mathbb{Z}_n,$$

we have $0 \not\in [1,t]\hat{}_{\pm}A$. Suppose that we have

$$\epsilon_0, \epsilon_1, \ldots, \epsilon_{t-2} \in \{-1, 0, 1\}$$

for which

$$\epsilon_0 \cdot 1 + \epsilon_1 \cdot 2 + \cdots + \epsilon_{t-2} \cdot 2^{t-2} = 0$$

in $\mathbb{Z}_n$. Since the left-hand side has absolute value at most $2^{t-1} - 1 < n$, the same equation must hold in $\mathbb{Z}$ as well. Now if there is an index $0 \leq i \leq t-2$ for which $\epsilon_i \neq 0$, then let

$$k = \min\{i \mid \epsilon_i \neq 0\}.$$

Dividing our equation (in $\mathbb{Z}$) by 2^k we get

$$\epsilon_k + \epsilon_{k+1} \cdot 2 + \cdots + \epsilon_{t-2} \cdot 2^{t-2-k} = 0,$$

which is a contradiction, since the left-hand side is odd. Therefore, $\epsilon_i = 0$ for all $0 \leq i \leq t-2$, proving our claim. □

Proof of Proposition G.22

Let

$$A = \{a, a+d, \ldots, a+(m-1)d\}$$

be a sum-free arithmetic progression in $\mathbb{Z}_n$ of size

$$m = v_1(n,3) = \begin{cases} \left(1+\frac{1}{p}\right)\frac{n}{3} & \text{if } n \text{ has prime divisors congruent to 2 mod 3,} \\ & \text{and } p \text{ is the smallest such divisor,} \\ \left\lfloor \frac{n}{3} \right\rfloor & \text{otherwise.} \end{cases}$$

Note that $m \geq (n-1)/3$ (since in the case when $n \equiv 2 \mod 3$, n must have a prime divisor $p \equiv 2 \mod 3$).

Let $\delta = \gcd(d,n)$, $d = d_1\delta$, $n = n_1\delta$. Since d_1 and n_1 are relatively prime, there are integers b and c for which

$$b \cdot d_1 = 1 + c \cdot n_1.$$

Then

$$b \cdot d = \delta + c \cdot n,$$

so

$$b \cdot A = \{ba, ba+\delta, \ldots, ba+(m-1)\delta\}.$$

Note that the fact that A is sum-free implies that $b \cdot A$ is as well.

Observe that $b \cdot A$ has size m, since otherwise we would have an $i \in \{1, 2, \ldots, m-1\}$ for which $i\delta = 0$ in $\mathbb{Z}_n$, but then $id = 0$ in $\mathbb{Z}_n$ for the same i, contradicting $|A| = m$. Furthermore, $b \cdot A$ has size at most $n_1 = n/\delta$, since it is contained in a coset of the subgroup of G that has order n_1. Thus $m \leq n_1$; but since $m \geq (n-1)/3$, we get $n/\delta \geq (n-1)/3$. Therefore, either $\delta \leq 3$, or $\delta = 4$ and $n = 4$, but the latter case is impossible since for even n we have $m = n/2$, contradicting $m \leq n_1$.

We consider the cases $\delta = 3$, $\delta = 2$, and $\delta = 1$ separately.

When $\delta = 3$, we have $n_1 = n/3$, so $n/3 \geq m \geq (n-1)/3$ implies that $m = n/3$ and thus $b \cdot A$ is a full coset of the subgroup of G that has order $n/3$. This coset cannot be the subgroup itself, since then $0 \in A$, contradicting sum-freeness. Therefore,

$$b \cdot A = \{1, 4, 7, \ldots, n-2\}$$

or

$$b \cdot A = \{2, 5, 8, \ldots, n-1\},$$

but note that in the second case we have

$$(2b) \cdot A = \{1, 4, 7, \ldots, n-2\}.$$

Since $m = v_1(n,3) = n/3$, we see that n cannot have any prime divisors congruent to 2 mod 3; in particular, n is odd. We also see that b and n are relatively prime, since $1 \in b \cdot A$ or $-1 \in b \cdot A$. Since n is odd, $2b$ and n must be relatively prime too. This yields case 2 (a).

If $\delta = 2$, then n is even and thus $m = v_1(n,3) = n/2$. Therefore,

$$b \cdot A = ba + \{0, 2, \ldots, n-2\} = \{1, 3, \ldots, n-1\}.$$

Since then b is relatively prime to n, it has an inverse, with which we get

$$A = a + \{0, 2, \ldots, n-2\} = \{1, 3, \ldots, n-1\}.$$

This yields case 1.

Finally, assume that $\delta = 1$. In this case,

$$b \cdot A = \{ba, ba + 1, \ldots, ba + (m - 1)\},$$

and so

$$2(b \cdot A) = b \cdot A + b \cdot A = \{2ba, 2ba + 1, \ldots, 2ba + 2(m - 1)\},$$

and thus

$$2(b \cdot A) - b \cdot A = \{ba - m + 1, ba - m, ba - m + 1, \ldots, ba + 2m - 2\}.$$

Therefore, we see that $b \cdot A$ is sum-free, that is, $0 \notin 2(b \cdot A) - b \cdot A$, exactly when there is an integer k for which

$$kn + 1 \leq ba - m + 1$$

and

$$ba + 2m - 2 \leq (k + 1)n - 1,$$

or, equivalently,

$$m \leq ba - kn \leq n - 2m + 1.$$

Therefore, $m \leq (n + 1)/3$. We separate three subcases.

If $n \equiv 2 \bmod 3$, then

$$m = v_1(n, 3) = \left(1 + \frac{1}{p}\right)\frac{n}{3},$$

where p is the smallest prime divisor of n with $p \equiv 2 \bmod 3$. So $m \leq (n + 1)/3$ is possible only when

$$m = (n + 1)/3 = \left(1 + \frac{1}{p}\right)\frac{n}{3},$$

which gives $n = p$. So $m = (p + 1)/3$,

$$(p + 1)/3 \leq ba - kp \leq n - 2(p + 1)/3 + 1,$$

so

$$ba - kp = (p + 1)/3,$$

and thus $ba = (p + 1)/3$ in $\mathbb{Z}_p$. This yields case 3.

If $3|n$, then $m \leq (n + 1)/3$ implies that $m \leq n/3$, but we also know that $m \geq (n - 1)/3$, so $m = n/3$. Therefore, n cannot have a prime divisor congruent to 2 mod 3. We have

$$n/3 \leq ba - kn \leq n - 2n/3 + 1,$$

so $ba - kn \in \{n/3, n/3 + 1\}$ and thus $ba = n/3$ or $ba = n/3 + 1$ in $\mathbb{Z}_n$, with which

$$b \cdot A = \{n/3, n/3 + 1, \ldots, 2n/3 - 1\}$$

or

$$b \cdot A = \{n/3 + 1, n/3 + 2, \ldots, 2n/3\}.$$

Note that

$$\{n/3, n/3 + 1, \ldots, 2n/3 - 1\} = -\{n/3 + 1, n/3 + 2, \ldots, 2n/3\},$$

so the second possibility is superfluous. This yields case 2 (b).

Finally, suppose that $n \equiv 1 \bmod 3$. Then $m \leq (n+1)/3$ implies that $m \leq (n-1)/3$, but we also know that $m \geq (n-1)/3$, so $m = (n-1)/3$. Therefore, n cannot have a prime divisor congruent to 2 mod 3. We have

$$(n-1)/3 \leq ba - kn \leq n - 2(n-1)/3 + 1,$$

so

$$ba - kn \in \{(n-1)/3, (n-1)/3+1, (n-1)/3+2\}$$

and thus $ba = (n-1)/3$, $ba = (n-1)/3+1$, or $ba = (n-1)/3+2$ in $\mathbb{Z}_n$, with which

$$b \cdot A = \{(n-1)/3, (n-1)/3+1, \ldots, 2(n-1)/3-1\},$$

$$b \cdot A = \{(n-1)/3+1, (n-1)/3+2, \ldots, 2(n-1)/3\},$$

or

$$b \cdot A = \{(n-1)/3+2, (n-1)/3+3, \ldots, 2(n-1)/3+1\}.$$

Note that

$$\{(n-1)/3, (n-1)/3+1, \ldots, 2(n-1)/3-1\} = -\{(n-1)/3+2, (n-1)/3+3, \ldots, 2(n-1)/3+1\},$$

so the third possibility is superfluous. This yields case 4. □

Proof of Theorem G.27

Let A be a (k,l)-sum-free set of size $M+1$ in $\mathbb{Z}_p$. According to Theorem G.26, A is an arithmetic progression (of common difference d); since p is prime, (multiplying by the inverse of d in $\mathbb{Z}_p$) we may assume that

$$A = \{a, a+1, \ldots, a+M\}$$

for some $a \in \mathbb{Z}_p$. Then

$$0 \notin kA - lA = \{(k-l)a - lM, (k-l)a - lM + 1, \ldots, (k-l)a + kM\},$$

so there is an integer c for which

$$cp + 1 \leq (k-l)a - lM$$

and

$$(k-l)a + kM \leq (c+1)p - 1.$$

Combining these two inequalities we get

$$lM + 1 \leq (k-l)a - cp \leq p - kM - 1$$

or, equivalently,

$$(k-l)a = lM + i$$

in $\mathbb{Z}_p$ for some

$$1 \leq i \leq (p - kM - 1) - (lM + 1) + 1 = r + 1.$$

Thus A is (a dilate of) one of the $r+1$ sets

$$\{a_i, a_i + 1, \ldots, a_i + M\}.$$

Due to symmetry, we don't need $\lfloor (r+1)/2 \rfloor$ of these choices; more precisely, for any i with $1 \leq i \leq \lfloor (r+1)/2 \rfloor$, $A_i = -A_{r+2-i}$, since

$$a_i = -(a_{r+2-i} + M).$$

To see this, note that this equation holds if, and only if,

$$(k-l)a_i = -(k-l)(a_{r+2-i} + M).$$

This equation holds, since

$$-(k-l)(a_{r+2-i} + M) = -(lM + r + 2 - i) - (k-l)M = lM + i = (k-l)a_i$$

in $\mathbb{Z}_p$. Therefore, it suffices to assume that $i \leq r + 1 - \lfloor (r+1)/2 \rfloor = \lfloor r/2 \rfloor + 1$. □

Proof of Proposition G.64

Let

$$A = [a, a+m-1] = \{a, a+1, \ldots, a+m-1\}$$

be an interval of length $m-1$ (size m) in $\mathbb{Z}_n$. Note that if A is an interval, then so is $h\hat{\ }A$ for all integers h with $1 \leq h \leq m$; in particular,

$$h\hat{\ }A = [ha + h(h-1)/2, h(a+m-1) - h(h-1)/2].$$

Therefore, for positive integers k and l with $l < k \leq M+1$, $k\hat{\ }A - l\hat{\ }A$ is also an interval: its "smallest" element is

$$ka + k(k-1)/2 - l(a+m-1) + l(l-1)/2,$$

and its "largest" element equals

$$k(a+m-1) - k(k-1)/2 - la - l(l-1)/2;$$

using the notation

$$J = k(k-1) + l(l-1) = k^2 + l^2 - k - l,$$

we have

$$k\hat{\ }A - l\hat{\ }A = [ka - l(a+m-1) + J/2, k(a+m-1) - la - J/2].$$

Now A is a weak (k,l)-sum-free set in $\mathbb{Z}_n$ if, and only if, $0 \not\in k\hat{\ }A - l\hat{\ }A$, and this occurs if, and only if, there is an integer b for which

$$ka - l(a+m-1) + J/2 \geq bn + 1$$

and

$$k(a+m-1) - la - J/2 \leq (b+1)n - 1,$$

or, equivalently,

$$l(m-1) - J/2 + 1 \leq (k-l)a - bn \leq n - k(m-1) + J/2 - 1.$$

We can divide by $\delta = \gcd(n, k-l)$ and write

$$\frac{l(m-1) - J/2 + 1}{\delta} \leq \frac{k-l}{\delta}a - \frac{n}{\delta}b \leq \frac{n - k(m-1) + J/2 - 1}{\delta}.$$

Note that $(k-l)/\delta$ and n/δ are relatively prime integers, so any integer can be written as their linear combination (with integer coefficients). We thus see that $\mathbb{Z}_n$ contains a weak (k,l)-sum-free interval of size m if, and only if, there is an integer C for which

$$\frac{l(m-1)-J/2+1}{\delta} \leq C \leq \frac{n-k(m-1)+J/2-1}{\delta}.$$

Therefore, a necessary condition for the existence of a weak (k,l)-sum-free interval of size m is that

$$\frac{l(m-1)-J/2+1}{\delta} \leq \frac{n-k(m-1)+J/2-1}{\delta},$$

from which

$$m \leq \left\lfloor \frac{n+J-2}{k+l} \right\rfloor + 1 = M+1$$

follows. On the other hand, a sufficient condition for the existence of a weak (k,l)-sum-free interval of size m is that

$$\frac{l(m-1)-J/2+1}{\delta} + 1 \leq \frac{n-k(m-1)+J/2-1}{\delta},$$

for which it suffices that $m \leq M$ since then

$$m \leq \frac{n+J-2}{k+l} \leq \frac{n+J-2-\delta}{k+l} + 1,$$

from which our condition follows. Therefore, $\gamma\hat{}(\mathbb{Z}_n,\{k,l\})$ equals either M or $M+1$.

To see when $\gamma\hat{}(\mathbb{Z}_n,\{k,l\}) = M+1$, note that the existence of an integer C for which

$$\frac{L}{\delta} = \frac{lM-J/2+1}{\delta} \leq C \leq \frac{n-kM+J/2-1}{\delta} = \frac{n-K}{\delta}$$

is equivalent to saying that

$$\frac{L}{\delta} \leq \left\lfloor \frac{n-K}{\delta} \right\rfloor.$$

This completes our proof. □

Proof of Corollary G.65

We provide two proofs: one using Proposition G.64 and another that is more direct.

Proof I: Here we use Proposition G.64; it suffices to prove that

$$\gamma\hat{}(\mathbb{Z}_n,\{k,l\}) \geq \left\lfloor \frac{n+k^2+l^2-\gcd(n,k-l)-1}{k+l} \right\rfloor.$$

Recall the notations

$$\delta = \gcd(n,k-l),$$

$$J = k^2+l^2-(k+l),$$

$$M = \lfloor (n+J-2)/(k+l) \rfloor,$$

$$K = kM - J/2 + 1,$$

$$L = lM - J/2 + 1.$$

Since

$$M+1=\left\lfloor \frac{n+k^2+l^2-2}{k+l} \right\rfloor \geq \left\lfloor \frac{n+k^2+l^2-\gcd(n,k-l)-1}{k+l} \right\rfloor,$$

by Proposition G.64 we may assume that

$$L/\delta > \lfloor (n-K)/\delta \rfloor,$$

and we need to show that

$$M \geq \left\lfloor \frac{n+k^2+l^2-\gcd(n,k-l)-1}{k+l} \right\rfloor$$

in this case.

Let R be the remainder of $n+k^2+l^2-2$ when divided by $k+l$; we can then write

$$n+k^2+l^2-2=(k+l)(M+1)+R,$$

and so

$$n-K=(k+l)(M+1)+R-(k^2+l^2-2)-(kM-J/2+1)=L+R.$$

Therefore, our hypothesis that

$$L/\delta > \lfloor (n-K)/\delta \rfloor$$

can only hold if $R<\delta-1$. But then

$$\left\lfloor \frac{n+k^2+l^2-\gcd(n,k-l)-1}{k+l} \right\rfloor = M+1+\left\lfloor \frac{R-\delta+1}{k+l} \right\rfloor \leq M+1+\left\lfloor \frac{-1}{k+l} \right\rfloor = M,$$

as claimed. □

Proof II: We set $\delta=\gcd(n,k-l)$ and

$$c=\left\lfloor \frac{n+k^2+l^2-\delta-1}{k+l} \right\rfloor.$$

We need to prove that $\mu\hat{}(\mathbb{Z}_n,\{k,l\}) \geq c$.

Note that $c<n$ since

$$c \leq \frac{n+k^2+l^2-2}{k+l} = n - \frac{n(k+l-1)-(k^2+l^2-2)}{k+l},$$

and, since $0<l<k\leq n$, we have

$$\begin{aligned} n(k+l-1)-(k^2+l^2-2) &\geq k(k+l-1)-(k^2+l^2-2) \\ &= k(l-1)-(l^2-2) \\ &> k(l-1)-(l^2-1) \\ &= (k-l-1)(l-1) \\ &\geq 0. \end{aligned}$$

Now

$$\overline{h}=\frac{k^2+l^2-k+l-2}{2}$$

is an integer; let r be the remainder of $\overline{h}-lc$ when divided by δ; we then have $0\leq r\leq \delta-1$.

Since δ is a divisor of $\overline{h} - lc - r$, we can find integers a and b for which

$$a \cdot (k-l) - b \cdot n = -(\overline{h} - lc - r);$$

furthermore, we may assume w.l.o.g. that $b \geq 0$.

Now define the set A as

$$A = \{a, a+1, \ldots, a+c-1\}.$$

(Throughout this proof, we will talk about integers and elements of the group $\mathbb{Z}_n$ interchangeably: when an integer is considered as an element of $\mathbb{Z}_n$, we regard it mod n; on the other hand, an element t of $\mathbb{Z}_n$ is treated as the integer t in $\mathbb{Z}$.)

Clearly, $|A| = c$; we will show that A is a weak (k,l)-sum-free set in $\mathbb{Z}_n$ by showing that every element of $k\hat{\ }A - l\hat{\ }A$ is strictly between bn and $(b+1)n$.

The smallest element of $k\hat{\ }A - l\hat{\ }A$ is

$$\begin{aligned} ka + \frac{k(k-1)}{2} - la - lc + \frac{l(l+1)}{2} &= (k-l)a - lc + \overline{h} + 1 \\ &= bn - \overline{h} + r + \overline{h} + 1 \\ &= bn + 1 + r \\ &\geq bn + 1. \end{aligned}$$

For the largest element of $k\hat{\ }A - l\hat{\ }A$ we have

$$\begin{aligned} ka + kc - \frac{k(k+1)}{2} - la - \frac{l(l-1)}{2} &= (k-l)a + kc - \frac{k^2 + l^2 + k - l}{2} \\ &= bn + lc - \overline{h} + r + kc - \frac{k^2 + l^2 + k - l}{2} \\ &= bn + (k+l)c - k^2 - l^2 + 1 + r \\ &\leq bn + (n + k^2 + l^2 - \delta - 1) - k^2 - l^2 + 1 + r \\ &= (b+1)n - \delta + r \\ &\leq (b+1)n - 1. \end{aligned}$$

Therefore, no element of $k\hat{\ }A - l\hat{\ }A$ is 0 in $\mathbb{Z}_n$. □

Proof of Theorem G.67

Let A be a weak sum-free set in $\mathbb{Z}_n$. We claim that

$$|A| \leq \begin{cases} \left(1 + \frac{1}{p}\right) \frac{n}{3} & \text{if } n \text{ has prime divisors congruent to 2 mod 3,} \\ & \text{and } p \text{ is the smallest such divisor,} \\ \left\lfloor \frac{n}{3} \right\rfloor + 1 & \text{otherwise.} \end{cases}$$

If $A = \{0\}$, the claim is obviously true. If A contains a nonzero element a, then it cannot contain 0, since otherwise $0 + a = a$ would contradict A being weakly sum-free. So we may assume that $0 \not\in A$.

By Theorem G.4, the claim is clear when A is sum-free, so we may assume that there is an element $a_0 \in A$ for which $2a_0 \in A$. We have $a_0 \neq 0$ and thus $a_0 \neq 2a_0$.

Let

$$A_1 = a_0 + (A \setminus \{a_0\}) = \{a_0 + a \mid a \in A \setminus \{a_0\}\}$$

and

$$A_2 = 2a_0 + (A \setminus \{a_0, 2a_0\}) = \{2a_0 + a \mid a \in A \setminus \{a_0, 2a_0\}\}.$$

Note that $A_1 \subseteq 2\hat{}A$ and $A_2 \subseteq 2\hat{}A$, so A is disjoint from both A_1 and A_2. Furthermore, A_1 and A_2 are disjoint too, since otherwise we would have elements $a_1 \in A \setminus \{a_0\}$ and $a_2 \in A \setminus \{a_0, 2a_0\}$ for which

$$a_0 + a_1 = 2a_0 + a_2,$$

but then

$$a_1 = a_0 + a_2,$$

contradicting that A and A_1 are disjoint. Since A, A_1, and A_2 are pairwise disjoint, we have

$$|A| + |A_1| + |A_2| = 3|A| - 3 \leq n,$$

from which $|A| \leq \lfloor n/3 \rfloor + 1$, proving our claim. □

Proof of Proposition G.73

Since

$$A = \{a, a+1, \dots, a+c\} + H$$

consists of $c+1$ consecutive cosets of H, and each coset is of size $k-1$, we have

$$k\hat{}A = \{ka+1, ka+2, \dots, ka+kc-1\} + H,$$

and

$$l\hat{}A \subseteq \{la, la+1, \dots, la+lc\} + H,$$

with equality holding if $l \leq k-2$; however, for $l = k-1$, only a single element from each of the cosets $la + H$ and $la + lc + H$ is in $l\hat{}A$.

Using our specified values of a and c, we find that

$$la + lc = \frac{l^2 n}{(k-1)(k^2-l^2)} + \frac{ln}{(k-1)(k+l)} = \frac{kln}{(k-1)(k^2-l^2)} = ka,$$

and

$$\begin{aligned} ka + kc - 1 &= \frac{kln}{(k-1)(k^2-l^2)} + \frac{kn}{(k-1)(k+l)} - 1 \\ &= \frac{k^2 n}{(k-1)(k^2-l^2)} - 1 \\ &= \frac{n}{k-1} + \frac{l^2 n}{(k-1)(k^2-l^2)} - 1 \\ &= \frac{n}{k-1} + la - 1. \end{aligned}$$

Therefore, $k\hat{}A \cap l\hat{}A = \emptyset$, as claimed. We also see that, in the case of $l \leq k-2$, we have $k\hat{}A \cup l\hat{}A = \mathbb{Z}_n$, so A is a complete weak (k,l)-sum-free set in $\mathbb{Z}_n$. □

Proof of Proposition G.74

Using our notations, for any positive integer h, we have:

$$h\hat{\ }A \subseteq \{h_1, h_1+1, \dots, h_2\} + H,$$

where

$$h_1 = \max\{0, h - d/2 + 1\},$$

and

$$h_2 = \min\{h, n/d\}.$$

Therefore,

$$k\hat{\ }A \subseteq \{k - d/2 + 1, k - d/2 + 2, \dots, k\} + H,$$

and

$$l\hat{\ }A \subseteq \{l - d/2 + 1, l - d/2 + 2, \dots, l\} + H.$$

So both $k\hat{\ }A$ and $l\hat{\ }A$ are within $d/2$ consecutive cosets of H; furthermore, these d cosets are all distinct, since

$$(k - d/2) + H = l + H.$$

This proves that $k\hat{\ }A$ and $l\hat{\ }A$ are disjoint.

Furthermore, we see that

$$k\hat{\ }A \cup l\hat{\ }A = \mathbb{Z}_n$$

if, and only if, the inequalities $k < n/d$, $k > d/2 - 1$, $l < n/d$, and $l > d/2 - 1$ all hold; so A is a complete weak (k,l)-sum-free set in $\mathbb{Z}_n$ if, and only if,

$$d/2 - 1 < l < k < n/d.$$

□

Proof of Proposition G.82

By Proposition G.81, we may assume that $n \geq k+1$. For $i = 0, 1, 2, \dots, k$, consider the subsets

$$A_i = \{0, 1, 2, \dots, k\} \setminus \{i\}$$

of $\mathbb{Z}_n$.

Consider first the case when $n = k+1$ and $k \equiv 1$ mod 4. Note that every k-subset of $\mathbb{Z}_n$ is of the form A_i for some $i = 0, 1, 2, \dots, k$; to prove our claim, we will show that

$$s_i = k\hat{\ }A = 0 + 1 + 2 + \cdots + k - i \in A_i$$

for each i. Indeed, if for some i we were to have

$$s_i = k(k+1)/2 - i = i$$

in $\mathbb{Z}_n$, then

$$k(k+1)/2 = 2i,$$

which is impossible, since n is even, $2i$ is even, but $k(k+1)/2$ is odd.

We will now show that in all other cases, $\mathbb{Z}_n$ contains a k-subset that is weakly $(k,1)$-sum-free. Assuming that there is no i for which A_i is weakly $(k,1)$-sum-free yields that, for each i, $s_i \in A_i$, so $s_i \in \{0, 1, \dots, k\}$ and thus

$$\{s_k, s_{k-1}, \dots, s_1, s_0\} \subseteq \{0, 1, 2, \dots, k\}.$$

Observe that

$$(s_k, s_{k-1}, \dots, s_1, s_0) = (s_k, s_k + 1, \dots, s_k + k),$$

so we must have $s_i = k - i$ for each i. In particular,

$$s_0 = k(k+1)/2 = k.$$

We consider three cases. Assume first that $n \geq k + 3$. Then for

$$B = \{0, 1, 2, \dots, k, k+1, k+2\} \setminus \{1, k, k+1\} = A_0 \cup \{0, k+2\} \setminus \{1, k\}$$

we have

$$k\hat{\ }B = s_0 + (k+2) - 1 - k = k + 1.$$

Since $k + 1 \notin B$, B is weakly $(k, 1)$-sum-free in $\mathbb{Z}_n$.

Assume now that $n = k + 2$. In this case, we have

$$6 = (k^2 - k) - (k+2)(k-3) = k^2 - k = 2 \cdot k(k+1)/2 - 2k = 2s_0 - 2s_0 = 0.$$

This can only happen in $\mathbb{Z}_{k+2}$ if $k + 2 = 6$ and thus $k = 4$. However, in this case,

$$A_2 = \{0, 1, 3, 4\}$$

is weakly $(4, 1)$-sum-free in $\mathbb{Z}_6$.

Next, assume that $n = k + 1$ and that k is even. As we have seen above, we have

$$s_{k/2} = k - k/2 = k/2,$$

so $A_{k/2}$ is weakly $(k, 1)$-sum-free in $\mathbb{Z}_n$.

Finally, assume that $n = k + 1$ and $k \equiv 3 \mod 4$. In this case, we have

$$s_{(k+1)/4} = k(k+1)/2 - (k+1)/4 = (k+1) \cdot (k-1)/2 + (k+1)/4 = (k+1)/4,$$

which means that $A_{(k+1)/4}$ is weakly $(k, 1)$-sum-free in $\mathbb{Z}_n$.

To prove our second claim, first note that if $n \leq 2k + 1$ and $|A| = k + 1$, then A and $k\hat{\ }A$ cannot be disjoint, since $|k\hat{\ }A| = k + 1$. Suppose then that $n \geq 2k + 2$. We will consider two cases depending on the parity of k.

When k is even, we set

$$A = \{1\} \cup \{\pm 2, \pm 4, \dots, \pm k\}.$$

Then $|A| = k + 1$, and

$$k\hat{\ }A = 1 - A = \{0\} \cup \{-1\} \cup \{\pm 3, \pm 5, \dots, \pm (k-1)\} \cup \{k+1\}.$$

Since $k + 1 < n - k$, A is weakly $(k, 1)$-sum-free in $\mathbb{Z}_n$.

When k is odd, we let

$$A = \{0, -1, -2, 4\} \cup \{\pm 5, \pm 7, \dots, \pm k\}.$$

(We simply let $A = \{0, -1, -2, 4\}$ if $k = 3$.) Again, $|A| = k + 1$, and

$$k\hat{\ }A = 1 - A = \{1, 2, \pm 3, -4\} \cup \{\pm 6, \pm 8, \dots, \pm (k-1)\} \cup \{k+1\},$$

and A is weakly $(k, 1)$-sum-free in $\mathbb{Z}_n$.

We can also observe that if $n = 2k + 2$, then A is a complete weak $(k, 1)$-sum-free set. □

Bibliography

[1] N. Alon, Subset sums. *J. Number Theory* **27** (1987), no. 2, 196–205.

[2] N. Alon and D. J. Kleitman, Sum-free subsets. *A tribute to Paul Erdős*, 13–26, *Cambridge Univ. Press, Cambridge*, 1990.

[3] N. Alon, M. B. Nathanson, and I. Ruzsa, Adding distinct congruence classes modulo a prime. *Amer. Math. Monthly* **102** (1995), no. 3, 250–255.

[4] N. Alon, M. B. Nathanson, and I. Ruzsa, The polynomial method and restricted sums of congruence classes. *J. Number Theory* **56** (1996), no. 2, 404–417.

[5] L. Babai, Email communication (July, 2017).

[6] L. Babai and V. T. Sós, Sidon sets in groups and induced subgraphs of Cayley graphs. *European J. Combin.* **6** (1985), no. 2, 101–114.

[7] B. Bajnok, Construction of spherical 4- and 5-designs. *Graphs Combin.* **7** (1991), no. 3, 219–233.

[8] B. Bajnok, Construction of spherical t-designs. *Geom. Dedicata.* **43** (1992), no. 2, 167–179.

[9] B. Bajnok, Constructions of spherical 3-designs. *Graphs Combin.* **14** (1998), no. 2, 97–107.

[10] B. Bajnok, Spherical designs and generalized sum-free sets in abelian groups. Special issue dedicated to Dr. Jaap Seidel on the occasion of his 80th birthday (Oisterwijk, 1999). *Des. Codes Cryptogr.* **21** (2000), no. 1-3, 11–18.

[11] B. Bajnok, The spanning number and the independence number of a subset of an abelian group. *Number theory (New York, 2003)*, 1–16, *Springer, New York,* 2004.

[12] B. Bajnok, Research classes at Gettysburg College. In *Proceedings of the Conference on Promoting Undergraduate Research in Mathematics.* J. A. Gallian (Ed.), American Mathematical Society (2007), 223–226.

[13] B. Bajnok, On the maximum size of a (k,l)-sum-free subset of an abelian group. *Int. J. Number Theory* **5** (2009), no. 6, 953–971.

[14] B. Bajnok, *An Invitation to Abstract Mathematics.* Undergraduate Texts in Mathematics, Springer–Verlag, New York (2013).

[15] B. Bajnok, On the minimum size of restricted sumsets in cyclic groups. *Acta Math. Hungar.* **148** (2016), no. 1, 228–256.

[16] B. Bajnok, The h-critical number of finite abelian groups. *Unif. Distrib. Theory* **10** (2015), no. 2, 93–15.

[17] B. Bajnok, Comments on Y. O. Hamidoune's paper "Adding distinct congruence classes." *Combin. Probab. Comput.* **25** (2016), no. 5, 641–644.

[18] B. Bajnok, More on the h-critical numbers of finite abelian groups. *Ann. Univ. Sci. Budapest. Eötvös Sect. Math.* **59** (2016), 113–122.

[19] B. Bajnok, Corrigendum to "The h-critical number of finite abelian groups." *Unif. Distrib. Theory* **12** (2017), no. 2, 119–124.

[20] B. Bajnok, Open problems about Sumsets in finite abelian groups: minimum sizes and critical numbers. To appear in *Combinatorial and Additive Number Theory II: CANT, New York, NY, USA, 2015 and 2016, Springer, New York,* 2017.

[21] B. Bajnok and S. Edwards, On two questions about restricted sumsets in finite abelian groups. *Australas. J. Combin.* **68** (2017), no. 2, 229–244.

[22] B. Bajnok and R. Matzke, The minimum size of signed sumsets. *Electron. J. Combin.* **22** (2015), no. 2, Paper 2.50, 17 pp.

[23] B. Bajnok and R. Matzke, On the minimum size of signed sumsets in elementary abelian groups. *J. Number Theory* **159** (2016), 384–401.

[24] B. Bajnok and I. Ruzsa, The independence number of a subset of an abelian group. *Integers* **3** (2003), A2, 23 pp.

[25] É. Balandraud, An addition theorem and maximal zero-sum free sets in $\mathbb{Z}/p\mathbb{Z}$. *Israel J. Math.* **188** (2012), 405–429.

[26] É. Balandraud, Erratum to "An addition theorem and maximal zero-sum free sets in $\mathbb{Z}/p\mathbb{Z}$". *Israel J. Math.* **192** (2012), no. 2, 1009–1010.

[27] É. Balandraud, Addition theorems in $\mathbb{F}_p$ via the polynomial method. arXiv:1702.06419v1 [math.CO].

[28] R. Balasubramanian, G. Prakash, and D. S. Ramana, Sum-free subsets of finite abelian groups of type III. *European J. Combin.* **58** (2016), 181–202.

[29] Ei. Bannai and Et. Bannai, A survey on spherical designs and algebraic combinatorics on spheres. *European J. Combin.* **30** (2009), no. 6, 1392–1425.

[30] Ei. Bannai and R. M. Damerell, Tight spherical designs I. *J. Math. Soc. Japan,* **31** (1979), no.1, 199–207.

[31] Ei. Bannai and R. M. Damerell, Tight spherical designs II. *J. London Math. Soc. (2)* **21** (1980), no.1, 13–30.

[32] M. Bateman and N. H. Katz, New bounds on cap sets. *J. Amer. Math. Soc.* **25** (2012), no. 2, 585–613.

[33] G. Bhowmik, I. Halupczok, and J-C. Schlage-Puchta, Zero-sum free sets with small sumset. *Math. Comp.* **80** (2011), no. 276, 2253–2258.

[34] G. Bhowmik and J-C. Schlage-Puchta, Davenport's constant for groups of the form $\mathbb{Z}_3 \times \mathbb{Z}_3 \times \mathbb{Z}_{3d}$. *Additive Combinatorics*, 307–326, CRM Proc. Lecture Notes, **43**, *Amer. Math. Soc., Providence, RI,* 2007.

[35] G. Bhowmik and J-C. Schlage-Puchta, Inductive methods and zero-sum free sequences. *Integers* **9** (2009), A40, 515–536.

[36] G. Bhowmik and J-C. Schlage-Puchta, An improvement on Olson's constant for $\mathbb{Z}_p \times \mathbb{Z}_p$. *Acta Arith.* **141** (2010), no. 4, 311–319.

[37] G. Bhowmik and J-C. Schlage-Puchta, Corrigendum to "An improvement on Olson's constant for $\mathbb{Z}_p \times \mathbb{Z}_p$." *Acta Arith.* **149** (2011), no. 4, 99–100.

[38] T. Bier and A. Y. M. Chin, On (k,l)-sets in cyclic groups of odd prime order. *Bull. Austral. Math. Soc.* **63** (2001), no. 1, 115–121.

[39] J. Bierbrauer and Y. Edel, Bounds on affine caps. *J. Comb. Des.* **10** (2002), no. 2, 111–115.

[40] N. Blyler, Examination of the maximum size of weak zero-h-sum-free sets. *Research Papers in Mathematics*, B. Bajnok, ed., Gettysburg College, Vol. 10 (2010).

[41] R. C. Bose, An affine analogue of Singer's theorem. *J. Indian Math. Soc. (N.S.)* **6** (1942), 1–15.

[42] R. C. Bose and S. Chowla, Theorems in the additive theory of numbers. *Comment. Math. Helv.* **37** (1962/1963), 141–147.

[43] J. J. Bravo, D. F. Ruiz, and C. A. Trujillo, Cardinality of sets associated to B_3 and B_4 sets. *Rev. Colombiana Mat.* **46** (2012), no. 1, 27–37.

[44] M. Buell, What's nu? Exploring when $\nu_{\pm}(G,m,h) < \min\{n, c(m,h)\}$. *Research Papers in Mathematics*, B. Bajnok, ed., Gettysburg College, Vol. 16 (2014).

[45] C. Bui, Weakly zero-3-sum-free sets in the group $\mathbb{Z}_3^k$. *Research Papers in Mathematics*, B. Bajnok, ed., Gettysburg College, Vol. 9 (2009).

[46] J. Butterworth, Examining the arithmetic function $v_g(n,h)$. *Research Papers in Mathematics*, B. Bajnok, ed., Gettysburg College, Vol. 8 (2008).

[47] J. Butterworth, Finding the maximum size of a (k,l)-sum-free subset of an abelian group. *Research Papers in Mathematics*, B. Bajnok, ed., Gettysburg College, Vol. 9 (2009).

[48] W. C. Calhoun, Counting subgroups of some finite groups. *Amer. Math. Monthly* **94** (1987), no. 1, 54–59.

[49] K. Campbell, An exploration of h-critical numbers. *Research Papers in Mathematics*, B. Bajnok, ed., Gettysburg College, Vol. 16 (2014).

[50] P. Candela and H. A. Helfgott, On the dimension of additive sets. *Acta Arith.* **167** (2015), no. 1, 91–100.

[51] E. Carrick, Evaluating $\rho\hat{}(G,5,2)$. *Research Papers in Mathematics*, B. Bajnok, ed., Gettysburg College, Vol. 17 (2015).

[52] A–L. Cauchy, Recherches sur les nombres. *J. École Polytechnique* **9** (1813), 99–123.

[53] G. Chiaselotti, Sums of distinct elements in finite abelian groups. *Boll. Un. Mat. Ital. A (7)* **7** (1993), no. 2, 243–251.

[54] G. Chiaselotti, On additive bases with two elements. *Acta Arith.* **101** (2002), no. 2, 115–119.

[55] A. Y. M. Chin, Maximal sum-free sets and block designs. *Math. Slovaca* **51** (2001), no. 3, 295–299.

[56] J. Cilleruelo, Combinatorial problems in finite fields and Sidon sets. *Combinatorica* **32** (2012), no. 5, 497–511.

[57] W. E. Clark and J. Pedersen, Sum-free sets in vector spaces over $GF(2)$. *J. Comb. Theory Ser. A* **61** (1992), no. 2, 222–229.

[58] K. Collins, Finding restricted zero-h-sum-free sets. *Research Papers in Mathematics*, B. Bajnok, ed., Gettysburg College, Vol. 16 (2014).

[59] H. Davenport, On the addition of residue classes. *J. London Math. Soc.* **10** (1935), 30–32.

[60] H. Davenport, A historical note. *J. London Math. Soc.* **22** (1947), 100–101.

[61] R. Day, Findings on the size of $\nu_{\pm}(\mathbb{Z}_n, m, h)$ for all n, m, and h. *Research Papers in Mathematics*, B. Bajnok, ed., Gettysburg College, Vol. 11 (2011).

[62] H. Delannoy, Emploi de d'échiquier pour la résolution de certains problèmes de probabilités. *Assoc. Franc. Bordeaux* **24** (1895), 70–90.

[63] P. Delsarte, J. M. Goethals, and J. J. Seidel, Spherical codes and designs. *Geom. Dedicata* **6** (1977), no. 3, 363–388.

[64] J–M. Deshouillers and G. Prakash, Large zero-free subsets of $\mathbb{Z}_p$. *Integers* **11** (2011), no. 3, 363–388.

[65] P. H. Diananda and H. P. Yap, Maximal sum-free sets of elements of finite groups. *Proc. Japan Acad.* **45** (1969), 1–5.

[66] J. A. Dias Da Silva and Y. O. Hamidoune, Cyclic space for Grassmann derivatives and additive theory. *Bull. London Math. Soc.* **26** (1994), no. 2, 140–146.

[67] G. T. Diderrich, An addition theorem for abelian groups of order pq. *J. Number Theory* **7** (1975), 33–48.

[68] G. T. Diderrich and H. B. Mann, Combinatorial problems in finite abelian groups. *Survey of combinatorial theory (Proc. Internat. Sympos., Colorado State Univ., Fort Collins, Colo., 1971)*, pp. 95-100. *North-Holland, Amsterdam*, 1973.

[69] N. Doskov, P. Pokhrel, and S. Singh, A few insights on s-spanning sets. *Research Papers in Mathematics*, B. Bajnok, ed., Gettysburg College, Vol. 0 (2003).

[70] Y. Edel, Extensions of generalized product caps. *Des. Codes Cryptogr.* **31** (2004), no. 1, 5–14.

[71] S. Edwards, How now brown tau: finding weakly sum-free sets. *Research Papers in Mathematics*, B. Bajnok, ed., Gettysburg College, Vol. 17 (Fall 2015).

[72] R. B. Eggleton and P. Erdős, Two combinatorial problems in group theory. *Acta Arith.* **21** (1972), 111–116.

[73] S. Eliahou and M. Kervaire, Sumsets in vector spaces over finite fields. *J. Number Theory* **71** (1998), no. 1, 12–39.

[74] S. Eliahou and M. Kervaire, Restricted sums of sets of cardinality $1 + p$ in a vector space over $\mathbf{F}_p$. Combinatorics (Prague, 1998). *Discrete Math.* **235** (2001), no. 1–3, 199–213.

[75] S. Eliahou and M. Kervaire, Restricted sumsets in finite vector spaces: the case $p = 3$, *Integers* **1** (2001), A2, 19 pp.

[76] S. Eliahou, M. Kervaire, and A. Plagne, Optimally small sumsets in finite abelian groups. *J. Number Theory* **101** (2003), 338–348.

[77] P. Erdős, Extremal problems in number theory. *Proc. Sympos. Pure Math., Vol. VIII* pp. 181–189 *Amer. Math. Soc., Providence, R.I.*, 1965.

[78] P. Erdős, Problems and results on combinatorial number theory. *A survey of combinatorial theory (Proc. Internat. Sympos., Colorado State Univ., Fort Collins, Colo., 1971)*, pp. 117–138. *North-Holland, Amsterdam*, 1973.

[79] P. Erdős and R. L. Graham, Old and new problems and results in combinatorial number theory. Monographies de L'Enseignement Mathématique [Monographs of L'Enseignement Mathématique], 28. *Université de Genéve, L'Enseignement Mathématique, Geneva*, 1980. 128 pp.

[80] P. Erdős and H. Heilbronn, On the addition of residue classes mod p. *Acta Arith.* **9** (1964), 149–159.

[81] C. Finn, Some don't sum to zero: examination of the maximum size of zero-sum-free subsets of cyclic and noncyclic groups. *Research Papers in Mathematics.* B. Bajnok, ed., Gettysburg College, Vol. 14 (2013).

[82] M. A. Fiol, Comments on "Extremal Cayley digraphs of finite abelian groups" [Intercon. Networks 12 (2011), no. 1–2, 125–135]. *J. Inter. Net.* **14** (2013), no. 4, 1350016.

[83] M. A. Fiol, J. L. A. Yebra, I. Alegre, and M. Valero, A discrete optimization problem in local networks and data alignment. *IEEE Trans. Comput.* **36** (1987), no. 6, 702–713.

[84] P. Frankl, R. L. Graham, and V. Rödl, On subsets of abelian groups with no 3-term arithmetic progression. *J. Comb. Theory Ser. A* **45** (1987), no. 1, 157–161.

[85] M. Freeze, W. Gao, and A. Geroldinger, The critical number of finite abelian groups. *J. Number Theory* **129** (2009), no. 11, 2766–2777.

[86] M. Freeze, W. Gao, and A. Geroldinger, Coorigendum to "The critical number of finite abelian groups." *J. Number Theory* **152** (2015) 205–207.

[87] G. A. Freiman, Foundations of a structural theory of set addition. Translated from the Russian. Translations of Mathematical Monographs, Vol 37. *American Mathematical Society, Providence, R. I.*, 1973. vii+108 pp.

[88] K. Fried, Rare bases for finite intervals of integers. *Acta Sci. Math. (Szeged)* **52** (1988), no. 3–4, 303–305.

[89] L. Gallardo, G. Grekos, L. Habsieger, F. Hennecart, B. Landreau, and A. Plagne, Restricted addition in $\mathbb{Z}/n\mathbb{Z}$ and an application to the Erdős–Ginzburg–Ziv problem. *J. London Math. Soc. (2)* **65** (2002), no. 3, 513–523.

[90] W. Gao and A. Geroldinger, On long minimal zero sequences in finite abelian groups. *Period. Math. Hungar.* **38** (1999), no. 3, 179–211.

[91] W. Gao and A. Geroldinger, Zero-sum problems in finite abelian groups: a survey, *Expo. Math.* **24** (2006), no. 4, 337–369.

[92] W. Gao, A. Geroldinger, and W. Schmid, Inverse zero-sum problems. *Acta Arith.* **128** (2007), no. 3, 245–279.

[93] W. Gao and Y. O. Hamidoune, On additive bases. *Acta Arith.* **88** (1999), no. 3, 233–237.

[94] W. Gao, Y. O. Hamidoune, A. S. Lladó, and O. Serra, Covering a finite abelian group by subset sums. *Combinatorica* **23** (2003), no. 4, 599–611.

[95] W. Gao, Y. Li, J. Peng, and F. Sun, Subsums of a zero-sum free subset of an abelian group. *Electron. J. Combin.* **15** (2008), no. 1, Research Paper 116, 36 pp.

[96] W. D. Gao, I. Z. Ruzsa, and R. Thangadurai, Olson's constant for the group $\mathbb{Z}_p \times \mathbb{Z}_p$. *J. Combin. Theory Ser. A* **107** (2004), no. 1, 49–67.

[97] W. D. Gao and R. Thangadurai, A variant of Kemnitz conjecture. *J. Combin. Theory Ser. A* **107** (2004), no. 1, 69–86.

[98] A. Geroldinger and I. Ruzsa, Combinatorial number theory and additive group theory. Courses and seminars from the DocCourse in Combinatorics and Geometry held in Barcelona, 2008. Advanced Courses in Mathematics. CRM Barcelona. *Birkhäuser Verlag, Basel*, 2009. xii+330 pp.

[99] B. Girard, S. Griffiths, and Y. O. Hamidoune, k-sums in abelian groups. *Combin. Probab. Comput.* **21** (2012), no. 4, 582–596.

[100] R. L. Graham and N. J. A. Sloane, On additive bases and harmonious graphs. *SIAM J. Algebraic Discrete Methods* **1** (1980), no. 4, 382–404.

[101] B. Green and I. Ruzsa, Sum-free sets in abelian groups. *Israel J. Math.* **147** (2005), 157–188.

[102] J. R. Griggs, Spanning subset sums for finite abelian groups. Combinatorics, graph theory, algorithms, and applications. *Discrete Math.* **229** (2001), no. 1–3, 89–99.

[103] D. J. Grynkiewicz, Structural additive theory. *Developments in Mathematics*, **30** *Springer, Cham*, 2013. xii + 426 pp.

[104] D. J. Grynkiewicz, Email communication. (July, 2017).

[105] R. Guy, Unsolved problems in number theory. Third edition. Problem books in mathematics. *Springer–Verlag, New York*, 2004. xviii+437 pp.

[106] H. Haanpää, Minimum sum and difference covers of abelian groups. *J. Integer Seq.* **7** (2004), no. 2, Article 04.2.6, 10 pp.

[107] H. Haanpää, A. Huima, and P. Östergård, Sets in $\mathbb{Z}_n$ with distinct sums of pairs. Optimal discrete structures and algorithms (ODSA 2000). *Discrete Applied Math.* **138** (2004), no. 1–2, 99–106.

[108] H. Haanpää and P. Östergård, Sets in abelian groups with distinct sums of pairs. *J. Number Theory* **123** (2007), no. 1, 144–153.

[109] H. Halberstam and K. F. Roth, Sequences. Second Edition. *Springer-Verlag, New York-Berlin*, 1983. xviii+292 pp.

[110] S. Hallfors, On the maximum size of weakly (k,l)-sum-free sets of finite abelian groups. *Research Papers in Mathematics*, B. Bajnok, ed., Gettysburg College, Vol. 11 (2011).

[111] S. Hallfors, Maximum size of weak (k,l)-sum-free sets in abelian groups. *Research Papers in Mathematics*, B. Bajnok, ed., Gettysburg College, Vol. 12 (2012).

[112] Y. O. Hamidoune, Adding distinct congruence classes. *Combin. Probab. Comput.* **7** (1998), no. 1, 81–87.

[113] Y. O. Hamidoune, A. S. Lladó, and O. Serra, On restricted sums. *Combin. Probab. Comput.* **9** (2000), no. 6, 513–518.

[114] Y. O. Hamidoune and A. Plagne, A new critical pair theorem applied to sum-free sets in abelian groups. *Comment. Math. Helv.* **79** (2004), no. 1, 183–207.

[115] Y. O. Hamidoune and G. Zémor, On zero-free subset sums. *Acta Arith.* **78** (1996), no. 3, 143–152.

[116] M. Hampejs, N. Holighaus, L. Tóth, and C. Wiesmeyr, On the subgroups of the group $\mathbb{Z}_m \times \mathbb{Z}_n$. arXiv:1211.1797v1 [math.GR].

[117] I. Haviv and D. Levy, Symmetric complete sum-free sets in cyclic groups. arXiv:1703.04118 [math.CO].

[118] H. A. Helfgott, The ternary Goldbach problem. arXiv:1501.05438 [math.NT].

[119] D. F. Hsu and X. Jia, Extremal problems in the construction of distributed loop networks. *SIAM J. Discrete Math.* **7** (1994), no. 1, 57–71.

[120] I. Ilinkin, Sumset generator. Published at http://addcomb.gettysburg.edu.

[121] T. Jankowski, Explorations of spanning sets in noncyclic groups. *Research Papers in Mathematics*, B. Bajnok, ed., Gettysburg College, Vol. 12 (2012).

[122] X. Jia, Thin bases for finite abelian groups. *J. Number Theory* **36** (1990), no. 2, 254–256.

[123] X. Jia and J. Shen, Extremal bases for finite cyclic groups. *SIAM J. Discrete Math.* **31** (2017), no. 2, 796–804.

[124] Gy. Károlyi, On restricted set addition in abelian groups. *Ann. Univ. Sci. Budapest, Eötvös Sect. Math.* **46**, (2003) 47–54 (2004).

[125] Gy. Károlyi, The Erdős–Heilbronn problem in abelian groups. *Israel J. Math.* **139** (2004) 349–359.

[126] Gy. Károlyi, An inverse theorem for the restricted set addition in abelian groups. *J. Algebra* **290** (2005), no. 2, 557–593.

[127] P. M. Kayll, Well-spread sequences and edge-labellings with constant Hamilton-weight. *Discrete Math. Theor. Comput. Sci.* **6** (2004), no. 2, 401–408.

[128] A. Kemnitz, On a lattice point problem. *Ars Combin.* **16** (1983), B, 151–160.

[129] J. H. B. Kemperman, On small sumsets in an abelian group. *Acta Math.* **103** (1960), 63–88.

[130] C. Kiefer, Examining the maximum size of zero-h-sum-free subsets. *Research Papers in Mathematics*, B. Bajnok, ed., Gettysburg College, Vol. 19 (2016).

[131] B. Klopsch and V. F. Lev, How long does it take to generate a group? *J. Algebra* **261** (2003), no. 1, 145–171.

[132] B. Klopsch and V. F. Lev, Generating abelian groups by addition only. *Forum Math.* **21** (2009), no. 1, 23–41.

[133] M. Kneser, Abschätzungen der asymptotichen Dichte von Summenmengen. *Math. Z.* **58** (1953). 459–484.

[134] D. Knuth, http://sunburn.stanford.edu//simknuth/programs.html.

[135] J. Kohonen, An improved lower bound for finite additive 2-bases. *J. Number Theory* **174** (2017), 518–524.

[136] A. Kotzig, On well spread sets of integers. Tech. Report CRM–161, Centre Res. Math., Université de Montréal (1972), 83 pp.

[137] O. Krasny, Finding minimum spanning sets with limited number of terms in cyclic groups. *Research Papers in Mathematics*, B. Bajnok, ed., Gettysburg College, Vol. 10 (2010).

[138] N. Laza, Personal communication (2000–2003).

[139] M. Lee, Sets with few differences in abelian groups. arXiv:1508.05524 [Math.CO].

[140] V. F. Lev, Restricted set addition in groups. I. The classical setting. *J. London Math. Soc. (2)* **62** (2000), no. 1, 27–40.

[141] V. F. Lev, Three-fold restricted set addition in groups. *European J. Combin.* **23** (2002), no. 5, 613–617.

[142] V. F. Lev, Generating binary spaces. *J. Combin. Theory Ser. A* **102** (2003), no. 1, 94–109.

[143] V. F. Lev, Progression-free sets in finite abelian groups. *J. Number Theory* **104** (2004), no. 1, 162–169.

[144] V. F. Lev, Stability result for sets with $3A \neq \mathbb{Z}_5^n$. arXiv:1602.06715 [Math.NT].

[145] V. F. Lev and R. Yuster, On the size of dissociated bases. *Electron. J. Combin.* **18** (2011), no. 1, Paper 117, 5 pp.

[146] MAA CUPM Subcommittee on Research by Undergraduates, Mathematics research by undergraduates: costs and benefits to faculty and the institution. In *Proceedings of the Conference on Promoting Undergraduate Research in Mathematics*, J. A. Gallian (Ed.), American Mathematical Society (2007), 325–328.

[147] M. Malec, An algorithmic approach to computing sumset sizes. *Research Papers in Mathematics*. B. Bajnok, ed., Gettysburg College, Vol. 14 (2013).

[148] D. Manandhar, Decoding minimum h-fold restricted sumset size of a cyclic group. *Research Papers in Mathematics*, B. Bajnok, ed., Gettysburg College, Vol. 13 (2012).

[149] H. B. Mann and Y. F. Wou, Addition theorem for the elementary abelian group of type (p, p), *Monatsh. Math.* **102** (1986), no. 2, 273–308.

[150] L. E. Marchan, O. Ordaz, D. Ramos, and W. Schmid, Some exact values of the Harborth constant and its plus-minus weighted analogue. *Arch. Math. (Basel)* **101** (2013), no. 6, 501–512.

[151] L. E. Marchan, O. Ordaz, D. Ramos, and W. Schmid, Inverse results for weighted Harborth constants. *Int. J. Number Theory* **12** (2016), no. 7, 1845–1861.

[152] M. T. Margotta, Examining the number and nature of spanning sets for $\mathbb{Z}_n$. *Research Papers in Mathematics*, B. Bajnok, ed., Gettysburg College, Vol. 10 (2010).

[153] A. Maturo, Finding the folding number of abelian groups. *Research Papers in Mathematics*, B. Bajnok, ed., Gettysburg College, Vol. 9 (2009).

[154] A. Maturo and D. Yager-Elorriaga, Finding Sidon sets in abelian groups. *Research Papers in Mathematics*, B. Bajnok, ed., Gettysburg College, Vol. 7 (2008).

[155] R. Matzke, It's the least we can do. *Research Papers in Mathematics*, B. Bajnok, ed., Gettysburg College, Vol. 14 (2013).

[156] E. Matys, Tauer of terror: not as scary as one might think. *Research Papers in Mathematics*, B. Bajnok, ed., Gettysburg College, Vol. 16 (2014).

[157] Z. Miller, Give me liberty or give me death: on t-independent sets in finite cyclic groups. *Research Papers in Mathematics*, B. Bajnok, ed., Gettysburg College, Vol. 15 (2013).

[158] Y. Mimura, A construction of spherical 2-designs. *Graphs Combin.* **6** (1990), no. 4, 369–372.

[159] P. Morillo, M. A. Fiol, and J. Fàbrega, The diameter of directed graphs associated to plane tessellations. Tenth British combinatorial conference (Glasgow, 1985). *Ars Combin.* **20-A** (1985), A, 17–27.

[160] A. Mrose, Untere Schranken für die Reichweiten von Extremalbasen fester Ordnung. *Abh. Math. Sem. Univ. Hamburg* **48** (1979), 118–124.

[161] M. B. Nathanson, Additive number theory. Inverse problems and the geometry of sumsets. Graduate Texts in Mathematics, **165**. *Springer–Verlag, New York*, 1996. xiv+293 pp.

[162] A. Navarro, On the subsets of size $\chi(G, h) - 1$. *Research Papers in Mathematics.* B. Bajnok, ed., Gettysburg College, Vol. 18 (2016).

[163] N. H. Nguyen, E. Szemerédi, and V. H. Vu, Subset sums modulo a prime. *Acta Arith.* **131** (2008), no. 4, 303–316.

[164] N. H. Nguyen and V. H. Vu, Classification theorems for sumsets modulo a prime. *J. Combin. Theory Ser. A* **116** (2009), no. 4, 936–959.

[165] T. Olans, Exploring $\nu\hat{_\pm}(\mathbb{Z}_n, m, h)$. *Research Papers in Mathematics.* B. Bajnok, ed., Gettysburg College, Vol. 15 (2013).

[166] J. E. Olson, An addition theorem modulo p. *J. Combin. Theory Ser. A* **5** (1968), 45–52.

[167] J. E. Olson, Sums of sets of group elements. *Acta Arith.* **28** (1975/76), no. 2, 147–156.

[168] O. Ordaz, A. Philipp, I. Santos, and W. A. Schmid, On the Olson and the strong Davenport constants. *J. Théor. Nombres Bordeaux* **23** (2011), no. 3, 715–750.

[169] K. Phillips, Examining the maximum size of a zero-4-sum-free set over $\mathbb{Z}$. *Research Papers in Mathematics*. B. Bajnok, ed., Gettysburg College, Vol. 11 (2011).

[170] N. C. K. Phillips and W. D. Wallis, Well-spread sequences. *J. Combin. Math. Combin. Comput.* **31** (1999), 91–96.

[171] A. Plagne, Maximal (k,l)-free sets in $\mathbb{Z}/p\mathbb{Z}$ are arithmetic progressions. *Bull. Austral. Math. Soc.* **65** (2002), no. 3, 137–144.

[172] A. Plagne, Additive number theory sheds extra light on the Hopf–Stiefel $\circ$ function. *Enseign. Math. (2)* **49** (2003), no. 1–2, 109–116.

[173] A. Plagne, Optimally small sumsets in groups, I. The supersmall sumset property, the $\mu_G^{(k)}$ and the $\nu_G^{(k)}$ functions. *Unif. Distrib. Theory* **1** (2006), no. 1, 27–44.

[174] A. Plagne, Optimally small sumsets in groups, II. The hypersmall sumset property and restricted addition. *Unif. Distrib. Theory* **1** (2006), no. 1, 111–124.

[175] A. Potechin, Maximal caps in AG(6, 3). *Des. Codes Cryptogr.* **46** (2008), no. 3, 243–259.

[176] Y. Qu, G. Wang, Q. Wang, and D. Guo, Extremal incomplete sets in finite abelian groups. *Ars Combin.* **116** (2014), 457–475.

[177] T. Reckner, On the minimum size of spanning sets in finite cyclic groups. *Research Papers in Mathematics*, B. Bajnok, ed., Gettysburg College, Vol. 15 (2013).

[178] A. H. Rhemtulla and A. P. Street, Maximal sum-free sets in elementary abelian p-groups. *Canad. Math. Bull.* **14** (1971), no. 1, 73–80.

[179] R. M. Roth and A. Lempel, t-sum generators of finite abelian groups. *Discrete Math.* **103** (1992), no. 3, 279–292.

[180] D. F. Ruiz and C. A. Trujillo, Construction of $B_h[g]$ sets in product of groups. *Rev. Colombiana Mat.* **50** (2016), no. 2, 163–172.

[181] I. Ruzsa, Solving a linear equation in a set of integers. I. *Acta Arith.* **65** (1993), no. 3, 259–282.

[182] W. Samotij and B. Sudakov, The number of additive triples in subsets of abelian groups. *Math. Proc. Cambridge Philos. Soc.* **160** (2016), no. 3, 495–512.

[183] W. Samotij and B. Sudakov, The number of additive triples in subsets of abelian groups. arXiv:1507.03764 [math.NT].

[184] W. Schmid, Email communication. (July, 2017).

[185] T. Schoen and I. D. Shkredov, Additive dimension and a theorem of Sanders. *J. Aust. Math. Soc.* **100** (2016), no. 1, 124–144.

[186] SET Enterprises, http://www.setgame.com.

[187] J. Singer, A theorem in finite projective geometry and some applications to number theory. *Trans. Amer. Math. Soc.* **43** (1938), no. 3, 377–385.

[188] N. J. A. Sloane, ed., The On-Line Encyclopedia of Integer Sequences. Published at https://oeis.org.

[189] S. Smale, Mathematical problems for the next century. *Math. Intelligencer* **20** (1998), no. 2, 7–15.

[190] T. Soma, A "nu" result in additive combinatorics. *Research Papers in Mathematics*, B. Bajnok, ed., Gettysburg College, Vol. 11 (2011).

[191] A. P. Street, Maximal sum-free sets in abelian groups of order divisible by three. *Bull. Austral. Math. Soc.* **6** (1972), 439–441.

[192] A. P. Street, Corrigendum: "Maximal sum-free sets in abelian groups of order divisible by three". *Bull. Austral. Math. Soc.* **7** (1972), 317–318.

[193] J. C. Subocz G., Some values of Olson's constant. *Divulg. Mat.* **8** (2000), no. 2, 121–128.

[194] T. Tao, Open question: best bounds for cap sets. Blog post https://terrytao.wordpress.com/2007/02/23/open-question-best-bounds-for-cap-sets/.

[195] T. Tao and V. Vu, Additive combinatorics. *Cambridge Studies in Advanced Mathematics*, **105**. *Cambridge University Press, Cambridge*, 2010. xviii+512 pp.

[196] T. Tao and V. Vu, Sum-free sets in groups: a survey. arXiv:1603.03071 [math.CO].

[197] A. G. Vosper, The critical pairs of subsets of a group of prime order. *J. London Math. Soc.* **31** (1956), 200–205.

[198] A. G. Vosper, Addendum to "The critical pairs of subsets of a group of prime order". *J. London Math. Soc.* **31** (1956), 280–282.

[199] W. D. Wallis, A. P. Street, and J. S. Wallis, Combinatorics: room squares, sum-free sets, Hadamard matrices. Lecture Notes in Mathematics, **292**, *Springer–Verlag, Berlin-New York*, 1972. iv+508 pp.

[200] D. Yager-Elorriaga, Weakly zero-sum-free sets in finite abelian groups. *Research Papers in Mathematics*, B. Bajnok, ed., Gettysburg College, Vol. 8 (2008).

[201] H. P. Yap, Maximal sum-free sets in finite abelian groups. II. *Bull. Austral. Math. Soc.* **5** (1971), 43–54.

[202] Y. Yin, Maximum size of weakly zero-sum-free subsets of abelian groups. *Research Papers in Mathematics*, B. Bajnok, ed., Gettysburg College, Vol. 18 (2016).

[203] P. Yuan and X. Zeng, On zero-sum free subsets of length 7, *Electron. J. Combin.* **17** (2010), no. 1, Research Paper 104, 13 pp.

[204] C. Zajaczkowski, Addition without sums: analyzing the size of sum-free sets. *Research Papers in Mathematics*, B. Bajnok, ed., Gettysburg College, Vol. 15 (2013).

[205] U. Zannier, Email communication (August, 2015).

[206] D. Ziegler, A further exploration of weakly (k,l)-sum-free sets in finite abelian groups. *Research Papers in Mathematics*, B. Bajnok, ed., Gettysburg College, Vol. 9 (2009).

Author Index